Modern Birkhäuser Classics

Many of the original research and survey monographs, as well as textbooks, in pure and applied mathematics published by Birkhäuser in recent decades have been groundbreaking and have come to be regarded as foundational to the subject. Through the MBC Series, a select number of these modern classics, entirely uncorrected, are being re-released in paperback (and as eBooks) to ensure that these treasures remain accessible to new generations of students, scholars, and researchers.

T0202971

The Self-Avoiding Walk

Neal Madras
Gordon Slade

Reprint of the 1996 Edition

 Birkhäuser

Neal Madras
Department of Mathematics and Statistics
York University
Toronto, Ontario
Canada

Gordon Slade
Department of Mathematics
The University of British Columbia
Vancouver, British Columbia
Canada

Originally published in the series *Probability and Its Applications*

ISBN 978-1-4614-6024-4   ISBN 978-1-4614-6025-1 (eBook)
DOI 10.1007/978-1-4614-6025-1
Springer New York Heidelberg Dordrecht London

Library of Congress Control Number: 2012952314

Printed on acid-free paper

Springer is part of Springer Science+Business Media (www.birkhauser-science.com)

Neal Madras
Gordon Slade

The Self-Avoiding Walk

Birkhäuser
Boston · Basel · Berlin

Neal Madras
Department of Mathematics
and Statistics
York University
Downsview, Ontario
Canada   M3J 1P3

Gordon Slade
Department of Mathematics
and Statistics
McMaster University
Hamilton, Ontario
Canada   L8S 4K1

Printed on acid-free paper

© 1996 Birkhäuser Boston

Birkhäuser ®

Hardcover edition, printed in 1993 by Birkhäuser Boston

ISBN 0-8176-3891-1
ISBN 3-7643-3891-1

Camera-ready text prepared in L^ATEX by the authors.
Printed and bound by Quinn-Woodbine, Woodbine, NJ.
Printed in the U.S.A.
9 8 7 6 5 4 3 2 1

To Joyce and Joanne

Contents

Preface

A self-avoiding walk is a path on a lattice that does not visit the same site more than once. In spite of this simple definition, many of the most basic questions about this model are difficult to resolve in a mathematically rigorous fashion. In particular, we do not know much about how far an n-step self-avoiding walk typically travels from its starting point, or even how many such walks there are. These and other important questions about the self-avoiding walk remain unsolved in the rigorous mathematical sense, although the physics and chemistry communities have reached consensus on the answers by a variety of nonrigorous methods, including computer simulations. But there has been progress among mathematicians as well, much of it in the last decade, and the primary goal of this book is to give an account of the current state of the art as far as rigorous results are concerned.

A second goal of this book is to discuss some of the applications of the self-avoiding walk in physics and chemistry, and to describe some of the nonrigorous methods used in those fields. The model originated in chemistry several decades ago as a model for long-chain polymer molecules. Since then it has become an important model in statistical physics, as it exhibits critical behaviour analogous to that occurring in the Ising model and related systems such as percolation. It is also of interest in probability theory as a basic example which does not respond well to standard probabilistic methods. Methods originating in mathematical physics and combinatorics have been more successful. Computer simulations have played an important role in formulating conjectures, and interesting computational and algorithmic issues have arisen in the process.

We have attempted to make this book as self-contained as possible. It should be accessible to graduate students in mathematics and to graduate students in physics and chemistry who are mathematically inclined.

Chapter 1 gives a general introduction to the basic questions and conjectures about the self-avoiding walk. The important notion of subadditivity is introduced in Section 1.2. Its relevance was pointed out by Hammersley and Morton (1954), and its interplay with concatenation is a recurring theme throughout the book.

Chapter 2 is devoted to a discussion of some nonrigorous and applied topics, namely scaling theory, the relation to polymers and the Flory argument, and the identification of the self-avoiding walk as a "zero-component" ferromagnet.

In 1962, Hammersley and Welsh proved an upper bound on the number of n-step self-avoiding walks which remains the best available bound in two dimensions. Their proof is given in Section 3.1. Shortly afterward, Kesten (1964) improved the Hammersley–Welsh bound in three or more dimensions. Kesten's bound is proven in Section 3.3; it remains the best bound in dimensions three and four. Bounds on the number of self-avoiding polygons are proven in Section 3.2. Subadditivity is the driving force in these arguments, implemented at times with considerable sophistication.

Chapter 4 is concerned with the decay of the subcritical two-point function. In particular, existence of a mass (or inverse correlation length) is proven, as well as existence of Ornstein–Zernike decay near a coordinate axis. This chapter also makes use of subadditivity in a fundamental way, using bridges and irreducible bridges. It has close connections with probabilistic renewal theory.

Chapters 5 and 6 are concerned with the self-avoiding walk above four dimensions and the recent proof by Hara and Slade (1992a, 1992b) of mean-field behaviour in five or more dimensions. The main tool in the proof is the lace expansion of Brydges and Spencer (1985), which is the subject of Chapter 5. Section 5.5 indicates how the lace expansion can also be applied to lattice trees, lattice animals and percolation, and attempts to describe the expansions for the various models in a manner which emphasizes their similarities. The results in high dimensions are summarized in Section 6.1, before proving convergence of the lace expansion in Section 6.2. The convergence proof uses a number of estimates for ordinary random walk; these are given in Appendix A. In order to keep the convergence proof as simple as possible, we do not prove mean-field behaviour for the nearest-neighbour model in five or more dimensions, but rather consider the nearest-neighbour model in sufficiently high dimensions and the "spread-out" model above four dimensions; this allows us to present the simplest proof of convergence of the lace expansion to appear in print to date. Sections 6.3 to 6.8 show how convergence of the lace expansion leads to existence of critical exponents and other results stated in Section 6.1.

Chapter 7 is devoted to a proof of the pattern theorem of Kesten (1963) and a discussion of some of its consequences. These consequences are primarily in the form of ratio limit theorems for the number of n-step self-avoiding walks and related quantities.

Chapter 8 contains a short potpourri of additional results: upper bounds on the critical exponent α_{sing}, comments on self-avoiding walks in restricted geometries, construction of the infinite bridge, and some comments on the occurrence of knots in self-avoiding polygons.

Chapter 9 gives an extensive survey of various Monte Carlo algorithms, both static and dynamic, that have been used to simulate self-avoiding walks. Special attention is paid to the rigorous analysis of ergodicity properties and autocorrelation times of the algorithms.

Finally in Chapter 10 a brief discussion is given of four related topics: the Edwards model and weakly self-avoiding walk, the loop-erased self-avoiding random walk, intersection properties of simple random walks, and the "myopic" or "true" self-avoiding walk.

With the exception of Chapter 10 and the Appendices, most references to the literature are postponed to Notes which follow each chapter.

Enumerations of self-avoiding walks are proceeding at a rapid pace as computer technology advances. Appendix C gives tables of the number of self-avoiding walks, the square displacement, and the number of polygons, on the hypercubic lattice in dimensions two through six.

An overview of the key concepts, results, and methods of the book can be obtained from a reading of all of Chapter 1, together with Sections 3.1, 3.2, 4.1, 4.2, 5.1, 5.2, 5.4, 6.1, 6.2, 7.1, and 9.1. Any section not on the above list is rarely referenced outside its own chapter. In fact, the chapters of this book to a large extent can be read in any order, with the following exceptions: Chapter 1 should be read first; Chapter 3 should precede Chapter 4; Chapter 5 should precede Chapter 6; and Chapters 3, 4, and 7 should precede Chapter 8.

In view of our emphasis on rigorous results, we have omitted any description of such important ideas as the exact solution in two dimensions arising from connections with the Coulomb gas due to Nienhuis, and with conformal invariance in work of Duplantier and others. Nor have we attempted to describe the work on renormalization in the physics and polymer literature.

We have benefitted from the help of various people throughout the course of writing this book. Takashi Hara, Alan Sokal and Stu Whittington each read various portions of the manuscript and made numerous suggestions for improvements. Harry Kesten provided extensive notes for Section 3.3. Greg Lawler clarified several issues for us in Sections 10.2 and 10.3. Tony Guttmann kept us up-to-date with the latest world records in self-avoiding walk enumerations, and guided us through the related literature. We extend our thanks to all these, and to the many others who have offered advice in correspondence or conversation. NM is indebted to Alan Sokal and Stu Whittington for teaching him much about this field while collaborating on various enjoyable and fascinating projects. GS expresses special thanks to David Brydges for inspiring his initial interest in the lace expansion and suggesting that it might converge in five dimensions, and to Takashi Hara for the pleasure of four years of collaboration and for permission to include unpublished joint work. We gratefully acknowledge financial support from the Natural Sciences and Engineering Research Council of Canada. Finally we offer our thanks and deep appreciation to Joyce Kruskal and Joanne Nakonechny for their encouragement, support, and tolerance.

Toronto and Hamilton
July 7, 1992

Chapter 1

Introduction

1.1 The basic questions

Imagine that you are standing at an intersection in the centre of a large city whose streets are laid out in a square grid. You choose a street at random and begin walking away from your starting point, and at each intersection you reach you choose to continue straight ahead or to turn left or right. There is only one rule: you must not return to any intersection already visited in your journey. In other words, your path should be self-avoiding. It is possible that you will lead yourself into a trap, reaching an intersection whose neighbours have all been visited already, but barring this disaster you continue walking until you have walked some large number N of blocks. There are two basic questions:

- How many possible paths could you have followed?

- Assuming that any one path is just as likely as any other, how far will you be on the average from your starting point?

These questions are straightforward enough, but the answers are only known for small values of N. It is widely accepted that a search for general exact formulas is an enormously difficult problem which lies beyond the reach of current methods. A less difficult question would be to ask for the asymptotic behaviour of the answers as N becomes very large, but this too is very hard. Physicists and chemists who are interested in this and related problems have applied a variety of methods and have produced many intriguing results, but a great deal of work is still needed to settle these issues in a mathematically rigorous way. In this book we will state some of the

1

results of nonrigorous work in the field, and describe the rigorous work in some detail.

At first glance one might expect that the easiest way to answer the above questions, at least approximately, would be to use a computer. Much numerical work has been done in this direction, and in Chapter 9 some of it will be discussed. Here too, however, the situation is not so easy: exact enumeration of all possible routes has been done to date only for $N \leq 34$, with further enumerations made difficult because of the exponential growth in the number of paths as N increases. Larger values of N can be studied by extrapolation of the exact enumeration data, or by Monte Carlo simulations.

There is no need to restrict the walk to a two-dimensional grid, and it is easy to generalize the above questions to general dimension d. It is also possible to generalize the problem by changing from a rectangular to a triangular or other type of grid. There is at least one case where the above questions can be easily answered, and this is the case of a one-dimensional walk. A self-avoiding walker in one dimension has no alternative but to continue travelling in the direction initially chosen, so there are exactly two paths for every value of N and the distance travelled is exactly N blocks. That was easy, but not very interesting. Higher dimensions provide a vastly richer structure.

In general, a self-avoiding walk takes place on a graph. A graph (more precisely, an undirected graph) is a collection of points, together with a collection of pairs of points known as *edges*. The basic example that will concern us most is the *d-dimensional hypercubic lattice* $\mathbf{Z}^d$. The points of this graph are the points of the d-dimensional Euclidean space $\mathbf{R}^d$ whose components are all integers, and the edges are given by the set of all unit line segments joining neighbouring points. The points will be referred to as *sites*, and the unit line segments as *nearest-neighbour bonds*. Sites will typically be denoted by letters such as u, v, x, y, and their components by subscripts: $x = (x_1, x_2, \ldots, x_d)$. The usual Euclidean dot product on $\mathbf{Z}^d$ will be written $x \cdot y = \sum_{i=1}^{d} x_i y_i$, and the Euclidean norm will be written $|x| = \sqrt{x \cdot x}$. We will also use the notation $\|x\|_p = (\sum_{i=1}^{d} x_i^p)^{1/p}$, and $\|x\|_\infty = \max\{|x_i|: i = 1, \ldots, d\}$.

An N-step self-avoiding walk ω on $\mathbf{Z}^d$, beginning at the site x, is defined as a sequence of sites $(\omega(0), \omega(1), \ldots, \omega(N))$ with $\omega(0) = x$, satisfying $|\omega(j+1) - \omega(j)| = 1$, and $\omega(i) \neq \omega(j)$ for all $i \neq j$. We write $|\omega| = N$ to denote the length of ω, and we denote the components of $\omega(j)$ by $\omega_i(j)$ $(i = 1, \ldots, d)$. Let c_N denote the number of N-step self-avoiding walks beginning at the origin. By convention, $c_0 = 1$. Then the first of our basic questions above is asking for the value of c_N. More modestly, we could ask

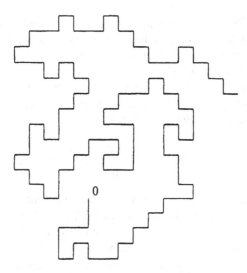

Figure 1.1: A two-dimensional self-avoiding walk with 115 steps.

for the asymptotic form of c_N as $N \to \infty$. It is easy to find the exact values of c_N (as a function of d) for very small values of N, for example $c_1 = 2d$, $c_2 = 2d(2d - 1)$, $c_3 = 2d(2d - 1)^2$, and $c_4 = 2d(2d - 1)^3 - 2d(2d - 2)$ (for c_4 the second term subtracts the contribution of squares to the first term). However, the combinatorics quickly become difficult as N increases and then soon become intractable. Tables in Appendix C give enumerations of c_N for dimensions two through six.

The simplest bounds on the behaviour of c_N are obtained as follows. An upper bound on c_N is given by the number of walks which have no immediate reversals, or in other words which never visit the same site at times i and $i + 2$. Avoiding immediate reversals allows $2d$ choices for the initial step, and $2d - 1$ choices for the $N - 1$ remaining steps, for a total of $2d(2d - 1)^{N-1}$. For a lower bound we simply count the number of walks in which each step is in one of the d positive coordinate directions. Such walks are necessarily self-avoiding. Thus we have

$$d^N \leq c_N \leq 2d(2d - 1)^{N-1}. \qquad (1.1.1)$$

To discuss the average distance from the origin after N steps, we need to introduce a probability measure on N-step self-avoiding walks. The measure that we shall use throughout this book is the uniform measure, which assigns equal weight c_N^{-1} to each N-step self-avoiding walk. It is worth noting that although we originally introduced the self-avoiding walk in terms of a walker moving in time, the uniform measure is a measure on paths of length N and does not define a stochastic process evolving in time (for example, a walk may be trapped and impossible to extend without introducing a self-intersection).

Denoting expectation with respect to the uniform measure by angular brackets, the average distance (squared) from the origin after N steps is then given by the *mean-square displacement*

$$\langle |\omega(N)|^2 \rangle = \frac{1}{c_N} \sum_{\omega:|\omega|=N} |\omega(N)|^2. \tag{1.1.2}$$

The sum over ω is the sum over all N-step self-avoiding walks beginning at the origin. Like c_N, the mean-square displacement can also be calculated by hand for very small values of N, but the combinatorics quickly become intractable as N increases. Enumerations are tabulated in Appendix C.

It is instructive to compare the behaviour of the self-avoiding walk with that of the simple random walk. An N-step simple random walk on $\mathbf{Z}^d$, starting at the origin, is a sequence $\omega = (\omega(0), \omega(1), \ldots, \omega(N))$ of sites with $\omega(0) = 0$ and $|\omega(j+1) - \omega(j)| = 1$, with the uniform measure on the set of all such walks. Without the self-avoidance constraint the situation is rather easy. Indeed, since each site has $2d$ nearest neighbours, the number of N-step simple random walks is exactly $(2d)^N$. To analyse the mean-square displacement, we represent the simple random walk in the following way. Let $\{X^{(i)}\}$ be independent and identically distributed random variables with $X^{(i)}$ uniformly distributed over the $2d$ (positive and negative) unit vectors. Then the position after N steps can be represented as the sum $S_N = X^{(1)} + X^{(2)} + \ldots + X^{(N)}$. Expanding $|S_N|^2$, the mean-square displacement is given by

$$\langle |S_N|^2 \rangle = \sum_{i,j=1}^{N} \langle X^{(i)} \cdot X^{(j)} \rangle. \tag{1.1.3}$$

For $i \neq j$, $\langle X^{(i)} \cdot X^{(j)} \rangle = 0$, using independence and the fact that $\langle X^{(i)} \rangle = 0$. Since $\langle X^{(i)} \cdot X^{(i)} \rangle = 1$, it follows that the mean-square displacement is equal to N. Similarly, if we consider a random walk in $\mathbf{Z}^d$ in which steps lie in a symmetric finite set $\Omega \subset \mathbf{Z}^d$ of cardinality $|\Omega|$, with each possible step equally likely, then the number of N-step walks is $|\Omega|^N$ and the mean-square displacement is $N\sigma^2$, where σ^2 is the mean-square displacement of a single step.

For the self-avoiding walk it is believed that there is exponential growth of c_N with power law corrections, unlike the pure exponential growth of the simple random walk. It is also believed that the mean-square displacement will not always be linear in the number of steps, in contrast to the diffusive behaviour of the simple random walk. These beliefs are in harmony with known properties of other models of statistical mechanics, and are supported by numerical and nonrigorous calculations. The conjectured behaviour of c_N and $\langle |\omega(N)|^2 \rangle$ is thus

$$c_N \sim A\mu^N N^{\gamma-1} \tag{1.1.4}$$

and

$$\langle |\omega(N)|^2 \rangle \sim DN^{2\nu}, \tag{1.1.5}$$

where A, D, μ, γ and ν are dimension-dependent positive constants. We shall refer to μ as the *connective constant*, and γ and ν are examples of *critical exponents*. In four dimensions the above two relations should be modified by logarithmic factors; see (1.1.13) and (1.1.14) below. Here $f(N) \sim g(N)$ means that f is asymptotic to g as $N \to \infty$:

$$\lim_{N \to \infty} \frac{f(N)}{g(N)} = 1.$$

For ordinary random walk (1.1.4) and (1.1.5) hold with $\gamma = 1$ and $\nu = 1/2$, both for the nearest-neighbour and more general walks.

In the next section the existence of the limit

$$\mu = \lim_{N \to \infty} c_N^{1/N} \tag{1.1.6}$$

will be proven, which is the first step in justifying (1.1.4). The simple bounds of (1.1.1) then immediately imply that

$$d \le \mu \le 2d - 1. \tag{1.1.7}$$

The exact value of μ is not known for the hypercubic lattice in any dimension $d \ge 2$, although for the honeycomb lattice in two dimensions there is nonrigorous evidence that $\mu = \sqrt{2 + \sqrt{2}}$. Improvements to (1.1.7) will be discussed in the next section. For high dimensions it is known that as $d \to \infty$

$$\mu = 2d - 1 - \frac{1}{2d} - \frac{3}{(2d)^2} + O\left(\frac{1}{(2d)^3}\right); \tag{1.1.8}$$

references are given in the Notes. In fact Fisher and Sykes (1959) established the coefficients in the $1/d$ expansion up to and including order d^{-4}, although there is no rigorous control of their error term. Intuitively (1.1.8) says that in high dimensions the principal effect of the self-avoidance constraint is to rule out immediate reversals.

Concerning γ, we will show in Section 1.2 that $c_N \geq \mu^N$ and hence $\gamma \geq 1$ in all dimensions. There is still no proof, however, that γ is finite in two, three or four dimensions, where the best bounds are

$$c_N \leq \begin{cases} \mu^N \exp[KN^{1/2}] & d = 2 \\ \mu^N \exp[KN^{2/(2+d)} \log N] & d = 3, 4 \end{cases} \qquad (1.1.9)$$

for a positive constant K; these bounds will be discussed in Sections 3.1 and 3.3. In Chapter 6 we will describe a proof that (1.1.4) holds with $\gamma = 1$ for $d \geq 5$. In addition to characterizing the asymptotic behaviour of c_N, the exponent γ provides a measure of the probability that two N-step self-avoiding walks starting at the same point do not intersect. In fact, this probability is equal to c_{2N}/c_N^2, and assuming (1.1.4) we have

$$\frac{c_{2N}}{c_N^2} \sim \frac{2^{\gamma-1}}{A} N^{1-\gamma}. \qquad (1.1.10)$$

If $\gamma > 1$ then this probability goes to zero as $N \to \infty$, while if $\gamma = 1$ it remains positive. For the simple random walk the analogous probability is known to remain positive as $N \to \infty$ for $d > 4$, and roughly speaking to go to zero like $(\log N)^{-1/2}$ for $d = 4$ and as an inverse power of N for $d = 2, 3$. A survey of the simple random walk results is given in Section 10.3.

Intuitively it is to be expected that the repulsive interaction of the self-avoiding walk will tend to drive the endpoint of the walk away from the origin faster than for simple random walk, or in other words that $\nu \geq 1/2$. However it is still an open question to prove that this "obvious" inequality $\langle |\omega(N)|^2 \rangle \geq CN$ holds in all dimensions. On the other hand, bounding $|\omega(N)|^2$ above by N^2 in (1.1.2) gives the upper bound $\langle |\omega(N)|^2 \rangle \leq N^2$, or $\nu \leq 1$. This bound is optimal in one dimension, but seems far from optimal in two or more dimensions. No upper bound of the form $CN^{2-\epsilon}$ ($C, \epsilon > 0$), or in other words $\nu < 1$, has been proven for dimensions two, three or four, however. For $d \geq 5$ it has been proved that $\nu = 1/2$; this proof will be described in Chapter 6. It will also be shown that for high dimensions the diffusion constant D is strictly greater than the simple random walk value of 1. Thus in high dimensions the self-avoiding walk does move away from the origin more quickly than the simple random walk, but only at the level of the diffusion constant and not at the level of the exponent ν. The tendency of the self-avoiding walk to move away from the origin more quickly than the simple random walk should become less pronounced as the dimension increases, and hence it is to be expected that ν is a nonincreasing function of the dimension.

The critical exponents γ and ν are believed to be dimension dependent, but independent of the type of allowed steps (as long as there are only finitely many possible steps and the allowed steps are symmetric) or even of the type of lattice—the exponents are believed, for example, to be the same for the square and triangular lattices. This lack of dependence on the detailed definition of the model is known as *universality*, and models with the same exponents are said to be in the same *universality class*. The *connective constant* μ appearing in (1.1.4) represents the effective coordination number of the lattice and is not universal—it depends on the details of the allowed steps and the underlying lattice, as well as the dimension d.

It seems clear that in high dimensions the self-avoiding walk should be closer to the simple random walk than in low dimensions, since a simple random walk is less likely to intersect itself in high dimensions. Four dimensions plays a special role: for simple random walk the expected time of the first return to the origin, conditioned on the event that this return occurs, is finite for $d > 4$; this suggests that above four dimensions self-avoidance is a short-range effect rather than a long-range one, and hence that it will not affect the critical exponents. In addition, as mentioned above, the probability that two independent simple random walks of length N do not intersect remains bounded away from zero as $N \to \infty$ for $d > 4$, but not for $d \leq 4$.

The conjectured values of γ and ν are as follows:

$$\gamma = \begin{cases} \frac{43}{32} & d = 2 \\ 1.162... & d = 3 \\ 1 \text{ with logarithmic corrections} & d = 4 \\ 1 & d \geq 5 \end{cases} \tag{1.1.11}$$

$$\nu = \begin{cases} \frac{3}{4} & d = 2 \\ 0.59... & d = 3 \\ \frac{1}{2} \text{ with logarithmic corrections} & d = 4 \\ \frac{1}{2} & d \geq 5 \end{cases} \tag{1.1.12}$$

Currently the only rigorous results which prove power law behaviour and confirm the conjectured values of γ and ν are for $d \geq 5$. These are discussed in detail in Chapter 6. The conjectured logarithmic corrections to γ and ν in four dimensions, predicted by the renormalization group, are given by:

$$c_N \sim A\mu^N [\log N]^{1/4}, \quad d = 4 \tag{1.1.13}$$

$$\langle |\omega(N)|^2 \rangle \sim DN[\log N]^{1/4}, \quad d = 4. \tag{1.1.14}$$

Equations (1.1.11) to (1.1.14) are typical of what is found for other statistical mechanical models, such as the Ising model or percolation. A common feature is the existence of a certain dimension, the so-called *upper critical dimension*, at which there are logarithmic corrections to critical exponents and above which all critical exponents are dimension independent and are given by the corresponding critical exponents for a simpler model, known as the *mean-field*[1] model. For the self-avoiding walk the mean-field model is the simple random walk and the simple random walk critical exponents are sometimes referred to as the mean-field exponents.

The rational values for two dimensions given in (1.1.11) and (1.1.12) come from a nonrigorous exact solution of the $O(N)$ spin model which includes the self-avoiding walk as the special case $N = 0$ (see Section 2.3). This remarkable work exploits a connection between the $O(N)$ model and the Coulomb gas and uses the renormalization group. From a different approach, nonrigorous conformal invariance arguments reproduce the same rational values. There is no analogous exact solution in three dimensions, and the $d = 3$ values given in (1.1.11) and (1.1.12) are from numerical results and field-theoretic calculations using the ϵ-expansion. References for these topics are given in the Notes.

An early conjecture for the values of ν was made by Flory, and will be discussed in Section 2.2. The Flory exponents are given by $\nu_{Flory} = 3/(2 + d)$ for $d \leq 4$ and $\nu_{Flory} = 1/2$ for $d > 4$. This agrees with Equation (1.1.12) for $d = 2$ and $d \geq 4$ (apart from the logarithmic correction when $d = 4$), and comes very close for $d = 3$. The exact Flory value $\nu_{Flory} = 3/5$ in three dimensions has been ruled out by numerical work, however.

1.2   The connective constant

If (1.1.4) correctly represents the behaviour of c_N for large N, then the limit

$$\mu = \lim_{N \to \infty} c_N^{1/N} \tag{1.2.1}$$

must exist. One purpose of this section is to prove the existence of this limit as a simple consequence of a subadditive property of $\log c_N$. It then follows immediately from (1.1.1) that

$$d \leq \mu \leq 2d - 1. \tag{1.2.2}$$

The proof involves the notion of concatenation of two self-avoiding walks.

---

[1]This terminology has its origin in the Ising model. For the Ising model the upper critical dimension is also four, and above four dimensions critical exponents are given by the exactly solvable model in which a spin interacts with the *average* of all the other spins. References are given in the Notes.

Definition 1.2.1 *The concatenation $\omega^{(1)} \circ \omega^{(2)}$ of an M-step self-avoiding walk $\omega^{(2)}$ to an N-step self-avoiding walk $\omega^{(1)}$ is the $(N + M)$-step walk ω, which in general need not be self-avoiding, given by*

$$
\begin{aligned}
\omega(k) &= \omega^{(1)}(k), & k &= 0,\ldots,N \\
\omega(k) &= \omega^{(1)}(N) + \omega^{(2)}(k - N) - \omega^{(2)}(0), & k &= N+1,\ldots,N+M.
\end{aligned}
$$

The product $c_N c_M$ is equal to the cardinality of the set of $(N + M)$-step simple random walks which are self-avoiding for the initial N steps and the final M steps, but which may not be completely self-avoiding. This can be seen by concatenations of M-step walks to N-step walks, and implies that

$$c_{N+M} \le c_N c_M. \tag{1.2.3}$$

In fact equality holds in (1.2.3) only if N or M is zero, since otherwise there will be at least one M-step walk whose concatenation with a given N-step walk fails to be self-avoiding. Taking logarithms in (1.2.3) shows that the sequence $\{\log c_n\}$ is *subadditive*:

$$\log c_{N+M} \le \log c_N + \log c_M. \tag{1.2.4}$$

The existence of the limit (1.2.1) is a consequence of (1.2.4) and the following standard result; this was first observed by Hammersley and Morton (1954).

Lemma 1.2.2 *Let $\{a_n\}_{n \ge 1}$ be a sequence of real numbers which is subadditive, i.e., $a_{n+m} \le a_n + a_m$. Then the limit $\lim_{n \to \infty} n^{-1} a_n$ exists in $[-\infty, \infty)$ and is equal to*

$$\lim_{n \to \infty} \frac{a_n}{n} = \inf_{n \ge 1} \frac{a_n}{n}. \tag{1.2.5}$$

Proof. It suffices to show that

$$\limsup_{n \to \infty} \frac{a_n}{n} \le \frac{a_k}{k} \tag{1.2.6}$$

for every k, since taking the $\liminf_{k \to \infty}$ in (1.2.6) gives existence of the limit, and then (1.2.5) can be seen by taking the $\inf_{k \ge 1}$ in (1.2.6).

To prove (1.2.6), we fix k and let

$$A_k = \max_{1 \le r \le k} a_r. \tag{1.2.7}$$

Given a positive integer n we let j denote the largest integer which is strictly less than n/k. Then $n = jk + r$ for some integer r with $1 \le r \le k$. Using subadditivity, we have

$$a_n \le j a_k + a_r \le \frac{n}{k} a_k + A_k. \tag{1.2.8}$$

Dividing by n and taking the $\limsup_{n\to\infty}$ then gives (1.2.6).

Equation (1.2.5) shows that $\lim_{n\to\infty} n^{-1} a_n < \infty$. In general, the possibility that the limit equals $-\infty$ cannot be excluded, as is illustrated by the example of $a_n = -n^2$. For many applications, however, this is ruled out by an *a priori* bound such as $a_n \geq 0$.   □

Together with (1.2.4), Lemma 1.2.2 implies the existence of the limit $\log \mu \equiv \lim_{N\to\infty} N^{-1} \log c_N$, and hence gives (1.2.1). In fact (1.2.5) shows more:

$$\log \mu = \inf_{N\geq 1} N^{-1} \log c_N, \tag{1.2.9}$$

and hence

$$\mu^N \leq c_N, \quad N \geq 1. \tag{1.2.10}$$

This inequality can be summarized by the statement $\gamma \geq 1$, where γ is as introduced in (1.1.4), although strictly speaking we do not know that γ exists. Equation (1.2.10) also yields $\mu \leq c_N^{1/N}$. This gives a sequence of upper bounds for μ, but they converge to μ very slowly. A better bound is

$$\mu \leq \left(\frac{c_N}{c_1}\right)^{1/(N-1)}, \quad N \geq 2. \tag{1.2.11}$$

References for this and other improvements are given in the Notes.

Another sequence of upper bounds for μ can be obtained by considering walks which are self-avoiding only over a finite time scale or *memory* τ. We define $c_{N,\tau}$ to be the number of N-step walks ω beginning at the origin, for which $\omega(i) \neq \omega(j)$ whenever $0 < |i - j| \leq \tau$. Self-intersections occurring after an interval of more than τ steps are permitted. For example, $c_{N,2} = 2d(2d-1)^{N-1}$ for $N \geq 1$, since memory $\tau = 2$ simply rules out immediate reversals. For $\tau \geq N$, $c_{N,\tau} = c_N$. Memory $\tau = 0$ corresponds to the simple random walk.

The sequence $\{\log c_{N,\tau}\}_{N=1}^{\infty}$ is subadditive for every τ (for the same reason that $\{\log c_N\}_{N=1}^{\infty}$ is), and hence by Lemma 1.2.2 there is a μ_τ such that

$$\mu_\tau = \lim_{N\to\infty} c_{N,\tau}^{1/N} = \inf_{N\geq 1} c_{N,\tau}^{1/N}. \tag{1.2.12}$$

Since $c_{N,\tau} \geq c_N$, μ_τ provides an upper bound for μ. The next lemma shows that this sequence of upper bounds converges monotonically to μ.

Lemma 1.2.3 $\mu_\tau \searrow \mu$ *as* $\tau \to \infty$.

Proof. For $\sigma \leq \tau$, $c_{N,\sigma} \geq c_{N,\tau}$ and hence $\mu_\sigma \geq \mu_\tau$. By (1.2.12), $\mu_\tau \leq c_{N,\tau}^{1/N}$ for all N, τ. Taking $N = \tau$ gives

$$\mu \leq \mu_\tau \leq c_{\tau,\tau}^{1/\tau} = c_\tau^{1/\tau}. \tag{1.2.13}$$

Taking the limit $\tau \to \infty$ and using (1.2.1) gives the desired result. $\square$

The connective constant for the walk with memory $\tau = 4$ was shown in Fisher and Sykes (1959) to be given by the largest root of the cubic equation

$$\theta^3 - 2(d-1)\theta^2 - 2(d-1)\theta - 1 = 0. \tag{1.2.14}$$

For $d = 2$ this gives $\mu_4(2) = 2.8312$, where we have made the dimension dependence explicit by writing $\mu_\tau(d)$.

A number of investigations into the self-avoiding walk have approached the problem via the limit of finite memory walks as the memory goes to infinity. This approach was used in particular by Brydges and Spencer (1985) in applying their lace expansion to study weakly self-avoiding walk for $d > 4$, and will be adopted in Section 6.8 to obtain an upper bound in high dimensions on $c_N(0, x)$, the number of N-step self-avoiding walks which begin at the origin and end at x.

A lower bound on μ can be obtained in terms of *bridges*.

Definition 1.2.4 *An N-step bridge is defined to be an N-step self-avoiding walk ω whose first components satisfy the inequality*

$$\omega_1(0) < \omega_1(i) \le \omega_1(N)$$

for $1 \le i \le N$. The number of N-step bridges starting at the origin is denoted b_N. By convention, $b_0 = 1$.

The concatenation of two bridges will always yield another bridge, so

$$b_M b_N \le b_{M+N}. \tag{1.2.15}$$

Hence $\{-\log b_n\}$ is subadditive and so by Lemma 1.2.2 the limit

$$\mu_{Bridge} \equiv \lim_{n \to \infty} b_n^{1/n} = \sup_{n \ge 1} b_n^{1/n} \tag{1.2.16}$$

exists. Clearly $b_n \le c_n$. Therefore $\mu_{Bridge} \le \mu$, and so by (1.2.16)

$$b_N^{1/N} \le \mu_{Bridge} \le \mu. \tag{1.2.17}$$

In Section 3.1 it will be shown that in fact $\mu_{Bridge} = \mu$. Although the lower bound (1.2.17) is very slowly convergent, a more sophisticated use of bridges leads to better lower bounds. References can be found in the Notes at the end of this chapter.

We conclude this section with a table showing the current best rigorous upper and lower bounds on μ, together with estimates of the precise value, for the hypercubic lattice in dimensions $d = 2, 3, 4, 5, 6$.

d	lower bound	estimate	upper bound
2	2.61987^a	$2.6381585 \pm 0.0000010^d$	2.69576^b
3	4.43733^c	4.6839066 ± 0.0002^e	4.756^b
4	6.71800^c	6.7720 ± 0.0005^f	6.832^b
5	8.82128^c	8.83861^g	8.881^b
6	10.871199^c	10.87879^g	10.903^b

Table 1.1: Current best rigorous upper and lower bounds on the hypercubic lattice connective constant μ, together with estimates of actual values.
a) Conway and Guttmann (to be published), b) Alm (1992), c) Hara and Slade (1992b), d) Guttmann and Enting (1988), e) Guttmann (1987), f) Guttmann (1978), g) Guttmann (1981).

1.3   Generating functions

A common tool for understanding the behaviour of a sequence is its generating function. The generating function of the sequence $\{c_N\}$ is defined by

$$\chi(z) = \sum_{N=0}^{\infty} c_N z^N = \sum_{\omega} z^{|\omega|}. \qquad (1.3.1)$$

The sum over ω is the sum over all self-avoiding walks, of arbitrary length $|\omega|$, which begin at the origin. The parameter z is known as the *activity*. Physically the activity occurs in the study of a canonical ensemble of polymers of variable length, and in this context is nonnegative. From a mathematical point of view, however, it will sometimes be useful to consider χ to be an analytic function of complex z.

Given two sites x and y, let $c_N(x, y)$ be the number of N-step self-avoiding walks ω with $\omega(0) = x$ and $\omega(N) = y$. The *two-point function* is the generating function for the sequence $c_N(x, y)$, i.e.,

$$G_z(x, y) = \sum_{N=0}^{\infty} c_N(x, y) z^N = \sum_{\omega:x \to y} z^{|\omega|}. \qquad (1.3.2)$$

On the right side, the sum over ω is the sum over all self-avoiding walks, of arbitrary length, which begin at x and end at y. This is clearly translation invariant, so $G_z(x, y) = G_z(0, y - x)$. The two-point function is the self-avoiding walk analogue of the simple random walk Green function with killing rate $1 - 2dz$:

$$C_z(x,y) = \sum_{N=0}^{\infty} p_N(x,y)(2d\dot{z})^N, \tag{1.3.3}$$

where $p_N(x,y)$ is the probability that an N-step simple random walk beginning at x ends at y.

The generating function for c_N can be written in terms of the two-point function as

$$\chi(z) = \sum_{x \in Z^d} G_z(0,x). \tag{1.3.4}$$

In analogy with spin systems (see Section 2.3) we will refer to the generating function $\chi(z)$ as the *susceptibility*. The power series defining the susceptibility has radius of convergence

$$z_c \equiv \left[\lim_{N \to \infty} c_N^{1/N} \right]^{-1} = \frac{1}{\mu}, \tag{1.3.5}$$

and hence defines an analytic function in the *complex* parameter z if $|z| < z_c$. Since $c_N(0,x) \leq c_N$, the two-point function has radius of convergence at least z_c. It will be shown in Section 3.2 that in fact the radius of convergence is equal to z_c, for all $x \neq 0$. We will refer to z_c as the *critical point*, since it plays a role analogous to the critical point in statistical mechanical systems such as the Ising model or percolation.

It follows from (1.2.10) that

$$\chi(z) \geq \sum_{N=0}^{\infty} (\mu z)^N = \frac{1}{1 - \mu z} \tag{1.3.6}$$

and hence χ is "continuous" at the critical point, in the sense that $\chi(z) \to \infty$ as $z \nearrow z_c$. The manner of divergence of $\chi(z)$ at the critical point is related to the behaviour of the coefficients c_N for large N. To see this, we proceed as follows.

First we introduce the notation

$$f(x) \simeq g(x) \quad \text{as } x \to x_0 \tag{1.3.7}$$

to mean that there are positive constants C_1 and C_2 such that

$$C_1 g(x) \leq f(x) \leq C_2 g(x) \tag{1.3.8}$$

uniformly for x near its limiting value. Assuming that there is a γ such that

$$c_N \simeq \mu^N N^{\gamma-1} \quad \text{as } N \to \infty, \tag{1.3.9}$$

it can be concluded that

$$\chi(z) \simeq (z_c - z)^{-\gamma} \quad \text{as } z \nearrow z_c, \tag{1.3.10}$$

as follows. We write $z = \mu^{-1}e^{-t}$, so that $t \simeq z_c - z$. By the definition of $\chi(z)$,

$$\chi(z) \simeq \sum_{N=1}^{\infty} N^{\gamma-1} e^{-tN} \simeq \int_1^{\infty} x^{\gamma-1} e^{-tx} dx$$

$$= t^{-\gamma} \int_t^{\infty} y^{\gamma-1} e^{-y} dy \simeq t^{-\gamma}.$$

In the above the sum can be replaced by the integral using Riemann sum approximations. The second integral converges as $t \searrow 0$, since by (1.2.10) $\gamma \geq 1$. Thus it is conjectured that

$$\chi(z) \sim A'(z_c - z)^{-\bar{\gamma}} \quad \text{as } z \nearrow z_c, \tag{1.3.11}$$

with $\bar{\gamma} = \gamma$.

As for the converse, it does not follow directly from (1.3.10) that (1.3.9) holds, without further assumptions. In general, the problem of extracting the large-n asymptotics of a sequence from the manner of divergence of its generating function is a Tauberian problem. An example of a Tauberian theorem providing a converse to the above argument will be given in Lemma 6.3.4.

Power law behaviour such as (1.3.10) is also observed for spin systems and percolation, and is characteristic of critical phenomena. It follows from (1.3.6) that $\bar{\gamma} \geq 1$, assuming that $\bar{\gamma}$ exists. In four dimensions, where it is believed that $c_N \sim A\mu^N (\log N)^{1/4}$, we expect similarly that $\chi(z) \sim A'(z_c - z)^{-1} |\log(z_c - z)|^{1/4}$.

The analogue $\chi_0(z)$ of $\chi(z)$ for simple random walk can be calculated explicitly:

$$\chi_0(z) = \sum_{N=0}^{\infty} (2dz)^N = \frac{1}{1 - 2dz}.$$

Thus the mean-field value of $\bar{\gamma}$ is 1, which not surprisingly is equal to the mean-field value for γ. The inequality $\bar{\gamma} \geq 1$ is an example of a *mean-field bound*. There is a sufficient condition for the opposite bound $\bar{\gamma} \leq 1$ known as the bubble condition, which is known to hold for $d \geq 5$ (and is believed not to hold for $d \leq 4$), and which will be discussed in detail in Section 1.5. There are many examples in critical phenomena of rigorous mean-field bounds, but, as mentioned in Section 1.1, no general proof is known of the mean-field bound $\nu \geq 1/2$.

We now turn our attention to the long distance behaviour of the two-point function. Below the critical point the two-point function decays exponentially. To see this, we note that $c_N(0, x) = 0$ for $N < ||x||_\infty$, and hence

$$G_z(0, x) = \sum_{N=||x||_\infty}^{\infty} c_N(0, x) z^N \leq \sum_{N=||x||_\infty}^{\infty} c_N z^N. \qquad (1.3.12)$$

Since $c_N^{1/N} \to \mu$ by (1.2.1), for any $\epsilon > 0$ there is a positive K_ϵ such that

$$c_N \leq K_\epsilon (\mu + \epsilon)^N \qquad (1.3.13)$$

for all $N \geq 1$. Given a positive $z < z_c = \mu^{-1}$, we choose $\epsilon(z) > 0$ such that $\theta_z \equiv (\mu + \epsilon(z))z < 1$. Then substitution of (1.3.13) into (1.3.12) gives

$$G_z(0, x) \leq C_z \exp[-|\log \theta_z| \, ||x||_\infty], \qquad (1.3.14)$$

with $C_z = K_{\epsilon(z)}(1 - \theta_z)^{-1}$. This shows the desired exponential decay of the subcritical two-point function.

We define the *mass* $m(z)$ to be the rate of exponential decay of the two-point function along a coordinate axis:

$$m(z) = \liminf_{n \to \infty} \frac{-\log G_z(0, (n, 0, \ldots, 0))}{n}. \qquad (1.3.15)$$

In Theorem 4.1.3 it will be shown that in fact the lim inf in the definition of m can be replaced by the limit. The *correlation length* $\xi(z)$, defined by $\xi(z) = m(z)^{-1}$, provides a characteristic length scale for the model.

The mass $m(z)$ is clearly not infinite for $0 < z < z_c$: considering only the shortest self-avoiding walk from 0 to $(n, 0, \ldots, 0)$ gives

$$G_z(0, (n, 0, \ldots, 0)) \geq z^n, \qquad (1.3.16)$$

and hence $m(z) \leq -\log z$. By (1.3.14), $m(z) > 0$ for $z \in (0, z_c)$. By definition the mass is a nonincreasing function of positive z, and in Section 4.1 it will be shown that $m(z) \searrow 0$ as $z \nearrow z_c$. Given that the radius of convergence of $G_z(0, x)$ is z_c for all $x \neq 0$, it follows that $m(z) = -\infty$ for $z > z_c$. It has been proven that $m(z_c) = 0$ for $d \geq 5$; see Corollary 6.1.7. Although this is believed to be true in all dimensions, a negative mass at the critical point has not yet been ruled out rigorously in dimensions 2, 3 or 4.

Since the mass $m(z)$ approaches zero as $z \nearrow z_c$, it follows that the correlation length $\xi(z) = m(z)^{-1}$ diverges as $z \nearrow z_c$. It is believed that the manner of divergence of $\xi(z)$ is via a power law of the form

$$\xi(z) \sim \text{const.}(z_c - z)^{-\nu} \quad \text{as } z \nearrow z_c. \qquad (1.3.17)$$

Formal scaling theory predicts that $\bar{\nu} = \nu$; this will be discussed in Section 2.1. The equality of these two critical exponents is part of a general belief that all length scales for the self-avoiding walk should be governed by the same critical exponent. The same belief generally applies to other statistical mechanical models as well.

Another correlation length, ξ_p, known as the *correlation length of order p*, is defined for each $p > 0$ by

$$\xi_p(z) = \left[\frac{\sum_\omega |\omega(|\omega|)|^p z^{|\omega|}}{\sum_\omega z^{|\omega|}} \right]^{1/p} = \left[\frac{\sum_x |x|^p G_z(0, x)}{\sum_x G_z(0, x)} \right]^{1/p} . \qquad (1.3.18)$$

By Hölder's inequality ξ_p is increasing in p. A formal argument similar to that showing $\bar{\nu} = \nu$ gives

$$\xi_p(z) \sim \text{const.}(z_c - z)^{-\nu_p} \text{ as } z \nearrow z_c,$$

with $\nu_p = \nu$ for all p.

For $p = 2$ there is no need to appeal to scaling theory to argue that $\nu_2 = \nu$. Instead we can argue as we did for the equality of γ and $\bar{\gamma}$, in the following way. We will assume that there exist exponents γ and ν such that $c_N \simeq \mu^N N^{\gamma-1}$ and $\langle |\omega|^2 \rangle \simeq N^{2\nu}$, and show that this implies that $\xi_2(z) \simeq (z_c - z)^{-\nu}$. Given the assumptions, we have

$$\sum_x |x|^2 G_z(0, x) = \sum_N z^N \sum_{\omega:|\omega|=N} |\omega(N)|^2 \qquad (1.3.19)$$

$$\simeq \sum_N z^N N^{2\nu} c_N \simeq \sum_N z^N N^{2\nu+\gamma-1} \mu^N .$$

Again writing $z = \mu^{-1} e^{-t}$, we obtain

$$\sum_x |x|^2 G_z(0, x) \simeq \sum_N N^{2\nu+\gamma-1} e^{-tN}$$

$$\simeq \int_1^\infty x^{2\nu+\gamma-1} e^{-tx} dx \simeq t^{-2\nu-\gamma}. \qquad (1.3.20)$$

This implies that

$$\xi_2(z)^2 \simeq \frac{t^{-2\nu-\gamma}}{\chi(z)} \simeq t^{-2\nu-\gamma+\bar{\gamma}}. \qquad (1.3.21)$$

Using $\bar{\gamma} = \gamma$ it follows that $\xi_2(z) \simeq (z_c - z)^{-\nu}$, so $\nu_2 = \nu$.

1.4 Critical exponents

So far we have introduced the five critical exponents $\gamma, \bar{\gamma}, \nu, \bar{\nu}, \nu_p$. It was shown in Section 1.3 that if γ exists then $\gamma = \bar{\gamma}$. Heuristic arguments that $\nu = \bar{\nu} = \nu_p$ (for $0 < p < \infty$) will be given in the Section 2.1. The exponents were defined as follows:

$$c_N \sim A\mu^N N^{\gamma-1} \tag{1.4.1}$$

$$\chi(z) \sim A'(z_c - z)^{-\bar{\gamma}} \tag{1.4.2}$$

$$\langle |\omega(N)|^2 \rangle \sim DN^{2\nu} \tag{1.4.3}$$

$$\xi(z) \sim B(z_c - z)^{-\bar{\nu}} \tag{1.4.4}$$

$$\xi_p(z) \sim B_p(z_c - z)^{-\nu_p}. \tag{1.4.5}$$

We have written the above relations as if the various quantities involved are *asymptotically* given by power laws. This is consistent with the existing rigorous results, but some authors prefer a more conservative definition of the exponents. For example, one could require only that $c_N \simeq \mu^N N^{\gamma-1}$ [see (1.3.7)], with corresponding statements for the other exponents. A weaker definition, appearing sometimes in the literature, is to define the exponents by equations such as

$$\bar{\gamma} = -\lim_{z \nearrow z_c} \frac{\log \chi(z)}{\log(z_c - z)}, \tag{1.4.6}$$

but we will not need this definition. We shall take the optimistic view that the power law behaviour is asymptotic, although none of (1.4.1)–(1.4.5) has been proven in dimensions 2, 3, or 4 for any of these definitions of the exponents.

We will use the notation

$$f(x) \approx g(x) \tag{1.4.7}$$

in informal (nonrigorous) discussions to mean that $f(x)$ and $g(x)$ appear to have the same asymptotic behaviour in some sense which we will not attempt to specify.

In this section three additional critical exponents η, α_{sing}, and Δ_4 will be introduced. All these critical exponents are believed to be universal in the sense that they depend only on the dimension d of the lattice. In particular, the exponents are believed to be the same for the nearest-neighbour self-avoiding walk on $\mathbf{Z}^d$ as for a self-avoiding walk on $\mathbf{Z}^d$ in which steps can be within a fixed finite set $\Omega \subset \mathbf{Z}^d$ which is symmetric with respect to the symmetries of the lattice. Moreover the exponents ought even to be the same for a self-avoiding walk which can take unboundedly long steps, provided the weight of a step decays rapidly enough with its length (e.g., exponentially). This independence of the step set Ω is partially borne out in the rigorous results in high dimensions in Chapter 6.

We begin with the exponent η, which describes the conjectured long-distance behaviour of the two-point function at the critical point. Given that $m(z) \to 0$ as $z \nearrow z_c$, and given the belief that $m(z_c) = 0$ in all dimensions, it might be expected that the two-point function decays via a power law at the critical point. For simple random walk (with $d > 2$) the mass is certainly zero at the critical point, as it is well-known that the critical simple random walk two-point function $C_{1/2d}(0, x)$ decays like $|x|^{2-d}$ at large distances [see for example Lawler (1991)]. The conjectured large distance behaviour of the critical self-avoiding walk two-point function is

$$G_{z_c}(0, x) \sim \frac{C}{|x|^{d-2+\eta}} \quad \text{as } |x| \to \infty, \qquad (1.4.8)$$

where C is a constant. This is believed to hold in all dimensions $d \geq 2$, including $d = 2$. Comparison with the simple random walk decay yields the mean field value 0 for η. Unfortunately it has not yet been proved rigorously that $G_{z_c}(0, x)$ is even finite for $d = 2, 3$ or 4, for any value of $x \neq 0$. For $d \geq 5$ somewhat weaker decay than (1.4.8) has been proved; see Theorem 6.1.6.

Assuming that (1.4.8) does provide the correct behaviour, it follows from the fact that the susceptibility is infinite at the critical point that $\eta \leq 2$. The value of η is believed to be determined from the values of γ and ν according to Fisher's scaling relation

$$\gamma = (2 - \eta)\nu. \qquad (1.4.9)$$

The hypotheses leading to (1.4.9) will be discussed in Section 2.1. Inserting the conjectured values for γ and ν given in (1.1.11) and (1.1.12) into (1.4.9) gives the values for η appearing in Table 1.2. In contrast to γ and ν, the renormalization group predicts no logarithmic corrections to η in four dimensions. Logarithmic corrections are however expected in higher order terms in the asymptotic expansion of the critical two-point function in four dimensions.

One way to gain insight into the long distance behaviour of the critical two-point function is to examine the behaviour of its Fourier transform near the origin. In general, given a function $f(x)$ on the lattice whose absolute value is summable, we define its Fourier transform by

$$\hat{f}(k) = \sum_{x \in Z^d} f(x)e^{ik \cdot x}, \quad k \in [-\pi, \pi]^d. \qquad (1.4.10)$$

d	2	3	≥ 4
γ	$\frac{43}{32}$	$1.162\ldots$	1
ν	$\frac{3}{4}$	$0.59\ldots$	$\frac{1}{2}$
η	$\frac{5}{24}$	$0.03\ldots$	0

Table 1.2: Conjectured values of γ, ν, η.

It is generally expected in critical phenomena that (1.4.8) is associated with behaviour of the form

$$\hat{G}_{z_c}(k) \sim \frac{C'}{k^{2-\eta}} \quad \text{as } k \to 0 \tag{1.4.11}$$

for some constant C'. [However (1.4.8) and (1.4.11) are not mathematically equivalent — an example of a function satisfying (1.4.11) but not (1.4.8) is given in the Notes at the end of the chapter.] Equation (1.4.11) has been established for $d \geq 5$ with $\eta = 0$ (see Theorem 6.1.6), but not yet for $d = 2, 3$ or 4. The conjectured values of η are all nonnegative. It is thus suggestive to conjecture the *infrared bound*

$$\hat{G}_z(k) \leq \frac{C}{k^2}, \tag{1.4.12}$$

with C independent of $k \in [-\pi, \pi]^d$ and $z \leq z_c$. For the nearest-neighbour Ising model and other reflection-positive spin systems the infrared bound is known rigorously to hold and was of considerable importance in the proof of mean-field behaviour of such models above four dimensions. For the self-avoiding walk it is still an open problem to prove the infrared bound in dimensions 2, 3 or 4, but in higher dimensions it has been proved (see Theorem 6.1.6). It is worth noting that the infrared bound is believed by some to be false for percolation and lattice animals below dimensions six and eight respectively (see Section 5.5 for more details about these models).

The exponent α_{sing} describes the behaviour of the number $c_N(0, x)$ of N-step self-avoiding walks which begin at the origin and end at x, as $N \to \infty$ with x fixed. For x equal to a nearest-neighbour e of the origin, $c_N(0, e)$ is closely related to the number of self-avoiding polygons. Self-avoiding polygons will be studied in detail in Section 3.2. It will be shown in Corollary 3.2.6 that the leading asymptotic behaviour of $c_N(0, x)$ as $N \to \infty$ is μ^N. As is the case for c_N, this leading behaviour is believed to have a power law correction of the form

$$c_N(0, x) \sim B\mu^N N^{\alpha_{sing}-2}. \qquad (1.4.13)$$

Here x is fixed and nonzero, and $N \to \infty$ through a sequence of values with the same parity as $\|x\|_1$. It is believed that α_{sing} is independent of x, and we formalize this conjecture for future reference as follows.

Conjecture 1.4.1 *For every pair of nonzero points x and y in $\mathbf{Z}^d$, there exist positive constants A_1 and A_2 and an integer N_0 (all depending on x and y) such that*

$$A_1 c_N(0, y) \leq c_N(0, x) \leq A_2 c_N(0, y) \qquad \text{for all } N \geq N_0$$

if $\|x - y\|_1$ is even, and

$$A_1 c_{N+1}(0, y) \leq c_N(0, x) \leq A_2 c_{N+1}(0, y) \qquad \text{for all } N \geq N_0$$

if $\|x - y\|_1$ is odd.

A special case of this conjecture is proven in Proposition 7.4.4. The value of B is also believed to be independent of x (as it is for simple random walk). For simple random walk the local central limit theorem states that the probability $p_N(0, x)$ that a simple random walk starting at 0 ends after N steps at x is given asymptotically by $\text{const.}N^{-d/2}\exp[-d|x|^2/2N] \sim \text{const.}N^{-d/2}$, as $N \to \infty$. Hence the mean-field value of $\alpha_{sing} - 2$ is $-d/2$. The value of α_{sing} is believed to be determined from the value of ν and the dimension d via the hyperscaling relation

$$\alpha_{sing} - 2 = -d\nu. \qquad (1.4.14)$$

This hyperscaling relation will be discussed in Section 2.1. If (1.4.14) and the values given for ν in Table 1.2 are true, then it would follow that $\alpha_{sing} - 2 < -1$ in all dimensions and hence that the critical two-point function $G_{z_c}(0, x) = \sum_N c_N(0, x)\mu^{-N}$ is finite in all dimensions, including $d = 2$. This is in contrast to the situation for simple random walk, where in two dimensions the Green function is infinite at the critical point.

The strongest bounds on $c_N(0, x)$ are for high dimensions. It is proved in Theorem 6.1.3 that for d sufficiently large, or for $d > 4$ for a walk allowed to take long enough steps, that

$$c_N(0, x) \leq B\mu^N N^{-d/2} \qquad (1.4.15)$$

for some constant B. Although this bound has not yet been extended to all $d \geq 5$ for the nearest-neighbour model, the weaker result that for all

$a < -1 + d/2$

$$\sup_{x} \sum_{N=0}^{\infty} N^{a} c_{N}(0, x) \mu^{-N} < \infty \tag{1.4.16}$$

has been proved for all $d \geq 5$; see Theorem 6.1.4. Either of (1.4.15) or (1.4.16) could be summarized by the inequality $\alpha_{sing} - 2 \leq -d/2$. For dimensions 2, 3 and 4, the best results are for x a nearest neighbour of the origin. These results are described in Section 8.1 and can be summarized by the inequalities

$$\alpha_{sing} \leq \frac{5}{2} \quad (d = 2) \tag{1.4.17}$$

$$\alpha_{sing} \leq 2 \quad (d = 3) \tag{1.4.18}$$

$$\alpha_{sing} < 2 \quad (d \geq 4). \tag{1.4.19}$$

Finally, we introduce the critical exponent Δ_4. Let $c_{N_1,N_2}(x)$ denote the number of pairs of self-avoiding walks of lengths N_1 and N_2 and respective starting points 0 and x which intersect each other, and let

$$c_{N_1,N_2} = \sum_{x} c_{N_1,N_2}(x). \tag{1.4.20}$$

This quantity occurs in the study of interacting polymer chains. The asymptotic behaviour of c_{N_1,N_2} is believed to be given by

$$c_{N_1,N_2} \sim \text{const.} \mu^{N_1 + N_2} N_1^{2\Delta_4 + \gamma - 2} f(N_1/N_2) \quad \text{as } N_1, N_2 \to \infty \tag{1.4.21}$$

for some critical exponent Δ_4 and universal scaling function f. The quantity

$$g(z) = \xi(z)^{-d} \chi(z)^{-2} \sum_{N_1,N_2=0}^{\infty} c_{N_1,N_2} z^{N_1 + N_2} \tag{1.4.22}$$

represents a kind of average intersection probability. In quantum field theory, an analogue of $g(z)$ is referred to as the renormalized coupling constant. An informal calculation in which (1.4.21) is substituted into (1.4.22) leads to

$$g(z) \sim \text{const.} (z_c - z)^{d\nu - 2\Delta_4 + \gamma} \quad \text{as } z \nearrow z_c. \tag{1.4.23}$$

For simple random walk it is known that the analogue of $g(z)$ satisfies (1.4.23), with $d/2 - 2\Delta_4 + 1 = 0$ for $d = 2, 3, 4$ (with a logarithmic correction in four dimensions), and $\Delta_4 = 3/2$ for $d \geq 5$; see Section 10.3. Similar behaviour is believed to hold for the self-avoiding walk. In particular, for the self-avoiding walk it is believed that in dimensions 2, 3 and 4 the hyperscaling relation

$$d\nu - 2\Delta_4 + \gamma = 0 \tag{1.4.24}$$

is satisfied. Heuristic arguments in support of this hyperscaling relation will be given in Section 2.1. It has been proved that $\Delta_4 = 3/2$ for the self-avoiding walk in dimensions $d \geq 6$ (see Theorem 1.5.5 and the Remark following its statement); it is believed that $\Delta_4 = 3/2$ for all $d > 4$.

Elementary bounds on Δ_4 can be obtained as follows. Consider all pairs of N-step self-avoiding walks $\omega^{(1)}$ and $\omega^{(2)}$ which intersect somewhere, with $\omega^{(1)}$ beginning at the origin and $\omega^{(2)}$ beginning anywhere. There are $c_{N,N}$ such pairs. Since there are $N+1$ possible sites on each of $\omega^{(1)}$ and $\omega^{(2)}$ where an intersection can occur, $c_{N,N} \leq (N+1)^2 c_N^2$. On the other hand if we count only those pairs for which $\omega^{(2)}(0) = \omega^{(1)}(j)$ for some $j = 0, \ldots, n$, we obtain $c_{N,N} \geq (N+1)c_N^2$. Together these bounds give

$$\frac{\gamma + 1}{2} \leq \Delta_4 \leq \frac{\gamma + 2}{2}. \tag{1.4.25}$$

This can be rewritten as

$$1 \leq 2\Delta_4 - \gamma \leq 2. \tag{1.4.26}$$

The upper bound implies that the hyperscaling relation (1.4.24) fails if $d\nu > 2$. Since it is known that $\nu = 1/2$ for $d \geq 5$ (see Section 6.1), this implies failure of hyperscaling for $d > 4$.

1.5   The bubble condition

The lower bound on the susceptibility (1.3.6) can be rewritten in terms of $z_c = \mu^{-1}$ as

$$\chi(z) \geq \frac{z_c}{z_c - z} \tag{1.5.1}$$

for $0 \leq z < z_c$. The bubble condition is a sufficient condition for the complementary bound

$$\chi(z) \leq \frac{C}{z_c - z} \tag{1.5.2}$$

for some constant C and for $0 \leq z < z_c$. Thus the bubble condition implies that $\bar\gamma = 1$ in the sense that

$$\chi(z) \simeq (z_c - z)^{-1} \quad \text{as } z \nearrow z_c. \tag{1.5.3}$$

The bubble condition was proven to hold in five or more dimensions in Hara and Slade (1992b) (see Section 6.1), and is believed not to hold for $d \leq 4$.

To state the bubble condition we first introduce the *bubble diagram*

$$\mathsf{B}(z) = \sum_{x \in Z^d} G_z(0, x)^2. \tag{1.5.4}$$

The name "bubble diagram" comes from a Feynman diagram notation in which the two-point function or *propagator* evaluated at sites x and y is denoted by a line terminating at x and y. In this notation

$$\mathsf{B}(z) \;=\; \sum_x 0 \;\bigcirc\; x \;=\; \bigcirc$$

where in the diagram on the right it is implicit that one vertex is fixed at the origin and the other is summed over the lattice. The bubble diagram can be rewritten in terms of the Fourier transform of the two-point function, using (1.5.4) and the Parseval relation, as

$$\mathsf{B}(z) = \|G_z(0,\cdot)\|_2^2 = \|\hat{G}_z\|_2^2 = \int_{[-\pi,\pi]^d} \hat{G}_z(k)^2 \frac{d^d k}{(2\pi)^d}. \tag{1.5.5}$$

Definition 1.5.1 *The* bubble condition *states that the bubble diagram is finite at the critical point, i.e.*

$$\mathsf{B}(z_c) < \infty.$$

In view of the definition of η in (1.4.8) or (1.4.11), it follows from (1.5.5) that the bubble condition is satisfied provided $\eta > (4-d)/2$. Hence the bubble condition for $d > 4$ is implied by the infrared bound $\eta \geq 0$. If the values for η given in Table 1.2 are correct, then the bubble condition will not hold in dimensions $2, 3$ or 4, with the divergence of the bubble diagram being only logarithmic in four dimensions.

The next lemma provides the principal step in proving that the bubble condition implies (1.5.2) and hence implies (1.5.3).

Lemma 1.5.2 *For any $z \in [0, z_c)$, the derivative of the susceptibility satisfies*

$$\frac{\chi(z)^2}{\mathsf{B}(z)} - \chi(z) \leq z\chi'(z) \leq \chi(z)^2 - \chi(z). \tag{1.5.6}$$

Proof. Below the critical point the derivative of χ can be obtained by term by term differentiation:

$$z\chi'(z) = \sum_\omega |\omega| z^{|\omega|} = \sum_\omega (|\omega| + 1) z^{|\omega|} - \chi(z), \tag{1.5.7}$$

where the sums are over self-avoiding walks of arbitrary length which begin at the origin. The summation on the right side can be written

$$\sum_y \sum_{\omega:0 \to y} \sum_x I[\omega(j) = x \text{ for some } j] z^{|\omega|}$$

$$= \sum_{x,y} \sum_{\substack{\omega^{(1)}\,:\,0\,\to\,x \\ \omega^{(2)}\,:\,x\,\to\,y}} z^{|\omega^{(1)}|+|\omega^{(2)}|} I[\omega^{(1)} \cap \omega^{(2)} = \{x\}]$$

$$\equiv \; Q(z), \tag{1.5.8}$$

where I denotes the indicator function and the last summation is over self-avoiding walks $\omega^{(1)}$ and $\omega^{(2)}$ of arbitrary length and having the prescribed endpoints. Then

$$z\chi'(z) = Q(z) - \chi(z). \tag{1.5.9}$$

The upper bound in (1.5.6) then follows since the indicator function in the middle member of (1.5.8) is bounded above by one.

The first step toward obtaining the lower bound is to use the inclusion-exclusion relation in the form

$$I[\omega^{(1)} \cap \omega^{(2)} = \{x\}] = 1 - I[\omega^{(1)} \cap \omega^{(2)} \neq \{x\}].$$

This gives

$$Q(z) = \chi(z)^2 - \sum_{x,y} \sum_{\substack{\omega^{(1)}\,:\,0\,\to\,x \\ \omega^{(2)}\,:\,x\,\to\,y}} z^{|\omega^{(1)}|+|\omega^{(2)}|} I[\omega^{(1)} \cap \omega^{(2)} \neq \{x\}]. \tag{1.5.10}$$

In the last term on the right side of (1.5.10), let $w = \omega^{(2)}(l)$ be the site of the last intersection of $\omega^{(2)}$ with $\omega^{(1)}$, where time is measured along $\omega^{(2)}$ beginning at its starting point x. Then the portion of $\omega^{(2)}$ corresponding to times greater than l must avoid all of $\omega^{(1)}$. Relaxing the restrictions that this portion of $\omega^{(2)}$ avoid both the remainder of $\omega^{(2)}$ and the part of $\omega^{(1)}$ linking w to x gives the upper bound

$$\sum_{x,y} \sum_{\substack{\omega^{(1)}\,:\,0\,\to\,x \\ \omega^{(2)}\,:\,x\,\to\,y}} z^{|\omega^{(1)}|+|\omega^{(2)}|} I[\omega^{(1)} \cap \omega^{(2)} \neq \{x\}] \leq Q(z)[B(z) - 1].$$

$$\tag{1.5.11}$$

Here the factor $B(z) - 1$ arises from the two paths joining w and x. The upper bound involves $B(z) - 1$ rather than $B(z)$ since there will be no contribution here from the $x = 0$ term in (1.5.4). This type of distinction will be crucial in similar bounds on the lace expansion used in Chapter 6.

Combining (1.5.10) and (1.5.11) gives

$$Q(z) \geq \chi(z)^2 - Q(z)[B(z) - 1]. \tag{1.5.12}$$

This inequality is illustrated in Figure 1.2. Solving for $Q(z)$ gives

$$Q(z) \geq \frac{\chi(z)^2}{B(z)}.$$

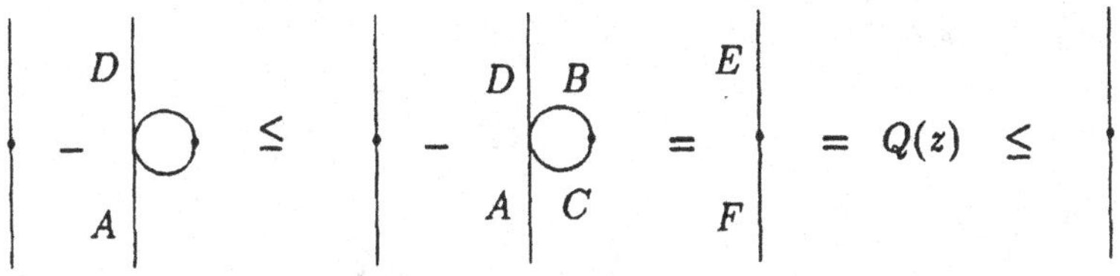

$$\left|\begin{array}{c} D \\ - \quad \bigcirc \\ A \end{array}\right| \quad \leq \quad \left|\begin{array}{cc} D & B \\ - \quad \bigcirc \\ A & C \end{array}\right| \quad = \quad \left|\begin{array}{c} E \\ \\ F \end{array}\right| \quad = \quad Q(z) \quad \leq \quad \left|\begin{array}{c} \\ \\ \end{array}\right|$$

$$[AD] \qquad\qquad [AD, AB, CD, BD] \qquad [EF]$$

Figure 1.2: A diagrammatic representation of the inequality $\chi(z)^2 - Q(z)[B(z) - 1] \leq Q(z) \leq \chi(z)^2$ occurring in the proof of Lemma 1.5.2. The list of pairs of lines indicates interactions between the propagators, in the sense that the corresponding walks must avoid each other.

Combining this inequality with (1.5.9) completes the proof of the lemma.

□

The quantity $\chi(z)^{-2}Q(z)$ can be interpreted as the probability that two self-avoiding walks of arbitrary length, which start at the origin, do not intersect. Lemma 1.5.2 can be restated as saying that this probability lies in the interval $[B(z)^{-1}, 1]$, and hence remains strictly positive at the critical point if the bubble condition is satisfied.

In the next theorem it is shown that the lower bound of Lemma 1.5.2 implies that if the bubble condition is satisfied, then (1.5.3) holds.

Theorem 1.5.3 *If the bubble condition is satisfied, and hence in particular if the infrared bound holds and $d > 4$, then there is a positive function $\epsilon(z)$ with $\lim_{z \nearrow z_c} \epsilon(z) = 0$ such that for z less than but near z_c*

$$\frac{z_c}{z_c - z} \leq \chi(z) \leq \frac{z_c[B(z_c) + \epsilon(z)]}{z_c - z}.$$

Hence if in fact there is a constant A such that $\chi(z) \sim A z_c(z_c - z)^{-1}$, then $1 \leq A \leq B(z_c)$.

Proof. The lower bound in the statement of the theorem is just (1.5.1). For the upper bound, let $z_1 \in (0, z_c)$. It follows from the lower bound in (1.5.6) that for $z \in [z_1, z_c)$

$$z\left(-\frac{d\chi^{-1}}{dz}\right) \geq \frac{1}{B(z)} - \frac{1}{\chi(z)}$$

$$\geq \frac{1}{B(z_c)} - \frac{1}{\chi(z_1)}. \qquad (1.5.13)$$

We bound the factor of z on the left side by z_c and then integrate from z_1 to z_c. Using the fact that $\chi(z_c)^{-1} = 0$ by (1.5.1), this gives

$$z_c \chi(z_1)^{-1} \geq [\mathsf{B}(z_c)^{-1} - \chi(z_1)^{-1}](z_c - z_1). \qquad (1.5.14)$$

Rewriting gives

$$\chi(z_1) \leq \frac{\mathsf{B}(z_c)}{1 - \mathsf{B}(z_c)\chi(z_1)^{-1}} \frac{z_c}{z_c - z_1}. \qquad (1.5.15)$$

This gives the desired upper bound on the susceptibility, since by (1.5.1) the inverse susceptibility on the right side can be made arbitrarily small by taking z_1 sufficiently close to z_c. □

Although the bubble condition is expected not to hold in four dimensions, it is nevertheless possible to draw some conclusions from the lower bound of Lemma 1.5.2 if we assume the infrared bound (1.4.12). While not sharp compared to the expected behaviour

$$\chi(z) \sim \frac{A}{z_c - z} |\log(z_c - z)|^{1/4},$$

the upper bound that we obtain on χ shows that the deviation from mean-field behaviour is at worst logarithmic in four dimensions, if the infrared bound is satisfied.

Theorem 1.5.4 *Let $d = 4$. If the infrared bound (1.4.12) is satisfied then for z less than but near z_c,*

$$\frac{z_c}{z_c - z} \leq \chi(z) \leq C \frac{|\log(z_c - z)|}{z_c - z}$$

for some constant C which does not depend on z.

Proof. The lower bound in the statement of the theorem is just (1.5.1), which holds in all dimensions. It remains to prove the upper bound. In the following, C represents a constant whose value may change from one occurrence to another.

Let $0 < z < z_c$. Since

$$\chi(z) = \hat{G}_z(0) \geq |\hat{G}_z(k)|$$

for all k, it follows from the infrared bound that

$$|\hat{G}_z(k)| \leq \frac{2}{|\hat{G}_z(k)|^{-1} + \chi(z)^{-1}} \leq \frac{2C}{k^2 + C\chi(z)^{-1}}. \qquad (1.5.16)$$

Using the fact that

$$B(z) = \int_{[-\pi,\pi]^4} \hat{G}_z(k)^2 \frac{d^4k}{(2\pi)^4},$$

a routine calculation using (1.5.16) gives the bound

$$B(z) \le C[1 + \log\chi(z)]. \tag{1.5.17}$$

By (1.5.6), (1.5.1) and (1.5.17), for z sufficiently close to z_c we have

$$-z\frac{d\chi^{-1}}{dz} \ge \frac{1}{B(z)} - \frac{1}{\chi(z)}$$

$$\ge \frac{1}{2B(z)}$$

$$\ge \frac{C}{1 + \log\chi(z)}$$

and therefore

$$-[1 + \log\chi(z)]\frac{d\chi^{-1}}{dz} \ge C. \tag{1.5.18}$$

The left side of (1.5.18) is the derivative of $-\chi(z)^{-1}[2 + \log\chi(z)]$. Hence for z close to z_c integration of (1.5.18) over the interval (z, z_c) gives

$$\chi(z)^{-1}[2 + \log\chi(z)] \ge C(z_c - z),$$

where we used (1.5.1) to see that the contribution from the upper limit of integration on the left side is zero. Decreasing C slightly we obtain

$$\frac{1}{C(z_c - z)} \ge \chi(z)[\log\chi(z)]^{-1}. \tag{1.5.19}$$

Taking logarithms, and taking z sufficiently close to z_c, gives

$$C|\log(z_c - z)| \ge \log\chi(z) - \log\log\chi(z) \ge \frac{1}{2}\log\chi(z). \tag{1.5.20}$$

Inserting the lower bound for $[\log\chi(z)]^{-1}$ given by (1.5.20) into (1.5.19) gives

$$C(z_c - z)^{-1} \ge \chi(z)|\log(z_c - z)|^{-1}.$$

This gives the upper bound on χ in the statement of the theorem. □

Finally we turn to a connection between the bubble diagram and the critical exponent Δ_4 for the renormalized coupling constant $g(z)$, which was defined in (1.4.22) by

$$g(z) = \xi(z)^{-d}\chi(z)^{-2}\sum_{N_1,N_2=0}^{\infty} c_{N_1,N_2}z^{N_1+N_2}. \tag{1.5.21}$$

Here c_{N_1,N_2} is the sum over sites x of the number of intersecting pairs of self-avoiding walks of length N_1 and N_2 starting at 0 and x respectively. The critical behaviour of $g(z)$ is believed to be of the form $(z_c - z)^{d\nu - 2\Delta_4 + \gamma}$.

The next theorem gives sufficient conditions for Δ_4 to take its mean-field value 3/2. The theorem is most efficiently stated in terms of the *repulsive* bubble diagram $R(z) < B(z)$, which is defined by taking only those contributions to the bubble from pairs of walks which are mutually avoiding apart from their common endpoints:

$$R(z) = \sum_{x \in Z^d} \sum_{\substack{\omega^{(1)}:\, 0 \to x \\ \omega^{(2)}:\, 0 \to x}} z^{|\omega^{(1)}| + |\omega^{(2)}|} I[\omega^{(1)} \cap \omega^{(2)} = \{0, x\}]. \qquad (1.5.22)$$

Theorem 1.5.5 *If* $B(z_c) < \infty$ *and in addition* $R(z_c) - 1 < 1/4$, *then* $g(z) \simeq \xi(z)^{-d}(z_c - z)^{-2}$. *If also* $\xi(z) \simeq (z_c - z)^{-\nu}$, *then* $\Delta_4 = 3/2$ *in the sense that*

$$g(z) \simeq (z_c - z)^{d\nu - 3 + \gamma} = (z_c - z)^{d\nu - 2}$$

(assuming that the exponent γ for c_n is equal to the exponent for the susceptibility).

Remark. The best current bound on $R(z_c) - 1$ in five dimensions is $0.434636 > 0.25$ [Hara and Slade (1991b)]. For $d = 6$ the same reference reports $B(z_c) - 1 \leq 0.25974$. However the repulsive bubble in six dimensions satisfies $R(z_c) - 1 \leq 0.2343 < 0.25$, and is smaller still in more than six dimensions [Hara and Slade (unpublished)]. Together with Theorem 1.5.5 and the result of Hara and Slade (1991a) that for $d \geq 5$ the correlation length exhibits the mean-field behaviour $\xi(z) \sim \text{const.}(z_c - z)^{-1/2}$ (and that the exponent for c_n is $\gamma = 1$), this implies that

$$g(z) \simeq (z_c - z)^{(d-4)/2} \qquad (1.5.23)$$

for $d \geq 6$. Although the same conclusion cannot yet be made for $d = 5$, it will be shown in Chapter 6 (see Theorem 6.2.5 and the remark preceding it, and Theorem 6.1.5) that for a sufficiently spread-out self-avoiding walk in more than four dimensions, $B(z_c) - 1 < 1/4$ and $\xi(z) \sim \text{const.}(z_c - z)^{-1/2}$, and hence $g(z) \simeq (z_c - z)^{(d-4)/2}$.

Proof of Theorem 1.5.5. By Theorem 1.5.3, the bubble condition implies that $\chi(z) \simeq (z_c - z)^{-1}$. Hence to prove the theorem it suffices to show that

$$\sum_{N_1, N_2 = 0}^{\infty} c_{N_1, N_2} z^{N_1 + N_2} \simeq (z_c - z)^{-4}. \qquad (1.5.24)$$

The left side is equal to

$$\sum_{\substack{x,y,v}} \sum_{\substack{\omega\,:\,0\,\rightarrow\,v \\ \rho\,:\,x\,\rightarrow\,y}} z^{|\omega|+|\rho|} I[\omega \cap \rho \neq \phi]. \qquad (1.5.25)$$

In a nonzero contribution to this sum, let u be the first site along ω where ω and ρ intersect. Then the portion of ω before u avoids ρ as well as the latter part of ω, while the latter part of ω avoids only the former part of ω and may intersect ρ. This gives the following diagrammatic interpretation of the left side of (1.5.24) (in which the list of pairs indicates mutually avoiding walks):

$$[AB, CD, AC, AD] \qquad (1.5.26)$$

Neglecting all mutual avoidance between the four lines of the diagram gives the upper bound $\chi^4 \leq \text{const.}(z_c - z)^{-4}$ for the left side of (1.5.24).

For a lower bound on (1.5.26) we apply inclusion-exclusion, as follows. The indicator function for the event that the various mutual avoidances shown in (1.5.26) occur can be written as one minus the event that at least one of the required mutual avoidances is violated. This leads to the lower bound

$$\sum_{\substack{u,v,x,y}} \sum_{\substack{\omega^{(1)}\,:\,0\,\rightarrow\,u \\ \omega^{(2)}\,:\,u\,\rightarrow\,v}} \sum_{\substack{\rho^{(1)}\,:\,x\,\rightarrow\,u \\ \rho^{(2)}\,:\,u\,\rightarrow\,y}} z^{|\omega^{(1)}|+|\omega^{(2)}|+|\rho^{(1)}|+|\rho^{(2)}|}$$

$$\times \left\{ 1 - I[\omega^{(1)} \cap \omega^{(2)} \neq \{u\}] - I[\omega^{(1)} \cap \rho^{(1)} \neq \{u\}] \right.$$

$$\left. - I[\omega^{(1)} \cap \rho^{(2)} \neq \{u\}] - I[\rho^{(1)} \cap \rho^{(2)} \neq \{u\}] \right\}. \qquad (1.5.27)$$

This bound is equal to

$$\chi^4 - 4\chi^2 \sum_{\substack{u,x}} \sum_{\substack{\gamma^{(1)}\,:\,0\,\rightarrow\,u \\ \gamma^{(2)}\,:\,u\,\rightarrow\,x}} z^{|\gamma^{(1)}|+|\gamma^{(2)}|} I[\gamma^{(1)} \cap \gamma^{(2)} \neq \{u\}]. \qquad (1.5.28)$$

We now argue as in (1.5.11), but this time we let w be the site of the *first* intersection (measured along $\gamma^{(2)}$) of $\gamma^{(2)}$ with $\gamma^{(1)}$. This gives the lower bound

$$\chi^4 - 4\chi^4[\mathsf{R}(z) - 1] \geq \text{const.}\chi^4 \qquad (1.5.29)$$

for (1.5.26), assuming that $\mathsf{R}(z_c) - 1 < 1/4$. $\qquad\qquad\qquad\qquad\qquad$ $\square$

1.6 Notes

Section 1.1. Existence of the connective constant $\mu = \lim_{N\to\infty} c_N^{1/N}$ was first proven in Hammersley and Morton (1954); this paper essentially marks the beginning of rigorous results for the self-avoiding walk. The (nonrigorous) derivation of $\mu = \sqrt{2+\sqrt{2}}$ for the honeycomb lattice is due to Nienhuis (1982); see also Nienhuis (1984) and Nienhuis (1987). For high dimensions, it was shown in Kesten (1964) that $\mu = 2d-1-(2d)^{-1}+O(d^{-2})$, and this has recently been improved to $\mu = 2d - 1 - (2d)^{-1} - 3(2d)^{-2} + O(d^{-3})$ using the lace expansion [Hara and Slade (unpublished)].

The conjectured values for γ and ν in two dimensions arise from an exact solution which is described in the articles by Nienhuis cited above. An alternate approach, based on conformal invariance, is discussed in Duplantier (1989), Duplantier (1990), and references therein. A rigorous argument leading to these two-dimensional critical exponents remains an open problem of major importance, and a solution would likely have far-reaching implications. For $d = 3$, field theoretic computations of the critical exponents are given in Le Guillou and Zinn-Justin (1989). Monte Carlo computations of the exponents are given for example in Madras and Sokal (1988), and numerical computations using extrapolation of exact enumerations are given in Guttmann and Wang (1991). The logarithmic corrections in four dimensions are obtained in Larkin and Khmel'Nitskii (1969), Wegner and Riedel (1973) and Brezin, Le Guillou and Zinn-Justin (1973). For recent progress on rigorous results in four dimensions, see Brydges, Evans and Imbrie (1992) and Arnaudon, Iagolnitzer and Magnen (1991). Existence of critical exponents for $d \geq 5$ is proven in Hara and Slade (1992a,1992b).

A necessary and sufficient condition for a bound of the form $c_N \leq \text{const.}\mu^N N^H$ for some finite H, i.e. for the finiteness of the critical exponent γ, is given in Hammersley (1991, 1992). We note the presence of a minor error in Hammersley (1991): the right side of (30) does not follow from the inequality that precedes it. This is easy to fix, however, as follows. In Hammersley's notation, the bound $f(m) \leq Gm^H\mu^m$ implies that $f(m,r) \leq \sum G^r n_1^H n_2^H \cdots n_r^H \mu^m$, where the sum is over all $n_1, \ldots, n_r \geq 1$ that sum to m. By the arithmetic-geometric inequality we have $n_1 n_2 \cdots n_r \leq (m/r)^r$, which implies $f(m,r) \leq \binom{m-1}{r-1} G^r (m/r)^{rH} \mu^m$. This gives us (30) with $Hu \log H$ replaced by $-Hu \log u$, and Hammersley's Equation (3) follows.

A rigorous understanding of the self-avoiding walk on finitely ramified fractals has recently emerged; see Hattori (1992) for a review.

For the Ising model (and also for φ^4 field theory), the following references prove results concerning mean-field behaviour above four dimensions: Sokal (1979), Aizenman (1982), Fröhlich (1982), Aizenman and Fernández (1986), Fernández, Fröhlich and Sokal (1992).

Section 1.2. The bound $(c_N/c_1)^{1/(N-1)}$ (for all $N \geq 2$) is attributed to Alm in Ahlberg and Janson (1980). The latter reference obtains an improvement when $c_N/c_{N-1} > c_1 - 2$: they show that μ is bounded above by the unique positive root of the polynomial

$$c_1 x^{N-1} = [c_N - (c_1 - 2)c_{N-1}]x + (c_1 - 2)[(c_1 - 1)c_{N-1} - c_N] \quad (1.6.1)$$

(for all $N \geq 3$). Currently the best upper bounds available are due to Alm (1992).

A method for obtaining lower bounds on μ using bridges was given in Guttmann (1983). The current best lower bound in two dimensions, due to Conway and Guttmann (to be published), also uses bridges. For $d \geq 3$, the best lower bounds are due to Hara and Slade (1992b), who use a different approach involving loop erasure.

The numerical estimates for μ cited in Table 1.1 are from exact enumeration data.

Section 1.3. Exponential decay of the subcritical two-point function was proven in Fisher (1966), as part of a study of the form of the distribution of $c_N(0, x)$.

Section 1.4. We make no attempt here to refer to the original literature on critical exponents; the ideas in this section are part of the standard physics picture of critical phenomena.

The infrared bound was proven for reflection-positive spin systems in all dimensions in Fröhlich, Simon and Spencer (1976). For branched polymers and for percolation, there are arguments that the infrared bound does not hold below eight and six dimensions respectively; see Bovier, Fröhlich and Glaus (1986) and Adler (1984) respectively.

For $d \geq 5$ it has been proven that $\hat{G}_{z_c}(k) \sim const.k^{-2}$ as $k \to 0$, but although it is believed that $G_{z_c}(x)$ is asymptotic to a multiple of $|x|^{2-d}$, this has not yet been proven (see Theorem 6.1.6 for a weaker result). It is thus of interest to know under what conditions behaviour of the form

$k^{-2+\eta}$ for a Fourier transform $\hat{g}(k)$ implies behaviour of the form $|x|^{2-d-\eta}$ for $g(x)$. For the case $\eta = 0$, the following sufficient condition was pointed out to us by S. Kotani (private communication); we omit the proof.

Theorem 1.6.1 *Let $d \geq 3$, and let $\mathbf{T}^d \equiv (\mathbf{R}/2\pi\mathbf{Z})^d$. Let $\hat{g}$ be a function in $C^{d-2}(\mathbf{T}^d \backslash \{0\})$, let $\hat{h}(k) = k^2\hat{g}(k)$, and for $x \in \mathbf{Z}^d$ let $g(x) = (2\pi)^{-d} \int_{\mathbf{T}^d} \hat{g}(k)e^{-ik\cdot x}d^dk$. Suppose that there is a neighbourhood $U \subset \mathbf{T}^d$ of 0 such that*

$$\hat{h} \in \left\{ \begin{array}{ll} C^{d-1}(U) & \text{if} \quad d = 3, 4 \\ C^{d-2}(U) & \text{if} \quad d \geq 5. \end{array} \right.$$

Then as $|x| \to \infty$,

$$g(x) \sim \hat{h}(0)\frac{\Gamma(d/2)}{2(d-2)\pi^{d/2}}|x|^{-(d-2)}.$$

The following shows that in general the hypothesis of existence of $d-2$ derivatives for $\hat{h}$ cannot be relaxed: we give an example[2] of a function $\hat{g}$ on $\mathbf{T}^d$, for $d \geq 3$, which is asymptotic to a multiple of k^{-2} as $k \to 0$, with $\hat{h}(k) = k^2\hat{g}(k)$ having $d-3$ but not $d-2$ derivatives in a neighbourhood of $k = 0$, but for which $g(x)$ is not bounded above by a multiple of $|x|^{2-d}$ for large x.

Example 1.6.2 Let $d \geq 3$, and let $C(x)$ be the critical simple random walk two-point function (or in other words the Green function) studied in Appendix A. Then $C(x)$ is asymptotic to a multiple of $|x|^{2-d}$ for large x [see, e.g., Lawler (1991)]. Also, $\hat{C}(k)^{-1} = 1 - d^{-1}\sum_{\mu=1}^{d} \cos k_\mu$ is asymptotic to $(2d)^{-1}k^2$ as $k \to 0$. Fix q such that $d-3 < q < d-2$, and for $-\pi \leq t \leq \pi$ define

$$\hat{f}(t) = \sum_{m=-\infty}^{\infty} 2^{-q|m|} \exp[it(\text{sgn}\,m)2^{|m|}], \tag{1.6.2}$$

where $\text{sgn}\,m = +1$ if $m > 0$; $= 0$ if $m = 0$; $= -1$ if $m < 0$. For $k \in [-\pi, \pi]^d$, let

$$\hat{F}(k) = \epsilon \prod_{\mu=1}^{d} \hat{f}(k_\mu) \tag{1.6.3}$$

where ϵ is chosen small enough that $1 + \hat{C}(k)^{-1}\hat{F}(k)$ is strictly positive uniformly in all $k \in [-\pi, \pi]^d$. (This is possible since $\hat{C}(k)^{-1}$ and the product in (1.6.3) are both bounded uniformly in k.) Observe that $\hat{F} \in C^s(\mathbf{T}^d)$ for $s < q$, but that for $s > q$, $\partial_\mu^s \hat{F}(0)$ does not exist. Now let

$$\hat{g}(k) = \hat{C}(k) + \hat{F}(k) = \hat{C}(k)[1 + \hat{C}(k)^{-1}\hat{F}(k)] \tag{1.6.4}$$

---

[2] The example was arrived at in conversation with T. Hara.

and

$$\hat{h}(k) = k^2 \hat{g}(k). \tag{1.6.5}$$

Then $\hat{g}(k)$ is asymptotic to $(2d)k^{-2}$ as $k \to 0$, and $\hat{h}(k) \in C^{d-3}(\mathbf{T}^d)$. However $g(x)$ is not bounded above by a multiple of $|x|^{2-d}$ for large x, because $F(x) = \epsilon |x|^{-q}$ for x having one component of the form $\pm 2^{|m|}$ (for any integer m) and all other components zero.

See Appendix A of Sokal (1982) for a discussion of some related issues.

Section 1.5. For reflection positive spin systems the infrared bound was proven in Fröhlich, Simon and Spencer (1976). As a consequence the bubble diagram for such systems is finite at the critical point above four dimensions, and diverges logarithmically in four dimensions. This was used to prove mean-field behaviour for spin systems for dimensions greater than four in Aizenman (1982) and Fröhlich (1982). In Bovier, Felder and Fröhlich (1984) Theorem 1.5.3 was proved, although at that time for the self-avoiding walk neither the infrared bound nor the bubble condition were known to hold in any dimension. In the same paper it was observed that if the infrared bound holds in four dimensions then the deviation from mean-field behaviour for the susceptibility is at most logarithmic. Our proof of Theorem 1.5.4 yields this conclusion in a slightly stronger form, following the methods used for spin systems in Aizenman and Graham (1983). Results analogous to Theorem 1.5.5 were obtained for spin systems in Aizenman (1982) and Fröhlich (1982). The proof that $\Delta_4 = 3/2$ for $d \geq 6$ is new, and is due to Hara and Slade (unpublished).

For percolation and branched polymers (lattice trees and lattice animals) the role of the bubble diagram is played by the triangle and the square diagram respectively; see Section 5.5. For percolation see Aizenman and Newman (1984), Nguyen (1987), Barsky and Aizenman (1991), Hara and Slade (1990a) and Nguyen and Yang (1991). For lattice trees and lattice animals see Bovier, Fröhlich and Glaus (1986), Tasaki and Hara (1987) and Hara and Slade (1990b).

Chapter 2

Scaling, polymers and spins

2.1 Scaling theory

This chapter is concerned with some of the nonrigorous work on the self-avoiding walk: the scaling theory which leads to the scaling relations stated in Section 1.4, the connection with polymers and the derivation of the Flory values for the critical exponent ν, and finally, the interpretation of the self-avoiding walk as a "zero-component" ferromagnet.

We begin in this section with a discussion of scaling theory, giving heuristic derivations of Fisher's scaling relation (1.4.9) and of the hyperscaling relations (1.4.14) and (1.4.24). There are a variety of approaches which can be used to derive these relations, and here we content ourselves with giving a representative sample of the types of arguments which are frequently used. Although the sort of arguments we will describe are part of the standard lore of theoretical physics (applicable to a wide variety of models), from a mathematical point of view they may appear to be on rather shaky ground. There will be no rigorous results in this section, and we will make frequent use of the symbol $\approx$, which implies a leap of faith.

Our starting point will be an assumption about the behaviour of the two-point function in the limit as both $z \nearrow z_c$ and $x \to \infty$. The correlation length $\xi(z)$ is to be interpreted as the important length scale of the system. For $x \to \infty$ at fixed $z < z_c$, the two-point function is believed to obey the *Ornstein-Zernike* decay

$$G_z(0, x) \sim C_z |x|^{-(d-1)/2} e^{-|x|/\xi(z)}, \qquad (2.1.1)$$

35

where C_z depends only on z, and strictly speaking the norm on the right side is equivalent to but not equal to the Euclidean norm. A proof of (2.1.1) for x on a coordinate axis will be given in Theorem 4.4.7. However the Ornstein-Zernike decay describes the behaviour of the two-point function on a length scale $|x| \gg \xi(z)$, and is not believed to be accurate on the important length scale where x is of the order of $\xi(z)$. Instead, the decay of the *critical* two-point function is considered to be fundamental on the scale of the correlation length, and we define a function $h(z; x)$ by

$$G_z(0, x) = \frac{1}{|x|^{d-2+\eta}} h(z; x). \qquad (2.1.2)$$

The assumption now is that the important contribution to $h(z; x)$ will come from x of the order of the correlation length $\xi(z)$, and that we can write

$$G_z(0, x) \approx \frac{1}{|x|^{d-2+\eta}} g(|x|/\xi(z)), \qquad (2.1.3)$$

for some universal function g of a single variable. The function g will be assumed to decay at infinity sufficiently rapidly that its product with any power of x is integrable, for example an exponential function.

Given (2.1.3), the following argument can be put forth in support of Fisher's relation $\gamma = (2 - \eta)\nu$. We assume that the main contribution to the susceptibility

$$\chi(z) = \sum_x G_z(0, x)$$

is due to x of the order of the correlation length, and that we may therefore substitute (2.1.3) into the sum over x. By definition of $\bar{\gamma}$ we then have

$$(z_c - z)^{-\bar{\gamma}} \quad \approx \quad \sum_x \frac{g(|x|/\xi)}{|x|^{d-2+\eta}} \approx \int_0^\infty r^{1-\eta} g(r/\xi) dr$$

$$= \quad \text{const.}\xi^{2-\eta} \sim \text{const.}(z_c - z)^{-(2-\eta)\bar{\nu}}. \qquad (2.1.4)$$

This gives

$$\bar{\gamma} = (2 - \eta)\bar{\nu}. \qquad (2.1.5)$$

It has already been argued in Section 1.3 that $\bar{\gamma} = \gamma$, and we will shortly argue that $\bar{\nu} = \nu$, which then gives $\gamma = (2 - \eta)\nu$.

The following is an alternate derivation of Fisher's relation which does not rely on (2.1.3). In the sum

$$G_{z_c}(0, x) = \sum_{N=0}^{\infty} c_N(0, x)\mu^{-N} \qquad (2.1.6)$$

we assume that $c_N(0, x)$ is significant only when $|x|$ is of the order of N^ν. There are about $N^{d\nu}$ such sites x, and we assume that an N-step self-avoiding walk is equally likely to end at any one of them, or in other words,

$$c_N(0, x) \approx c_N N^{-d\nu} \approx \mu^N N^{\gamma-1-d\nu}. \qquad (2.1.7)$$

Restricting the sum in (2.1.6) to N between $c_1|x|^{1/\nu}$ and $c_2|x|^{1/\nu}$, for some positive constants c_1 and c_2, and then using (2.1.7), gives

$$G_{z_c}(0, x) \approx \sum_{N=c_1|x|^{1/\nu}}^{c_2|x|^{1/\nu}} N^{\gamma-1-d\nu} \approx |x|^{(\gamma-d\nu)/\nu}. \qquad (2.1.8)$$

This implies $-(d - 2 + \eta) = (\gamma - d\nu)/\nu$, which can be rewritten as $\gamma = (2 - \eta)\nu$.

Continuing in the spirit of the calculation leading to (2.1.5), we now argue that $\bar{\nu} = \nu_p = \nu$. For any $p \in (0, \infty)$,

$$\sum_x |x|^p G_z(0, x) \approx \sum_x |x|^p \frac{g(|x|/\xi)}{|x|^{d-2+\eta}} \approx \int_0^\infty r^{p+1-\eta} g(r/\xi) dr$$
$$= \text{const.} \xi^{p+2-\eta} \sim \text{const.} (z_c - z)^{-(p+2-\eta)\bar{\nu}}. \qquad (2.1.9)$$

Using the definition of ξ_p and (2.1.5), this gives

$$\xi_p(z)^p \approx (z_c - z)^{\bar{\gamma}-(2-\eta)\bar{\nu}-p\bar{\nu}} = (z_c - z)^{-p\bar{\nu}} \qquad (2.1.10)$$

and hence $\bar{\nu} = \nu_p$. To show that $\bar{\nu} = \nu$, we first observe that by (1.3.20),

$$\sum_x |x|^2 G_z(0, x) \approx (z_c - z)^{-(2\nu+\gamma)}. \qquad (2.1.11)$$

Comparing with (2.1.9), with $p = 2$, gives

$$(4 - \eta)\bar{\nu} = 2\nu + \gamma. \qquad (2.1.12)$$

Now by (2.1.5) and the equality of $\bar{\gamma}$ and γ we conclude that $\bar{\nu} = \nu$.

We now turn to the hyperscaling relations (1.4.14) and (1.4.24). These cannot be derived from the scaling hypothesis (2.1.3) and require additional assumptions. Less numerical testing has been done of the hyperscaling relations than on the calculation of γ and ν, but both the Monte Carlo and series extrapolation computations which have been done are consistent with them.

The hyperscaling relation involving α_{sing} may at first glance seem somewhat surprising. It would perhaps seem natural to assume that the probability that an N-step self-avoiding walk ends at x would be proportional to the characteristic volume $N^{-d\nu}$ in the limit as $N \to \infty$, as is the case for simple random walk. However this leads to the conclusion $\alpha_{sing} - \gamma - 1 = -d\nu$ rather than to the hyperscaling relation $\alpha_{sing} - 2 = -d\nu$. This incorrect argument fails to take into account the fact that for fixed x it is difficult for a long self-avoiding walk to return near to its starting point at the origin, and as the length of the walk goes to infinity x must be regarded as being close to the origin. The argument should be reasonable for x of the order of the typical length scale N^ν, and indeed this is what we assumed in the second derivation of Fisher's relation given above.

To obtain the hyperscaling relation $\alpha_{sing} - 2 = -d\nu$, we proceed as follows. First we assume that $c_N(0, x)$ will have the same scaling behaviour for any fixed x as $N \to \infty$, and consider the case $x = e$, with e a nearest neighbour of the origin. This assumption is mild in comparison with the assumptions we will make next. By adding an extra step to a walk ending at e we obtain a closed self-avoiding loop. Let $n = (N + 1)/2$. Then by summing over the position of the walk after n steps, and using symmetry, we have

$$c_N(0,e) = (2d)^{-1} \sum_x \sum_{\substack{\omega^{(1)} : \omega^{(1)}(n) = x \\ \omega^{(2)} : \omega^{(2)}(n) = x}} I[\omega^{(1)} \cap \omega^{(2)} = \{0, x\}]. \quad (2.1.13)$$

Here both of the self-avoiding walks $\omega^{(i)}$ begin at the origin and consist of n steps. We now make three assumptions. First, we assume that the main contribution to the above sum will be from x of the order of n^ν. Thus there are of the order of $n^{\nu d}$ relevant terms in the sum. Second, we assume that the effect of the avoidance constraint between $\omega^{(1)}$ and $\omega^{(2)}$ can be incorporated by replacing the quantity being summed over x by $c_N(0, x)^2$ multiplied by the square of the probability that two n-step self-avoiding walks beginning at the same point avoid each other; this probability is $c_{2n}/c_n^2 \sim 2^{\gamma-1}A^{-1}n^{1-\gamma}$. Here we use the square of this probability to account for the avoidance both near 0 and near x. Third, we assume that for x of the order of n^ν the probability that an n-step walk ends at x is of the order of the inverse of the characteristic volume $n^{\nu d}$, so that $c_n(0, x)$ is of the order of $\mu^n n^{\gamma-1} n^{-\nu d}$. With these three assumptions we have from (2.1.13) that

$$c_N(0,e) \approx n^{\nu d}[\mu^n n^{\gamma-1-\nu d}]^2[n^{1-\gamma}]^2 = \mu^{2n} n^{-\nu d}. \quad (2.1.14)$$

Comparison with the definition of α_{sing} gives $\alpha_{sing} - 2 = -d\nu$.

We next turn to the hyperscaling relation (1.4.24) for Δ_4, which is believed to hold only for $d \leq 4$. For simplicity, we take $N_1 = N_2 = n$ in the definition (1.4.21) of Δ_4. To begin, we assume that since a self-avoiding walk of length n goes a distance of about n^ν in each direction, such a self-avoiding walk will primarily lie in a hypercube of side n^ν and volume $n^{\nu d}$.

Given $0 \le D \le d$, consider a subset of the hypercubic lattice such that the cube of side R contains of the order of R^D points as $R \to \infty$. Such a subset can in some sense be thought of as being D-dimensional. From this point of view a long self-avoiding walk, which consists of n points in volume approximately equal to $n^{\nu d}$, is a $1/\nu$-dimensional set. Typically two D-dimensional subsets will intersect in d dimensions if $2D \ge d$, but otherwise will not. This suggests that two n-step self-avoiding walks with a common point will typically have additional intersections if $d\nu \le 2$, and typically will not if $d\nu > 2$.

Consider first the case of $d\nu > 2$, for which we have already seen in (1.4.26) that the hyperscaling relation fails. According to the values of ν given in Table 1.2, this inequality says $d > 4$. Two n-step self-avoiding walks lying in the cube of volume $n^{\nu d}$ will typically not intersect each other, and so there should be no overcounting in writing

$$c_{n,n} \approx c_n^2 n^2 \sim A^2 \mu^{2n} n^{2\gamma}. \tag{2.1.15}$$

Here the factor n^2 comes from choosing a point on each walk at which the two walks can be joined. By definition of Δ_4, this gives $2\Delta_4 + \gamma - 2 = 2\gamma$, or $\Delta_4 = 1 + \gamma/2$. Using the mean-field value $\gamma = 1$ known to be correct for $d \ge 5$, we obtain $\Delta_4 = 3/2$ (which of course is consistent with the rigorous result for $d \ge 6$ obtained in Section 1.5).

We next consider the case $d\nu \le 2$, which corresponds to $d \le 4$. Here the factor of n^2 in (2.1.15) would overcount. Given one n-step walk, which will lie roughly within a cube of volume $n^{\nu d}$, a second n-step walk will typically intersect the first if it is started at any one of the $n^{\nu d}$ points in the cube. This leads to

$$c_{n,n} \approx c_n^2 n^{\nu d} \sim A^2 \mu^{2n} n^{2\gamma-2+\nu d}. \tag{2.1.16}$$

By definition of Δ_4, this gives $2\Delta_4 + \gamma - 2 = 2\gamma - 2 + \nu d$, which simplifies to the hyperscaling relation $d\nu - 2\Delta_4 + \gamma = 0$.

2.2 Polymers

One of the most important applications of the self-avoiding walk is as a model for linear polymer molecules in chemical physics. In this section we shall briefly describe some aspects of this role, including a nonrigorous derivation of the "Flory values" for the critical exponent ν.

A *polymer* is a molecule that consists of many "monomers" (groups of atoms) joined together by chemical bonds. The *functionality* of a monomer is the number of available chemical bonds that it has, i.e. the number of other monomers with which it must bond. If each monomer has functionality two, then a *linear* polymer is formed. If we denote a monomer by (A), then a linear polymer may be represented schematically as

$$\cdots -(A)-(A)-(A)-(A)-(A)- \cdots.$$

One simple example is polyethylene, where each monomer is CH_2 (one carbon atom and two hydrogen atoms). The pattern terminates either by bonding with a monomer of functionality one, such as CH_3, at each end, or else by closing on itself to form a "ring polymer". When we speak of linear polymers, we shall be referring to the former. By way of contrast, if a polymer includes monomers of functionality three or more, then a *branched polymer* is formed; these are often modelled by lattice trees or lattice animals (see Section 5.5.1).

The preceding paragraph deals only with the topological structure of a polymer. Properties of its spatial configuration are no less important. Polymers can be very large; some linear polymers consist of more than 10^5 monomers. Thus the length scale of the entire polymer is macroscopic with respect to the length scale of the individual monomers. Consider a linear polymer consisting of $N+1$ monomers, and label the monomers $0, 1, \ldots, N$ from one end to the other. Let $x(i) \in \mathbf{R}^3$ denote the location of the i-th monomer. Then the i-th (monomer-monomer) bond may be represented by the line segment joining $x(i-1)$ to $x(i)$. Typically, the length of each bond is essentially constant throughout the chain, as is the angle between each pair of consecutive monomer-monomer bonds. However, there is some rotational freedom for the i-th bond around the axis determined by the $(i-1)$-th bond. In some cases, a reasonably good approximation may be obtained by allowing the rotational angle of the i-th bond around the $(i-1)$-th bond to take on three different values, say $0°$ and $\pm120°$, perhaps with different probabilities (an angle of $0°$ means that the i-th, $(i-1)$-th, and $(i-2)$-th bonds all lie in one plane). These angles correspond to local configurations of minimal energy, and depend on the details of the monomers.

We see that one possible model for the spatial configuration of a linear polymer is simply a random walk in $\mathbf{R}^3$, and in fact this model is known as the *ideal polymer chain*. Alternatively, one can work with a lattice approximation, say a random walk on $\mathbf{Z}^3$. The model can be embellished by turning it into a Markov chain (or random walk with some finite memory), and it works reasonably well in some situations. However, there is a fundamental limitation of the ideal polymer chain, namely the *excluded volume* effect.

Two monomers cannot occupy the same position in space: the presence of a monomer at position x prohibits any other part of the polymer from getting too close to x, that is, other monomers are excluded from a certain volume of space. This is the excluded volume effect. When we take this effect into account, it becomes apparent that a self-avoiding walk is a more appropriate model for a linear polymer than is a random walk. The self-avoiding walk model is best for the case of a dilute polymer solution (where polymers are far apart, so that there is little interaction between distinct molecules) and a good solvent (which minimizes attractive forces between monomers).

We remark that there are some situations in which polymers really do behave ideally on large length scales, even though excluded volume effects are present. One is in a dense system (or "melt") of many polymers, where monomers fill three-dimensional space uniformly and a given polymer interacts with many other monomers besides its own. Another is at certain values of temperature and solvent quality where roughly speaking the attractive forces between monomers exactly balance the excluded volume repulsion (the "Θ point"). For more details, see the general polymer references listed in the Notes at the end of the chapter.

For the remainder of this section, we shall only discuss linear polymers in dilute solutions with good solvents. These are believed to be in the same "universality class" as the self-avoiding walk, which means in particular that they have the same critical exponents. For example, consider the *radius of gyration* of a polymer, which is the average distance of the monomers from the centre of mass of the polymer. The radius of gyration of polymers can be determined experimentally, for example from light scattering properties. For a polymer consisting of N monomers, the radius of gyration is expected to be asymptotic to DN^ν as $N \to \infty$, where D and ν are constants. The exponent ν is believed to be *universal*: it should be the same for all linear polymers (in dilute solution with good solvents), and for the self-avoiding walk as well. Moreover the exponent ν for the radius of gyration is believed to be the same as the critical exponent ν defined in (1.1.5) for the mean-square displacement, since polymers are expected to have only one macroscopic length scale. In contrast, the amplitude D is non-universal: it depends on microscopic details of the monomers and the solvent molecules.

The chemist Paul J. Flory developed an effective (but nonrigorous) method for computing the exponent ν [Flory (1949)]. We give a brief description of this method in general dimension d; for simplicity, we ignore all multiplicative constants. (A more probabilistic description of the method will be given afterward.) Fix N and consider a linear polymer with $N + 1$ monomers, represented by an N-step walk $\omega = (\omega(0), \ldots, \omega(N))$ in $\mathbf{Z}^d$ (not necessarily self-avoiding). Let L be the radius of gyration of ω, or any other "effective radius" of the walk. Then ω consists of $N + 1$ monomers (sites) spread through a box of volume L^d. Assuming uniformity, this gives a density of

$$\rho = \frac{N}{L^d} \qquad (2.2.1)$$

monomers per unit volume. The repulsive energy per unit volume depends on the number of pairs of monomers per unit volume, which we approximate by ρ^2. This is a "mean-field" approximation: it uses the assumption of uniformity very heavily, ignoring the strong correlations in the locations of consecutive monomers along the polymer. If we accept this approximation, then the total repulsive energy of the polymer is given by

$$E_{rep} = L^d \rho^2 = \frac{N^2}{L^d}. \qquad (2.2.2)$$

Naturally the repulsive energy is lower for highly extended chains, i.e. large values of L.

Now consider the free energy F of the polymer of radius L, in the absence of the repulsion. This is given (up to constants) by (-1) times the entropy[1], and the entropy in turn is just the logarithm of the number of walks of radius L. Without repulsions, this can be found from the Gaussian behaviour of the ideal chain, as follows. Taking L now to denote the end-to-end distance and fixing $\omega(0) = 0$, we have

$$\Pr\{\omega(N) = x\} \approx N^{-d/2} \exp(-|x|^2/N) \qquad (2.2.3)$$

for every $x \in \mathbf{Z}^d$, and hence

$$\Pr\{|\omega(N)| = L\} \approx \frac{L^{d-1}}{N^{d/2}} \exp(-L^2/N). \qquad (2.2.4)$$

The total number of N-step walks is $(2d)^N$ in the nearest-neighbour case, so the free energy is

$$\begin{aligned} F &= -\log[(2d)^N \Pr\{|\omega(N)| = L\}] \\ &= -(d-1)\log L + \frac{L^2}{N} + \text{terms independent of } L. \quad (2.2.5) \end{aligned}$$

The term F may also be viewed as an "elastic energy" term, which prevents L from getting too large. The total energy of the polymer is now given by

---

[1] In thermodynamics we have $F = U - TS$, where U is internal energy, T is temperature, and S is entropy. Here U depends on the number of monomers but not on L, so for our purposes it is constant and hence we ignore it.

the sum of the two energy terms (2.2.2) and (2.2.5):

$$E_{rep} + F = \frac{N^2}{L^d} + \frac{L^2}{N} - (d-1)\log L + K, \qquad (2.2.6)$$

where K is independent of L. Now put $L = N^\nu$. Then the total energy (2.2.6) becomes

$$E_{rep} + F = N^{2-d\nu} + N^{2\nu-1} - \nu(d-1)\log N + K. \qquad (2.2.7)$$

The value of ν that minimizes the energy (2.2.7) may be found by first equating the first two powers of N: solving $2 - d\nu = 2\nu - 1$ gives

$$\nu = \frac{3}{d+2}. \qquad (2.2.8)$$

Substituting this back into (2.2.7), the first two terms become $N^{(4-d)/(d+2)}$, and these are the dominant terms if and only if $d < 4$. Therefore this argument predicts that (2.2.8) gives the correct value of ν whenever $d < 4$. When $d = 4$, this argument also predicts $\nu = 3/(4+2) = 1/2$ since this is the only value for which the first two terms of (2.2.7) remain bounded. However, when $d > 4$, any value of ν in the interval $[2/d, 1/2]$ keeps the first two terms of (2.2.7) bounded. Pushing this argument further suggests that we should take the largest value in this interval so as to minimize the $-\nu(d-1)\log N$ term in (2.2.7), obtaining $\nu = 1/2$ for $d > 4$. This answer makes sense: since ν equals $1/2$ in the ideal case, the addition of a repulsive energy term should not decrease ν below $1/2$, and so we conclude that $\nu = 1/2$ whenever $d > 4$.

To summarize, the above argument makes the following predictions for ν:

$$\nu_{Flory} = \begin{cases} 1 & \text{if } d = 1 \\ 3/4 & \text{if } d = 2 \\ 3/5 & \text{if } d = 3 \\ 1/2 & \text{if } d \geq 4. \end{cases} \qquad (2.2.9)$$

These predictions are known as the *Flory values* for ν. As described in Section 1.1, they are known to be correct for $d = 1$ and $d \geq 5$, and they are believed to be correct for $d = 2$ and $d = 4$ as well. The Flory value for $d = 3$ is generally believed to be slightly too large: numerical and field theory calculations indicate that the actual value is probably close to 0.59 (some references are given in the Notes for Section 1.1). The success of Flory's argument is all the more remarkable when one realizes that it benefits greatly from the cancellation of two errors: both E_{rep} and F are greatly overestimated (see p. 46 of de Gennes (1979) for a brief discussion).

To conclude this section, we shall recast the Flory argument in a more probabilistic language. In (2.2.4) we calculated the probability that an N-step random walk ω (starting at the origin) has $|\omega(N)| = L$. Now let us estimate the probability that ω is self-avoiding given that $|\omega(N)| = L$. We shall write $L = N^\nu$ and choose ν to maximize this probability. As above, we assume that the $N + 1$ sites of ω are spread uniformly through a box of volume L^d. Given that $\omega(0), \ldots, \omega(k-1)$ are all distinct, the probability that $\omega(k)$ does not coincide with any one of the previous k sites is approximately $1 - kL^{-d}$ (this is the "mean-field" approximation). Hence the probability that ω is self-avoiding given that $|\omega(N)| = L$ is approximately

$$\prod_{k=1}^{N}(1 - kL^{-d}) \approx \exp\left(-\sum_{k=1}^{N} kL^{-d}\right) \approx \exp(-N^2/L^d). \tag{2.2.10}$$

Multiplying (2.2.10) by (2.2.4) yields

$$\Pr\{\omega \text{ is self-avoiding and } |\omega(N)| = L\} \approx \frac{L^{d-1}}{N^{d/2}} \exp\left[-\frac{L^2}{N} - \frac{N^2}{L^d}\right].$$
$$\tag{2.2.11}$$

To find the most likely value of L, we maximize the above probability for fixed N. Since the logarithm of this probability is just the negative of the total energy (2.2.6), we are again led to the Flory exponents.

2.3   The $N \to 0$ limit

In this section we describe a connection, discovered by de Gennes, between the self-avoiding walk and the spin systems of classical statistical mechanics: the self-avoiding walk can be considered to be a "zero-component" ferromagnet. Although this connection has not yet provided methods for obtaining rigorous results for the self-avoiding walk, it has been an important tool for physicists and has been used for example to compute the values for the critical exponents γ and ν for $d = 2, 3, 4$ given in (1.1.11) – (1.1.14). To make the discussion more self-contained, we first describe very briefly the basic set-up of spin models. The prototype of these models is the Ising model, and we begin with this fundamental model of ferromagnetism.

For simplicity we restrict attention to the hypercubic lattice $\mathbf{Z}^d$, although this is not essential. Let Λ denote the sites in $\mathbf{Z}^d$ which are in the cube $[-L, L]^d$, for $L \geq 1$. Eventually we will want to take the limit as $L \to \infty$. In the Ising model, a spin variable $S^{(x)}$ taking the value plus one or minus one is associated to each site $x \in \Lambda$. These spin variables interact via a Hamiltonian

$$\mathcal{H} = - \sum_{(x,y)} S^{(x)} S^{(y)}, \tag{2.3.1}$$

where the sum represents the sum over all nearest-neighbour pairs of sites in Λ. The Hamiltonian represents the energy of a spin configuration (choice of ± 1 for each spin), and is lowest when neighbouring spins agree. The expected value of any function F of the spins in Λ is then given by

$$\langle F \rangle = \frac{1}{Z} E \left(F e^{-\beta \mathcal{H}} \right), \tag{2.3.2}$$

where the expectation E on the right side is with respect to the product of the Bernoulli measures assigning probability one-half to each of the possible values ± 1 for the spin variables, and the *partition function*

$$Z = E \left(e^{-\beta \mathcal{H}} \right) \tag{2.3.3}$$

is a normalization factor. The nonnegative parameter β corresponds to inverse temperature. The partition function and expectations depend on the volume Λ, but to simplify the notation no subscripts Λ will be used to keep track of this.

An important example is $F = S^{(0)} S^{(x)}$, the product of the values of the spins at the origin and at x. For any finite volume Λ it follows from the symmetry of the Hamiltonian under the global spin flip, in which each spin is multiplied by minus one, that $\langle S^{(y)} \rangle = 0$ for any site $y \in \Lambda$. Hence the *two-point function* $\langle S^{(0)} S^{(x)} \rangle$ represents the correlation between the spins at the origin and at x. It follows from the fact that the two-point function lies in the compact interval $[0, 1]$ that there is a subsequence of volumes tending to infinity such that the limit of the two-point function exists along the subsequence. (The same subsequence can be used for all x by a diagonal argument.) In fact it can be shown using correlation inequalities that the infinite volume limit of the two-point function exists, without recourse to subsequences. The infinite volume limit is often referred to as the *thermodynamic limit*. Here we are using *free boundary conditions*, in which spins on the inside boundary of Λ interact only with their nearest neighbours inside Λ. It is known that in the thermodynamic limit, for high temperatures (or in other words for low β) the two-point function decays exponentially as $|x| \to \infty$. The inverse of the decay rate defines a correlation length $\xi(\beta)$. For dimensions $d \geq 2$ there is a critical value β_c (corresponding to the Curie point) such that the correlation length diverges to infinity as $\beta \nearrow \beta_c$. This corresponds to the onset of long range order.

Associated with the critical point β_c, a number of critical exponents can be defined which are analogous to the exponents defined for the self-avoiding walk. For example a critical exponent, known as ν as for the self-avoiding walk, defines the power law according to which the correlation length diverges:

$$\xi(\beta) \sim \text{const.}(\beta_c - \beta)^{-\nu} \text{ as } \beta \nearrow \beta_c. \qquad (2.3.4)$$

The susceptibility is defined by

$$\chi(\beta) = d^{-1} \sum_x \langle S^{(0)} \cdot S^{(x)} \rangle, \qquad (2.3.5)$$

for the infinite volume theory with $\beta < \beta_c$. The susceptibility diverges as $\beta \nearrow \beta_c$ and the power law at which the divergence takes place defines a critical exponent γ:

$$\chi(\beta) \sim \text{const.}(\beta_c - \beta)^{-\gamma} \text{ as } \beta \nearrow \beta_c. \qquad (2.3.6)$$

These qualitative analogies between the critical behaviours of spin systems and the self-avoiding walk can be made more quantitative. For this we need to introduce a generalization of the Ising model, known as the N-vector or $O(N)$ model. In this generalization the Ising spins are replaced by spins taking values in the N-dimensional sphere of radius $\sqrt{N}$, for some positive integer N, and the Hamiltonian becomes

$$\mathcal{H} = - \sum_{\langle x,y \rangle} S^{(x)} \cdot S^{(y)}, \qquad (2.3.7)$$

where the dot product is the usual Euclidean one. The two-point function for the N-vector model is then defined in finite volume as for the Ising model, with the change that now the single spin distribution is the uniform measure on $S(N, \sqrt{N})$, where

$$S(n, r) = \{(a_1, \ldots, a_n) \in \mathbf{R}^n : a_1^2 + \cdots + a_n^2 = r^2\} \qquad (2.3.8)$$

is the sphere of radius r in $\mathbf{R}^n$. For $N = 1$ this is just the Ising model. For $N \geq 2$ the N-vector model also has a critical point, and shares many common features with the Ising model (although the change from discrete to continuous symmetry group introduces new elements). Critical exponents can be defined, which will in general depend on N as well as on the dimension d. In a manner to be described in more detail below, the N-vector model can be defined in the limit as $N \to 0$, and this limit gives the self-avoiding walk. The N-vector model can be analyzed, at least nonrigorously, using renormalization methods, and this analysis yields values for the critical exponents in which N appears as a parameter which may be assigned values other than positive integers. Taking $N = 0$ in the expression for the critical exponents then gives values which are believed to correspond to the self-avoiding walk exponents, and indeed the values given for two dimensions in (1.1.11) and (1.1.12) were obtained by Nienhuis in this way. The self-avoiding walk exponents for three dimensions and the logarithmic corrections for $d = 4$ can be arrived at similarly.

To take the $N \to 0$ limit, consider a fixed finite volume Λ with free boundary conditions (i.e. we consider only sites in Λ in the sum defining the Hamiltonian, and do not take the infinite volume limit). Our aim is to show that the two-point function for the N-vector model converges to that of the self-avoiding walk, i.e. for any $\beta \geq 0$ and for any fixed i, j and sites x, y,

$$\lim_{N \to 0} \langle S_i^{(x)} S_j^{(y)} \rangle = \delta_{i,j} \sum_{\omega : x \to y} \beta^{|\omega|} = \delta_{i,j} G_\beta(x, y), \qquad (2.3.9)$$

where the subscript i denotes the i-th spin component and the sum is over all self-avoiding walks (in Λ) of any length, from x to y. In the process, it will be necessary to define what is meant by the limit on the left side of (2.3.9), since the N-vector model two-point function has only been defined when N is a positive integer. To begin with some notation, for a function F of the spins in Λ and for N a positive integer, we write

$$\langle F \rangle = \frac{1}{Z} E\left(F e^{-\beta \mathcal{H}}\right) = \frac{1}{Z} \int F e^{-\beta \mathcal{H}} d\mu_N, \qquad (2.3.10)$$

where $d\mu_N$ denotes the product over the spins of uniform measures on $\mathcal{S}(N, \sqrt{N})$, and the partition function Z is the normalization

$$Z = E(e^{-\beta \mathcal{H}}) = \int e^{-\beta \mathcal{H}} d\mu_N. \qquad (2.3.11)$$

To obtain (2.3.9) it will be argued that

$$\lim_{N \to 0} Z = \lim_{N \to 0} E(e^{-\beta \mathcal{H}}) = 1 \qquad (2.3.12)$$

and

$$\lim_{N \to 0} E\left(S_i^{(x)} S_j^{(y)} e^{-\beta \mathcal{H}}\right) = \delta_{i,j} \sum_{\omega : x \to y} \beta^{|\omega|}. \qquad (2.3.13)$$

The analysis will not proceed by extending the definitions of the expectations whose limits are being taken in the above two equations to positive real values of N, and then taking the limit in the strict mathematical sense. Rather, we will show that a certain plausible interpretation of the limit leads to (2.3.12) and (2.3.13); thus our arguments do not lead to these equations as rigorous mathematical statements.

Both (2.3.12) and (2.3.13) will be obtained in the same way. The first step is to expand the exponential in a power series:

$$e^{-\beta\mathcal{H}} = \prod_{(x,y)} \exp[\beta S^{(x)} \cdot S^{(y)}] = \prod_{(x,y)} \sum_{m_{xy}=0}^{\infty} \frac{\beta^{m_{xy}}}{m_{xy}!}(S^{(x)} \cdot S^{(y)})^{m_{xy}}. \quad (2.3.14)$$

Then we label the nearest-neighbour (undirected) bonds of Λ by $b_1, \ldots b_B$, and for each bond b_α label one of its endpoints b_α^- and the other b_α^+. In this notation (2.3.14) can be written

$$e^{-\beta\mathcal{H}} = \sum_{m_1,\ldots,m_B=0}^{\infty} \frac{\beta^{\Sigma_\alpha m_\alpha}}{\prod_\alpha m_\alpha!} \prod_\alpha (S^{(b_\alpha^-)} \cdot S^{(b_\alpha^+)})^{m_\alpha}. \quad (2.3.15)$$

Hence Z or the two-point function be computed in terms of expectations of products of powers of spin components. Such expectations can be evaluated using the following lemma.

Lemma 2.3.1 *Fix an integer $N \geq 1$. Let $S = (S_1, \ldots, S_N)$ denote a vector which is uniformly distributed on $\mathcal{S}(N, \sqrt{N})$. Given nonnegative integers $k_1, \ldots, k_N$,*

$$E(S_1^{k_1} \cdots S_N^{k_N}) = \begin{cases} \dfrac{2\Gamma(\frac{N+2}{2})\prod_{l=1}^{N}\Gamma(\frac{k_l+1}{2})}{\pi^{N/2}\Gamma(\frac{k_1+\cdots+k_N+N}{2})}N^{(k_1+\cdots+k_N-2)/2} & \text{all } k_l \text{ even} \\ 0 & \text{otherwise,} \end{cases}$$

where Γ denotes the Gamma function.

Proof. The lemma is clearly true for $N = 1$, so we fix $N \geq 2$. Suppose $U = (U_1, \ldots, U_N)$ is uniformly distributed on the sphere $\mathcal{S}(N, 1)$. Using the fact that $\Gamma(\frac{N+2}{2}) = \frac{N}{2}\Gamma(\frac{N}{2})$, it suffices to show for all $k = 1, \ldots, N$ that if the integers m_l are all even then

$$E(U_1^{m_1} U_2^{m_2} \cdots U_k^{m_k}) = \frac{\Gamma(\frac{N}{2})\prod_{l=1}^{k}\Gamma(\frac{m_l+1}{2})}{\pi^{k/2}\Gamma(\frac{m_1+\cdots+m_k+N}{2})}, \quad (2.3.16)$$

and that this expectation is equal to 0 if any m_i is an odd integer. The latter follows by symmetry. We will prove the former by induction on k.

For $k = 1$, we use the fact [proved in Watson (1983), p.44] that the marginal density of U_i is

$$\frac{\Gamma(\frac{N}{2})}{\pi^{1/2}\Gamma(\frac{N-1}{2})}(1-a^2)^{(N-3)/2}, \quad -1 \leq a \leq 1. \quad (2.3.17)$$

It then follows from the identity

$$\int_0^1 t^c (1-t)^d \, dt = \frac{\Gamma(c+1)\Gamma(d+1)}{\Gamma(c+d+2)} \qquad (2.3.18)$$

(for $c, d > -1$) that

$$E(U_1^m) = 0 \qquad \text{if } m \text{ is odd, and} \qquad (2.3.19)$$

$$E(U_1^m) = \frac{\Gamma\left(\frac{N}{2}\right)\Gamma\left(\frac{m+1}{2}\right)}{\pi^{1/2}\Gamma\left(\frac{N+m}{2}\right)} \qquad \text{if } m \text{ is even} \qquad (2.3.20)$$

which gives (2.3.16) for $k = 1$.

Suppose now that $k > 1$ and assume that (2.3.16) is true for $k - 1$. Conditioned on $U_1 = a$, the distribution of $(U_2, \dots, U_N)$ is uniform on the set $S(N-1, (1-a^2)^{1/2})$. The inductive hypothesis then gives the conditional expectation

$$E(U_2^{m_2} \cdots U_k^{m_k} | U_1 = a) \qquad (2.3.21)$$

$$= (1-a^2)^{(m_2 + \cdots + m_k)/2} \frac{\Gamma\left(\frac{N-1}{2}\right) \prod_{l=2}^k \Gamma\left(\frac{m_l+1}{2}\right)}{\pi^{(k-1)/2}\Gamma\left(\frac{m_2 + \cdots + m_k + N - 1}{2}\right)}.$$

Inserting this into

$$E(U_1^{m_1} U_2^{m_2} \cdots U_k^{m_k})$$

$$= \int_{-1}^{+1} a^{m_1} E(U_2^{m_2} \cdots U_k^{m_k} | U_1 = a) \frac{\Gamma\left(\frac{N}{2}\right)}{\pi^{1/2}\Gamma\left(\frac{N-1}{2}\right)} (1-a^2)^{(N-3)/2} \, da$$

then gives the desired result (2.3.16). □

We now use Lemma 2.3.1 to define what we mean by the limit as $N \to 0$ of expectations like those in the statement of the lemma. It follows from the lemma that for any positive integer N and any index i,

$$E(S_i^2) = 1;$$

in fact this can be seen more easily by symmetry and the fact that $E(S_1^2 + \cdots + S_N^2) = N$. We will therefore assert, by way of definition, that

$$\lim_{N \to 0} E(S_i^2) = 1. \qquad (2.3.22)$$

Also, we will define the limit as $N \to 0$ of any expectation as in the statement of the lemma to be zero if $k_1 + \dots + k_N > 2$. This is consistent with the result of the lemma; e.g. if two k_l's equal 2 and the others are 0, then

$$E(S_i^2 S_j^2) = \frac{2\Gamma(\frac{N+2}{2})\left(\frac{1}{2}\right)^2}{\Gamma(\frac{N+4}{2})} N, \qquad (2.3.23)$$

which converges to zero as $N \to 0$. According to this definition,

$$\lim_{N \to 0} E(S_1^{k_1} \cdots S_N^{k_N}) = \begin{cases} 1 & \text{all } k_l = 0, \text{ or one } k_l = 2 \text{ and } k_j = 0, j \neq l \\ 0 & \text{otherwise.} \end{cases}$$

$$(2.3.24)$$

Consider now the partition function

$$Z = \sum_{m_1,\ldots,m_B=0}^{\infty} \frac{\beta^{\Sigma_\alpha m_\alpha}}{\prod_\alpha m_\alpha!} E\left(\prod_\alpha (S^{(b_\alpha^-)} \cdot S^{(b_\alpha^+)})^{m_\alpha}\right). \qquad (2.3.25)$$

Equation (2.3.24) provides a means of extending Z to $N = 0$, by taking the limit as $N \to 0$ termwise in the above sum. A graphical interpretation of the sum in (2.3.25) can be obtained by associating to each term in the sum a graph whose edges are given by m_i undirected edges joining the endpoints of the bond b_i. It then follows from (2.3.24) that any term whose corresponding graph has a vertex from which other than two or zero edges emanate will approach zero in the limit as $N \to 0$. This can be seen by considering a specific example. Consider the graph consisting of the four nearest-neighbour edges $\{z, x\}, \{z, x\}, \{z, y\}, \{z, w\}$. The expectation arising from the corresponding term in (2.3.25) is

$$\sum_{i,j,k,l} E\left(S_i^{(y)} S_i^{(z)} S_j^{(w)} S_j^{(z)} S_k^{(x)} S_k^{(z)} S_l^{(x)} S_l^{(z)}\right), \qquad (2.3.26)$$

where the sum is over spin components. Since spins at different sites are independent, the expectation in the above sum factors into a product of four expectations. The factor corresponding to the site z is

$$E\left(S_i^{(z)} S_j^{(z)} S_k^{(z)} S_l^{(z)}\right), \qquad (2.3.27)$$

which will go to zero in the limit as $N \to 0$, for any choice of i, j, k, l, by (2.3.24). There is further N-dependence arising from the number of terms in the sum over spin components i, j, k, l, but this will be interpreted as only helping to drive the limit to zero.

The relevant graphs in the limit are therefore the graphs consisting of a finite number (possibly zero) of nonintersecting self-avoiding polygons in Λ, where the degenerate polygons consisting of two edges linking a pair of nearest-neighbour sites are allowed possibilities. The graph with no edges corresponds to the term in the sum with all $m_i = 0$, and contributes an amount 1. A two-edge polygon with nearest-neighbour vertices x, y contributes an amount

$$\frac{\beta^2}{2} E\left([S^{(x)} \cdot S^{(y)}]^2\right). \qquad (2.3.28)$$

Since an expectation involving an odd power of $S_i^{(x)}$ is zero, this is equal to

$$\frac{\beta^2}{2} \sum_{i=1}^{N} E\left([S_i^{(x)} S_i^{(y)}]^2\right) = \frac{\beta^2}{2} N \qquad (2.3.29)$$

and hence does not contribute in the limit.

A nondegenerate polygon, in other words a polygon consisting of at least four bonds, contributes an amount

$$\beta^k E\left((S^{(y_1)} \cdot S^{(y_2)})(S^{(y_2)} \cdot S^{(y_3)}) \ldots (S^{(y_r)} \cdot S^{(y_1)})\right) \qquad (2.3.30)$$

where y_l is a neighbour of y_{l+1} for each $l = 1, \ldots, r-1$, y_r is a neighbour of y_1, and $y_1, \ldots, y_r$ are distinct. This expression is equal to

$$\beta^k \sum_{i=1}^{N} E\left((S_i^{(y_1)} S_i^{(y_2)} \ldots S_i^{(y_r)})^2\right) = \beta^k N, \qquad (2.3.31)$$

which also converges to 0 as $N \to 0$. We are thus led to conclude that

$$\lim_{N \to 0} Z = 1. \qquad (2.3.32)$$

For the two-point function the analysis is similar. We would like to compute the limit as $N \to 0$ of the expectation

$$\sum_{m_1, \ldots, m_B = 0}^{\infty} \frac{\beta^{\sum_\alpha m_\alpha}}{\prod_\alpha m_\alpha!} E\left(S_i^{(x)} S_j^{(y)} \prod_\alpha (S^{(b_\alpha^-)} \cdot S^{(b_\alpha^+)})^{m_\alpha}\right). \qquad (2.3.33)$$

For $x = y$ the limit of the above expression is equal to 1 by an analysis similar to that used to analyze the partition function. Suppose now that $x \neq y$. Again there is a correspondence between the terms in the sum and graphs on Λ, but now there can be a nonzero contribution to the limit only from those graphs in which exactly one edge emanates from each of the vertices x and y, and either two or zero edges emanate from every other vertex. Such a graph must consist of a self-avoiding walk from x to y together with a finite number of (possibly degenerate) self-avoiding polygons. The contribution to the limit from the polygons is equal to zero as it is for the partition function. The contribution due to the self-avoiding walk with vertices $(x, v_1, \ldots, v_{k-1}, y)$ is

$$\beta^k E\left(S_i^{(x)}(S^{(x)} \cdot S^{(v_1)})(S^{(v_1)} \cdot S^{(v_2)}) \ldots (S^{(v_{k-1})} \cdot S^{(y)}) S_j^{(y)}\right). \qquad (2.3.34)$$

This expression is equal to $\beta^k \delta_{i,j}$, since upon expanding the dot products everything has expectation 0 except $(S_i^{(x)} S_i^{(v_1)} \cdots S_i^{(v_{k-1})} S_i^{(y)})^2$. Since there is one such term for every self-avoiding walk from x to y,

$$\lim_{N \to 0} \langle S_i^{(x)} S_j^{(y)} \rangle = \delta_{i,j} \sum_{\omega: x \to y} \beta^{|\omega|} = \delta_{i,j} G_\beta(x, y). \qquad (2.3.35)$$

This correspondence of two-point functions is responsible for the general belief that the critical exponents γ and ν also correspond in the $N \to 0$ limit.

We end this section with a nonrigorous discussion of the equality in the $N \to 0$ limit of the self-avoiding walk critical exponent α_{sing}, defined in (1.4.13), and the critical exponent for the singular part of the specific heat of the N-vector model. To distinguish between these two exponents we shall denote the latter by α_s.

To define the specific heat we first introduce the expected energy per unit volume, which is given by

$$\mathcal{E}(\beta) = \langle \sum_{\langle 0,x \rangle} S^{(0)} \cdot S^{(x)} \rangle = 2dN \langle S_i^{(0)} S_i^{(e)} \rangle, \qquad (2.3.36)$$

for any fixed i and any nearest neighbour e of the origin. The prefactor $2dN$ is irrelevant as far as the behaviour of $\mathcal{E}(\beta)$ near the critical point is concerned, so we introduce

$$u(\beta) = \langle S_i^{(0)} S_i^{(e)} \rangle. \qquad (2.3.37)$$

The specific heat is defined as the rate of change of the energy with respect to temperature β^{-1}, i.e.

$$C(\beta) = -\beta^2 \frac{du}{d\beta}. \qquad (2.3.38)$$

Typically the specific heat either diverges as β increases to β_c, or there is a nonnegative integer M such that it has M but not $M+1$ derivatives at β_c^-, with

$$C(\beta) \approx \sum_{j=0}^{M} \frac{C^{(j)}(\beta_c)}{j!} (\beta - \beta_c)^j + (\beta_c - \beta)^{-\alpha_s} \qquad (2.3.39)$$

for some exponent $-\alpha_s \in (M, M+1]$. In principle both α_s and M can depend on N. Here we are using the symbol $\approx$ which indicates a crude correspondence between the right and left sides; in particular in the correction term we are dropping sign and constant factors, and also possible logarithmic factors which can be expected to be present when $-\alpha_s$ is an integer.

We also allow $M = -1$ in (2.3.39), with the empty sum interpreted as zero, to deal simultaneously with both finite and infinite $C(\beta_c)$. For $M = -1$, requiring that the energy $u(\beta)$ remains bounded as $\beta \nearrow \beta_c$ implies that $-\alpha_s \in (-1, 0]$.

In view of (2.3.38), the behaviour (2.3.39) of the specific heat suggests that

$$u(\beta) \approx \sum_{j=0}^{M+1} \frac{u^{(j)}(\beta_c)}{j!}(\beta - \beta_c)^j + (\beta_c - \beta)^{-\alpha_s+1}. \qquad (2.3.40)$$

Assuming that the form of the above relation persists in the limit as $N \to 0$, from (2.3.37) and (2.3.35) we obtain

$$G_\beta(0, e) = \sum_{n=1}^{\infty} c_n(0, e)\beta^n \approx \sum_{j=0}^{M+1} g_j(\beta - \beta_c)^j + (\beta_c - \beta)^{-\alpha_s+1}, \qquad (2.3.41)$$

where

$$g_j = \frac{1}{j!}\frac{d^j}{d\beta^j}G_\beta(0, e)\Big|_{\beta_c} = \sum_{n=1}^{\infty} c_n(0, e) \binom{n}{j} \beta_c^{n-j}. \qquad (2.3.42)$$

The exponent α_s in (2.3.41) is interpreted as the $N \to 0$ limit of an N-dependent exponent. Our goal now is to argue that this limiting value of α_s is equal to α_{sing}. We further assume that in (2.3.41) M is the largest integer such that g_{M+1} is finite, so that $g_{M+2} = \infty$. Assuming now that $c_n(0, e) \approx n^{\alpha_{sing}-2}\mu^n = n^{\alpha_{sing}-2}\beta_c^{-n}$, g_j will be finite if and only if $\alpha_{sing}-2+j < -1$. Thus we have $-\alpha_{sing} \in (M, M+1]$, which is consistent with the restriction $-\alpha_s \in (M, M+1]$ in (2.3.39).

Writing $\beta = \beta_c e^{-t}$, so that $\beta_c - \beta \sim \beta_c t$, we have

$$G_\beta(0, e) - \sum_{j=0}^{M+1} g_j(\beta - \beta_c)^j \approx \sum_{n=1}^{\infty} n^{\alpha_{sing}-2}\left[e^{-nt} - \sum_{j=0}^{M+1} \binom{n}{j}(-t)^j\right]. \qquad (2.3.43)$$

Approximating the sum over n on the right side by an integral and then making the change of variables $y = xt$, the right side of (2.3.43) is given approximately by

$$\int_1^{\infty} x^{\alpha_{sing}-2}\left[e^{-xt} - \sum_{j=0}^{M+1}\binom{x}{j}(-t)^j\right] dx$$

$$= t^{-\alpha_{sing}+1}\int_t^{\infty} y^{\alpha_{sing}-2}\left[e^{-y} - \sum_{j=0}^{M+1}\binom{y/t}{j}(-t)^j\right] dy.$$

As $t \to 0$, the right side behaves like

$$t^{-\alpha_{sing}+1} \int_t^\infty y^{\alpha_{sing}-2} \left[e^{-y} - \sum_{j=0}^{M+1} \frac{1}{j!}(-y)^j \right] dy. \qquad (2.3.44)$$

In view of the fact that $-\alpha_{sing} \in (M, M+1]$, the above integral is convergent both for large y and for $y \to 0$ (apart from a logarithmic divergence as $y \to 0$ when $-\alpha_{sing} = M+1$). The integral is clearly nonzero, since the quantity in square brackets in the integrand is of the same sign for all positive y, by Taylor's Theorem with remainder. Hence the overall behaviour is $t^{-\alpha_{sing}+1}$. Comparing now with (2.3.41), we conclude that $\alpha_s = \alpha_{sing}$.

2.4 Notes

Section 2.1. Scaling theory is discussed in many theoretical physics texts on critical phenomena, for example Amit (1984), and we shall make no attempt here to refer to the original literature.

Section 2.2. Some general references on polymers which elaborate on the topics mentioned here include Flory (1971), de Gennes (1979), Doi and Edwards (1986), and des Cloiseaux and Jannink (1990). A readable survey is given in Flory's 1974 Nobel lecture [Flory (1976)]. Whittington (1982) discusses several additional topics in the statistical mechanics of polymers and self-avoiding walks.

Flory (1949) originally discussed only the three-dimensional case of the argument presented in this section. The extension to other dimensions was first observed by Fisher (1969). There are many other arguments which derive the Flory exponents; for example, see Edwards (1965), Freed (1981), and Bouchaud and Georges (1989).

Section 2.3. For a general introduction to rigorous results for spin systems, see for example Fernández, Fröhlich and Sokal (1992), Thompson (1988), Glimm and Jaffe (1987), Ellis (1985), Ruelle (1969). More physics-oriented accounts are given in for example Itzykson and Drouffe (1989), Parisi (1988), Amit (1984).

The fact that the $N \to 0$ limit of the N-vector model gives the self-avoiding walk was first observed in de Gennes (1972); see also de Gennes (1979). The use of Lemma 2.3.1 in deriving the $N \to 0$ limit appears to be new. Other approaches can be found in Aragão de Carvalho, Caracciolo and Fröhlich (1983), Halley and Dasgupta (1983), Domb (1976), and Bowers and McKerrell (1973). The calculation of critical exponents for the two-dimensional N-vector model, and corresponding identification of the exponents for $N = 0$, was carried out in Nienhuis (1982); see also Nienhuis (1984) and Nienhuis (1987). For $d = 3$, the $N = 0$ critical exponents are calculated in Le Guillou and Zinn-Justin (1989). Logarithmic corrections in four dimensions were computed in Larkin and Khmel'Nitskii (1969), Wegner and Riedel (1973) and Brezin, Le Guillou and Zinn-Justin (1973).

There is an intimate relation between spin systems and interacting random walks of various types (going far beyond the $N \to 0$ limit). This was emphasized by Symanzik (1969), and developed further in Brydges, Fröhlich and Spencer (1982). A detailed account of the random walk representations of spin systems is given in Fernández, Fröhlich and Sokal (1992).

Chapter 3

Some combinatorial bounds

3.1 The Hammersley-Welsh method

As was mentioned in Section 1.1, there is still no rigorous proof of the finiteness of the critical exponent γ for the number of self-avoiding walks [see Equation (1.1.4)] in dimensions two, three and four. The best rigorous upper bounds on c_N/μ^N are essentially of the form $\exp(O(N^p))$ for some constant $0 < p < 1$. It is a major open problem to replace this bound by a polynomial in N. We remark that subadditivity (Section 1.2) by itself gives no information about such subexponential behaviour.

Theorem 3.1.1 below, with its elegant proof, is due to Hammersley and Welsh (1962). Although this result (which holds for all $d \geq 2$) has subsequently been improved for $d > 2$, after three decades it remains the best rigorous upper bound on c_N in two dimensions. Improved bounds for $d > 2$ can be obtained, with considerably more work, using an extension of the Hammersley-Welsh method. These improved bounds, which are given in Theorem 3.3.1, remain the best available in three and four dimensions. In five or more dimensions, entirely different methods have been used to prove that c_N/μ^N is asymptotically constant (and hence bounded); these methods will be described in Chapter 6.

Theorem 3.1.1 *Let $d \geq 2$. For any constant $B > \pi(2/3)^{1/2}$, there exists an $N_0(B)$ independent of d such that*

$$c_N \leq \mu^{N+1} e^{BN^{1/2}} \text{ for all } N \geq N_0. \tag{3.1.1}$$

The proof relies on bridges (see Definition 1.2.4), and yields as a bonus the fact that $\mu_{Bridge} = \mu$. In particular, it uses the fact that walks are "subadditive" [Equation (1.2.3)] while bridges are "superadditive" [Equation (1.2.15)] and plays these two off against each other. The basic idea is that every self-avoiding walk can be "unfolded" into a bridge, and that this transformation is at most $\exp(O(N^{1/2}))$-to-one. Before we give the details, we require a few definitions, as well as a classical theorem of number theory which we will quote without proof.

Definition 3.1.2 *An N-step* half-space walk *is an N-step self-avoiding walk ω whose first components satisfy the inequality*

$$\omega_1(0) < \omega_1(i) \text{ for all } i = 1, \ldots, N.$$

The number of N-step half-space walks starting at the origin is denoted h_N. By convention, $h_0 = 1$.

In particular, every bridge is a half-space walk.

Definition 3.1.3 *The* span *of an N-step self-avoiding walk ω is*

$$\max_{0 \leq j \leq N} \omega_1(j) - \min_{0 \leq j \leq N} \omega_1(j).$$

The number of N-step half-space walks (respectively, bridges) starting at the origin and having span A is denoted $h_{N,A}$ (respectively, $b_{N,A}$).

Note that $h_{N,0}$ is 1 if $N = 0$ and is 0 otherwise.

Theorem 3.1.4 *For each integer $A \geq 1$, let $P_D(A)$ denote the number of partitions of A into distinct integers (i.e. the number of ways to write $A = A_1 + \cdots + A_k$ where $A_1 > \ldots > A_k$). Then*

$$\log P_D(A) \sim \pi \left(\frac{A}{3}\right)^{1/2} \text{ as } A \to \infty. \tag{3.1.2}$$

This theorem is proved in Hardy and Ramanujan (1917).

The following proposition contains the first part of the proof of Theorem 3.1.1, in which half-space walks are "unfolded" into bridges by a sequence of reflections.

Proposition 3.1.5 *For every $N \geq 1$,*

$$h_N \leq P_D(N)b_N, \tag{3.1.3}$$

where $P_D(N)$ is defined in Theorem 3.1.4.

Proof. Let $N \geq 1$, and let ω be an N-step half-space walk that starts at the origin. Let $n_0 = 0$. For each $j = 1, 2, \ldots$, recursively define $A_j(\omega)$ and $n_j(\omega)$ so that

$$A_j = \max_{n_{j-1} < i \leq N} (-1)^j (\omega_1(n_{j-1}) - \omega_1(i))$$

and n_j is the largest value of i for which this maximum is attained. The recursion is stopped at the smallest integer k such that $n_k = N$; this means that $A_{k+1}(\omega)$ and $n_{k+1}(\omega)$ are not defined. (See Figure 3.1.) Observe that

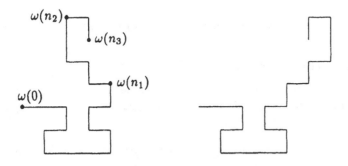

Figure 3.1: A half-space walk ω in $H_{20}[4, 2, 1]$ and the transformed walk ω' in $H_{20}[6, 1]$.

$A_1(\omega)$ is the span of ω; in general, $A_{j+1}(\omega)$ is the span of the self-avoiding walk $(\omega(n_j), \ldots, \omega(N))$, which is either a half-space walk or the reflection of one. Moreover, each of the subwalks $(\omega(n_j), \ldots, \omega(n_{j+1}))$ is either a bridge or the reflection of one. Also observe that $A_1 > A_2 > \ldots > A_k > 0$.

For every decreasing sequence of k positive integers $a_1 > a_2 \ldots > a_k > 0$, let $H_N[a_1, \ldots, a_k]$ be the set of N-step half-space walks ω with $\omega(0) = 0$ and $A_1(\omega) = a_1, \cdots, A_k(\omega) = a_k$, and $n_k(\omega) = N$ (and hence $A_{k+1}(\omega)$ is not defined). Note that in particular $H_N[a]$ is the set of N-step bridges of span a.

Given an N-step half-space walk ω, define a new N-step walk ω' as follows: for $0 \leq i \leq n_1(\omega)$, define $\omega'(i) = \omega(i)$; and for $n_1(\omega) < i \leq N$, define $\omega'(i)$ to be the reflection of the point $\omega(i)$ in the hyperplane $x_1 = A_1(\omega)$. Observe that if ω is in $H_N[a_1, a_2, \cdots, a_k]$, then ω' is in $H_N[a_1 + a_2, a_3, \cdots, a_k]$; moreover, this transformation is one-to-one, so

$$|H_N[a_1, a_2, \cdots, a_k]| \leq |H_N[a_1 + a_2, a_3, \cdots, a_k]|.$$

Therefore, summing over all finite integer sequences $a_1 > \ldots > a_k > 0$,

$$h_N = \sum |H_N[a_1, \cdots, a_k]|$$

$$\leq \sum |H_N[a_1 + \cdots + a_k]|$$
$$= \sum b_{N,a_1 + \cdots + a_k},$$

which tells us that

$$h_N \leq \sum_{A=1}^{N} P_D(A) b_{N,A}. \tag{3.1.4}$$

Since $P_D(A) \leq P_D(N)$ for $A \leq N$, it follows from (3.1.4) that

$$h_N \leq P_D(N) \sum_{A=1}^{N} b_{N,A} = P_D(N) b_N, \tag{3.1.5}$$

which proves the proposition.   □

We now complete the proof of the Hammersley-Welsh bound on c_N. The idea is to split each self-avoiding walk into two half-space walks, and then to use Proposition 3.1.5.

Proof of Theorem 3.1.1. Fix $B > \pi(2/3)^{1/2}$, and choose $\epsilon > 0$ so that $B - \epsilon > \pi(2/3)^{1/2}$. By Theorem 3.1.4, there exists a constant K such that

$$P_D(A) \leq K \exp\left[(B - \epsilon)(A/2)^{1/2}\right] \text{ for all } A. \tag{3.1.6}$$

Given an arbitrary n-step self-avoiding walk ω, let $M = \min_i \omega_1(i)$ and let m be the largest i such that $\omega_1(i) = M$. Then $(\omega(m), \ldots, \omega(n))$ is a half-space walk, as is

$$(\omega(m) - (1, 0, 0, \ldots, 0), \omega(m), \omega(m-1), \ldots, \omega(0)).$$

Using this decomposition, as well as Proposition 3.1.5, the inequality $b_i b_j \leq b_{i+j}$ [from (1.2.15)], (3.1.6), and the inequality $x^{1/2} + y^{1/2} \leq (2x + 2y)^{1/2}$, we obtain

$$c_n \leq \sum_{m=0}^{n} h_{n-m} h_{m+1}$$

$$\leq \sum_{m=0}^{n} b_{m+1} b_{n-m} P_D(m+1) P_D(n-m)$$

$$\leq b_{n+1} \sum_{m=0}^{n} K^2 \exp\left((B - \epsilon)\left[\left(\frac{m+1}{2}\right)^{1/2} + \left(\frac{n-m}{2}\right)^{1/2}\right]\right)$$

$$\leq b_{n+1}(n+1) K^2 \exp\left[(B - \epsilon)(n+1)^{1/2}\right] \tag{3.1.7}$$

for all n. Therefore, there exists an $N_0(B)$ (independent of d) such that

$$c_n \leq b_{n+1} e^{Bn^{1/2}} \text{ for all } n \geq N_0. \tag{3.1.8}$$

Since $b_{n+1} \leq \mu^{n+1}$ [Equation (1.2.17)], Theorem 3.1.1 is now proven. □

Corollary 3.1.6 *Let B be as in Theorem 3.1.1. Then, for all sufficiently large N,*

$$\mu^{N-1} e^{-BN^{1/2}} \leq b_N \leq \mu^N. \tag{3.1.9}$$

In particular,

$$\lim_{N \to \infty} (b_N)^{1/N} = \mu. \tag{3.1.10}$$

Proof. The right inequality of (3.1.9) is just (1.2.17); the left inequality comes from the bound $\mu^n \leq c_n$ [recall (1.2.10)] and from (3.1.8) (with n replaced by $N - 1$). Equation (3.1.10) is then immediate. □

Definition 3.1.7 *The generating function for the number of bridges is denoted B_z and is given by*

$$B_z = \sum_{N=0}^{\infty} b_N z^N.$$

Equation (3.1.10) says that the radius of convergence of B_z is $z_c = \mu^{-1}$. The following corollary says that B_z actually diverges at $z = z_c$.

Corollary 3.1.8

$$\lim_{z \nearrow z_c} B_z = +\infty;$$

that is,

$$B_{z_c} = \sum_{N=1}^{\infty} b_N \mu^{-N} = +\infty. \tag{3.1.11}$$

Proof. The proof of Theorem 3.1.1 shows that every N-step half-space walk may be decomposed into a finite sequence of bridges $\{\omega^{(i)}\}$ having spans A_i and lengths m_i, where $A_1 > A_2 > \ldots$ and $\sum m_i = N$; moreover, the sequence of bridges uniquely determines the original half-space walk. Therefore for $N \geq 1$

$$h_N \leq \sum \left(\prod_{i=1}^{k} b_{m_i, A_i} \right) \tag{3.1.12}$$

where the sum is over all integers $k \geq 1$, all integers $A_1 > \ldots > A_k > 0$, and all integers $m_1, \ldots, m_k \geq 1$ that sum to N. Consequently, for $z > 0$

$$\sum_{N=0}^{\infty} h_N z^N \leq \prod_{A=1}^{\infty} \left(1 + \sum_{m=1}^{\infty} b_{m,A} z^m \right);$$

this can be seen by comparing z^N terms on both sides and using (3.1.12). Combining this inequality with the elementary inequality $1 + x \leq e^x$, we find

$$\sum_{N=0}^{\infty} h_N z^N \leq \exp \left(\sum_{A=1}^{\infty} \sum_{n=1}^{\infty} b_{n,A} z^n \right) = \exp(B_z - 1).$$

This and the first inequality of (3.1.7) imply

$$\sum_{N=0}^{\infty} c_N z^N \leq z^{-1} \left(\sum_{n=0}^{\infty} h_n z^n \right)^2 \leq z^{-1} e^{2(B_z - 1)}. \tag{3.1.13}$$

By Equation (1.3.6), the leftmost term of (3.1.13) diverges at $z = z_c$, hence so does the rightmost term. This proves the corollary. $\qquad\square$

We remark that the above proof gives an explicit bound on the rate of divergence of B_z: indeed, combining (3.1.13) with (1.3.6) yields

$$B_z \geq 1 + \frac{1}{2} \log \frac{z_c z}{z_c - z} \qquad \text{for } 0 < z < z_c. \tag{3.1.14}$$

3.2 Self-avoiding polygons

Intuitively, a *self-avoiding polygon* may be thought of as a simple (i.e. non-self-intersecting) closed curve embedded in the lattice, with neither starting point nor orientation specified. The precise definition is as follows.

Definition 3.2.1 *Let N be an integer greater than 2. An N-step self-avoiding polygon is a set $\mathcal{P}$ of N nearest-neighbour bonds with the following property: there exists a corresponding $(N - 1)$-step self-avoiding walk ω having $|\omega(N - 1) - \omega(0)| = 1$ such that $\mathcal{P}$ consists of precisely the bond joining $\omega(N - 1)$ to $\omega(0)$ and the $N - 1$ bonds joining $\omega(i - 1)$ to $\omega(i)$ $(i = 1, \ldots, N - 1)$.*

Observe that ω is not uniquely determined by $\mathcal{P}$; in fact, each N-step self-avoiding polygon has precisely $2N$ corresponding self-avoiding walks (there are N choices of starting point and two choices of orientation). However, no $(N - 1)$-step self-avoiding walk corresponds to more than one N-step polygon.

We want to count self-avoiding polygons by ignoring translations and only counting different shapes; thus, in $\mathbf{Z}^2$, there should be only one 4-step self-avoiding polygon (a unit square) and two 6-step self-avoiding polygons (rectangles, one being a 90° rotation of the other). This leads us to the following definition.

Definition 3.2.2 *Two N-step self-avoiding polygons are said to be* equivalent up to translation *if there is a vector v in $\mathbf{R}^d$ such that translation by v defines a one-to-one correspondence from the set of bonds of one polygon to the set of bonds of the other polygon. Also, we denote by q_N the number of distinct equivalence classes up to translation of N-step self-avoiding polygons.*

Thus, if e is one of the $2d$ nearest neighbours of the origin in $\mathbf{Z}^d$, then the observations following Definition 3.2.1 tell us that

$$2Nq_N = 2dc_{N-1}(0, e) \qquad (3.2.1)$$

for every $N > 2$ (recall from Section 1.3 that $c_n(x, y)$ is the number of n-step self-avoiding walks from x to y). In particular, for $d = 2$, we have $q_4 = 1$, $c_3(0, e) = 2$, $q_6 = 2$, and $c_5(0, e) = 6$. Observe that $q_N = 0$ for every odd N.

Two self-avoiding polygons can be concatenated to form a larger self-avoiding polygon. The procedure is clear in two dimensions (see Figure 3.2): join a "rightmost" bond of one to a "leftmost" bond of the other. In higher

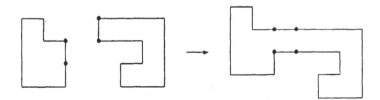

Figure 3.2: Concatenation of a 10-step polygon and a 14-step polygon to produce a 24-step polygon in $\mathbf{Z}^2$. The dots are the endpoints of the bonds that are changed during the concatenation.

dimensions, however, the procedure is slightly more involved, because such a pair of edges need not be parallel. In general, the concatenation effectively occurs in a $(d-1)$-dimensional hyperplane, and so there will be an additional factor of $d - 1$ to account for the number of possible orientations of the "leftmost" bond. The correct form of the subadditivity relation for polygons is the following.

Theorem 3.2.3 *For even integers* $M, N \geq 4$,

$$\frac{q_M q_N}{d-1} \leq q_{N+M} \tag{3.2.2}$$

and

$$q_N \leq q_{N+2}. \tag{3.2.3}$$

Proof. We prove (3.2.2) first; the proof of (3.2.3), which is similar, will follow. First, we define the *lexicographic ordering* on $\mathbf{Z}^d$, as follows. We say that $(a_1, \ldots, a_d) \prec (b_1, \ldots, b_d)$ if for some j (with $1 \leq j \leq d$) we have: $a_i = b_i$ whenever $1 \leq i < j$, and $a_j < b_j$. For even integers $N \geq 4$, let $Q[N]$ be the set of N-step self-avoiding polygons whose lexicographically smallest point is the origin. Then $Q[N]$ has exactly q_N members.

For each $i = 1, \ldots, d$, let $e^{(i)}$ be the neighbour of the origin with $e_i^{(i)} = 1$ and $e_j^{(i)} = 0$ for $j \neq i$. For $i = 2, \ldots, d$ and for even $M \geq 4$, let $Q_i[M]$ be the set of M-step self-avoiding polygons that lie in the half-space $x_1 \geq 0$ and that contain the bond joining the origin to $e^{(i)}$. Then $Q[M]$ is contained in the union of $Q_2[M], \ldots, Q_d[M]$, and so, by symmetry,

$$|Q_2[M]| = \cdots = |Q_d[M]| \geq \frac{q_M}{d-1}. \tag{3.2.4}$$

Choose an arbitrary N-step polygon $\mathcal{P}$ in $Q[N]$, and let p be its lexicographically largest point. There are two values of i ($1 \leq i \leq d$) such that $\mathcal{P}$ contains the bond joining p to $p - e^{(i)}$; let I be the larger of these two values. (In particular, we have $I \geq 2$.) Then let Q be an arbitrary self-avoiding polygon in $Q_I[M]$.

We now concatenate $\mathcal{P}$ and Q. First translate Q by the vector $p - e^{(I)} + e^{(1)}$ (so the resulting polygon lies in the half-space $x_1 \geq p_1 + 1$ and contains the bond joining $p - e^{(I)} + e^{(1)}$ to $p + e^{(1)}$). Then take all of the bonds in the translated Q *except* the bond joining $p - e^{(I)} + e^{(1)}$ to $p + e^{(1)}$, and all of the bonds of $\mathcal{P}$ *except* the bond joining p to $p - e^{(I)}$, and also take the two bonds that join $p - e^{(I)}$ to $p - e^{(I)} + e^{(1)}$ and p to $p + e^{(1)}$. Since $\mathcal{P}$ is contained in the half-space $x_1 \leq p_1$, the result is a self-avoiding polygon in $Q[N + M]$. Conversely, given an $(N + M)$-step polygon constructed in this fashion, we can reconstruct $\mathcal{P}$ and Q, because the N sites with smallest first coordinate are precisely the points of $\mathcal{P}$. (Of course, not every polygon in $Q[N + M]$ can be obtained by such a concatenation.)

Since there were q_N ways to choose $\mathcal{P}$, and at least $q_M/(d-1)$ ways to choose $\mathcal{Q}$ given $\mathcal{P}$ [by (3.2.4)], inequality (3.2.2) follows immediately.

Finally we prove (3.2.3). Choose $\mathcal{P}$, p, and I as above. Then remove the bond joining p to $p - e^{(I)}$ from $\mathcal{P}$, and add the three bonds of the walk $(p, p+e^{(1)}, p+e^{(1)}-e^{(I)}, p-e^{(I)})$ to $\mathcal{P}$. The result is a self-avoiding polygon in $\mathcal{Q}[N+2]$ from which $\mathcal{Q}$ can be unambiguously determined as above. This proves (3.2.3). □

Now let $a_1 = 0$ and $a_n = -\log(q_{2n}/(d-1))$ for $n \geq 2$. Then Theorem 3.2.3 says that $\{a_n\}_{n\geq 1}$ is a subadditive sequence. Therefore Lemma 1.2.2 implies that $\lim_{n\to\infty}(q_{2n}/(d-1))^{1/2n}$ exists and equals some number $\mu_{Polygon} \leq \mu$, and that

$$q_N \leq (d-1)(\mu_{Polygon})^N \qquad (3.2.5)$$

for all even $N > 2$. In fact, $\mu_{Polygon} = \mu$; this will be a corollary of the next theorem, independent of Theorem 3.2.3.

Theorem 3.2.4 *Let e be a nearest neighbour of the origin in $\mathbf{Z}^d$. There exists a constant K, depending only on the dimension d, such that for every integer $M \geq 1$,*

$$c_{2M+1}(0, e) \geq KM^{-d-2}(b_M)^2. \qquad (3.2.6)$$

Proof. For a point x in $\mathbf{Z}^d$, let $B[M, x]$ denote the set of M-step bridges which begin at the origin and end at x, and let $|B[M, x]|$ denote the number of bridges in this set.

Consider a point x for which $B[M, x]$ is not empty, and let ω and v be bridges in $B[M, x]$ (not necessarily different). See Figure 3.3. Choose any vector $\mathbf{v} \equiv \mathbf{v}(x)$ in $\mathbf{R}^d$ which is orthogonal to the line containing 0 and x. Let i (respectively j) be chosen from among those values of $\{0, 1, \ldots, M\}$ that maximize (respectively, minimize) the dot product $\omega(i)\cdot\mathbf{v}$ (respectively, $v(j) \cdot \mathbf{v}$). Define

$$\overline{\omega} = (\omega(i), \ldots, \omega(M), \omega(1) + \omega(M), \ldots, \omega(i) + \omega(M)),$$
$$\overline{v} = (v(j), \ldots, v(M), v(1) + v(M), \ldots, v(j) + v(M)).$$

It is not hard to check that $\overline{\omega}$ and $\overline{v}$ are both self-avoiding walks (since ω and v were bridges), that $\overline{\omega}(M) - \overline{\omega}(0)$ and $\overline{v}(M) - \overline{v}(0)$ both equal x, and that

$$\overline{\omega}(0) \cdot \mathbf{v} = \overline{\omega}(M) \cdot \mathbf{v} \geq \overline{\omega}(k) \cdot \mathbf{v}$$
$$\overline{v}(0) \cdot \mathbf{v} = \overline{v}(M) \cdot \mathbf{v} \leq \overline{v}(k) \cdot \mathbf{v}$$

for all $k = 0, \ldots, M$. To interpret these inequalities, think of two hyperplanes orthogonal to $\mathbf{v}$, one passing through $\overline{\omega}(0)$ and the other through $\overline{v}(0)$; then $\overline{\omega}$ and $\overline{v}$ lie on opposite sides of their respective hyperplanes.

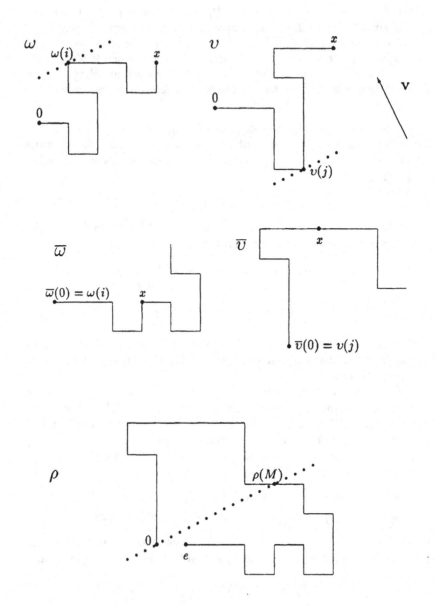

Figure 3.3: Proof of Theorem 3.2.4. Here $M = 12$. Top: the M-step bridges ω and v, and the vector $\mathbf{v}$. Middle: the derived walks $\overline{\omega}$ and $\overline{v}$. Bottom: the $(2M + 1)$-step walk ρ. The dotted lines are orthogonal to $\mathbf{v}$.

Now let e be the nearest neighbour of the origin such that $e \cdot \mathbf{v} < 0$. Let ϱ be the $(2M + 1)$-step walk starting at the origin and consisting of $\bar{v}$, followed by one step in the e direction, followed by the reversal of $\bar{w}$; that is,

$$\varrho(k) = \begin{cases} \bar{v}(k) - \bar{v}(0) & \text{for } 0 \le k \le M, \\ \bar{w}(2M + 1 - k) - \bar{w}(0) + e & \text{for } M + 1 \le k \le 2M + 1. \end{cases}$$

Then ϱ is a self-avoiding walk since the hyperplane with normal vector $\mathbf{v}$ that passes through the origin separates the first $M + 1$ points of ϱ from the last $M + 1$. Also, $\varrho(2M + 1) - \varrho(0) = e$ and $\rho(M) = x$.

Given a self-avoiding walk ϱ that has been constructed as above, we could reconstruct the original bridges ω and υ if we only knew i and j. There are $M + 1$ possible values for each of i and j. Therefore, if S denotes the set of $(2M + 1)$-step self-avoiding walks ϱ with $\varrho(0) = 0$ and $|\varrho(2M + 1)| = 1$, then the number of walks in S having $\varrho(M) = x$ is at least $|B[M, x]|^2/(M + 1)^2$. Since there are fewer than $M(2M + 1)^{d-1}$ values of x for which $|B[M, x]| > 0$, it follows from the above argument and the Schwarz inequality that

$$|S| \ge \frac{\sum_x |B[M, x]|^2}{(M + 1)^2} \ge \frac{(\sum_x |B[M, x]|)^2}{(M + 1)^2 M(2M + 1)^{d-1}}. \tag{3.2.7}$$

The theorem is a direct consequence of (3.2.7). $\quad\square$

Corollary 3.2.5 *There exists a constant C depending only on the dimension d such that*

$$\mu^{2M} e^{-CM^{1/2}} \le c_{2M+1}(0, e) \le \frac{2(M + 1)(d - 1)}{d} \mu^{2M+2} \tag{3.2.8}$$

for all $M \ge 1$. In particular we have

$$\mu_{Polygon} = \lim_{n \to \infty} (q_{2n})^{1/2n} = \mu. \tag{3.2.9}$$

Proof. The first inequality of (3.2.8) is a direct consequence of Theorem 3.2.4 and Equation (3.1.9) (the constant C can absorb all factors of polynomial order). The second inequality follows from (3.2.1), (3.2.5), and the obvious bound $\mu_{Polygon} \le \mu$. Finally, Equation (3.2.9) follows immediately from Equations (3.2.8) and (3.2.1). $\quad\square$

Remark. It is possible to prove that $\mu_{Polygon} = \mu_{Bridge}$ directly, without using the results of Section 3.1. This can be done using Theorem 3.2.4 to prove $\mu_{Polygon} \ge \mu_{Bridge}$, and the bound $q_N \le d(d - 1)b_N$, which follows from Proposition 8.1.2, to prove $\mu_{Polygon} \le \mu_{Bridge}$.

Corollary 3.2.6 *Let $\{x^{(N)}\}$ be a sequence of sites in $\mathbf{Z}^d \setminus \{0\}$ such that $|x^{(N)}| = o(N)$ as $N \to \infty$. To avoid trivialities, we also assume that $\|x^{(N)}\|_1$ has the same parity as N. Then*

$$\lim_{N \to \infty} \left(c_N(0, x^{(N)}) \right)^{1/N} = \mu. \tag{3.2.10}$$

In particular, for every $x \neq 0$, the two-point function $G_z(0, x)$ has the same radius of convergence $z_c = \mu^{-1}$ as the susceptibility $\chi(z)$.

Proof. For fixed N, let ϕ be a fixed self-avoiding walk from the origin to $x^{(N)}$ of length $\|x^{(N)}\|_1$ (or possibly $\|x^{(N)}\|_1 + 2$) whose lexicographically largest point, p, is neither 0 nor $x^{(N)}$. Let L_N be the length of ϕ. Choose $I \geq 2$ so that the bond joining p and $p - e^{(I)}$ is a bond of ϕ. We now consider concatenation of ϕ and self-avoiding polygons Q in $Q_I[N - L_N]$, as in the fourth paragraph of the proof of Theorem 3.2.3. In detail: Given such a polygon Q, translate it by the vector $p - e^{(I)} + e^{(1)}$. Then take all of the bonds in the translated Q *except* the bond joining $p - e^{(I)} + e^{(1)}$ to $p + e^{(1)}$, and all of the bonds of ϕ *except* the bond joining p to $p - e^{(I)}$, and also take the two bonds that join $p - e^{(I)}$ to $p - e^{(I)} + e^{(1)}$ and p to $p + e^{(1)}$. Since ϕ is contained in the half-space $x_1 \leq p_1$, and since the translated polygon lies in the half-space $x_1 \geq p_1 + 1$, the result determines an N-step self-avoiding walk from 0 to $x^{(N)}$. We conclude from (3.2.4) that

$$c_N(0, x^{(N)}) \geq \frac{q^{N-L_N}}{d-1}. \tag{3.2.11}$$

Since $L_N = o(N)$, the result now follows immediately from Corollary 3.2.5 and the trivial bound $c_N(0, x) \leq c_N$. $\qquad\square$

We remark that for fixed x (i.e. $x^{(N)}$ independent of N), the lower bound (3.2.11) can be improved by a factor of order N; see Proposition 7.4.4.

3.3 Kesten's bound on c_N

In this section, we shall prove the following upper bound on the number of self-avoiding walks:

Theorem 3.3.1 *Let $d \geq 2$. Then there exists a constant Q, depending only on d, such that for every $N \geq 2$*

$$c_N \leq \mu^N \exp[Q N^{2/(d+2)} \log N]. \tag{3.3.1}$$

As we observed in Section 1.1, this is the best bound that is known rigorously in three and four dimensions. In two dimensions, it is not quite as good as the Hammersley-Welsh bound (Theorem 3.1.1), while above four dimensions we know that $c_N \sim \text{const.} \mu^N$ (see Section 6.1).

This theorem first appeared in Kesten (1964). However, that paper only presented a proof of the weaker bound

$$c_N \leq \mu^N \exp[Q N^{2/(d+1)} \log N]. \tag{3.3.2}$$

The proof of Theorem 3.3.1 builds on the proof of (3.3.2) and draws on the ideas of the Hammersley-Welsh argument. The full proof was not given in Kesten (1964) because it was hoped that someone would find a better bound, but almost thirty years later this has still not come to pass for $d = 3, 4$. The proof of Theorem 3.3.1 that we present here is due to Kesten (private communication).

To begin with, observe that by the first inequality of (3.1.7) and the inequality $x^a + y^a \leq 2^{1-a}(x + y)^a$ for $0 < a < 1$ and $x, y \geq 0$, it suffices to prove the same bound for half space walks, i.e. that there exists a Q depending only on d such that

$$h_n \leq \mu^n \exp[Q n^{2/(d+2)} \log n]. \tag{3.3.3}$$

for every $n \geq 2$. Therefore we shall work with half-space walks for much of the proof. We first need an extension of Definitions 3.1.2 and 3.1.3.

Definition 3.3.2 *For integers N and S, let $h_{N,S}^* = \sum_{i=0}^{S} h_{N,i}$ be the number of half-space walks starting at the origin and having span at most S.*

Consider an integer $n \geq 1$. If ω is an n-step half-space walk that starts at the origin, then let $K(\omega)$ denote the span of ω and let $I(\omega)$ be the largest value of i such that $\omega_1(i) = K(\omega)$. Since $|\omega| > 0$, both $K(\omega)$ and $I(\omega)$ are nonzero. Observe that the first $I(\omega)$ steps of ω is a bridge of span $K(\omega)$, and the remainder of ω is (the reflection of) a half-space walk whose span is less than $K(\omega)$. Since there are at most n^2 possibilities for the pair $I(\omega)$ and $K(\omega)$, there exist integers $i[0]$ and $k[0]$ in $\{1, \ldots, n\}$ such that the number of half-space walks ω having $I(\omega) = i[0]$ and $K(\omega) = k[0]$ is at least $n^{-2} h_n$. Therefore the above decomposition shows that

$$h_n \leq n^2 b_{i[0],k[0]} h_{n-i[0],k[0]}^*. \tag{3.3.4}$$

At this point, we shall state a lemma which is crucial for the proof of Theorem 3.3.1. Its proof will be deferred to the end of the section. To help the reader appreciate the role of the lemma, we shall show how the weaker bound (3.3.2) may be obtained as an immediate corollary of the lemma and (3.3.4).

Lemma 3.3.3 *Let k, l, and m be strictly positive integers, and let B be a real number satisfying $0 < B < 1$. Let $V = (m^{1-B}l)^{1/d}$. Then there exists a constant D, depending only on the dimension d, such that*

$$b_{l,k}h^*_{m,k} \leq \mu^{m+l+dV}[D(m+l)]^{12m^B+3dV}. \qquad (3.3.5)$$

Corollary 3.3.4 *Let $d \geq 2$. There exists a constant Q, depending only on d, such that (3.3.2) holds for every $n \geq 2$.*

Proof. As explained prior to (3.3.3), it suffices to prove (3.3.2) with c_n replaced by h_n. Let $B = 2/(d+1)$, and let $n \geq 2$. If $i[0] = n$, then the result follows from (3.3.4) together with the basic relations $b_n \leq \mu^n$ and $h^*_{0,k} = 1$ for every k. Therefore, assume that $i[0] < n$. By (3.3.4) and Lemma 3.3.3, we have

$$h_n \leq n^2\mu^{n+dV}[Dn]^{12n^B+3dV}, \qquad (3.3.6)$$

where

$$V = ((n - i[0])^{1-B}i[0])^{1/d} \leq n^{(2-B)/d}. \qquad (3.3.7)$$

Since $(2 - B)/d = 2/(d+1) = B$, we see that

$$h_n \leq \mu^n(\mu^d)^{n^B}[Dn]^{(12+3d)n^B+2}, \qquad (3.3.8)$$

and the result follows.   □

We now proceed with the proof of Theorem 3.3.1. The idea is to iterate (3.3.4) until certain auxiliary conditions are satisfied, and then to make some estimates and apply Lemma 3.3.3. Fix real numbers A and B in the interval $(0, 1)$, and fix an integer $n \geq 1$. (As we shall see by the end of the proof, we are specifically interested in the values $A = d/(d+2)$ and $B = 2/(d+2)$.)

We shall now give a procedure for defining an integer $u \geq 0$ and integers $i[0], \ldots, i[u]$, $k[0], \ldots, k[u] > 0$ (all depending on A and n) having certain properties. We have already seen how to define $i[0]$ and $k[0]$. Next, if $n^2b_{i[0],k[0]} > \mu^{i[0]}$ and $i[0] < n^A$, or if $i[0] = n$, then set $u = 0$ and stop; otherwise we reapply the decomposition of (3.3.4) with n replaced by $n-i[0]$ to choose $i[1]$ and $k[1]$ from $\{1, \ldots, n - i[0]\}$ such that

$$h^*_{n-i[0],k[0]} \leq (n - i[0])^2 b_{i[1],k[1]}h^*_{n-i[0]-i[1],k[1]} \qquad (3.3.9)$$

(notice that $h^*_{n-i[0],k[0]} \leq h_{n-i[0]}$). Then (3.3.4) and (3.3.9) imply that

$$h_n \leq n^2b_{i[0],k[0]}(n - i[0])^2b_{i[1],k[1]}h^*_{n-i[0]-i[1],k[1]}. \qquad (3.3.10)$$

We now repeat this procedure inductively. Suppose that $i[j]$ and $k[j]$ have already been defined but $i[j+1]$ and $k[j+1]$ have not yet been defined. If

$$(n - i[0] - \cdots - i[j-1])^2 b_{i[j],k[j]} > \mu^{i[j]} \qquad (3.3.11)$$

and

$$i[j] < (n - i[0] - \cdots - i[j-1])^A, \qquad (3.3.12)$$

or if $i[0] + \cdots + i[j] = n$, then set u equal to this value of j and stop the procedure. Otherwise, choose $i[j+1]$ and $k[j+1]$ from $\{1, \ldots, n - i[0] - \cdots - i[j]\}$ such that

$$h^*_{n-i[0]-\cdots-i[j],k[j]} \qquad (3.3.13)$$
$$\leq \quad (n - i[0] - \cdots - i[j])^2 b_{i[j+1],k[j+1]} h^*_{n-i[0]-\cdots-i[j+1],k[j+1]}.$$

Thus we end up with the inequality

$$h_n \leq n^2 b_{i[0],k[0]} (n - i[0])^2 b_{i[1],k[1]} \cdots \qquad (3.3.14)$$
$$\times (n - i[0] - \cdots - i[u-1])^2 b_{i[u],k[u]} h^*_{n-i[0]-\cdots-i[u],k[u]}.$$

Let $\mathcal{I}$ be the set of j's in $\{0, 1, \ldots, u-1\}$ with the property that (3.3.11) holds (if $u = 0$, then $\mathcal{I}$ is the empty set). Therefore (3.3.12) fails for these values of j, i.e.

$$i[j] \geq (n - i[0] - \cdots - i[j-1])^A \quad \text{for every } j \text{ in } \mathcal{I}. \qquad (3.3.15)$$

If $0 \leq j < u$ and j is not in $\mathcal{I}$, then the reverse inequality of (3.3.11) holds (with $\leq$), while if j is in $\mathcal{I}$ or if j equals u, then we have the simple inequality

$$(n - i[0] - \cdots - i[j-1])^2 b_{i[j],k[j]} \leq n^2 \mu^{i[j]} \qquad (3.3.16)$$

(from $b_i \leq \mu^i$). Applying these inequalities to (3.3.14) yields

$$h_n \leq n^{2|\mathcal{I}|+2} \mu^{i[0]+\cdots+i[u]} h^*_{n-i[0]-\cdots-i[u],k[u]}. \qquad (3.3.17)$$

Next we claim that there exists a constant C depending on A but not on n such that

$$|\mathcal{I}| \leq C n^{1-A}. \qquad (3.3.18)$$

To see this, for each integer $a \geq 0$ we let $\mathcal{I}_a$ denote the subset of integers j in $\mathcal{I}$ with the property that

$$n 2^{-a} \geq n - i[0] - \cdots - i[j-1] \geq n 2^{-a-1}. \qquad (3.3.19)$$

If we can show that

$$|\mathcal{I}_a| \leq 1 + \left(n 2^{-a-1}\right)^{1-A} \qquad (3.3.20)$$

for every $a \geq 0$, then the claim (3.3.18) will follow from

$$|\mathcal{I}| \leq \sum_{a=0}^{\log_2 n} |\mathcal{I}_a| \leq 1 + \log_2 n + \sum_{a=0}^{\infty} \left(n2^{-a-1}\right)^{1-A}. \qquad (3.3.21)$$

If $|\mathcal{I}_a| \leq 1$, then (3.3.20) is trivial. Otherwise let f_a and F_a denote the smallest and largest members of $\mathcal{I}_a$ respectively, and let $\mathcal{I}_a'$ denote the set $\mathcal{I}_a$ with F_a removed. By (3.3.19),

$$n2^{-a} \geq n - i[0] - \cdots - i[f_a - 1] \geq n - i[0] - \cdots - i[F_a - 1] \geq n2^{-a-1}, \qquad (3.3.22)$$

and hence

$$\sum_{j \in \mathcal{I}_a'} i[j] \leq \sum_{j=f_a}^{F_a - 1} i[j] \leq n2^{-a} - n2^{-a-1} = n2^{-a-1}. \qquad (3.3.23)$$

Also, (3.3.15) and (3.3.19) imply that

$$\sum_{j \in \mathcal{I}_a'} i[j] \geq (|\mathcal{I}_a| - 1)(n2^{-a-1})^A. \qquad (3.3.24)$$

Combining (3.3.23) and (3.3.24) yields (3.3.20), and the claim (3.3.18) follows.

To prepare for the application of Lemma 3.3.3, we let $k = k[u]$, $l = i[u]$, and $m = n - i[0] - \cdots - i[u]$. (Recall that $k[u]$ and $i[u]$ are strictly positive.) Then (3.3.17) and (3.3.18) tell us that

$$h_n \leq \mu^{n-m} n^{2Cn^{1-A}+2} h_{m,k}^*. \qquad (3.3.25)$$

If $m = 0$, then since $h_{0,k}^* = 1$ for every k, the result (3.3.3) follows from (3.3.25) by simply taking $A = d/(d+2)$. So for the remainder of the proof, we shall assume that m is strictly positive. By the definition of u, we know from (3.3.11), (3.3.12), and the bound $m + l \leq n$ that

$$\mu^l < (m + l)^2 b_{l,k} \leq n^2 b_{l,k} \qquad (3.3.26)$$

and

$$l < (m + l)^A \leq n^A. \qquad (3.3.27)$$

Combining (3.3.26) with Lemma 3.3.3 yields

$$n^{-2} \mu^l h_{m,k}^* \leq \mu^{m+l+dV} [D(m + l)]^{12m^B + 3dV}, \qquad (3.3.28)$$

which implies

$$h_{m,k}^* \leq \mu^{m+dV} n^2 [Dn]^{12n^B + 3dV}. \qquad (3.3.29)$$

Applying (3.3.29) to (3.3.25), we obtain

$$h_n \leq \mu^{n+dV} n^{2Cn^{1-A}+4} [Dn]^{12n^B+3dV}. \tag{3.3.30}$$

Now, $V = (m^{1-B}l)^{1/d} \leq n^{(1-B+A)/d}$ by (3.3.27), so we see that there is a constant Q, depending only on A, B, and d, such that

$$h_n \leq \mu^n \exp[Q(n^{(1-B+A)/d} + n^{1-A} + n^B) \log n] \tag{3.3.31}$$

for every $n \geq 2$. Finally, set $A = d/(d+2)$ and $B = 2/(d+2)$, so that $1 - A = (1 - B + A)/d = 2/(d+2)$ (these are the optimal choices for A and B). This proves (3.3.3), and Theorem 3.3.1 follows.

Proof of Lemma 3.3.3. Let β be an arbitrary l-step bridge starting at the origin and having span k. Let η be an arbitrary m-step half-space walk starting at the origin and having span at most k. Now let

$$\mathcal{Y} = \{y \in \mathbf{Z}^d : y_1 \geq \eta_1(m) \text{ and } \|y - \eta(m)\|_\infty \leq V\}.$$

Then $\mathcal{Y}$ is a half-cube containing $(\lfloor V \rfloor + 1)(2\lfloor V \rfloor + 1)^{d-1}$ points of $\mathbf{Z}^d$, and hence

$$|\mathcal{Y}| > V^d = m^{1-B}l. \tag{3.3.32}$$

For each y in $\mathbf{Z}^d$, let $J(y)$ denote the number of pairs (i,j) ($0 \leq i \leq m$, $0 \leq j \leq l$) such that $\eta(i) - \beta(j) = y$. The sum of $J(y)$ over all $y \in \mathbf{Z}^d$ equals $(m+1)(l+1)$, and so the average value of $J(y)$ over $\mathcal{Y}$ is less than or equal to

$$\frac{(m+1)(l+1)}{|\mathcal{Y}|} \leq \frac{4ml}{m^{1-B}l} = 4m^B.$$

Thus we know that there exists a point Y in $\mathcal{Y}$ such that

$$J(Y) \leq 4m^B. \tag{3.3.33}$$

Let $g = \|Y - \eta(m)\|_1$; since $Y \in \mathcal{Y}$ we know that

$$0 \leq g \leq dV. \tag{3.3.34}$$

Now define a walk ρ, not necessarily self-avoiding, which consists of η, followed by a walk of minimal length from $\eta(m)$ to Y, followed by β (translated to begin at Y). Thus ρ has exactly $m + g + l$ steps. Observe that

$$0 < \rho_1(i) \leq Y_1 + k = Y_1 + \beta_1(l) = \rho_1(m+g+l) \quad \text{for all } i = 0, \ldots, m+g+l \tag{3.3.35}$$

(this is because (i) $\rho_1(i) = \eta_1(i) \in (0, k]$ whenever $0 < i \le m$, (ii) $\eta_1(m) \le \rho_1(i) \le Y_1$ whenever $m \le i \le m + g$, and (iii) $\rho_1(i) = Y_1 + \beta_1(i - m - g) \in [Y_1, Y_1 + k]$ whenever $m + g \le i \le m + g + l$). Let T be the number of self-intersections of ρ (if $\rho(i) = z$ for exactly n different values of i, then we count $n - 1$ self-intersections). There are exactly $J(Y)$ intersections of the first $m + 1$ sites of ρ with the last $l + 1$ sites, and at most $g - 1$ intersections of $(\rho(m + 1), \ldots, \rho(m + g - 1))$ with the rest of ρ; therefore

$$T \le J(Y) + g - 1 \le 4m^B + dV - 1 \tag{3.3.36}$$

by (3.3.33) and (3.3.34).

It follows from (3.3.35) that ρ can be obtained by taking a (self-avoiding) bridge of span $Y_1 + k$ and adjoining at most T self-avoiding polygons (including possibly "degenerate" two-step polygons). To understand this, think of traversing ρ one step at a time. When ρ first intersects itself, remove the segment of ρ between the two visits to the site where the self-intersection occurs. The removed segment is a self-avoiding polygon with a distinguished site (where the intersection occurs) and orientation (corresponding to the direction in which ρ traversed the polygon). Observe that some of these "polygons" may consist of only two steps, the direction of the second being the opposite of the first. Accordingly, we define the number of two-step polygons to be $q_2 = d$. Now continue to traverse ρ, and repeat this procedure: the next time that ρ visits a site that it has already visited (excluding visits that occurred on the removed segment), remove the resulting polygon, and so on. Let φ be the part of ρ that is never removed; then φ is a bridge of span $Y_1 + k$. (We remark that this procedure is essentially the same as the "loop-erasing" of Section 10.2.) Let $r = |\varphi|$, let t be the number of polygons that have been removed by this procedure, and let a_i be the number of steps in the i-th polygon. Then $r + a_1 + \cdots + a_t = |\rho|$.

Now, the number of different p-step walks ρ that can give rise to a particular choice of φ, t, and $a_1, \ldots, a_t$ is at most

$$\prod_{j=1}^{t} (2pa_j q_{a_j}), \tag{3.3.37}$$

because there are q_{a_j} choices for the j-th polygon, exactly a_j different points on the j-th polygon where it could be attached to the walk (i.e. to φ or one of the first $j - 1$ polygons), at most p places on the walk where the j-th polygon can be attached (in fact, at most $|\varphi| + a_1 + \cdots + a_{j-1}$ places), and two possible directions that the polygon can be traversed. [If a_j equals 2 for some j, then it in fact would have sufficed to have used $2pq_2$ for the j-th term in (3.3.37).] For every even $n \ge 2$ we have $q_n \le (d - 1)\mu^n$ [by (3.2.5) and (3.2.9) for $n \geq 4$; the case $n = 2$ is obvious since $\mu \geq d$], and so the product (3.3.37) can be bounded above by

$$[2(d-1)p^2]^t \mu^{a_1 + \cdots + a_t} = [2(d-1)p^2]^t \mu^{|\rho|-r}.$$

Therefore the number of possible walks ρ is at most

$$\sum_{p=m+l}^{m+l+dV} \sum_{r=1}^{p} b_r \sum_{t=0}^{H} \sum_{\substack{a_1, \cdots, a_t : \\ a_1 + \cdots + a_t = p - r}} [2(d-1)p^2]^t \mu^{p-r}, \qquad (3.3.38)$$

where $H = 4m^B + dV - 1$ [recall (3.3.36)]. Using the fact that there are at least $b_{l,k} h_{n,k}^*$ possible walks ρ, we conclude from (3.3.38) and the bound $b_r \leq \mu^r$ [Equation (1.2.17)] that

$$
\begin{aligned}
b_{l,k} h_{n,k}^* &\leq \sum_{p=m+l}^{m+l+dV} \sum_{r=1}^{p} \sum_{t=0}^{H} \sum_{\substack{a_1, \cdots, a_t : \\ a_1 + \cdots + a_t = p - r}} [2(d-1)p^2]^H \mu^p \\
&\leq \sum_{p=m+l}^{m+l+dV} p(H+1)p^H [2(d-1)p^2]^H \mu^p \qquad (3.3.39) \\
&\leq \mu^{m+l+dV} 4(m+l+dV)^3 [2(d-1)(m+l+dV)^3]^{4m^B+dV-1}.
\end{aligned}
$$

Finally, we obtain the inequality of the lemma from (3.3.39) and $V \leq (ml)^{1/2} \leq (m+l)/2$. □

3.4 Notes

Section 3.1. An earlier paper [Hammersley (1961b)] proved that $c_N \leq \mu^N \exp[O(N^{(d-1)/d} \log N)]$ for $d \geq 2$. The proof was more complicated than the proof of Theorem 3.1.1, but the methods were similar; they were also closely related to the methods of Theorem 3.3.1.

The asymptotics of the number of partitions of N can also be used to study "spiral" walks, which are self-avoiding walks in $\mathbf{Z}^2$ that cannot turn to the left. Guttmann and Wormald (1984) proved that the number of N-step spiral walks is $\exp[2\pi(N/3)^{1/2}]N^{-7/4}[C + O(N^{-1/2})]$, where $C = 4 \cdot 3^{5/4}/\pi$.

Corollary 3.1.8 is due to Kesten (1963).

Section 3.2. Theorem 3.2.3 and Corollary 3.2.6 are due to Hammersley (1961a), using essentially the same proofs as we present. This paper also contains a proof that $\mu_{Polygon} = \mu$, using very different methods from ours. A result similar to Theorem 3.2.4 appears in Equation (3.7) of Kesten (1963).

Dubins *et al.* (1988) proved the following result about self-avoiding polygons in $\mathbf{Z}^2$. Consider the set of all N-step polygons that have the origin as one of their sites. Then the probability that the point $(\frac{1}{2}, \frac{1}{2})$ lies in the inside region of a polygon chosen at random from this set equals $\frac{1}{2} - \frac{1}{N}$. (Here, the "inside region" is the bounded subset of $\mathbf{R}^2$ whose boundary is the simple closed curve determined by the polygon.) They conjecture that the analogous probability for any other point (a, b) of $\mathbf{R}^2$ (with a and b non-integer) should likewise increase to $1/2$ as $N \to \infty$, but nothing is known even for $(a, b) = (\frac{3}{2}, \frac{1}{2})$.

Chapter 4

Decay of the two-point function

4.1 Properties of the mass

In this section we shall develop some fundamental properties of the mass m, which we originally defined in Equation (1.3.15) as follows:

$$m(z) = \liminf_{n \to \infty} \frac{-\log G_z(0, (n, 0, \ldots, 0))}{n}. \tag{4.1.1}$$

Thus the mass describes the exponential decay rate of the two-point function. We shall see that the "lim inf" appearing in (4.1.1) is in fact a limit for every $z > 0$ except perhaps $z = z_c$, and that the two-point function decays (to leading order) like $\exp[-m(z)|x|_z]$ for $z < z_c$, for some norm[1] $|\cdot|_z$. We shall also show that $m(z)$ is a reasonably nice function, strictly positive below the critical point z_c, identically $-\infty$ above z_c, and decreasing to 0 as z approaches z_c from the left. Finally we shall prove that $m(z_c) = 0$ if the "bubble diagram" $B(z) \equiv \sum_x G_z(0, x)^2$ (see Section 1.5) is finite at the critical point $z = z_c$ (as it is for $d \geq 5$; see Corollary 6.1.7). It is expected that $m(z_c) = 0$ in all dimensions [in fact, $G_{z_c}(0, x)$ is believed to decay as a power law; see (1.4.8)], but it remains an open problem to prove this for $d = 2, 3, 4$. In particular, it is not even known rigorously that $G_{z_c}(0, x)$ is finite for any $x \neq 0$ in low dimensions.

---

[1] Alternatively, one can work with the Euclidean norm and a direction-dependent mass $m[v; z] \equiv m(z)|v|_z$ for vectors $v \in R^d$ such that $|v| = 1$. Then $G_z(0, x)$ decays like $\exp[-m[v; z]|x|]$ where $v = x/|x|$.

The term "mass" comes from quantum field theory, where the exponential decay rate of the theory's two-point function defines the physical mass of the particles in the theory. In statistical mechanics, the mass tending to 0 is equivalent to the correlation length $\xi(z) = 1/m(z)$ tending to ∞. In the context of the Ising model and the other N-vector models, for example, this says that spins are becoming correlated on larger and larger length scales. In a spin model, the divergence of the correlation length as $z \nearrow z_c$ is the precursor of the long range order (and spontaneous magnetization) that will occur for $z > z_c$. The self-avoiding walk corresponds to the $N = 0$ case, where the concept of long-range order does not really apply, but we are still interested in whether the mass tends to 0 by analogy to spin systems. In addition, it is believed (and known for $d \geq 5$) that the mass for the self-avoiding walk goes to zero as a power law

$$ m(z) \sim \text{const.}(z_c - z)^{\overline{\nu}} \quad \text{as} \quad z \nearrow z_c, \tag{4.1.2}$$

with $\overline{\nu} = \nu$ (recall Section 1.3). Proving that $m(z) \searrow 0$ as $z \nearrow z_c$ is a first step towards proving (4.1.2).

Many results of this section extend immediately to self-avoiding walks $(\omega(0), \ldots, \omega(N))$ whose steps $\omega(i) - \omega(i-1)$ all lie in a finite subset Ω of $\mathbf{Z}^d$ which is invariant under all symmetries of $\mathbf{Z}^d$. Besides the usual nearest-neighbour model ($\Omega = \{x : \|x\|_1 = 1\}$), in Chapter 6 we shall also be interested in the "spread-out" models which have $\Omega = \{x : 0 < \|x\|_\infty \leq L\}$ for some (large) integer L. In particular, everything in this section up to and including Theorem 4.1.6, as well as Theorem 4.1.18, hold for general symmetric Ω with only trivial changes in the proofs.

Our first proposition establishes some elementary properties.

Proposition 4.1.1 *(a) $m(z)$ is a concave function of $\log z$ for $z > 0$.*
(b) On the interval $(0, z_c)$, $m(z)$ is a nonincreasing, finite, strictly positive, and continuous function of z.
(c) If $z > z_c$, then $G_z(0, x) = +\infty$ for every $x \neq 0$, and hence $m(z) = -\infty$.

We delay the proof of this proposition just long enough to present the following lemma:

Lemma 4.1.2 *Let $\{a_n\}_{n \geq 0}$ be a sequence of nonnegative numbers. Then $-\log(\sum_n a_n e^{n\beta})$ is a concave function of β.*

Proof. This is a consequence of Hölder's inequality. For λ between 0 and 1,

$$ \sum_n a_n e^{n[\lambda\beta_1 + (1-\lambda)\beta_2]} \leq \left(\sum_n a_n e^{n\beta_1}\right)^\lambda \left(\sum_n a_n e^{n\beta_2}\right)^{1-\lambda}. $$

The lemma follows upon taking $-\log$ of both sides.   □

Proof of Proposition 4.1.1.   (a) Lemma 4.1.2 shows that $-\log G_z(0, x)$
is a concave function of $\log z$ for any x. Since the lim inf of a sequence of
concave functions is concave, the result follows.
(b) Since $G_z(0, x)$ is nondecreasing in z, it is apparent from (4.1.1) that
$m(z)$ is nonincreasing. We already saw in Section 1.3 that $0 < m(z) < +\infty$
whenever $0 < z < z_c$ [recall (1.3.14) and (1.3.16)]. Continuity follows from
the concavity and finiteness of $m(z)$ on the open interval $(0, z_c)$.
(c) This is an immediate consequence of Corollary 3.2.6.   □

Next, we shall show how to replace the "lim inf" in (4.1.1) by a limit,
obtaining as a by-product an explicit bound on the two-point function in
terms of the mass and the bubble diagram $B(z) = \sum_x G_z(0, x)^2$. We will
use the notation $(n, 0)$ to denote the point $(n, 0, \ldots, 0) \in \mathbf{Z}^d$; this notation
will be generalized in Definition 4.1.7 below.

Theorem 4.1.3 (a) If $0 < z < z_c$, then

$$\lim_{n \to \infty} \frac{-\log G_z(0, (n, 0))}{n} = m(z) = \inf_{n \geq 1} \frac{-\log[G_z(0, (n, 0))/B(z)]}{n}; \quad (4.1.3)$$

in particular, the limit exists and satisfies

$$G_z(0, (n, 0)) \leq B(z) e^{-m(z)n} \quad \text{for every } n \geq 1. \quad (4.1.4)$$

(b) If $B(z_c)$ is finite, then (4.1.3) and (4.1.4) also hold for $z = z_c$, and
$m(z)$ is left-continuous at z_c.

The proof depends on subadditivity and the following lemma.

Lemma 4.1.4 For any $z > 0$, and any x and y in $\mathbf{Z}^d$,

$$G_z(0, x)G_z(x, y) \leq B(z)G_z(0, y). \quad (4.1.5)$$

Proof.   For each nonnegative integer N, let $\mathcal{S}_N$ denote the set of all
ordered pairs of self-avoiding walks (ω_A, ω_B) such that: ω_A starts at 0 and
ends at x; ω_B starts at x and ends at y; and $|\omega_A| + |\omega_B| = N$. Also, let
$\mathcal{T}_N$ denote the set of all ordered triples of self-avoiding walks $(\omega_C, \omega_D, \omega_E)$
such that: ω_C starts at 0 and ends at y; ω_D and ω_E both start at x and end
at the same (arbitrary) point; and $|\omega_C| + |\omega_D| + |\omega_E| = N$. To prove the
lemma, it suffices to show that there is a one-to-one mapping from $\mathcal{S}_N$ into
$\mathcal{T}_N$, for this would imply an inequality between their respective generating
functions, which is precisely the inequality that we we want.

Let (ω_A, ω_B) be a member of $\mathcal{S}_N$. Let I be the smallest value of i such that $\omega_A(i)$ is a point of ω_B. Let $u = \omega_A(I)$. Let ω_C be the walk which follows ω_A from 0 to u and then follows ω_B from u to y; this is self-avoiding by our choice of u. Let ω_D (respectively, ω_E) be the part of ω_A (respectively, ω_B) between x and u. Then $(\omega_C, \omega_D, \omega_E)$ is in $\mathcal{T}_N$. This mapping is clearly one-to-one, so the lemma is proven.   □

Proof of Theorem 4.1.3.   Fix z with $B(z) < \infty$, so in particular we may choose any z in $(0, z_c)$ since $B(z) \leq \chi(z)^2$. For each integer $n \geq 1$, define $h_n(z) = -\log[G_z(0, (n, 0))/B(z)]$. Taking $x = (n, 0)$ and $y = (m + n, 0)$ in (4.1.5) and dividing by $B(z)^2$ shows that the sequence $\{h_n(z) : n \geq 1\}$ is subadditive. Therefore Lemma 1.2.2 implies that

$$\lim_{n \to \infty} \frac{h_n(z)}{n} = \inf_{n \geq 1} \frac{h_n(z)}{n}, \qquad (4.1.6)$$

which proves (4.1.3) in both cases (a) and (b). The bound (4.1.4) follows immediately from (4.1.3).

It only remains to prove that $m(z)$ is left-continuous at z_c if $B(z_c)$ is finite. Since $m(z)$ is nonincreasing, it suffices to show that

$$\limsup_{z \nearrow z_c} m(z) \leq m(z_c). \qquad (4.1.7)$$

Assume $B(z_c)$ is finite. This implies that $G_{z_c}(0, x)$ is finite (for every x); hence, since $B(z)$ and $G_z(0, x)$ are power series with nonnegative coefficients, they must be continuous on $(0, z_c]$. Therefore $h_n(z)$ is continuous on $(0, z_c]$ for every n. Together with the fact that $m(z) \leq h_n(z)/n$ for all z in $(0, z_c)$ [by (4.1.3)], this implies that

$$\limsup_{z \nearrow z_c} m(z) \leq \limsup_{z \nearrow z_c} \frac{h_n(z)}{n} = \frac{h_n(z_c)}{n}. \qquad (4.1.8)$$

Finally, we have seen that $m(z_c) = \inf_{n \geq 1} h_n(z_c)/n$ when $B(z_c)$ is finite, so taking the inf over $n \geq 1$ in (4.1.8) yields (4.1.7), which completes the proof.   □

We now turn our attention to the task of showing that the mass goes to 0 as z approaches z_c from the left. This turns out to be relatively easy if the bubble condition $B(z_c) < \infty$ holds. However, it is expected that $B(z_c)$ is infinite in 2, 3, and 4 dimensions, so we will have to work harder there. But first we shall take care of the high-dimensional case.

Lemma 4.1.5 *For any $z > 0$ and any x in $\mathbf{Z}^d$,*

$$G_z(0, x) \leq B(z)^{1/2} e^{-m(z)\|x\|_\infty}. \qquad (4.1.9)$$

Proof. Given x in $\mathbf{Z}^d$, let $K = \|x\|_\infty$ and let i be a coordinate such that $|x_i| = K$. Let u be the vector whose i-th coordinate is x_i and whose j-th coordinate, for every $j \neq i$, is $-x_j$. Then $x + u = \pm 2K e^{(i)}$, where $e^{(i)}$ is the unit vector whose i-th coordinate is 1. By Lemma 4.1.4 and symmetry considerations, we have

$$G_z(0, x) G_z(x, x + u) \leq \mathsf{B}(z) G_z(0, x + u) = \mathsf{B}(z) G_z(0, (2K, 0)). \quad (4.1.10)$$

Applying symmetry and (4.1.4), we obtain $G_z(0, x)^2 \leq \mathsf{B}(z) \exp[-m(z)2K]$, and (4.1.9) follows. $\qquad\square$

Theorem 4.1.6 *If* $\mathsf{B}(z_c) < \infty$, *then* $\lim_{z \nearrow z_c} m(z) = m(z_c) = 0$.

Proof. The first equality was proven in Theorem 4.1.3(b). Since $m(z) > 0$ for $z < z_c$ [by Proposition 4.1.1(b)], we see that $m(z_c) \geq 0$. Finally, if $m(z_c)$ were strictly positive, then Lemma 4.1.5 would imply that the critical two-point function decays exponentially, which would contradict the fact that the susceptibility is infinite at z_c [recall (1.3.6)]. Therefore $m(z_c)$ must equal 0. $\qquad\square$

For the rest of this chapter, we will not assume that the bubble condition holds. We will not be able to prove that the mass is 0 at z_c, but we will show that the mass decreases to 0 as z approaches z_c from the left. Some new ideas will be needed to accomplish this. In a nutshell, we would like a subadditivity relation in the spirit of Lemma 4.1.4 that holds nontrivially at the critical point. As in Section 3.1, we shall use bridges to get superadditivity relations instead [e.g. (4.1.13) and (4.1.14) below]. We first define generating functions and masses for classes of bridges, prove properties about these, and then show that these masses are the same as the one defined by (4.1.1). As the reader will discover, it is often easier to work with bridges than with general self-avoiding walks.

Definition 4.1.7 *Let* $y = (y_1, \ldots, y_{d-1})$ *be a point of* $\mathbf{Z}^{d-1}$, *and let* L *be a nonnegative integer. Then* (L, y) *denotes the point* $(L, y_1, \ldots, y_{d-1})$ *in* $\mathbf{Z}^d$, *and* $b_{N,L}(y)$ *denotes the number of* N-*step bridges* ω *with* $\omega(0) = 0$ *and* $\omega(N) = (L, y)$. *Recalling Definition 3.1.3, we see that*

$$b_{N,L} = \sum_{y \in \mathbf{Z}^{d-1}} b_{N,L}(y).$$

For each real $z > 0$, *we define the generating functions*

$$B_z(L, y) = \sum_{N=0}^{\infty} b_{N,L}(y) z^N$$

(the "point-to-point" bridge generating function) and

$$B_z(L) = \sum_{N=0}^{\infty} b_{N,L} z^N$$

(the "point-to-plane" bridge generating function). Observe that

$$B_z(L) = \sum_{y \in \mathbb{Z}^{d-1}} B_z(L, y).$$

Remark. To be fully consistent with our notation for the two-point function, we should be writing $B_z(0, (L, y))$ instead of $B_z(L, y)$. However, the shorter notation should not cause any confusion.

Proposition 4.1.8 *For every real $z > 0$, the limits*

$$M(z) = \lim_{L \to \infty} \frac{-\log B_z(L, 0)}{L} \quad \text{and} \quad \overline{M}(z) = \lim_{L \to \infty} \frac{-\log B_z(L)}{L} \qquad (4.1.11)$$

exist in $[-\infty, +\infty)$, and satisfy

$$B_z(L, 0) \le e^{-LM(z)} \quad \text{and} \quad B_z(L) \le e^{-L\overline{M}(z)} \qquad (4.1.12)$$

for every integer $L \ge 1$. Also, $M(z)$ and $\overline{M}(z)$ are nonincreasing functions of z, and they satisfy the obvious inequalities $M(z) \ge \overline{M}(z)$ and $M(z) \ge m(z)$.

Proof. Let L_1 and L_2 be nonnegative integers. The concatenation of a bridge of span L_1 with a bridge of span L_2 is a bridge of span $L_1 + L_2$, and the result uniquely determines the original pair. Thus it is apparent that

$$\sum_{n=0}^{N} b_{n,L_1} b_{N-n,L_2} \le b_{N,L_1+L_2}$$

and

$$\sum_{n=0}^{N} b_{n,L_1}(0) b_{N-n,L_2}(0) \le b_{N,L_1+L_2}(0)$$

for every nonnegative integer N, so

$$B_z(L_1) B_z(L_2) \le B_z(L_1 + L_2) \qquad (4.1.13)$$

and

$$B_z(L_1, 0) B_z(L_2, 0) \le B_z(L_1 + L_2, 0) \qquad (4.1.14)$$

for all $z > 0$. Therefore, by Lemma 1.2.2, the limits in (4.1.11) exist and satisfy (4.1.12) for all $L \geq 1$. □

We will now develop some properties of the mass M, and eventually we will show that M and $\overline{M}$ are identical. We begin with an analogue of Proposition 4.1.1.

Proposition 4.1.9 (a) $M(z) \leq -\log z$ for all $z > 0$.
(b) $M(z) \geq -\log(\mu z)$ for $0 < z < z_c$.
(c) M is a concave function of $\log z$ $(z > 0)$.
    In particular, M is finite and continuous on $(0, z_c)$.

Proof. (a) For every L, $b_{L,L}(0) = 1$, so $B_z(L, 0) \geq z^L$. The result follows now from (4.1.12).
(b) Fix z in $(0, z_c)$. Since $b_N \leq \mu^N$ by (1.2.17),

$$B_z(L, 0) \leq \sum_{N=L}^{\infty} \mu^N z^N = \frac{(\mu z)^L}{1 - \mu z}$$

for every $L > 0$. The result follows from (4.1.11).
(c) This follows by applying Lemma 4.1.2 with $a_n = b_{n,L}(0)$, dividing by L, and letting $L \to \infty$. □

We now describe "truncated" generating functions for bridges that are confined to a tube centred along the x_1-axis. The lemma which follows shows that the corresponding truncated mass converges to the mass $M(z)$ as the radius of the tube tends to infinity.

Definition 4.1.10 *For all positive integers N, L, and T, and for all points y in $\mathbf{Z}^{d-1}$, let $b_{N,L}^T(y)$ be the number of N-step bridges ω having $\omega(0) = 0$, $\omega(N) = (L, y)$, and $|\omega_i(k)| \leq T$ for every $i = 2, \ldots, d$ and $k = 0, \ldots, N$. For real $z > 0$, let*

$$B_z^T(L, y) = \sum_N b_{N,L}^T(y) z^N.$$

Observe that the Monotone Convergence Theorem implies that for every $z > 0$

$$\lim_{T \to \infty} B_z^T(L, y) = B_z(L, y) \text{ for all } L \text{ and } y. \tag{4.1.15}$$

Also, by the usual concatenation argument, for every T we have

$$B_z^T(L_1, 0) B_z^T(L_2, 0) \leq B_z^T(L_1 + L_2, 0) \text{ for all } L_1 \text{ and } L_2, \tag{4.1.16}$$

which implies, by Lemma 1.2.2, that we can define the "truncated masses"

$$M^T(z) = \lim_{L \to \infty} \frac{-\log B_z^T(L, 0)}{L} = \inf_{L \geq 1} \frac{-\log B_z^T(L, 0)}{L}. \tag{4.1.17}$$

Lemma 4.1.11 *Let T and L be positive integers, and let z be a positive real number.*
(a) Let $M_L^T(z) = -(\log B_z^T(L,0))/L$. Then

$$-\frac{(2T+1)^{d-1}}{z} \le \frac{d}{dz} M_L^T(z) \le -\frac{1}{z}.$$

(b) For $z_2 > z_1 > 0$:

$$-(2T+1)^{d-1}\log(z_2/z_1) \le M^T(z_2) - M^T(z_1) \le -\log(z_2/z_1).$$

In particular, M^T is a continuous decreasing function of z.
(c) $\lim_{T\to\infty} M^T(z) = \inf_{T\ge 1} M^T(z) = M(z)$.
(d) M is left-continuous; i.e., $\lim_{u\nearrow z} M(u) = M(z)$ for all $z > 0$.
   In (c) and (d), the limits may be $-\infty$.

Remark. Part (d) of this lemma is mainly of interest at the critical point, since $M(z)$ is already known to be continuous on $(0, z_c)$ by Proposition 4.1.9.

Proof of Lemma 4.1.11. (a) First observe that $B_z^T(L,0)$ is a polynomial with positive coefficients, and so $M^T(z)$ is differentiable at every $z > 0$:

$$\frac{d}{dz} M_L^T(z) = -\frac{\sum_N N b_{N,L}^T(0) z^{N-1}}{L \sum_N b_{N,L}^T(0) z^N}.$$

Since $b_{N,L}^T(0)$ is nonzero only if N is between L and $L(2T+1)^{d-1}$, the result follows.
(b) The result follows upon integrating the inequalities of part (a) from z_1 to z_2 and then letting $L \to \infty$ [using (4.1.17)].
(c) Since $M^T(z)$ is decreasing in T, it suffices to show that $\inf_{T\ge 1} M^T(z) = M(z)$. By subadditivity (recall Proposition 4.1.8) and (4.1.15),

$$M(z) = \inf_{L\ge 1} \frac{-\log B_z(L,0)}{L} = \inf_{L\ge 1} \inf_{T\ge 1} \frac{-\log B_z^T(L,0)}{L}.$$

The result now follows by interchanging the order of the infs in the last expression and using (4.1.17).
(d) This follows from parts (b) and (c), together with the general fact that the inf of a sequence of continuous decreasing functions is left-continuous. □

Lemma 4.1.12 $M(z) = \overline{M}(z)$ *for all $z > 0$.*

Proof. Since $M(z) \geq \overline{M}(z)$, it suffices to prove the reverse inequality. To do this, we use a slightly different form of truncation. Let $\overline{B}_z^T(L, y)$ denote the generating function of the collection of bridges ω from 0 to (L, y) with the property that

$$|\omega_i(j) - \omega_i(k)| \leq T \text{ for } 2 \leq i \leq d \text{ whenever } \omega_1(j) = \omega_1(k).$$

Let $\overline{B}_z^T(L) = \sum_y \overline{B}_z^T(L, y)$. The analogue of (4.1.16) holds for $\overline{B}_z^T(L)$, and hence the mass $\overline{M}^T(z)$ of $\overline{B}_z^T(L)$ exists and satisfies the analogue of (4.1.17). The same arguments as in the proof of Lemma 4.1.11(c) show that $\lim_{T \to \infty} \overline{M}^T(z) = \overline{M}(z)$. Observe that $\overline{B}_z^T(L, y) = 0$ if $\|y\|_\infty > LT$.

For any $L > 0$ and any y in $\mathbf{Z}^{d-1}$, we can get a bridge from 0 to $(2L + 1, 0)$ by the concatenation of a bridge from 0 to (L, y), a single step from (L, y) to $(L + 1, y)$, and another bridge from 0 to (L, y) that has been reflected through the hyperplane $x_1 = L + \frac{1}{2}$. This construction and the Schwarz inequality show that, for any $T > 0$,

$$
\begin{aligned}
B_z(2L + 1, 0) &\geq \sum_{y \in \mathbf{Z}^{d-1}:\|y\|_\infty \leq LT} z(\overline{B}_z^T(L, y))^2 \\
&\geq \frac{z(\sum_{y:\|y\|_\infty \leq LT} \overline{B}_z^T(L, y))^2}{\sum_{y:\|y\|_\infty \leq LT} 1^2} \\
&= \frac{z(\overline{B}_z^T(L))^2}{(2LT + 1)^{d-1}}.
\end{aligned}
$$

This implies that for every fixed T, $M(z) \leq \overline{M}^T(z)$. The lemma follows. $\square$

Theorem 4.1.13 $\lim_{z \nearrow z_c} M(z) = M(z_c) = 0.$

Proof. First, Lemma 4.1.11(d) says that M is left-continuous. By Proposition 4.1.9(b), we know that $M(z) > 0$ whenever $0 < z < z_c$, and so $M(z_c) \geq 0$. Next, Lemma 4.1.12 tells us that $M(z_c) = \overline{M}(z_c)$, so it only remains to prove that $\overline{M}(z_c) \leq 0$. But if it were true that $\overline{M}(z_c) > 0$, then it would follow from (4.1.12) that

$$B_{z_c} = \sum_{L=0}^{\infty} B_{z_c}(L) \leq \sum_{L=0}^{\infty} e^{-L\overline{M}(z_c)} < +\infty,$$

which contradicts Corollary 3.1.8. This completes the proof. $\square$

We are finally ready to prove that the various masses are identical below the critical point.

Theorem 4.1.14 *For all z in $(0, z_c)$, $m(z) = M(z) = \overline{M}(z)$.*

Proof.   Fix z in $(0, z_c)$. Recall $M(z) = \overline{M}(z)$ by Lemma 4.1.12. Since $G_z(0, (n, 0)) \geq B_z(n, 0)$, it is clear that $m(z) \leq M(z)$.

Any self-avoiding walk from 0 to $(n, 0)$ can be decomposed into three parts: cut the walk at the last time that it visits the hyperplane $x_1 = 0$ and at the first time after this that it visits the hyperplane $x_1 = n$. The middle piece is a bridge of span n; the other two pieces are self-avoiding walks (possibly having length 0). This decomposition shows that

$$G_z(0, (n, 0)) \leq (\chi(z))^2 B_z(n) \leq (\chi(z))^2 e^{-M(z)n}, \qquad (4.1.18)$$

where we have also used (4.1.12) and $\overline{M}(z) = M(z)$. Since $z < z_c$, $\chi(z)$ is finite, so $m(z) \geq M(z)$. The theorem follows.   □

Corollary 4.1.15 $\lim_{z \nearrow z_c} m(z) = 0$.

Proof.   This is an immediate consequence of Theorems 4.1.14 and 4.1.13.   □

Corollary 4.1.16 *The mass $m(z)$ is strictly decreasing on $(0, z_c)$, and $\lim_{z \searrow 0} m(z) = +\infty$.*

Proof.   By Lemma 4.1.11(b,c), the function $M(z) + \log z$ is nonincreasing on $(0, +\infty)$. Since $M(z)$ is finite and equals $m(z)$ on $(0, z_c)$, the corollary follows.   □

Corollary 4.1.17 *Define*

$$G_z(L) = \sum_{y \in \mathbb{Z}^{d-1}} G_z(0, (L, y)), \qquad (4.1.19)$$

the generating function of all self-avoiding walks from the origin to the hyperplane $x_1 = L$. Then

$$\lim_{L \to \infty} \frac{-\log G_z(L)}{L} = m(z) \text{ for all } z \text{ in } (0, z_c). \qquad (4.1.20)$$

In fact,

$$e^{-m(z)L} \leq G_z(L) \leq \chi(z)^2 e^{-m(z)L} \text{ for all } L \geq 1. \qquad (4.1.21)$$

Proof. Equation (4.1.20) follows from

$$B_z(L) \leq G_z(L) \leq \chi(z)^2 B_z(L), \tag{4.1.22}$$

which may be obtained by the same argument that gave (4.1.18). The upper bound of (4.1.21) follows from (4.1.22) and $B_z(L) \leq e^{-m(z)L}$. The lower bound of (4.1.21) follows from the subadditivity relation (1.2.5), since $G_z(L_1)G_z(L_2) \geq G_z(L_1 + L_2)$ whenever L_1 and L_2 are positive integers.
□

We now show that Theorem 4.1.3 can be generalized so that x can tend to infinity in any direction. We remark that the proof of the following theorem relies only on material from the first part of this section (prior to Definition 4.1.7). Recall that the bubble diagram $\mathsf{B}(z) = \sum_x G_z(0,x)^2$ is finite for $0 < z < z_c$ (since $\mathsf{B}(z) \leq \chi(z)^2$).

Theorem 4.1.18 *For any $0 < z < z_c$, there exists a norm $|\cdot|_z$ on $\mathbf{R}^d$, satisfying $\|u\|_\infty \leq |u|_z \leq \|u\|_1$ for every u in $\mathbf{R}^d$, such that*

$$\lim_{|x|_z \to \infty} \frac{-\log G_z(0,x)}{|x|_z} = m(z) \tag{4.1.23}$$

and

$$G_z(0,x) \leq \mathsf{B}(z)e^{-m(z)|x|_z} \quad \text{for every } x \text{ in } \mathbf{Z}^d. \tag{4.1.24}$$

Proof. Fix z in $(0, z_c)$. For each v in $\mathbf{Z}^d$, Lemma 4.1.4 tells us that

$$\frac{G_z(0,jv)}{\mathsf{B}(z)} \frac{G_z(0,kv)}{\mathsf{B}(z)} \leq \frac{G_z(0,(j+k)v)}{\mathsf{B}(z)}$$

for all nonnegative integers j and k. Therefore Lemma 1.2.2 implies that the limit

$$\lim_{n \to \infty} \frac{-\log G_z(0,nv)}{n}$$

exists, and, if we denote this limit by $m[v; z]$, that

$$G_z(0,nv) \leq \mathsf{B}(z)e^{-nm[v;z]} \quad \text{for every } n \geq 1. \tag{4.1.25}$$

We have $m[0; z] = 0$ since $G_z(0,0) = 1$, but for every nonzero v in $\mathbf{Z}^d$ we have $0 < m[v; z] \leq \|v\|_1 |\log z|$ (the lower bound follows from (1.3.14), while

$$G_z(0,v) \geq z^{\|v\|_1} \tag{4.1.26}$$

gives the upper bound). Also, it follows immediately that

$$m[kv; z] = |k|m[v; z] \tag{4.1.27}$$

for every integer k and every v in $\mathbf{Z}^d$, and (from Lemma 4.1.4) that

$$m[u + v; z] \leq m[u; z] + m[v; z]$$

for every u and v in $\mathbf{Z}^d$.

Define $|v|_z$ by

$$|v|_z = \frac{m[v; z]}{m(z)}$$

for v in $\mathbf{Z}^d$, and use (4.1.27) to extend the definition to v in $\mathbf{Q}^d$ (where $\mathbf{Q}$ denotes the rational numbers). Observe that $|u|_z = 1$ if u is one of the $2d$ unit vectors of $\mathbf{Z}^d$ (by Theorem 4.1.3), that $|kv|_z = |k||v|_z$ for every rational k and every v in $\mathbf{Q}^d$, and that $|u + v|_z \leq |u|_z + |v|_z$ for every u and v in $\mathbf{Q}^d$; in particular, then, we have $|v|_z \leq ||v||_1$ for every v in $\mathbf{Q}^d$. Consequently, we obtain

$$\big| |u|_z - |v|_z \big| \leq |u - v|_z \leq ||u - v||_1;$$

thus, $|\cdot|_z$ is uniformly continuous on $\mathbf{Q}^d$, and so it extends to a continuous function on all of $\mathbf{R}^d$, which will be a norm on $\mathbf{R}^d$. Next, Lemma 4.1.5 shows that $m[v; z] \geq m(z)||v||_\infty$ for every v in $\mathbf{Z}^d$. From this it follows that $||v||_\infty \leq |v|_z$ on $\mathbf{Z}^d$, and hence on all of $\mathbf{R}^d$.

From (4.1.25) and the definition of $|\cdot|_z$, we obtain

$$G_z(0, v) \leq \mathsf{B}(z) \exp(-|v|_z m(z)) \quad \text{for every } v \text{ in } \mathbf{Z}^d, \tag{4.1.28}$$

which is (4.1.24). It follows from this that if (4.1.23) is false then there exists a sequence $\{x_n\}$ of points in $\mathbf{Z}^d$, tending to infinity in norm, such that

$$\lim_{n \to \infty} \frac{-\log G_z(0, x_n)}{|x_n|_z} > m(z). \tag{4.1.29}$$

By choosing a subsequence if necessary, we can also assume that there exists a t in $\mathbf{R}^d$ such that $x_n/|x_n|_z$ converges to t. Let $\epsilon > 0$. Since $|t|_z = 1$, we can choose a v in $\mathbf{Z}^d$ and a positive integer J such that

$$||t - J^{-1}v||_1 \leq \epsilon \quad \text{and} \quad |J^{-1}v|_z \leq 1 + \epsilon.$$

Next, choose a sequence of integers $k(n)$ such that $k(n)/|x_n|_z$ tends to J^{-1}. When n is large, x_n is approximately $k(n)v$; this will lead to a contradiction of (4.1.29), as follows.

Lemma 4.1.4 tells us that

$$G_z(0, k(n)v)G_z(k(n)v, x_n) \leq \mathsf{B}(z)G_z(0, x_n).$$

Now take logs of this inequality, divide by $-|x_n|_z$, and let n tend to infinity. Using

$$\lim_{n\to\infty} \frac{-\log G_z(0, k(n)v)}{|x_n|_z} = J^{-1}m[v; z]$$
$$= |J^{-1}v|_z m(z)$$
$$\leq (1+\epsilon)m(z)$$

and [with the help of (4.1.26)]

$$\lim_{n\to\infty} \frac{-\log G_z(k(n)v, x_n)}{|x_n|_z} \leq -\lim_{n\to\infty} \frac{\|k(n)v - x_n\|_1}{|x_n|_z} \log z$$
$$= -\|J^{-1}v - t\|_1 \log z$$
$$\leq -\epsilon \log z,$$

we conclude that

$$\lim_{n\to\infty} \frac{-\log G_z(0, x_n)}{|x_n|_z} \leq (1+\epsilon)m(z) - \epsilon \log z.$$

Since ϵ can be made arbitrarily close to 0, this contradicts our choice of $\{x_n\}$ and completes the proof of the theorem. $\qquad\square$

4.2 Bridges and renewal theory

The main goal of the rest of this chapter is to give a more refined analysis of the asymptotics of $G_z(0, (L, 0, \ldots, 0))$ for large L when $z < z_c$. Most of the work can be done by focusing first on bridges and their two-point functions. The present section will give the asymptotic behaviour of the bridge generating functions. Specifically, we shall prove in Theorem 4.2.5 that $B_z(L)$ exhibits pure exponential decay, and in Theorem 4.2.6 that $B_z(L, y)$ exhibits exponential decay with a Gaussian power law correction, also known as *Ornstein-Zernike decay*. We shall also prove that the mass $m(z)$ is a real analytic function of z in the interval $(0, z_c)$. To obtain these results and develop some intuition about why they are true, we shall set up a correspondence between generating functions and certain probabilistic quantities in renewal theory.

We begin by discussing the decomposition of bridges. Given an N-step bridge ω, suppose that i satisfies $0 < i \leq N$ and

$$\omega_1(j) \leq \omega_1(i) < \omega_1(k) \quad \text{for all } j = 0, \ldots, i \text{ and } k = i+1, \ldots, N. \quad (4.2.1)$$

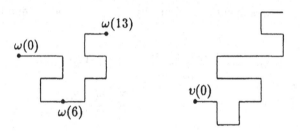

Figure 4.1: The bridge ω on the left can be decomposed into two smaller bridges: a 6-step bridge followed by a 7-step bridge. The bridge v on the right is irreducible.

Then ω can be decomposed into two smaller bridges, $(\omega(0), \ldots, \omega(i))$ and $(\omega(i), \ldots, \omega(N))$. (See Figure 4.1.) Observe that (4.2.1) always holds for $i = N$; in this case ω is trivially decomposed into ω and the 0-step bridge.

Definition 4.2.1 *We say that an N-step bridge is* irreducible *if the only i $(0 < i \le N)$ for which (4.2.1) holds is $i = N$. Let λ_N denote the number of irreducible N-step bridges, and let*

$$\Lambda_z = \sum_{N=1}^{\infty} \lambda_N z^N$$

be the corresponding generating function.

Observe that if ω is an irreducible N-step bridge, then for each $a = 1, \ldots, \omega_1(N)$, there exist at least *three* distinct values of i such that $\omega_1(i) = a$.

Given an N-step bridge ω (with $N > 0$), let s be the smallest index i for which (4.2.1) holds. Then $(\omega(0), \ldots, \omega(s))$ is an irreducible bridge and $(\omega(s), \ldots, \omega(N))$ is a bridge. It is thus straightforward to see that

$$b_N = \sum_{s=1}^{N} \lambda_s b_{N-s} + \delta_{N,0} \tag{4.2.2}$$

for every $N \ge 0$. From this equation we immediately obtain

$$B_z = \frac{1}{1 - \Lambda_z} \quad \text{for all complex } z \text{ with } |z| < z_c, \tag{4.2.3}$$

where B_z is the generating function for the number of bridges (recall Definition 3.1.7). For $0 < z < z_c$, B_z is finite, and so $\Lambda_z < 1$; therefore $\Lambda_{z_c} \le 1$

by the monotone convergence theorem. Also, B_z diverges at z_c (Corollary 3.1.8), so in fact $\Lambda_{z_c} = 1$; that is,

$$\sum_{k=1}^{\infty} \frac{\lambda_k}{\mu^k} = 1. \tag{4.2.4}$$

(It turns out that z_c is the radius of convergence of Λ_z; see Corollary 4.4.5.)

Equation (4.2.2) may now be transformed into a probabilistic renewal equation as follows. Let $p_k = \lambda_k \mu^{-k}$ (for $k \geq 1$) and $a_N = b_N \mu^{-N}$ (for $N \geq 0$). Observe that $a_0 = 1$, $a_k \leq 1$ by (1.2.17), and $\sum_k p_k = 1$ by (4.2.4). Multiplying (4.2.2) by μ^{-N} yields, for $N \geq 1$,

$$a_N = \sum_{k=1}^{N} p_k a_{N-k}. \tag{4.2.5}$$

To interpret this probabilistically, suppose that we have an independent sequence of random variables $X_1, X_2, \ldots$ with common distribution $\Pr\{X = k\} = p_k$. Then

$$a_N = \Pr\{X_1 + \ldots + X_k = N \text{ for some } k \geq 0\}; \tag{4.2.6}$$

i.e., a_N is the probability that there is a "renewal" at "time" N. (To understand this terminology, one can think of the X_i's as representing the lifetimes of light bulbs, where a new bulb immediately replaces one that burns out. Then a_N is the probability that a new bulb is installed on day N—i.e., that the system is renewed on day N.) Taking (4.2.6) as the definition of the sequence a_N, it is easy to verify (4.2.5). This probabilistic interpretation will be exploited later.

Equation (4.2.5) is a *discrete renewal equation*. The main theorem about these equations is the Renewal Theorem, which we state in the following form.

Theorem 4.2.2 *Assume that $\{f_n : n \geq 1\}$ and $\{g_n : n \geq 0\}$ are nonnegative sequences, and let*

$$f = \sum_{n=1}^{\infty} f_n \quad and \quad g = \sum_{n=0}^{\infty} g_n$$

denote their sums. Assume that $0 < g < +\infty$ and that $f_1 > 0$. Define the new sequence $v_0, v_1, \ldots$ by

$$v_0 = g_0$$
$$v_n = g_n + f_1 v_{n-1} + f_2 v_{n-2} + \cdots + f_n v_0, \text{ for all } n \geq 1.$$

(a) If $f < 1$, then $\lim_{n \to \infty} v_n = 0$ and $\sum_{n=0}^{\infty} v_n = g/(1 - f)$.

(b) If $f = 1$, then

$$\lim_{n \to \infty} v_n = \frac{g}{\sum_{k=1}^{\infty} k f_k}$$

(the limit is 0 if the sum in the denominator diverges). Also, $\sum_n v_n$ diverges.

(c) If $f > 1$, then $\limsup_{n \to \infty} v_n^{1/n} > 1$.

In the usual (more general) statements of the Renewal Theorem, the condition $f_1 > 0$ is replaced by the condition that the greatest common divisor of $\{n : f_n > 0\}$ is one. Also, there are more complete results available for part (c). For a full statement and proof, see Feller (1968, p. 330). Theorem 4.2.2 is sufficient for our needs; a proof appears in Appendix B.

In the present case, $f = \sum_k p_k = 1$, $g_n = \delta_{n,0}$, and $v_n = b_n \mu^{-n}$. Thus, the Renewal Theorem implies that $\lim_{N \to \infty} b_N/\mu^N$ exists and equals $(\sum_{k=1}^{\infty} k p_k)^{-1}$. If this limit were strictly positive, then it would say that the expected "time between renewals" would be finite, and hence that a typical N-step bridge would consist of at least ϵN irreducible bridges for some $\epsilon > 0$. But this would imply that the average value of $|\omega(N)|$ over the set of N-step bridges ω beginning at 0 would be proportional to N rather than to N^ν. But scaling theory predicts that only an exponentially small fraction of N-step self-avoiding walks have $|\omega(N)| > \epsilon N$ [e.g. Fisher (1966)], which contradicts the hypothesis that $\lim_N b_N/\mu^N > 0$. Therefore it is believed that $\lim_N b_N/\mu^N = 0$, but there is no known proof of this.

So far, we have only counted bridges according to the x_1 coordinates of their sites. To obtain more detailed information, we will also want to look at the remaining $d - 1$ coordinates. To this end, we introduce the following analogue of Definition 4.1.7 for irreducible bridges.

Definition 4.2.3 *For each point y in $\mathbf{Z}^{d-1}$ and each positive integer L and N, let $\lambda_{N,L}(y)$ denote the number of N-step irreducible bridges ω with $\omega(0) = 0$ and $\omega(N) = (L, y)$, and let*

$$\lambda_{N,L} = \sum_{y \in \mathbf{Z}^{d-1}} \lambda_{N,L}(y).$$

For each real $z > 0$, we define the generating functions

$$\Lambda_z(L, y) = \sum_{N=1}^{\infty} \lambda_{N,L}(y) z^N \quad \text{and} \quad \Lambda_z(L) = \sum_{N=1}^{\infty} \lambda_{N,L} z^N.$$

Using this definition, we have the following refinements of (4.2.2):

$$b_{N,L} = \sum_{k=1}^{N}\sum_{j=1}^{L} \lambda_{k,j} b_{N-k,L-j} + \delta_{N,0}\delta_{L,0} \tag{4.2.7}$$

$$b_{N,L}(y) = \sum_{k=1}^{N}\sum_{j=1}^{L}\sum_{v\in Z^{d-1}} \lambda_{k,j}(v) b_{N-k,L-j}(y-v) + \delta_{N,0}\delta_{L,0}\delta_{y,0}. \tag{4.2.8}$$

These imply the following equations for the generating functions:

$$B_z(L) = \sum_{j=1}^{L} \Lambda_z(j) B_z(L-j) + \delta_{L,0} \tag{4.2.9}$$

$$B_z(L,y) = \sum_{j=1}^{L}\sum_{v\in Z^{d-1}} \Lambda_z(j,v) B_z(L-j, y-v) + \delta_{L,0}\delta_{y,0}. \tag{4.2.10}$$

For the rest of this section, we shall only consider fixed $z < z_c$, so that $m(z) = M(z) = \overline{M}(z) > 0$ [by Theorem 4.1.14 and Proposition 4.1.1(b)]. First, let us turn (4.2.9) into a probabilistic renewal equation. We multiply both sides of the equation by $\exp(m(z)L)$, and set

$$p_L \equiv p_L(z) = \Lambda_z(L)e^{m(z)L} \qquad (L \geq 1) \tag{4.2.11}$$

and

$$a_L \equiv a_L(z) = B_z(L)e^{m(z)L} \qquad (L \geq 0). \tag{4.2.12}$$

Observe that $p_L \leq a_L \leq 1$ [by (4.1.12)]. Evidently the renewal equation (4.2.5) holds; to complete the probabilistic interpretation, we must show that $\sum_k p_k = 1$. Again, this can be accomplished by generating functions: take

$$P(s) = \sum_{k=1}^{\infty} p_k s^k \quad\text{and}\quad A(s) = \sum_{n=0}^{\infty} a_n s^n; \tag{4.2.13}$$

these are finite for $|s| < 1$. From (4.2.9) we then get

$$A(s) = \frac{1}{1 - P(s)} \text{ for } |s| < 1. \tag{4.2.14}$$

The sequence a_L is bounded away from zero [in fact, $a_L \geq (\chi(z))^{-2}$ by (4.2.12), (4.1.22), and (4.1.21)]. Therefore $A(s)$ diverges as s increases to 1, and so $P(1)$ must equal 1. That is,

$$\sum_{k=1}^{\infty} p_k = 1. \tag{4.2.15}$$

In addition, since we have just seen that a_L is bounded away from 0, the renewal theorem tells us that

$$\lim_{L \to \infty} a_L = \frac{1}{\sum_{k=1}^{\infty} k p_k} > 0. \tag{4.2.16}$$

Therefore

$$\sum_{k=1}^{\infty} k p_k < +\infty. \tag{4.2.17}$$

Using this kind of soft argument, there is not much more that can be said about the moments of the p_k sequence. However it turns out that

$$\limsup_{k \to \infty} p_k^{1/k} < 1. \tag{4.2.18}$$

This means that p_k has an exponential moment, or equivalently that the radius of convergence of $P(s)$ is strictly greater than 1. This can be expressed in terms of a "mass" for irreducible bridges:

Theorem 4.2.4 *For $0 < z < z_c$, define*

$$m_\Lambda(z) = \liminf_{L \to \infty} \frac{-\log \Lambda_z(L)}{L}. \tag{4.2.19}$$

Then

$$m_\Lambda(z) > m(z). \tag{4.2.20}$$

This theorem will be proven in the next section. It is clear that (4.2.20) is equivalent to (4.2.18).

We remark that it is not hard to see that (4.2.20) holds when z is sufficiently small. This is because an irreducible bridge of span $L \geq 2$ must have at least $3L$ steps, and when z is small the main contributions come from the shortest walks. Thus

$$\Lambda_z(L) = \sum_{N=3L}^{\infty} \lambda_{N,L} z^N \leq \sum_{N=3L}^{\infty} \mu^N z^N = \frac{(\mu z)^{3L}}{1 - \mu z}, \tag{4.2.21}$$

so

$$m_\Lambda(z) \geq -3 \log(\mu z). \tag{4.2.22}$$

But $m(z) \leq -\log z$ by Proposition 4.1.9(a), so $m_\Lambda(z) > m(z)$ whenever $z < \mu^{-3/2}$. Also note that these inequalities, together with Proposition 4.1.9(b) and the fact that

$$m_\Lambda(z) \leq -3 \log z \tag{4.2.23}$$

(which follows from the fact that $\lambda_{3L,L} \geq 1$ for $L \geq 2$), imply that $m_\Lambda(z) \sim 3m(z)$ as $z \searrow 0$. This is intuitively clear: for small z, the dominant contributions are from the shortest walks, and the shortest bridge of span L has length L while the shortest irreducible bridge has length $3L$. The $z \searrow 0$ limit is in fact where the greatest relative discrepancy between the two masses occurs, for we shall see in Corollary 4.4.4 that $m_\Lambda(z) \leq 3m(z)$ for every z in $(0, z_c)$.

For the rest of this section, we will concentrate on some of the consequences of Theorem 4.2.4. The first consequence is that the convergence in (4.2.16) is exponentially fast:

Theorem 4.2.5 *For* $0 < z < z_c$, *there exists a strictly positive constant* $\epsilon(z)$ *such that*

$$\left| B_z(L)e^{m(z)L} - C_z \right| \leq e^{-\epsilon(z)L}$$

for all $L \geq 1$, *where* $C_z = (\sum_{k=1}^\infty k\Lambda_z(k)e^{m(z)k})^{-1} > 0$.

Proof. Let s be complex, and define $P(s)$ and $A(s)$ as in (4.2.13). We know that $P(1) = 1$ by (4.2.15), and since the coefficients in the series defining $P(s)$ are all positive, we must have $|P(s)| < 1$ for all $s \neq 1$ such that $|s| \leq 1$. By Theorem 4.2.4, (4.2.18) holds, so $P(s)$ is analytic in some disc $|s| < R$ with R strictly greater than 1. Therefore there is an r between 1 and R such that $s = 1$ is the only zero of $P(s) - 1$ in the disc $|s| \leq r$. Next, observe that $P'(1) = 1/C_z$; in particular, $P'(1) \neq 0$, by (4.2.16). Denote the residue of the function $f(s)$ at a by $\mathrm{Res}(f(s), a)$. By the residue theorem and the integral theorem for the coefficients of a Laurent series, we have

$$a_L = -\mathrm{Res}\left(\frac{s^{-L-1}}{1 - P(s)}, 1\right) + \frac{1}{2\pi i}\oint_{z=r}\frac{s^{-L-1}}{1 - P(s)}ds.$$

The first term on the right is $1/P'(1)$ and the second is $O(r^{-L-1})$. This proves the theorem. $\qquad\qquad\square$

The probabilistic counterpart to the above theorem—that the probability of a renewal at time L converges exponentially fast to its limiting value when X_1 has an exponential moment—is well known (see Nummelin (1984), p.107). The next theorem is intrinsically probabilistic in nature: it gives a local central limit theorem for the endpoint of a bridge. In the terminology of mathematical physics, the $L^{-(d-1)/2}$ power law correction to the pure exponential decay in (4.2.25) below is known as *Ornstein-Zernike decay*. Such decay occurs in the subcritical two-point functions of a wide variety of statistical mechanical systems.

Theorem 4.2.6 *For $0 < z < z_c$, there exists a strictly positive constant δ_z such that*

$$\lim_{L \to \infty} \left| B_z(L,y) e^{m(z)L} L^{(d-1)/2} - C_z \frac{\exp(-|y|^2/\delta_z L)}{(\pi \delta_z)^{(d-1)/2}} \right| = 0 \qquad (4.2.24)$$

uniformly in y in $\mathbf{Z}^{d-1}$. Consequently,

$$B_z(L,y) \sim C_z e^{-m(z)L} \frac{\exp(-|y|^2/\delta_z L)}{(\pi \delta_z L)^{(d-1)/2}} \qquad (4.2.25)$$

as $L \to \infty$, uniformly in y in $\mathbf{Z}^{d-1}$ satisfying $|y| \le K L^{1/2}$ (for every fixed $K > 0$). Equivalently we have (by Theorem 4.2.5)

$$\frac{B_z(L,y)}{B_z(L)} \sim \frac{\exp(-|y|^2/\delta_z L)}{(\pi \delta_z L)^{(d-1)/2}} \qquad (4.2.26)$$

for the same range of y.

Intuitively, the relation (4.2.26) says that the endpoint of a bridge of span L has an asymptotically Gaussian probability distribution in the hyperplane $x_1 = L$.

We note here that Theorems 4.2.5 and 4.2.6 can be used to prove exactly analogous results for the full two-point function $G_z(0, x)$. This will be done in Section 4.4 (see Theorem 4.4.7).

We shall now provide a probabilistic context for interpreting and proving Theorem 4.2.6. Define the random vector (X, Y) such that X takes values in $\{1, 2, \ldots\}$ and Y takes values in $\mathbf{Z}^{d-1}$, and

$$\Pr\{(X,Y) = (L,y)\} = \Lambda_z(L,y) e^{m(z)L} \text{ for } L \ge 1, y \in \mathbf{Z}^{d-1}. \qquad (4.2.27)$$

Thus the marginal distribution of X is given by (4.2.11):

$$\Pr\{X = L\} = p_L(z), \text{ for } L = 1, 2, \ldots$$

(In particular, (4.2.15) guarantees that (4.2.27) describes a genuine probability distribution.) Let $\{(X_n, Y_n) : n \ge 1\}$ be a sequence of independent copies of (X, Y). Then we claim that

$$B_z(L,y) e^{m(z)L} = \Pr\{X_1 + \cdots + X_k = L \text{ and } Y_1 + \cdots + Y_k = y \text{ for some } k\} \qquad (4.2.28)$$

for $L \ge 1$ and y in $\mathbf{Z}^{d-1}$. (This also holds for $L = 0$ if we interpret the event on the right as occurring for $k = 0$ when $(L, y) = (0, 0)$.) Equation (4.2.28) follows by iteration of (4.2.10), which gives

$$B_z(L,y) = \sum_{k=1}^{L} \left[\sum \prod_{i=1}^{k} \Lambda_z(n_i, v_i) \right]$$

where the inner sum is over all $n_1, \ldots, n_k \geq 1$ and all $v_1, \ldots, v_k$ in $\mathbf{Z}^{d-1}$ such that $n_1 + \ldots + n_k = L$ and $v_1 + \ldots + v_k = y$.

Now, if we let $S_n = \sum_{i=1}^{n} (X_i, Y_i)$, then $\{S_n\}$ is a d-dimensional random walk with positive, finite drift in the x_1 direction, and $B_z(L, y)e^{m(z)L}$ is the probability that this walk ever hits the point (L, y). Thus Theorem 4.2.6 says that for large L, this probability factors into a $(d-1)$-dimensional Gaussian density, with variance proportional to L, times the inverse of the mean of X [which is the limiting probability that the hyperplane $x_1 = L$ is ever hit, according to (4.2.16)]. This just what one would expect probabilistically, if one knew that Y had finite variance. But in fact Theorem 4.2.4 implies that $|Y|$ has an exponential moment, as we now show.

Lemma 4.2.7 *(a) $m_\Lambda(z)$ is a concave function of $\log z$ ($z > 0$). In particular, it is finite and continuous for $0 < z < z_c$.*
(b) Fix $0 < z < z_c$. Then there exists an $s > 0$ for which $E(e^{s|Y|})$ is finite.

Proof. (a) By Lemma 4.1.2, $-L^{-1} \log \Lambda_z(L)$ is a concave function of $\log z$ for every $L \geq 1$. The desired concavity then follows from the fact that the lim inf of a concave sequence is concave. Inequalities (4.2.22) and (4.2.23) prove finiteness, and continuity follows immediately.
(b) By part (a), $m_\Lambda(z)$ is continuous at z. Therefore by (4.2.20) there exists an $s > 0$ such that $ze^s < z_c$ and $m_\Lambda(ze^s) > m(z)$. Then

$$
\begin{aligned}
E(e^{s|Y|}) &= \sum_{L,y} e^{s|y|} \Lambda_z(L, y) e^{m(z)L} \\
&\leq \sum_{L,y,N} e^{sN} \lambda_{N,L}(y) z^N e^{m(z)L} \\
&= \sum_{L} \Lambda_{ze^s}(L) e^{m(z)L} < \infty.
\end{aligned}
$$

$\square$

We now know exactly why Theorem 4.2.6 should be true—all that is left is to complete the technical details. This was done in Chayes and Chayes (1986a) via an explicit asymptotic analysis of the generating functions. We shall follow a different route here, appealing directly to a theorem from the probability literature.

Proof of Theorem 4.2.6. The following result is a (very) special case of Theorem 3.2 of Stam (1971); we shall not reproduce its proof here. Let $\{(X_n, Y_n) : n \geq 1\}$ be an independent, identically distributed sequence of $\mathbf{Z}^d$-valued random vectors (we follow our usual notation, so Y_n takes values in $\mathbf{Z}^{d-1}$). We write the coordinates as follows:

$$(X_n, Y_n) = (X_n, Y_{n2}, Y_{n3}, \ldots, Y_{nd}).$$

Assume the following: (i) that the expectation of X_1, which we denote θ, is finite and strictly positive; (ii) that the $((d-1)/2)$-th moment of X_1 is finite; (iii) that the covariance of Y_{1i} and Y_{1j} is 0 whenever $i \neq j$, and that for every $i = 2, \ldots, d$ the variance of Y_{1i} is a strictly positive finite constant v; and (iv) that the distribution of (X_n, Y_n) has no periodicities (for our purposes, it suffices to check that $\Pr\{(X_1, Y_1) = (L, y)\} > 0$ for every $L \geq 1$ and every y in $\mathbf{Z}^{d-1}$). Define

$$\varphi(L, y) = \theta^{-1} \left(\frac{\theta}{2\pi v} \right)^{(d-1)/2} \exp(-\theta|y|^2/2vL),$$

and let $U(L, y)$ denote the expected number of values of n such that

$$(X_1 + \cdots + X_n, Y_1 + \cdots + Y_n) = (L, y)$$

[i.e. the expected number of times that this d-dimensional random walk visits the point (L, y)]. Then Stam's theorem says that

$$\lim_{L \to \infty} |L^{(d-1)/2}U(L, y) - \varphi(L, y)| = 0,$$

uniformly in y in $\mathbf{Z}^{d-1}$. As a simple corollary, if we restrict $|y|$ to be no larger than $KL^{1/2}$ for some constant K, then $\varphi(L, y)$ remains bounded away from 0 and so

$$U(L, y) \sim L^{-(d-1)/2}\varphi(L, y)$$

as $L \to \infty$, uniformly for such y.

In our case, the distribution of the random vector is given by (4.2.27), for which assumption (iv) clearly holds. Since X_1 is always a positive integer, $\theta > 0$; in fact, $\theta = 1/C_z$, where C_z is the positive constant defined in Theorem 4.2.5. So assumption (i) holds. Assumption (ii) follows from the fact that X_1 has an exponential moment, by (4.2.18). For (iii): first, the covariances are 0 because the joint distribution of Y_{1i} and Y_{1j} is the same as the joint distribution of Y_{1i} and $-Y_{1j}$; secondly, symmetry implies that the variance of Y_{1i} does not depend on i; and thirdly, v is finite by Lemma 4.2.7(b). Finally, since X_1 is strictly positive, a point which is visited once by the random walk is never revisited; thus (4.2.28) tells us that $U(L, y) = B_z(L, y)e^{m(z)L}$. Now Stam's theorem proves Theorem 4.2.6, with $\delta_z = 2v/\theta$.   □

As a final consequence of Theorem 4.2.4, we shall prove that the mass is a real analytic function of z in $(0, z_c)$. This will be accomplished by applying the analytic implicit function theorem to (4.2.15), which is an equation relating z and $m(z)$.

Theorem 4.2.8 *The mass $m(z)$ is a real analytic function of z in the interval $(0, z_c)$.*

Proof. Fix a real z_0 in $(0, z_c)$ and let $u_0 = e^{m(z_0)}$. Define the function f of two (complex) arguments u and z by

$$f(u, z) \equiv \sum_{L=1}^{\infty} \sum_{N=1}^{\infty} \lambda_{N,L} z^N u^L = \sum_{L=1}^{\infty} \Lambda_z(L) u^L. \tag{4.2.29}$$

We must first show that $f(u, z)$ is convergent (and hence holomorphic) in some neighbourhood of (u_0, z_0). To this end, suppose that ϵ is a small positive number and suppose that (u, z) satisfies $|u - u_0| < \epsilon$ and $|z - z_0| < \epsilon$. Then

$$\sum_{L=1}^{\infty} \sum_{N=1}^{\infty} \lambda_{N,L} |z|^N |u|^L \leq \sum_{L=1}^{\infty} \Lambda_{z_0+\epsilon}(L) (e^{m(z_0)} + \epsilon)^L. \tag{4.2.30}$$

Applying the root test to the right hand sum and recalling the definition of m_Λ in (4.2.19), we see that the sum converges absolutely if

$$e^{-m_\Lambda(z_0+\epsilon)} \left(e^{m(z_0)} + \epsilon \right) < 1. \tag{4.2.31}$$

Since m_Λ is continuous at z_0 [by Lemma 4.2.7(a)] and since $m(z_0) - m_\Lambda(z_0) < 0$ (by Theorem 4.2.4), it is possible to choose ϵ small enough so that (4.2.31) holds. Therefore f is indeed holomorphic in a neighbourhood of (u_0, z_0).

We see from Equations (4.2.29), (4.2.11), and (4.2.15) that

$$f(e^{m(z)}, z) = 1 \quad \text{for all } z \text{ in } (0, z_c); \tag{4.2.32}$$

in particular, $f(u_0, z_0) = 1$. Also, $\partial f / \partial u$ is nonzero at (u_0, z_0), since it may be written as a series of positive terms. Therefore the analytic implicit function theorem [see for example Griffiths and Harris (1978), p. 19] tells us that there exists a function $w(z)$, holomorphic in a neighbourhood of z_0, with the following property: in a neighbourhood of (u_0, z_0), the equation $f(u, z) = 1$ holds if and only if $u = w(z)$. By (4.2.32) we see that $m(z) = \log w(z)$ for all real z in a neighbourhood of z_0. This proves the theorem. $\square$

4.3   Separation of the masses

This section is devoted to a proof of Theorem 4.2.4, which states that the "mass" $m_\Lambda(z)$ of irreducible bridges is strictly greater than the mass $m(z)$ whenever $0 < z < z_c$.

Definition 4.3.1 *Let $\omega = (\omega(0), \ldots, \omega(N))$ be a bridge. A backtrack of ω is a subwalk*

$$\omega[s; t] \equiv (\omega(s), \ldots, \omega(t))$$

$(0 \le s < t \le N)$ satisfying:
  (i) $\omega_1(t) \le \omega_1(i) < \omega_1(s)$ for all $i = s+1, \ldots, t-1$;
  (ii) $\omega_1(j) \le \omega_1(s)$ for all $j = 0, 1, \ldots, s-1$; and
  (iii) $\omega_1(t) < \omega_1(j)$ for all $j = t+1, \ldots, N$.
The span of the backtrack is defined to be $\omega_1(s) - \omega_1(t)$.

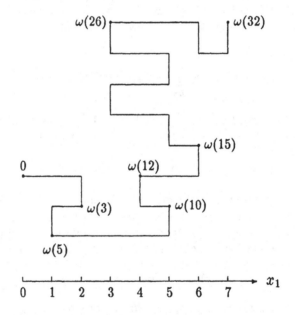

Figure 4.2: A bridge of length $N = 32$, with $\omega_1(0) = 0$ and $\omega_1(N) = 7$. There are three backtracks: $\omega[3; 5]$, $\omega[10; 12]$, and $\omega[15; 26]$.

Condition (i) says that a backtrack is itself a bridge, except that it goes right-to-left instead of left-to-right. Conditions (ii) and (iii) are maximality conditions: they guarantee that every subwalk $\omega[s; t]$ satisfying (i) is contained in a unique backtrack. In fact, the following is true:

Lemma 4.3.2 *Let ω be an N-step bridge. If $0 \leq s_0 < t_0 \leq N$ and $\omega_1(t_0) < \omega_1(s_0)$, then there is a unique backtrack $\omega[s, t]$ of ω that contains $\omega(s_0)$ and $\omega(t_0)$.*

Proof. Given such s_0 and t_0, define

$$
\begin{aligned}
A &= \min\{\omega_1(i) : t_0 \leq i \leq N\}, \\
B &= \max\{\omega_1(i) : 0 \leq i \leq s_0\}, \\
s &= \max\{i : 0 \leq i \leq s_0, \omega_1 = B\}, \\
t &= \max\{i : t_0 \leq i \leq N, \omega_1 = A\}.
\end{aligned}
$$

Then $\omega[s; t]$ is the unique backtrack containing $\omega(s_0)$ and $\omega(t_0)$. Details are left to the reader. $\qquad\square$

Corollary 4.3.3 *Two backtracks of the same bridge are either equal or else they have no sites in common.*

We now state two additional definitions related to the decomposition of a bridge into two smaller bridges, as can be done when (4.2.1) occurs for $i < N$.

Definition 4.3.4 *Let ω be an N-step bridge. A backtrack $\omega[s; t]$ is said to cover the integer j if $\omega_1(t) \leq j < \omega_1(s)$. We say that the integer j $[\omega_1(0) < j < \omega_1(N)]$ is a break point of ω if there exists an r in $\{1, \ldots, N-1\}$ such that $\omega_1(i) \leq j$ for all $i = 0, \ldots, r$ and $\omega_1(i) > j$ for all $i = r+1, \ldots, N$.*

For example, in Figure 4.2, the backtrack $\omega[15; 26]$ covers the integers 3, 4, and 5; and the integers 2 and 6 are break points (corresponding respectively to $r = 6$ and $r = 30$ in the definition). Notice that an integer can be covered by more than one backtrack (e.g. the integer 4 in Figure 4.2). Several remarks about Definition 4.3.4 are in order. Observe that j is a break point if and only if there exists an r such that $\omega[0; r]$ and $\omega[r; N]$ are both bridges, and $\omega_1(r) = j$. Also, j is a break point if and only if there is *only* one r such that $\omega_1(r) = j$ and $\omega_1(r+1) = j + 1$. A bridge is irreducible if and only if it has no break points. Finally, for every j strictly between $\omega_1(0)$ and $\omega_1(N)$, either j is a break point or j is covered by a backtrack.

For the rest of this section, we consider a fixed value of z, with $0 < z < z_c$, so we shall usually suppress z in our notation (we shall write m for $m(z)$, $B(L)$ for $B_z(L)$, etc.). We will use the following notation: If $\mathcal{S}$ is a set of self-avoiding walks, then $GF(\mathcal{S})$ denotes the generating function of $\mathcal{S}$:

$$
GF(\mathcal{S}) \equiv GF(\mathcal{S}, z) \equiv \sum_{\omega \in \mathcal{S}} z^{|\omega|}.
$$

We begin with a basic lemma which says that break points are common in the (subcritical) ensemble of bridges. In particular, it says that a long interval on the x_1-axis is unlikely to be free of break points. Later we will try to apply this bound to many long intervals simultaneously in order to prove that a complete absence of break points is exponentially rare, which is basically what Theorem 4.2.4 says.

Lemma 4.3.5 *For integers $c \geq 0$, $T \geq 1$ and $L \geq c + T$, let $B^*(L; c, T)$ denote the generating function of all bridges of span L with $\omega(0) = 0$ which have no break points among the integers $c + 1, \ldots, c + T - 1$. Then there exists a decreasing function $\epsilon(T)$ (independent of L and c) such that $\lim_{T \to \infty} \epsilon(T) = 0$ and*

$$B^*(L; c, T) \leq e^{-mL} \epsilon(T) \text{ for all } L \text{ and } c \ (0 \leq c \leq L - T). \qquad (4.3.1)$$

[For example, the bridge of Figure 4.2 is in $B^*(7; 2, 4)$ since 3, 4, and 5 are not break points, but it is not in $B^*(7; 1, 4)$ or $B^*(7; 2, 5)$.]

Proof. Considering the last break point j of a bridge before $c + 1$ and its first break point k after $c + T - 1$, we see that

$$B^*(L; c, T) = \sum_{j=0}^{c} \sum_{k=c+T}^{L} B(j) \Lambda(k - j) B(L - k).$$

Using the notation of the previous section [recall (4.2.11) and (4.2.12)],

$$B^*(L; c, T) e^{mL} = \sum_{j=0}^{c} \sum_{k=c+T}^{L} a_j p_{k-j} a_{L-k}$$

$$\leq \sum_{i=T}^{L} (i - T + 1) p_i$$

(where we have used $a_n \leq 1$ for all n). Therefore setting

$$\epsilon(T) = \sum_{i=T}^{\infty} i p_i$$

gives a function with the desired properties, by (4.2.17).   □

There is a "renewal theory" interpretation of the preceding lemma. The quantity $B^*(L; c, T) e^{mL}$ is the probability that there are no renewals between c and $c + T$ (and that there is a renewal at L). Since the mean time between renewals is finite, it is unlikely that a given long interval contains no renewals.

Our strategy now is the following. We want to bound the generating function of the set of irreducible bridges of span L (which start at the origin). We fix a large integer Q and for $L \gg Q$ we split the interval $[0, L]$ into blocks (subintervals) of size Q. Then we look at the backtracks that cover the endpoints of these blocks. For the subset of irreducible bridges in which many of these backtracks are small, the blocks approximately decouple into irreducible bridges, each of which has small probability (by Lemma 4.3.5), so many such blocks are exponentially rare. The remaining irreducible bridges have many large backtracks, so we use $B(k) \leq e^{-mk}$ to bound the contributions of these backtracks, again resulting in exponentially smaller quantities.

For the details, we proceed as follows. Let T and Δ be positive integers (to be specified in the proof of Theorem 4.2.4 below), and let $Q = 2\Delta + T$. For large L, let $k \equiv k(L)$ be the greatest integer less than or equal to $(L/Q) - 1$; we will split $[0, L]$ into $k + 1$ subintervals (observe that $k + 1$ is of order L). Given L, define the set

$$A = \{Q, 2Q, \ldots, kQ\}.$$

With 0 and L, the elements of A are the endpoints of the blocks. Consider any nonempty subset $S = \{n_1, \ldots, n_\tau\}$ of A, with $n_1 < \ldots < n_\tau$. Put $n_0 = 0$ and $n_{\tau+1} = L$. Then

$$n_{i+1} - n_i \geq Q \text{ for all } i = 0, 1, \ldots, \tau.$$

In the rest of this section, τ will always denote $|S|$. Let $\mathcal{I}_L$ denote the set of all irreducible bridges of span L. Let $\mathcal{I}_L(\leq \Delta; S)$ denote the set of all irreducible bridges of span L such that no point of S is covered by a backtrack having a span of more than Δ. For integers $\sigma_1, \ldots, \sigma_\tau \geq 1$, let $\mathcal{J}_L(S, \sigma_1, \ldots, \sigma_\tau)$ denote the set of all irreducible bridges of span L such that for each $i = 1, \ldots, \tau$ there is a backtrack of span σ_i which covers n_i but does not cover n_j for any other $j \neq i$.

We will now state three lemmas. We will then show how they can be used to prove that $m_\Lambda > m$, and finally we will prove the lemmas. The first lemma says that every irreducible bridge either has lots of small backtracks (i.e. of span Δ or less) or it has enough large backtracks. The second lemma bounds the generating function of the first kind of irreducible bridge, and the third lemma does the same for the second kind. The notation of the preceding paragraphs is assumed in each lemma.

Lemma 4.3.6 *The set $\mathcal{I}_L$ is contained in the union of*

$$\bigcup_{S \subseteq A, |S| \geq k/2} \mathcal{I}_L(\leq \Delta; S) \tag{4.3.2}$$

and

$$\bigcup_{SCA, |S| \geq 1} \bigcup_{\substack{\sigma_1 > \Delta, \ldots, \sigma_r > \Delta \\ \sigma_1 + \cdots + \sigma_r > k\Delta/2}} \mathcal{J}_L(S, \sigma_1, \ldots, \sigma_r). \qquad (4.3.3)$$

Lemma 4.3.7 $GF(\mathcal{I}_L(\leq \Delta; S)) \leq e^{-mL}(\chi \epsilon(T))^{r+1}.$

Here ϵ is from Lemma 4.3.5 and χ is the susceptibility [defined in (1.3.4)].

Lemma 4.3.8 $GF(\mathcal{J}_L(S, \sigma_1, \ldots, \sigma_r)) \leq e^{-mL} \chi^{2r} e^{-m(\sigma_1 + \cdots + \sigma_r)}.$

Proof of Theorem 4.2.4. Fix T so that

$$2\left(\chi \epsilon(T)\right)^{1/2} < \frac{1}{2}, \qquad (4.3.4)$$

and fix Δ so that

$$\left(1 + \frac{\chi^2}{1 - e^{-m/2}}\right) e^{-m\Delta/4} < \frac{1}{2}. \qquad (4.3.5)$$

Set $Q = 2\Delta + T$. Then, by the preceding three lemmas,

$$\begin{aligned}
\Lambda_z(L) &\equiv GF(\mathcal{I}_L) \\
&\leq 2^k e^{-mL} \left[\chi \epsilon(T)\right]^{1+k/2} \\
&\quad + e^{-mL} \sum_{\tau=1}^{k} \binom{k}{\tau} \chi^{2\tau} \sum_{\substack{\sigma_1 > \Delta, \ldots, \sigma_r > \Delta \\ \sigma_1 + \cdots + \sigma_r > k\Delta/2}} e^{-m(\sigma_1 + \cdots + \sigma_r)}.
\end{aligned}$$

Now, for any positive D and r such that $r < m$,

$$\begin{aligned}
\sum_{\substack{\sigma_1 > \Delta, \ldots, \sigma_r > \Delta \\ \sigma_1 + \cdots + \sigma_r > D}} e^{-m(\sigma_1 + \cdots + \sigma_r)} \\
\leq \sum_{\sigma_1 > \Delta, \ldots, \sigma_r > \Delta} e^{-m(\sigma_1 + \cdots + \sigma_r)} e^{r(\sigma_1 + \cdots + \sigma_r - D)} \\
\leq \left(\frac{e^{(r-m)\Delta}}{1 - e^{-(m-r)}}\right)^r e^{-rD}.
\end{aligned}$$

Putting $r = m/2$ and $D = k\Delta/2$, we obtain [using (4.3.4) and (4.3.5)]

$$\begin{aligned}
\Lambda_z(L) &\leq e^{-mL} \left(\left[2\left(\chi \epsilon(T)\right)^{1/2}\right]^k + \left[1 + \frac{\chi^2 e^{-m\Delta/2}}{1 - e^{-m/2}}\right]^k e^{-m\Delta k/4} \right) \\
&\leq e^{-mL} \left(\left(\frac{1}{2}\right)^k + \left(\frac{1}{2}\right)^k \right) \\
&\leq 2 e^{-mL} 2^{-(L/Q)}.
\end{aligned}$$

Therefore

$$m_\Lambda = \liminf_{L \to \infty} \frac{-\log \Lambda_z(L)}{L} \geq m + \frac{\log 2}{Q} > m.$$

This proves the theorem. □

Proof of Lemma 4.3.6. Suppose ω is in $\mathcal{I}_L$ but not in

$$\bigcup_{S \subset A, |S| \geq k/2} \mathcal{I}_L(\leq \Delta; S).$$

Then at least $k/2$ points of A are covered by backtracks of span greater than Δ. Some of these backtracks may cover more than one point of A. If a backtrack covers j points of A (where $j \geq 2$), then its span is at least $(j-1)Q$, which is greater than $j\Delta$ (recall $Q = 2\Delta + T$). Also, some points of A may be covered by several backtracks.

Choose an integer τ, as small as possible, having the following property: there exists a collection of τ backtracks of ω, each of span greater than Δ, such that at least $k/2$ of the points of A are covered by one or more of these τ backtracks. Then each backtrack covers some point of A which is not covered by any of the other $\tau - 1$ backtracks (otherwise, this backtrack could be removed from the collection, contradicting the minimality of τ). This shows that for some $S = \{n_1, \ldots, n_\tau\} \subset A$ and some $\sigma_1, \ldots, \sigma_\tau > \Delta$ satisfying $\sigma_1 + \cdots + \sigma_\tau > (k/2)\Delta$, ω is in $\mathcal{J}_L(S, \sigma_1, \ldots, \sigma_\tau)$. This proves the lemma. □

Proof of Lemma 4.3.7. Suppose that ω is in $\mathcal{I}_L(\leq \Delta; S)$, $S = \{n_1, \ldots, n_\tau\}$. Let

$$\begin{aligned} r_0 &= 0, \\ q_{\tau+1} &= |\omega|, \\ q_i &= \min\{j : \omega_1(j+1) = n_i + 1\}, \quad i = 1, \ldots, \tau, \\ r_i &= \max\{j : \omega_1(j) = n_i\}, \quad i = 1, \ldots, \tau. \end{aligned}$$

(See Figure 4.3.) Notice that since ω has no break points, q_i is always strictly less than r_i. By our choice of ω, we know that

$$\omega_1(j) \geq n_i - \Delta \qquad \text{for all } j \geq q_i$$

and

$$\omega_1(j) \leq n_i + \Delta \qquad \text{for all } j \leq r_i.$$

In particular, $r_{i-1} < q_i$ ($i = 1, \ldots, \tau+1$), since $n_{i+1} - n_i > 2\Delta$. Moreover, $\omega[r_{i-1}; q_i]$ is a bridge, and the rest of ω stays out of the strip

$$\{x \in \mathbf{R}^d : n_{i-1} + \Delta < x_1 < n_i - \Delta\}.$$

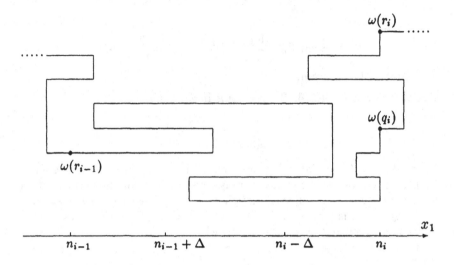

Figure 4.3: Proof of Lemma 4.3.7: part of a walk ω in $\mathcal{I}_L(\leq \Delta; S)$.

So, since ω has no break points, $\omega[r_{i-1}; q_i]$ has no break points between $n_{i-1} + \Delta$ and $n_i - \Delta$ (although it could have break points elsewhere). So, by cutting ω at each of $q_1, r_1, \ldots, q_\tau, r_\tau$, and looking at the $2\tau + 1$ subwalks that are obtained, we deduce (with the help of Lemma 4.3.5) that

$$GF(\mathcal{I}_L(\leq \Delta; S)) \leq \chi^\tau \prod_{i=1}^{\tau+1} B^*(n_i - n_{i-1}; \Delta, n_i - n_{i-1} - 2\Delta)$$

$$\leq \chi^\tau \prod_{i=1}^{\tau+1} e^{-m(n_i - n_{i-1})} \epsilon(n_i - n_{i-1} - 2\Delta).$$

The lemma now follows because $\chi > 1$, $\sum_{i=1}^{\tau+1}(n_i - n_{i-1}) = L$, $n_i - n_{i-1} - 2\Delta \geq T$, and ϵ is a decreasing function. $\quad\square$

Proof of Lemma 4.3.8. Suppose ω is in $\mathcal{J}_L(S, \sigma_1, \ldots, \sigma_\tau)$. For every $i = 1, \ldots, \tau$, there is at least one backtrack $\omega[s_i; t_i]$ of span σ_i such that

$$n_{i-1} < \omega_1(t_i) \leq n_i < \omega_1(s_i) \leq n_{i+1}.$$

(See Figure 4.4.) This implies that $t_i < s_{i+1}$ for every $i = 1, \ldots, \tau - 1$.
Define $l_0 = 0$ and, for $i = 1, \ldots, \tau$,

$$f_i = \min\{r > l_{i-1} : \omega_1(r+1) = n_i + 1\},$$
$$l_i = \max\{r : \omega_1(r) = n_i\}.$$

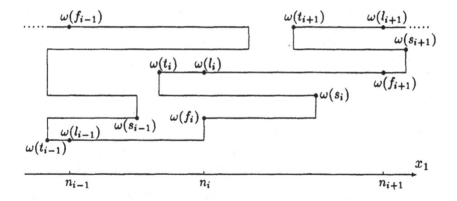

Figure 4.4: Proof of Lemma 4.3.8: part of a walk ω in $\mathcal{J}_L(S, \sigma_1, \ldots, \sigma_\tau)$.

Then, for every $i = 1, \ldots, \tau$,

$$l_{i-1} < f_i < s_i < t_i \leq l_i,$$

and $\omega[l_{i-1}; f_i]$ is a bridge of span $n_i - n_{i-1}$. If we consider cutting ω at every f_i, s_i, t_i, and l_i, then we deduce that

$$GF(\mathcal{J}_L(S, \sigma_1, \ldots, \sigma_\tau)) \leq \left(\prod_{i=1}^{\tau+1} B(n_i - n_{i-1}) \right) \prod_{i=1}^{\tau} (\chi B(\sigma_i) \chi)$$

$$\leq e^{-mL} \chi^{2\tau} e^{-m(\sigma_1 + \cdots + \sigma_\tau)},$$

since $\sum_{i=1}^{\tau+1}(n_i - n_{i-1}) = L$. $\square$

4.4 Ornstein-Zernike decay of $G_z(0, x)$

The preceding section showed that for subcritical z, the spatial decay rate for irreducible bridges is strictly larger than for bridges as a whole, or in other words that the ratio $\Lambda_z(L)/B_z(L)$ decays exponentially in L for any fixed $z < z_c$. This may be viewed intuitively as saying that in the subcritical ensemble of bridges with distant endpoints, irreducible bridges are exponentially rare. (In contrast, bridges with distant endpoints are *not* exponentially rare among all self-avoiding walks with the same endpoints, by Theorem 4.1.14). The first part of this section proves some results in the opposite direction: we get a lower bound on the scarcity of irreducible bridges in the form $m_\Lambda(z) \leq 3m(z)$ whenever $0 < z < z_c$ (Corollary 4.4.4).

Since $m(z)$ tends to 0 as z approaches z_c, this tells us intuitively that irreducible bridges are not exponentially rare in the critical ensemble. We shall also prove a more natural interpretation of this intuitive statement, namely that the "connective constant" for irreducible bridges is the same as for walks (Corollary 4.4.5). Finally, we shall use these results to help us prove detailed large-distance asymptotics (the "Ornstein-Zernike" decay) of the subcritical two-point function $G_z(0, x)$ (Theorem 4.4.7).

Definition 4.4.1 *For all nonnegative integers N, L, and r, let $\hat{B}_{N,L,r}$ be the set of bridges $(\omega(0), \ldots, \omega(N))$ of span L such that $\omega(0) = 0$, $\omega_2(N) = r$, and $0 \leq \omega_2(i) \leq r$ for every $i = 0, \ldots, N$. Also let $\hat{B}_{N,L} = \cup_{r \geq 0} \hat{B}_{N,L,r}$.*

Such bridges are useful for the following reason, which we shall explain more carefully below: we produce an irreducible bridge whenever we perform an appropriate concatenation of three of them, all with the same span but with the middle one reflected in the x_1 direction so that it goes right-to-left. Thus to get lower bounds on the number of irreducible bridges, we first derive a lower bound on the number of bridges in $\hat{B}_{N,L}$.

Proposition 4.4.2 *There exists an N_0 such that $b_{N,L} \leq e^{3N^{1/2}} |\hat{B}_{N,L}|$ for every $N \geq N_0$ and every $L \geq 1$.*

Proof. The idea is very similar to the proof of Theorem 3.1.1, but now we will "unfold" the walk in the x_2 direction. Let ω be any N-step bridge of span L. Let $n_0(\omega)$ be the largest value of i such that $\omega_2(i) = \min_j \omega_2(j)$. For each $j > 0$, recursively define $A_j(\omega)$ and $n_j(\omega)$ so that

$$A_j = \max_{n_0 < i \leq N} (-1)^j (\omega_2(n_{j-1}) - \omega_2(i))$$

and n_j is the largest value of i for which this maximum is attained. The recursion is stopped at the smallest integer k such that $n_k = N$. Also, let $m_0(\omega) = n_0(\omega)$, and for each $j > 0$ recursively define $\overline{A}_j(\omega)$ and $m_j(\omega)$ so that

$$\overline{A}_j = \max_{0 \leq i < m_{j-1}} (-1)^j (\omega_2(m_{j-1}) - \omega_2(i))$$

and m_j is the smallest value of i for which this maximum is attained. The recursion is stopped at the smallest integer l such that $m_l = 0$. (See Figure 4.5.)   Now, reflect the first m_{l-1} points of ω through the hyperplane $x_2 = \omega_2(m_{l-1})$, and continue reflecting through x_2 hyperplanes at $\omega(m_j)$, $j = l-2, \ldots, 0$, and at $\omega(n_j)$, $j = 1, \ldots, k-1$. The result of each reflection is still a bridge of span L, and the final result (after a translation in the x_2 direction and perhaps an overall reflection through $x_2 = 0$) will be in the set $\hat{B}_{N,L,A+\overline{A}}$, where $A = \sum A_j$ and $\overline{A} = \sum \overline{A}_j$. Since $A_1 > A_2 >$

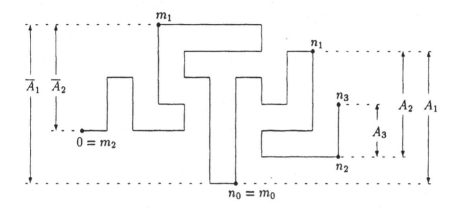

Figure 4.5: Proof of Proposition 4.4.2: The walk ω before "unfolding".

$\ldots > A_k$, there are $P_D(A)$ possible sequences of A_j's that sum to A; and since $\overline{A}_1 \geq \overline{A}_2 > \ldots > \overline{A}_l$, there are $P_D(\overline{A} + 1)$ possible sequences of $\overline{A}_j$'s that sum to $\overline{A}$. Observe that these two sequences together with the final bridge in $\hat{\mathcal{B}}_{N,L,A+\overline{A}}$ determine the original ω uniquely, except perhaps for a reflection through $x_2 = 0$. Therefore, since $A + \overline{A} \leq N$, we obtain

$$
\begin{aligned}
b_{N,L} &\leq 2 \sum_{r=0}^{N} \sum_{A=0}^{r} P_D(A) P_D(r - A + 1) |\hat{\mathcal{B}}_{N,L,r}| \\
&\leq 2 \sum_{r=0}^{N} (N + 1) K^2 \exp\left[(3 - \epsilon)(N + 1)^{1/2} \right] |\hat{\mathcal{B}}_{N,L,r}|,
\end{aligned}
$$

where the second inequality follows as in (3.1.7), having taken $B = 3$ in (3.1.6). The proposition follows.   □

We now establish several corollaries of this result. Recall that $M(z)$ is the mass for bridges, and that $M(z) = m(z)$ for all z in $(0, z_c)$.

Corollary 4.4.3 *Let $\hat{B}_z(L) = \sum_N |\hat{\mathcal{B}}_{N,L}| z^N$. Then*

$$
\lim_{L \to \infty} \frac{-\log \hat{B}_z(L)}{L} = M(z)
$$

for every $z > 0$.

Proof. By Proposition 4.4.2 and the observation that bridges of span L must have at least L steps,

$$\hat{B}_z(L) \geq \sum_{N=L}^{\infty} b_{N,L} e^{-3N^{1/2}} z^N$$

for every $L \geq N_0$ and every $z > 0$. Now let $\epsilon > 0$. Choose $N_1 \geq N_0$ such that $e^{-3N^{1/2}} > (1 - \epsilon)^N$ for every $N \geq N_1$. Then $\hat{B}_z(L) \geq B_{(1-\epsilon)z}(L)$ for every $L \geq N_1$, and hence

$$\limsup_{L \to \infty} \frac{-\log \hat{B}_z(L)}{L} \leq M((1 - \epsilon)z). \tag{4.4.1}$$

Since ϵ is arbitrary and M is left-continuous [Lemma 4.1.11(d)], we can replace $M((1 - \epsilon)z)$ by $M(z)$ on the right side of (4.4.1). The result then follows since the reverse inequality for the lim inf is a consequence of $\hat{B}_z(L) \leq B_z(L)$. $\quad\square$

Corollary 4.4.4 $m_\Lambda(z) \leq 3m(z)$ *for every z in* $(0, z_c)$*. Also,* $m_\Lambda(z_c) = 0$*.*

Proof. Fix L, and let ω_A, ω_B and ω_C be three walks in $\cup_N \hat{B}_{N,L}$. Let $\overline{\omega}_B$ be the reflection of ω_B through the hyperplane $x_1 = 0$. Let ω be the walk obtained by starting at the origin, taking one step in the $+x_1$ direction, followed by ω_A (appropriately translated), followed by one step in the $+x_2$ direction, followed by $\overline{\omega}_B$, followed by one step in the $+x_2$ direction, followed by ω_C. Then ω is self-avoiding; in fact, it is an irreducible bridge of span $L+1$. This proves that $\Lambda_z(L+1) \geq [z\hat{B}_z(L)]^3$, and so Corollary 4.4.3 implies that $m_\Lambda(z) \leq 3M(z)$ for every $z > 0$. With Theorem 4.1.14, this proves the first assertion of the corollary. The second assertion follows from $M(z) \leq m_\Lambda(z) \leq 3M(z)$ and the fact that $M(z_c) = 0$ (Theorem 4.1.13). $\quad\square$

Corollary 4.4.5 $\lim_{N \to \infty} (\lambda_N)^{1/N} = \mu$.

Proof. Since $\lambda_N \leq c_N$, it suffices to show that $\liminf_N (\lambda_N)^{1/N} \geq \mu$. In the proof of Corollary 4.4.4, suppose that ω_A, ω_B, and ω_C are all in $\hat{B}_{n-1,L}$; then the resulting irreducible bridge ω has $3n$ steps. Thus $\lambda_{3n,L+1}$ is bounded below by $|\hat{B}_{n-1,L}|^3$. Notice that we can get the same lower bound for $\lambda_{3n+1,L+1}$ and $\lambda_{3n+2,L+1}$, since we can add one or two steps in the $+x_2$ direction to the end of ω. Therefore

$$\lambda_{N,L+1} \geq |\hat{B}_{\lfloor N/3 \rfloor - 1, L}|^3$$

(where $\lfloor x \rfloor$ denotes the greatest integer less than or equal to x). Proposition 4.4.2 implies that

$$\lambda_{N,L+1} \geq (b_{\lfloor N/3 \rfloor - 1, L})^3 e^{-9(N/3)^{1/2}}.$$

We now sum this inequality over $L = 1, \ldots, N$ and apply Hölder's inequality in the form

$$N^2 \sum_{i=1}^{N} a_i^3 \geq \left(\sum_{i=1}^{N} a_i \right)^3$$

(for $a_i \geq 0$). This gives

$$\lambda_N \geq \frac{(b_{\lfloor N/3 \rfloor - 1})^3}{N^2} e^{-9(N/3)^{1/2}}$$

for every N. Finally we take N-th roots of both sides and let N tend to infinity; the desired result is a consequence of Corollary 3.1.6. □

For the remainder of this section, we need to extend the definition of *break point* in Definition 4.3.4 to an arbitrary N-step self-avoiding walk ω. Now we say that that the integer j is a break point of ω if there exists an r in $\{1, \ldots, N-1\}$ such that $\omega_1(i) \leq j$ for all $i = 0, \ldots, r$ and $\omega_1(i) > j$ for all $i = r+1, \ldots, N$. Observe that ω cannot have a break point unless $\omega_1(0) < \omega_1(N)$, and that each break point j must satisfy $\omega_1(0) \leq j < \omega_1(N)$. So the only real change in the definition is allowing $j = \omega_1(0)$. Notice that the two definitions coincide for bridges: if ω is a bridge then $\omega_1(0)$ cannot be a break point because we do not allow r to be 0.

Before proving the Ornstein-Zernike decay of $G_z(0,x)$, we require a lemma which establishes a decay rate for the point-to-plane generating functions of a new class of walks which is larger than the class of irreducible bridges. These walks end on a specified hyperplane $x_1 = L$, lie in the half-space $x_1 \leq L$, and have no break points.

Lemma 4.4.6 *Suppose $0 < z < z_c$. For each integer $L \geq 0$, let $H_z(L)$ be the generating function of the class of self-avoiding walks $(\omega(0), \ldots, \omega(N))$ such that: $\omega(0) = 0$; $\omega_1(N) = L$; $\omega_1(i) \leq L$ for every $i = 0, \ldots, N$; and ω has no break points. Then*

$$\liminf_{L \to \infty} \frac{-\log H_z(L)}{L} = m_\Lambda(z).$$

Proof. Since $H_z(L) \geq \Lambda_z(L)$ we have

$$\liminf_{L \to \infty} \frac{-\log H_z(L)}{L} \leq m_\Lambda(z),$$

so it remains to prove the reverse inequality.

Consider any ω in the class of walks that corresponds to $H_z(L)$. Let T be the largest i such that $\omega_1(i) = 0$. Let

$$J = \max\{\omega_1(i) : 0 \le i \le T\},$$

and let I be the first value of i for which this maximum is attained. Observe that the subwalk $(\omega(T), \ldots, \omega(N))$ is a bridge of span L, and that if it has any break points, then they must all be in $\{1, 2, \ldots, J-1\}$. We cut ω at I, T, and the last break point of $(\omega(T), \ldots, \omega(N))$; this decomposition shows that

$$H_z(L) \le \sum_{J=0}^{L} G_z(J)^2 \sum_{k=0}^{J-1} B_z(k)\Lambda_z(L-k),$$

where $G_z(J)$ was defined in (4.1.19).

Choose any number ρ such that $m(z) < \rho < m_\Lambda(z)$. By definition of $m_\Lambda(z)$, there exists an $R > 0$ such that $\Lambda_z(L) \le Re^{-\rho j}$ for every $j \ge 1$. Using this, (4.1.21), (4.1.12) and Theorem 4.1.14, we obtain

$$
\begin{aligned}
H_z(L) &\le \sum_{J=0}^{L} [\chi(z)^2 e^{-m(z)J}]^2 \sum_{k=0}^{J-1} e^{-m(z)k} Re^{-\rho(L-k)} \\
&\le \chi(z)^4 \sum_{J=0}^{L} e^{-2m(z)J} J Re^{(\rho - m(z))J} e^{-\rho L} \\
&\le \chi(z)^4 RL^2 e^{-\rho L}
\end{aligned}
$$

where we have used $\rho \le 3m(z)$ (Corollary 4.4.4) in the last line. The result now follows because ρ can be made arbitrarily close to $m_\Lambda(z)$. □

Finally we are ready to extend Theorems 4.2.5 and 4.2.6 to complete our picture of the long-distance asymptotics of the subcritical two-point function $G_z(0, x)$.

Theorem 4.4.7 *Fix z with $0 < z < z_c$.*
(a) There exist strictly positive, finite constants $\bar{\epsilon}(z)$ and $\overline{C}_z$ such that

$$\left| G_z(L)e^{m(z)L} - \overline{C}_z \right| \le e^{-\bar{\epsilon}(z)L}$$

for all sufficiently large L.
(b) (Ornstein-Zernike decay) For δ_z as in Theorem 4.2.6,

$$\lim_{L \to \infty} \left| G_z(0, (L, y))e^{m(z)L} L^{(d-1)/2} - \overline{C}_z \frac{\exp(-|y|^2/\delta_z L)}{(\pi \delta_z)^{(d-1)/2}} \right| = 0$$

uniformly in y in $\mathbf{Z}^{d-1}$.

The analogues of the asymptotic relations (4.2.25) and (4.2.26) also hold.

Proof. The basic idea is that in the subcritical ensemble, most walks with distant endpoints will have lots of break points; the difference between bridges and general walks in this respect is a matter of "boundary conditions" (at $x_1 = 0$ and $x_1 = L$). Our first job will be to show that walks with no break points are negligible. We shall then sum the remaining walks according to the locations of their first and last break points (which will typically be close to the endpoints). Then we apply Theorems 4.2.5 and 4.2.6 to the middle parts of the walks, which are bridges.

Throughout this proof we shall use m to denote $m(z)$.

(a) Let $G_z^*(L)$ be the generating function of all self-avoiding walks ω such that $\omega(0) = 0$, $\omega_1(N) = L$, and ω has no break points. We claim that the mass of $G_z^*(L)$ is strictly greater than m. The proof is similar in spirit to the proof of Lemma 4.4.6, so we shall be brief. For any walk ω contributing to $G_z^*(L)$, define T and J as in the proof of Lemma 4.4.6, and also define the analogous quantities $T' = \min\{i : \omega_1(i) = L\}$ and $J' = \min\{\omega_1(i) : T' \le i \le |\omega|\}$. The contribution of all walks which have either $J > L/3$ or $J' < 2L/3$ can be bounded by $2G_z(\lfloor L/3 \rfloor)^2 G_z(L)$, and hence has mass at least $5m/3$ by Corollary 4.1.17. The contribution of the remaining walks can be bounded by

$$\sum_{J=0}^{L/3} G_z(J)^2 \sum_{J'=2L/3}^{L} G_z(L - J')^2 \sum_{k=0}^{J} \sum_{k'=0}^{J'} B_z(k)\Lambda_z(k' - k)B_z(L - k'),$$

which also has mass strictly greater than m. This proves the claim. (We remark that with more care, one can show that the mass of $G_z^*(L)$ is $m_\Lambda(z)$.)

By considering the leftmost break point i and the rightmost break point $L - 1 - j$ of a given walk, we have

$$G_z(L) = \sum_{i=0}^{L-1} \sum_{j=0}^{L-1-i} H_z(i)B_z(L - 1 - i - j)zH_z(j) + G_z^*(L) \qquad (4.4.2)$$

(the factor of z is due to the single step of the walk from $x_1 = L - 1 - j$ to $x_1 = L - j$). Define $\overline{C}_z = ze^m C_z(\sum_{n \ge 0} H_z(n)e^{nm})^2$, where C_z is defined in Theorem 4.2.5, and the sum in parentheses converges by Lemma 4.4.6 and Theorem 4.2.4. From the above identity we have

$$G_z(L)e^{Lm} - \overline{C}_z =$$

$$\sum_{i=0}^{L-1}\sum_{j=0}^{L-1-i} zH_z(i)[B_z(L-1-i-j)e^{(L-1-i-j)m} - C_z]H_z(j)e^{(i+j+1)m}$$

$$-C_z \sum_{i+j\geq L} zH_z(i)H_z(j)e^{(i+j+1)m} \; + \; G_z^*(L)e^{Lm}.$$

Part (a) of the theorem now follows from Theorem 4.2.5 and Lemma 4.4.6.

(b) The proof is similar to part (a), but we need to pay more attention to technicalities. For every (L, y) in $\mathbf{Z}^d$, let $H_z(L, y)$ and $G_z^*(0, (L, y))$ denote the analogues of $H_z(L)$ and $G_z^*(L)$ when attention is restricted to walks that end at (L, y). Then

$$G_z(0, (L, y)) =$$
$$\sum_{i=0}^{L-1}\sum_{j=0}^{L-1-i}\sum_{u,v\in \mathbf{Z}^{d-1}} H_z(i, u)B_z(L-1-i-j, y-u-v)zH_z(j, v)$$
$$+ G_z^*(0, (L, y)). \tag{4.4.3}$$

Since the last term is less than $G_z^*(L)$ it decays faster than e^{-Lm}. Therefore, after substituting $k = i+j+1$ in (4.4.3), we are left with the task of showing that

$$\sup_{y\in \mathbf{Z}^{d-1}} |\sum_{k=1}^{L}\sum_{i=0}^{k-1}\sum_{u,v} zH_z(i, u)B_z(L-k, y-u-v)H_z(k-1-i, v)e^{Lm}L^{(d-1)/2}$$

$$- ze^{m}(\sum_{n=0}^{\infty} H_z(n)e^{nm})^2\varphi(L, y)| \tag{4.4.4}$$

converges to 0 as L tends to infinity, where as in the proof of Theorem 4.2.6

$$\varphi(L, y) = C_z(\pi\delta_z)^{-(d-1)/2}\exp(-|y|^2/\delta_z L).$$

As we shall see, this converges to 0 because the mass of the H_z terms is strictly greater than m and because

$$B_z(L, y)e^{Lm}L^{(d-1)/2} - \varphi(L, y) \rightarrow 0$$

uniformly in y as $L \rightarrow \infty$. To use these facts, we add and subtract several terms in (4.4.4) and use the triangle inequality, as well as the bound $B_z(L-k, y-u-v)e^{(L-k)m} \leq 1$ [which follows from (4.1.12) and Theorem 4.1.14]. As a result, (4.4.4) is bounded by

$$\sum_{k=1}^{L/2}\sum_{i=0}^{k-1}\sum_{u,v} zH_z(i, u)H_z(k-1-i, v)e^{km}(q_{L,k} + r_{L-k} + s_{L,k,u+v})$$

$$+ \sum_{k=L/2+1}^{L} \sum_{i=0}^{k-1} \sum_{u,v} z H_z(i, u) H_z(k-1-i, v) e^{km} L^{(d-1)/2}$$

$$+ \sum_{k=L/2+1}^{\infty} \sum_{i=0}^{k-1} z H_z(i) H_z(k-i-1) e^{km} \varphi(L, 0),$$

where we define

$$q_{L,k} = \sup_{x \in Z^{d-1}} |L^{(d-1)/2} - (L-k)^{(d-1)/2}| B(L-k, x) e^{(L-k)m},$$

$$r_n = \sup_{x \in Z^{d-1}} |n^{(d-1)/2} B_z(n, x) e^{nm} - \varphi(n, x)|,$$

and

$$s_{L,k,w} = \sup_{y \in Z^{d-1}} |\varphi(L-k, y-w) - \varphi(L, y)|.$$

By Lemma 4.4.6, there exist positive constants A and ϵ (depending only on z) such that

$$\sum_{i=0}^{k-1} \sum_{u,v} H_z(i, u) H_z(k-1-i, v) e^{km} = \sum_{i=0}^{k-1} H_z(i) H_z(k-1-i) e^{km}$$

$$\leq A e^{-\epsilon k} \tag{4.4.5}$$

for every $k \geq 1$. Thus, if we can prove that $q_{L,k}$, r_{L-k}, and $s_{L,k,w}$ (i) are bounded uniformly for all $L \geq 2k$ and all w, and (ii) converge to 0 as $L \to \infty$ for every fixed k and w, then the dominated convergence theorem will imply that (4.4.4) converges to 0, which is what we want.

To begin with, $\lim_{n \to \infty} r_n = 0$ by Theorem 4.2.6, so (i) and (ii) hold for r_{L-k}. Next we consider $q_{L,k}$. The uniform convergence in Theorem 4.2.6 implies that there is a finite constant Ψ (depending only on z and d) such that

$$(L-k)^{(d-1)/2} B_z(L-k, x) e^{(L-k)m} \leq \Psi$$

for every L, k, and x. This in turn implies that

$$q_{L,k} \leq \left| \left(\frac{L}{L-k} \right)^{(d-1)/2} - 1 \right| \Psi. \tag{4.4.6}$$

This proves (ii) for $q_{L,k}$. Since $0 \leq k \leq L/2$, the right side of (4.4.6) is bounded by $(2^{(d-1)/2} - 1)\Psi$, which proves (i).

Finally, φ is uniformly bounded, hence so is $s_{L,k,w}$. This proves (i) for this sequence. To check (ii), we need to consider bounds on

$$|\exp(-|y|^2/\delta L) - \exp(-|y-w|^2/\delta(L-k))|$$

that are uniform in y, where δ is a positive constant. First, the mean value theorem gives

$$
\begin{aligned}
|\exp(-|y|^2/\delta L) - \exp(-|y-w|^2/\delta L)| &\le \sup_{t\in R}\left|\frac{d}{dt}e^{-t^2/\delta}\right|\left|\frac{|y|}{\sqrt{L}} - \frac{|y-w|}{\sqrt{L}}\right| \\
&\le \sup_{t\in R}\left|\frac{2t}{\delta}e^{-t^2/\delta}\right|\frac{|w|}{\sqrt{L}} \\
&\le \text{const.}\frac{|w|}{\sqrt{L}}.
\end{aligned}
$$

Secondly, for arbitrary $\zeta \ge 0$ (so in particular for $\zeta = |y-w|^2/\delta$):

$$
\begin{aligned}
|\exp(-\zeta/L) - \exp(-\zeta/(L-k))| &= \exp(-\zeta/L)|1 - \exp(-\zeta k/L(L-k))| \\
&\le \exp(-\zeta/L)\frac{\zeta k}{L(L-k)} \\
&\le \left(\sup_{t\ge 0} te^{-t}\right)\frac{k}{L-k}
\end{aligned}
$$

where we have used $1 - e^{-a} \le a$ for $a \ge 0$. The above two inequalities show that

$$s_{L,k,w} \le \text{const.}\left(\frac{|w|}{\sqrt{L}} + \frac{k}{L-k}\right),$$

which proves (ii). $\qquad\square$

4.5 Notes

Section 4.1. Chayes and Chayes (1986a) perform a systematic study of the mass, deriving many of the results that are discussed in this section. In particular, our presentation of the material from Proposition 4.1.8 through Corollary 4.1.17 is largely based upon their work. They also prove that $M(z) = -\infty$ for all $z > z_c$.

Theorem 4.1.18 is modelled upon an analogous result for percolation in Alexander, Chayes and Chayes (1990). In place of the FKG inequality used in that paper, we require Lemma 4.1.4, which we believe has not been used before for this purpose. The proof of Lemma 4.1.4 is closely related to the proof of Lemma 1.5.2 (see in particular the first inequality depicted in Figure 1.2).

Section 4.2. Ornstein-Zernike decay of two-point correlation functions occurs in many lattice spin systems and Euclidean quantum field theories [see Chayes and Chayes (1986b) for references], as well as in percolation [Campanino, Chayes and Chayes (1991)]. The original reference [Ornstein and Zernike (1914)] was in the context of fluid mechanics.

The results of this section are due to Chayes and Chayes (1986a, 1986b). They give the Ornstein-Zernike decay with an explicit error term as a power of L^{-1} that is uniform over a certain region of $y \in \mathbf{Z}^{d-1}$ [see also the proof of Theorem 4.4 of Campanino, Chayes and Chayes (1991) for a correction of a misstatement in this respect in Chayes and Chayes (1986b)].

Section 4.3. Theorem 4.2.4 is due to Chayes and Chayes (1986b). Their proof splits the interval $[0, L]$ into blocks and uses a result similar to our Lemma 4.3.5, but then it works with various rescaled block generating functions and an "Ornstein-Zernike inequality". The proof that we give in this section is from Madras (1991a).

Section 4.4. Theorem 4.4.7 is due to Chayes and Chayes (1986b), which they state in a different form that includes an explicit error term, as they did for bridges. This paper also partially anticipates Corollary 4.4.4 and Lemma 4.4.6. The other results of this section are new.

Chapter 5

The lace expansion

5.1 Inclusion-exclusion

So far the lace expansion is the only method which has led to rigorous results proving existence of critical exponents for the self-avoiding walk. All results obtained so far are for dimensions greater than the conjectured upper critical dimension four.

The lace expansion was first introduced by Brydges and Spencer (1985), who used it to study *weakly* self-avoiding walk above four dimensions. Weakly self-avoiding walk will be defined precisely in the next section; roughly speaking it is a model of a random walk where walks which intersect themselves do have a nonzero weight, but this weight is smaller than for a walk which does not intersect itself. The size of the probability penalty imposed for a self-intersection (i.e. the weakness of the interaction) is the small parameter which provided convergence for the lace expansion.

Now consider a model of self-avoiding walk on a lattice with large coordination number $\mathcal{Z}$, or in other words a walk for which there are a large number of steps available at each site. The probability that the concatenation of a randomly chosen single step to a given walk produces an immediate reversal is $\mathcal{Z}^{-1}$, which is small. For example, consider the usual nearest-neighbour model in high dimensions, with coordination number $2d$. As we saw in Section 1.2, in high dimensions the main effect of the self-avoidance constraint is in some sense to rule out immediate reversals, and now it appears that immediate reversals are uncommon even in the absence of the constraint. This suggests that we can regard the interaction as being weak in high dimensions. Alternatively, we may consider a walk on a "low"-dimensional lattice with large coordination number. Our basic example of this type is a walk on the hypercubic lattice $\mathbf{Z}^d$, in which we allow all steps whose largest component has absolute value less than or equal to L, for some large parameter L. Then again the effect of ruling out immediate reversals will be small, but we must also worry about long-range effects. It will turn out that these are also small for $d > 4$, so that in this situation the self-avoidance is again weak. We shall see in Chapter 6 that in these two situations it is possible to obtain convergence of the lace expansion. In the former the small parameter responsible for convergence of the expansion is $(2d)^{-1}$, while in the latter it is L^{-1}. Remarkably, for the nearest-neighbour model in five dimensions the small parameter $1/10$ is sufficiently small to prove convergence.

The lace expansion was first derived by an expansion and resummation procedure reminiscent of the cluster expansions of statistical mechanics and constructive quantum field theory. Viewed differently, however, the lace expansion can be seen as resulting from repeated application of the inclusion-exclusion relation. In this section the derivation of the lace expansion via the inclusion-exclusion relation will be discussed, and in the next section the resummation approach will be described. The inclusion-exclusion approach is geometric and is useful for providing intuition as to the origin of the expansion, while the resummation procedure has the advantage of generating terms in the expansion in a systematic and algebraic manner. In the inclusion-exclusion approach, the name lace expansion may seem somewhat inappropriate, but the laces will appear in the next section. Both derivations lead to precisely the same expansion.

At the heart of the lace expansion method is a convolution equation for the two-point function which is a multi-dimensional analogue of the renewal equations of Section 4.2. Indeed the expansion amounts to identifying an "irreducible" two-point function, which will be denoted $\Pi_z(0, x)$, such that the two-point function is essentially given by its convolution with the irreducible two-point function. Fourier transform techniques will play a key role in the analysis of the convolution equation.

We now turn to the derivation of the expansion for the case of a walk which may take more general steps than just to a nearest neighbour. We fix a finite set $\Omega \not\ni 0$ in $\mathbf{Z}^d$ which is symmetric with respect to reflections in the coordinate hyperplanes and rotations about coordinate axes by $\pi/2$. The cardinality of Ω will also be denoted by Ω. We consider the ordinary random walk taking steps in Ω. More precisely, we consider $\omega = (\omega(0), \omega(1), \ldots, \omega(N))$, where each $\omega(i)$ is an element of $\mathbf{Z}^d$ and $\omega(i+1) - \omega(i) \in \Omega$. (If Ω is the set of nearest neighbours of the origin then this is just the simple random walk.) It is shown in (A.6) that for $z < \Omega^{-1}$ the

Fourier transform of the two-point function for this walk is given by

$$\hat{C}_z(k) = \frac{1}{1 - z\Omega\hat{D}(k)},$$

where

$$\hat{D}(k) = \frac{1}{\Omega}\sum_{x \in \Omega} e^{ik \cdot x}.$$

To simplify the notation we will not use a label Ω to keep track of the fact that we are not necessarily dealing with the nearest-neighbour walk; all walks in this section will take steps in Ω. Self-avoiding walks taking steps in Ω will satisfy the same subadditivity inequality as the nearest-neighbour self-avoiding walk, and hence will have a critical point $z_c = z_c(\Omega)$. We define $\hat{\Pi}_z(k)$, for $z < z_c$, implicitly by the equation

$$\hat{G}_z(k) = \frac{1}{1 - z\Omega\hat{D}(k) - \hat{\Pi}_z(k)}. \tag{5.1.1}$$

Then $\hat{\Pi}_z$ can be thought of as a measure of the difference between self-avoiding and simple random walk. The lace expansion is an expansion for $\hat{\Pi}_z$.

We denote by $\mathcal{C}_N(x, y)$ the set of all N-step self-avoiding walks from x to y (taking steps in Ω), and denote its cardinality by $c_N(x, y)$. The first step in deriving the expansion is to extract the term in $G_z(0, x)$ corresponding to $N = 0$:

$$G_z(0, x) = \delta_{0,x} + \sum_{N=1}^{\infty} c_N(0, x)z^N. \tag{5.1.2}$$

We shall now argue that for $N \geq 1$,

$$c_N(0, x) = \sum_{y \in \Omega} \left[c_1(0, y)c_{N-1}(y, x) - \sum_{\omega^{(1)} \in \mathcal{C}_{N-1}(y,x)} I[0 \in \omega^{(1)}] \right]. \tag{5.1.3}$$

Diagrammatically the right side of (5.1.3) can be represented by

In the first term on the right side the bold line is unconstrained, apart from the fact that it should be self-avoiding. The thin line in the first term represents a single step. Equation (5.1.3) is just the inclusion-exclusion relation: the first term on the right side counts all walks from 0 to x which are self-avoiding *after* the first step, and the second subtracts the contribution due to those which are not self-avoiding from the beginning, i.e., walks that return to the origin. Since $c_1(0, y) = 1$ for $y \in \Omega$, substitution of (5.1.3) into (5.1.2) gives

$$G_z(0, x) = \delta_{0,x} + z \sum_{y \in \Omega} G_z(y, x) - \sum_{y \in \Omega} \sum_{N=0}^{\infty} z^{N+1} \sum_{\omega^{(1)} \in \mathcal{C}_N(y,x)} I[0 \in \omega^{(1)}].$$

(5.1.4)

The inclusion-exclusion relation can now be applied to the last term on the right side of (5.1.4), as follows. Let S be the first (and only) time that $\omega^{(1)}(S) = 0$. Then

$$\sum_{\omega^{(1)} \in \mathcal{C}_N(y,x)} I[0 \in \omega^{(1)}] = \sum_{S=1}^{N} \sum_{\substack{\omega^{(2)} \in \mathcal{C}_S(y,0) \\ \omega^{(3)} \in \mathcal{C}_{N-S}(0,x)}} I[\omega^{(2)} \cap \omega^{(3)} = \{0\}]$$

$$= \sum_{S=1}^{N} \left[c_S(y, 0) c_{N-S}(0, x) - \sum_{\substack{\omega^{(2)} \in \mathcal{C}_S(y,0) \\ \omega^{(3)} \in \mathcal{C}_{N-S}(0,x)}} I[\omega^{(2)} \cap \omega^{(3)} \neq \{0\}] \right].$$

We can interpret $c_S(y, 0)$ as the number of $(S+1)$-step walks which step from the origin directly to y, then return to the origin in S steps, which have distinct vertices apart from the fact that they return to their starting point. Let $\mathcal{U}_S$ denote the set of all S-step self-avoiding loops at the origin (S-step walks which begin and end at the origin but which otherwise have distinct vertices), and let u_S be the cardinality of $\mathcal{U}_S$. Then

$$\sum_{y \in \Omega} \sum_{N=0}^{\infty} z^{N+1} \sum_{\omega^{(1)} \in \mathcal{C}_N(y,x)} I[0 \in \omega^{(1)}]$$

$$= \sum_{S=2}^{\infty} z^S u_S \cdot G_z(0, x) - \sum_{\substack{S = 2 \\ N = 0}}^{\infty} \sum_{\substack{\omega^{(2)} \in \mathcal{U}_S \\ \omega^{(3)} \in \mathcal{C}_N(0,x)}} z^{S+N} I[\omega^{(2)} \cap \omega^{(3)} \neq \{0\}].$$

Continuing in this fashion, in the last term on the right side of the above equation let $T_1 \geq 1$ be the first time along $\omega^{(3)}$ that $\omega^{(3)}(T_1) \in \omega^{(2)}$, and let $v = \omega^{(3)}(T_1)$. Then the inclusion-exclusion relation can be applied again to remove the avoidance between the portions of $\omega^{(3)}$ before and after T_1, and correct for this removal by the subtraction of a term involving a further intersection. For $z < z_c$ repetition of this procedure leads to the convolution equation

$$G_z(0, x) = \delta_{0,x} + z \sum_{y \in \Omega} G_z(y, x) + \sum_v \Pi_z(0, v) G_z(v, x), \qquad (5.1.5)$$

where the "irreducible" two-point function $\Pi_z(0, x)$ is given by

$$\Pi_z(0, v) = \sum_{N=1}^{\infty} (-1)^N \Pi_z^{(N)}(0, v), \qquad (5.1.6)$$

with the terms on the right side defined as follows. The $N = 1$ term is given by

$$\Pi_z^{(1)}(0, v) = \delta_{0,v} \sum_{S=2}^{\infty} z^S u_S \equiv \delta_{0,v} \; 0 \; \bigcirc \; .$$

The $N = 2$ term is

$$\Pi_z^{(2)}(0, v) = \prod_{i=1}^{3} \left[\sum_{T_i=1}^{\infty} z^{T_i} \sum_{\omega_i \in C_{T_i}(0, v)} \right] I(\omega_1, \omega_2, \omega_3),$$

where $I(\omega_1, \omega_2, \omega_3)$ is equal to 1 if the ω_i are pairwise mutually avoiding apart from their common endpoints, and otherwise equals 0. Diagrammatically this can be represented by

$$\Pi_z^{(2)}(0, v) = \; 0 \; \ominus \; v \; ,$$

where each line represents a sum over self-avoiding walks between the endpoints of the line, weighted by z^T, with mutual avoidance between the three pairs of lines in the diagram. Similarly

$$\Pi_z^{(3)}(0, v) = \; \underset{0 \qquad v}{\boxed{\diagup\!\!\!\diagdown}} \; ,$$

where now there is mutual avoidance between some but not all pairs of lines in the diagram; we defer any discussion of the details of this mutual avoidance until later in the chapter. The unlabelled vertex is summed over

$\mathbf{Z}^d$. A slashed propagator is used to indicate a walk which may have zero length, i.e., be a single site, whereas propagators without a slash correspond to walks of at least one step. All the higher order terms can be expressed as diagrams in this way, and with some care it is possible to keep track of the pattern of mutual avoidance between subwalks (individual lines in the diagram) which emerges. The algebraic derivation of the expansion, which will be given in the next section, keeps track of this mutual avoidance automatically. Equation (5.1.6) is the lace expansion.

In the above we have tacitly assumed that the lace expansion converges. To be more careful, we should have truncated the above procedure after some large finite number of terms had been generated, and then taken a limit as the number of terms grows to infinity. This convergence assumption will be made more explicit in the next section.

The Fourier transform of a function on $\mathbf{Z}^d$ was defined in (1.4.10). Using translation invariance, and the fact that the Fourier transform of a convolution is the product of Fourier transforms, taking the Fourier transform of (5.1.5) and solving for $\hat{G}_z(k)$ gives

$$\hat{G}_z(k) = \frac{1}{1 - z\Omega\hat{D}(k) - \hat{\Pi}_z(k)}. \qquad (5.1.7)$$

Here

$$\hat{\Pi}_z(k) = \sum_{N=1}^{\infty} (-1)^N \hat{\Pi}_z^{(N)}(k). \qquad (5.1.8)$$

In Section 5.4 we will show how $\hat{\Pi}_z^{(N)}(k)$ can by bounded using the diagrammatic representation of $\Pi_z^{(N)}(0, x)$ described above. But first we turn to the algebraic derivation of the lace expansion, in the next section.

5.2   Algebraic derivation of the lace expansion

In this section we give an algebraic derivation of the lace expansion for walks taking steps in a fixed set $\Omega \subset \mathbf{Z}^d$ which respects the symmetries of the lattice and does not contain the origin. As in the previous section we simplify the notation by omitting the label Ω. The number of sites in Ω is also denoted Ω.

Given a walk $\omega = (\omega(0), \omega(1), \dots, \omega(n))$ and two "times" s and t in $\{0, 1, \dots, n\}$, we define

$$U_{st}(\omega) = \begin{cases} -1 & \text{if } \omega(s) = \omega(t) \\ 0 & \text{if } \omega(s) \neq \omega(t). \end{cases} \qquad (5.2.1)$$

Then the two point function can be written

$$G_z(0,x) = \sum_{\omega:0\to x} z^{|\omega|} \prod_{0\le s<t\le|\omega|} (1+\mathcal{U}_{st}(\omega)). \qquad (5.2.2)$$

Here the activity z is any *complex* number for which $|z| < z_c = z_c(\Omega)$. The sum over ω is the sum over all ordinary (possibly self-intersecting) walks from 0 to x, although walks which do have self-intersections give zero contribution to (5.2.2). The walk ω takes steps (u,v) with $v-u \in \Omega$, and as usual $|\omega|$ denotes the number of steps in ω. The product in (5.2.2) is equal to one if ω is self-avoiding, and is equal to zero if not. The weakly self-avoiding walk, also known as the Domb–Joyce model, is defined by replacing the factor $1+\mathcal{U}_{st}$ by $1+\lambda\mathcal{U}_{st}$, with $\lambda \in (0,1)$. Taking $\lambda = 0$ gives the ordinary unconstrained random walk, while $\lambda \in (0,1)$ gives a walk for which self-intersections are suppressed but not prohibited. We take $\lambda = 1$.

We will have need of the self-avoiding walk with a memory τ. Its two-point function is defined by

$$G_z(0,x;\tau) = \sum_{\omega:0\to x} z^{|\omega|} \prod_{\substack{0\le s<t\le|\omega| \\ t-s\le\tau}} (1+\mathcal{U}_{st}(\omega)). \qquad (5.2.3)$$

If the memory τ is equal to zero, then (5.2.3) is just the two-point function of ordinary random walk. The case $\tau = \infty$ corresponds to the self-avoiding walk. Unless explicitly stated otherwise, the memory may take any value $0 \le \tau \le \infty$.

The Fourier transform of (5.2.3) is given by

$$\hat{G}_z(k;\tau) = \sum_{\omega} z^{|\omega|} e^{ik\cdot\omega(|\omega|)} \prod_{\substack{0\le s<t\le|\omega| \\ t-s\le\tau}} (1+\mathcal{U}_{st}(\omega)) \qquad (5.2.4)$$

for $k \in [-\pi,\pi]^d$; the sum over ω is the sum over all ordinary walks of arbitrary length beginning at the origin. The right hand side is a power series in z. We denote its radius of convergence by $z_c(k;\tau)$. For any k,

$$z_c(k;\tau) \ge z_c(0;\tau) = \mu_\tau^{-1}, \qquad (5.2.5)$$

where μ_τ was defined in (1.2.12). The self-avoiding walk critical point is then $z_c = z_c(0;\infty)$.

To obtain a formula for the inverse of $\hat{G}_z(k)$, we first introduce some terminology.

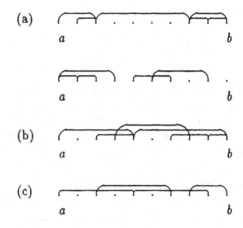

Figure 5.1: Graphs in which an edge st is represented by an arc joining s and t. (a) Examples of graphs which are not connected. (b) An example of a connected graph. (c) An example of a lace.

Definition 5.2.1 *Given an interval* $I = [a, b]$ *of positive integers, we refer to a pair* $\{s, t\}$ *($s < t$) of elements of* I *as an* edge. *To abbreviate the notation, we usually write* st *for* $\{s, t\}$. *The* length *of an edge* st *is* $t - s$. *A set of edges is called a* graph. *A graph* Γ *is said to be* connected *if both* a *and* b *are endpoints of edges in* Γ, *and if in addition, for any* $c \in (a, b)$, *there are* $s, t \in [a, b]$ *such that* $s < c < t$ *and* $st \in \Gamma$. *The set of all graphs on* $[a, b]$ *consisting of edges of length* τ *or less is denoted* $\mathcal{B}_\tau[a, b]$, *and the subset consisting of all connected graphs is denoted* $\mathcal{G}_\tau[a, b]$. *A* lace *is a minimally connected graph, i.e., a connected graph for which the removal of any edge would result in a disconnected graph. The set of laces on* $[a, b]$ *consisting of edges of length* τ *or less is denoted by* $\mathcal{L}_\tau[a, b]$, *and the set of laces on* $[a, b]$ *which consist of exactly* N *edges is denoted* $\mathcal{L}_{\tau, N}[a, b]$.

A convenient graphical representation of graphs and laces is illustrated in Figure 5.1.

Given a connected graph Γ, the following prescription associates to Γ a unique lace $\mathcal{L}_\Gamma$: The lace $\mathcal{L}_\Gamma$ consists of edges $s_1 t_1, s_2 t_2, \ldots$ where

$$s_1 = a, \quad t_1 = \max\{t : at \in \Gamma\}$$

$$t_{i+1} = \max\{t : st \in \Gamma, s < t_i\}$$

$$s_i = \min\{s : st_i \in \Gamma\}.$$

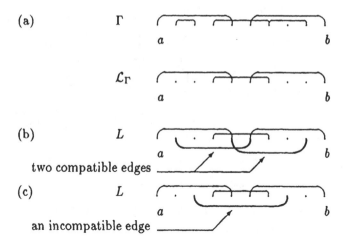

Figure 5.2: (a) An example of a connected graph Γ and its associated lace $L = \mathcal{L}_\Gamma$. (b) Examples of edges compatible with the lace L. (c) An example of an edge which is not compatible with the lace L.

Given a lace L, the set of all edges $st \notin L$ (of length τ or less) such that $\mathcal{L}_{L\cup\{st\}} = L$ is denoted $\mathcal{C}_\tau(L)$. Edges in $\mathcal{C}_\tau(L)$ are said to be *compatible* with L. Figure 5.2 illustrates these definitions.

For $a < b$ we define

$$K_\tau[a,b] = \prod_{\substack{s,t \in [a,b] \\ 0 < t-s \leq \tau}} (1 + \mathcal{U}_{st}). \tag{5.2.6}$$

We set $K_\tau[a,a] = 1$, and if $a > b$ then we set $K_\tau[a,b] = 0$. By expanding the product in (5.2.6) we obtain

$$K_\tau[a,b] = \sum_{\Gamma \in \mathcal{B}_\tau[a,b]} \prod_{st \in \Gamma} \mathcal{U}_{st}. \tag{5.2.7}$$

For $a < b$ we define an analogous quantity, in which the sum over graphs is restricted to connected graphs:

$$J_\tau[a,b] = \sum_{\Gamma \in \mathcal{G}_\tau[a,b]} \prod_{st \in \Gamma} \mathcal{U}_{st}. \tag{5.2.8}$$

We set $J_\tau[a,a] = 1$, and if $a > b$ then we set $J_\tau[a,b] = 0$. Partially resumming the right side of (5.2.8), we obtain

$$J_\tau[a, b] = \sum_{L \in \mathcal{L}_\tau[a,b]} \sum_{\Gamma : \mathcal{L}_\Gamma = L} \prod_{st \in L} \mathcal{U}_{st} \prod_{s't' \in \Gamma \setminus L} \mathcal{U}_{s't'}$$

$$= \sum_{L \in \mathcal{L}_\tau[a,b]} \prod_{st \in L} \mathcal{U}_{st} \prod_{s't' \in \mathcal{C}_\tau(L)} (1 + \mathcal{U}_{s't'}). \qquad (5.2.9)$$

For $a < b$ we define $J_{\tau,N}[a, b]$ to be the contribution to (5.2.9) from laces consisting of exactly N bonds:

$$J_{\tau,N}[a, b] = \sum_{L \in \mathcal{L}_{\tau,N}[a,b]} \prod_{st \in L} \mathcal{U}_{st} \prod_{s't' \in \mathcal{C}_\tau(L)} (1 + \mathcal{U}_{s't'}). \qquad (5.2.10)$$

Each term in the above sum is either 0 or $(-1)^N$. By (5.2.9) and (5.2.10),

$$J_\tau[a, b] = \sum_{N=1}^\infty J_{\tau,N}[a, b]. \qquad (5.2.11)$$

The sum over N in (5.2.11) is a finite sum, since the sum in (5.2.10) is empty for $N > b - a$ and hence $J_{\tau,N}[a, b] = 0$ if $N > b - a$. By definition $J_\tau[a, a + 1] = 0$, since the only lace on $[a, a + 1]$ consists of the single edge $\{a, a+1\}$, and $\mathcal{U}_{a,a+1}(\omega) = 0$ for all ω, because a walk cannot be at the same place at consecutive times.

Lemma 5.2.2 *For any $a < b$,*

$$K_\tau[a, b] = K_\tau[a + 1, b] + \sum_{j=a+2}^b J_\tau[a, j] K_\tau[j, b]. \qquad (5.2.12)$$

Proof. The contribution to the sum on the right side of (5.2.7) due to all graphs Γ for which a is not in an edge is exactly $K_\tau[a + 1, b]$. To resum the contribution due to the remaining graphs we proceed as follows. If Γ does contain an edge ending at a, let $j(\Gamma)$ be the largest value of j such that the set of edges in Γ with at least one end in the interval $[a, j]$ forms a connected graph on $[a, j]$. We lose nothing by taking $j \geq a + 2$, since as argued above the statement of the lemma, $\mathcal{U}_{a,a+1} = 0$. Then resummation over graphs on $[j, b]$ gives

$$K_\tau[a, b] = K_\tau[a + 1, b] + \sum_{j=a+2}^b \sum_{\Gamma \in \mathcal{G}_\tau[a,j]} \prod_{st \in \Gamma} \mathcal{U}_{st} \, K_\tau[j, b], \qquad (5.2.13)$$

which with (5.2.8) proves the lemma.   □

Now we define

$$\Pi_z^{(N)}(0, x; \tau) = (-1)^N \sum_{\substack{\omega \,:\, 0 \,\to\, x \\ |\omega| \geq 1}} z^{|\omega|} J_{\tau,N}[0, |\omega|] \qquad (5.2.14)$$

and

$$\Pi_z(0, x; \tau) = \sum_{N=1}^{\infty} (-1)^N \Pi_z^{(N)}(0, x; \tau) = \sum_{\substack{\omega \,:\, 0 \,\to\, x \\ |\omega| \geq 1}} z^{|\omega|} J_{\tau}[0, |\omega|], \quad (5.2.15)$$

for any z for which the right side converges. The factor $(-1)^N$ on the right side of (5.2.14) ensures that

$$\Pi_z^{(N)}(0, x; \tau) \geq 0 \quad \text{for nonnegative } z. \qquad (5.2.16)$$

Since $J_{\tau,N}[0, 1] = 0$, the sum on the right side of (5.2.14) or (5.2.15) could equally well be over walks ω of length greater than or equal to 2. Recall the notation

$$\hat{D}(k) = \frac{1}{\Omega} \sum_{x \in \Omega} e^{ik \cdot x}$$

Theorem 5.2.3 *For any value of z for which $\sum_{|\omega| \geq 2} z^{|\omega|} J_{\tau}[0, |\omega|]$ and $\sum_{|\omega| \geq 0} z^{|\omega|} K_{\tau}[0, |\omega|]$ converge absolutely,*

$$G_z(0, x; \tau) = \delta_{0,x} + z \sum_{u \in \Omega} G_z(u, x; \tau) + \sum_v \Pi_z(0, v; \tau) G_z(v, x; \tau) \quad (5.2.17)$$

and

$$\hat{G}_z(k; \tau) = \frac{1}{1 - z\Omega\hat{D}(k) - \hat{\Pi}_z(k; \tau)}. \qquad (5.2.18)$$

Proof. It suffices to obtain (5.2.17), since then (5.2.18) follows immediately upon taking the Fourier transform of (5.2.17) and using the fact that the Fourier transform of a convolution is the product of Fourier transforms. Existence of the Fourier transforms of $G_z(0, \cdot; \tau)$ and $\Pi_z(0, \cdot; \tau)$ is guaranteed by the hypotheses of the theorem.

To prove (5.2.17), we first extract the contribution to (5.2.2) due to the zero step walk:

$$G_z(0, x; \tau) = \delta_{0,x} + \sum_{\substack{\omega \,:\, 0 \,\to\, x \\ |\omega| \geq 1}} z^{|\omega|} K_{\tau}[0, |\omega|].$$

Substitution of (5.2.12) into this equation results in

$$G_z(0, x; \tau) = \delta_{0,x} + \sum_{\substack{\omega : 0 \to x \\ |\omega| \geq 1}} z^{|\omega|} K_\tau[1, |\omega|]$$

$$+ \sum_{\substack{\omega : 0 \to x \\ |\omega| \geq 1}} z^{|\omega|} \sum_{j=2}^{|\omega|} J_\tau[0, j] K_\tau[j, |\omega|]. \quad (5.2.19)$$

In the second term on the right side, the factor $K_\tau[1, |\omega|]$ is nonzero only if the walk is self-avoiding after the first step. Explicitly summing over the endpoint of the first step in ω, the second term is equal to

$$z \sum_{y \in \Omega} G_z(y, x; \tau).$$

The third term on the right side of (5.2.19) is equal to

$$\sum_{N=1}^{\infty} z^N \sum_{\substack{\omega : 0 \to x \\ |\omega| = N}} \sum_{j=2}^{N} J_\tau[0, j] K_\tau[j, N].$$

Interchanging the order of summation gives the following expression for this quantity:

$$\sum_{j=2}^{\infty} z^j \sum_{N=j}^{\infty} z^{N-j} \sum_{\substack{\omega : 0 \to x \\ |\omega| = N}} J_\tau[0, j] K_\tau[j, N].$$

In the sum over ω, there is no interaction between the initial j-step portion and final $(N - j)$-step portion of the walk. Factorizing the walk into these two pieces gives the desired result.

In the last step we interchanged two infinite sums. This is justified by the hypothesis that these two sums converge absolutely.   □

Next, we prove an identity which will be used to study the finite-dimensional distributions of the self-avoiding walk. It expresses the difference between a self-avoiding walk and two independent or decoupled self-avoiding walks, in the spirit of the inclusion-exclusion relation. First we need a definition.

Definition 5.2.4 *Any graph $B \in \mathcal{B}_\tau[0, b]$ breaks up into connected components in a natural way. Given an integer m in the open interval $(0, b)$, we let $C_m(B) = \{m\}$ if B does not contain a bond st with $s < m < t$. If B*

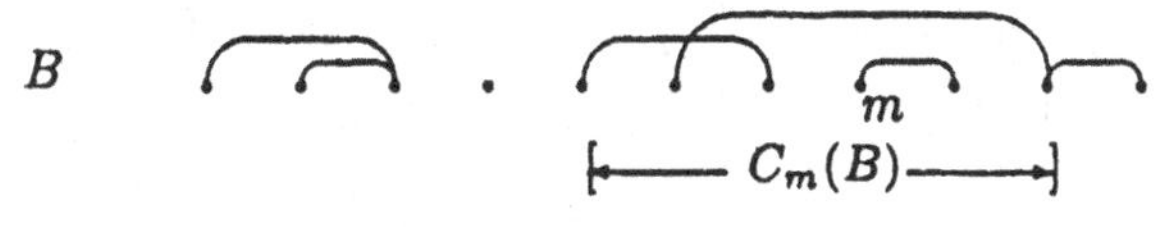

Figure 5.3: An example of a graph B and interval $C_m(B)$.

does contain a bond st with $s < m < t$, then there is a connected component Γ of B with bonds having endpoints less than and greater than m. In this case we let $C_m(B) = [i, j]$, where i is the smallest endpoint of all bonds in Γ, and j is the largest.

An example illustrating the definition is depicted in Figure 5.3.

Lemma 5.2.5 *For any integers $0 \leq m \leq b$,*

$$K_\tau[0, b] = \sum_{I \ni m} K_\tau[0, I_1] J_\tau[I_1, I_2] K_\tau[I_2, b],$$

where the sum over I is a sum over intervals $[I_1, I_2]$ of integers with either $0 \leq I_1 < m < I_2 \leq b$ or $I_1 = I_2 = m$.

Proof. By definition, $C_m(B)$ is an interval of the type being summed over in the statement of the lemma. Therefore a partial resummation of (5.2.7) gives

$$K_\tau[0, b] = \sum_{I \ni m} \sum_{B: C_m(B) = I} \prod_{st \in B} U_{st}.$$

Factoring the sum over B into three independent sums over graphs on $[0, I_1]$, connected graphs on $I = [I_1, I_2]$, and graphs on $[I_2, b]$, gives

$$K_\tau[0, b] = \sum_{I \ni m} K_\tau[0, I_1] \sum_{\Gamma \in \mathcal{G}_\tau(I)} \prod_{st \in \Gamma} U_{st} K_\tau[I_2, b].$$

The lemma then follows from (5.2.8). □

Finally we prove a lemma which will be used in Section 6.7 to prove existence of the infinite self-avoiding walk above four dimensions. To state the lemma we first introduce some notation. Since we will not need a finite memory in Section 6.7, we consider only the fully self-avoiding walk, and drop subscripts τ. Given $n \geq m \geq 0$ and an n-step self-avoiding walk ω, let ω_m denote the first m steps of ω. For $m \geq 1$, we write $\mathbf{k} = (k^{(1)}, \ldots, k^{(m)})$,

$k^{(i)} \in [-\pi, \pi]^d$, and $\mathbf{k} \cdot \omega_m = \sum_{j=1}^m k^{(j)} \cdot \omega(j)$. We define a quantity similar to the Fourier transform $\hat{G}_z(k)$ of the two-point function by

$$\Gamma_z(\mathbf{k}, m) = \sum_{n=m}^{\infty} \sum_{|\omega|=n} e^{i\mathbf{k} \cdot \omega_m} K[0, n] z^n. \qquad (5.2.20)$$

Since $|\Gamma_z(\mathbf{k}, m)| \leq \chi(|z|)$, this power series converges for $|z| < z_c$. We define a quantity similar to $\hat{\Pi}_z(k)$, again for $m \geq 0$, by

$$\Psi_z(\mathbf{k}, m) = \sum_{s=m}^{\infty} z^s \sum_{|\omega|=s} e^{i\mathbf{k} \cdot \omega_m} J[0, s]. \qquad (5.2.21)$$

For $j < m$ we define $\bar{\mathbf{k}}_j = (k^{(j+1)}, \ldots, k^{(m)})$.

Lemma 5.2.6 For $m \geq 1$ and for any z for which both sides make sense,

$$\begin{aligned}
\Gamma_z(\mathbf{k}, m) &= z\Omega\hat{D}(\sum_{j=1}^m k^{(j)})\Gamma_z(\bar{\mathbf{k}}_1, m-1) \\
&+ \sum_{s=2}^{m-1} z^s \sum_{|\omega|=s} \exp[i \sum_{j=1}^m k^{(j)} \cdot \omega(\min\{j, s\})] J[0, s]\Gamma_z(\bar{\mathbf{k}}_s, m-s) \\
&+ \Psi_z(\mathbf{k}, m)\chi(z).
\end{aligned}$$

Proof. The proof is similar to the proof of (5.2.17), but is complicated by the presence of the phase factor. We begin by replacing the factor $K[0, n]$ on the right side of (5.2.20), using Lemma 5.2.2, by

$$K[0, n] = K[1, n] + \sum_{s=2}^{m-1} J[0, s]K[s, n] + \sum_{s=m}^{n} J[0, s]K[s, n]. \qquad (5.2.22)$$

The three terms in the statement of the lemma correspond to the three terms on the right side of the above equation.

For example, the $K[1, n]$ term can be written

$$z \sum_{y \in \Omega} e^{ik^{(1)} \cdot y} \sum_{n'=m-1}^{\infty} \sum_{|\omega'|=n'} \exp\left[i \sum_{j=2}^m k^{(j)} \cdot (\omega'(j) + y)\right] K[0, n'] z^{n'}$$
$$= z\Omega\hat{D}(\sum_{j=1}^m k^{(j)})\Gamma_z(\bar{\mathbf{k}}_1, m-1).$$

Similarly, in the second and third terms the product $J[0, s]K[s, n]$ allows the sum over ω to be replaced by sums over independent walks of lengths s and $n - s$. For $s < m$ the phase factor makes a contribution to the second of these walks, while for $s \geq m$ it does not. $\qquad \square$

5.3 Example: the memory-two walk

As an example of a calculation using the lace expansion, we now solve the self-avoiding walk with memory equal to two by finding an explicit formula for $\hat{G}_z(k; 2)$. This allows for an explicit calculation of the mean-square displacement. Although the calculation of the mean-square displacement for the memory-two walk is much simpler than the corresponding calculation for the fully self-avoiding walk, the memory-two calculation does illustrate some of the basic features which will occur in Chapter 6.

We begin with a formula for $\hat{\Pi}_z(k; 2)$. The derivation of the formula requires $|z| < 1$, but the resulting expression has an analytic continuation to a meromorphic function, which in turn provides a meromorphic extension for $\hat{G}_z(k; 2)$.

Theorem 5.3.1 *For the self-avoiding walk with memory equal to two,*

$$\hat{\Pi}_z(k; 2) = \frac{z^2 \Omega}{z^2 - 1}[1 - z\hat{D}(k)].$$

Proof. We use (5.2.15), (5.2.11) and (5.2.10), with $\tau = 2$, to evaluate $\hat{\Pi}_z(k; 2)$. When the memory is equal to two, all contributing laces have all edges of length exactly equal to two. There is a unique N-edge lace L_N contributing to $J_{2,N}[0, |\omega|]$ if $|\omega| = N + 1$, illustrated in Figure 5.4. If $|\omega| \neq N + 1$, then $J_{2,N}[0, |\omega|] = 0$. There is therefore exactly one term in the sum (5.2.10) defining $J_{2,N}[0, |\omega|]$ if $|\omega| = N + 1$, and there are no terms otherwise.

Thus we can write the Fourier transform of (5.2.14) as

$$\begin{aligned}
\hat{\Pi}_z^{(N)}(k; 2) &= \sum_x \Pi_z^{(N)}(0, x; 2)e^{ik \cdot x} \\
&= (-1)^N z^{N+1} \sum_{|\omega|=N+1} J_{2,N}[0, N+1]e^{ik \cdot \omega(N+1)} \\
&= (-1)^N z^{N+1} \sum_{|\omega|=N+1} \prod_{st \in L_N} U_{st}(\omega)e^{ik \cdot \omega(N+1)}, \quad (5.3.1)
\end{aligned}$$

where L_N is the unique memory-two N-edge lace. The product over compatible edges in (5.2.10) is equal to 1 here, since all bonds compatible with L_N have length one. The product $\prod_{st \in L_N} U_{st}(\omega)$ is equal to $(-1)^N$ if it is nonzero, and it will be nonzero if and only if ω has the topology indicated in Figure 5.4. These walks end at the origin if N is odd, and end at a site in Ω if N is even. In either case they simply step back and forth repeatedly between the origin and a particular site in Ω.

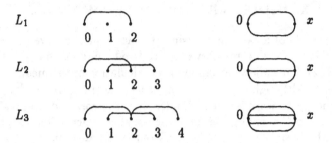

Figure 5.4: The laces for the memory-two walk, together with the corresponding walk topologies for $\prod_{st \in L_N} U_{st} \neq 0$. Each line in the diagrams on the right represents a single step of the walk, and x is a site in Ω.

Summing (5.3.1) times $(-1)^N$ over N, with the odd and even values of N summed separately, gives

$$
\begin{aligned}
\hat{\Pi}_z(k;2) &= -\sum_{n=1}^{\infty} z^{2n}\Omega + \sum_{n=1}^{\infty} z^{2n+1} \sum_{x \in \Omega} e^{ik \cdot x} \\
&= \frac{z^2}{z^2-1}\Omega[1-z\hat{D}(k)].
\end{aligned}
\tag{5.3.2}
$$

$\square$

Corollary 5.3.2 *For all $z \in \mathbb{C}$,*

$$
\hat{G}_z(k;2) = \frac{1-z^2}{1+(\Omega-1)z^2 - z\Omega\hat{D}(k)}.
\tag{5.3.3}
$$

Proof. Combining Theorem 5.3.1 with Theorem 5.2.3, we obtain (5.3.3) for all $|z| < \mu_2^{-1} = (\Omega-1)^{-1}$. But since the right side defines a function meromorphic in the plane, we have a meromorphic extension of $\hat{G}_z(k;2)$.
$\square$

The denominator of (5.3.3) has zeroes

$$
z_\pm(k) = \frac{\Omega\hat{D}(k) \pm \left[\Omega^2\hat{D}(k)^2 - 4\Omega + 4\right]^{1/2}}{2(\Omega-1)}.
\tag{5.3.4}
$$

At $k = 0$ these reduce to

$$
z_-(0) = \frac{1}{\Omega-1}, \quad z_+(0) = 1.
\tag{5.3.5}
$$

For $k = 0$ the singularity at $z_+(0) = 1$ in (5.3.3) is removable, and hence the susceptibility $\hat{G}_z(0; 2)$ has a unique singularity at the simple pole $z = (\Omega - 1)^{-1}$. For k near but not equal to 0, $\hat{G}_z(k; 2)$ has two simple poles. The location of the closest singularity of $\hat{G}_z(0; 2)$ to the origin could have been anticipated from the fact that $z_c(0; 2)$, given in (5.2.5), is equal to μ_2^{-1}.

Let $c_{T,2}(0, x)$ denote the number of T-step memory-two walks ending at x, and denote its Fourier transform by $\hat{c}_{T,2}(k)$. Then the mean-square displacement of the memory-two walk is given by

$$\langle |\omega(T)|^2 \rangle_{\tau=2} = -\frac{\nabla_k^2 \hat{c}_{T,2}(0)}{\Omega(\Omega - 1)^{T-1}}, \tag{5.3.6}$$

where ∇_k denotes the gradient in k-space.

Let C be a small circle centred at the origin of the complex plane. Since

$$\hat{G}_z(k; 2) = \sum_{T=0}^{\infty} \hat{c}_{T,2}(k) z^T, \tag{5.3.7}$$

we have

$$-\nabla_k^2 \hat{c}_{T,2}(0) = \frac{1}{2\pi i} \oint_C \frac{-\nabla_k^2 \hat{G}_z(0; 2)}{z^{T+1}} dz. \tag{5.3.8}$$

By (5.3.3),

$$-\nabla_k^2 \hat{G}_z(0; 2) = \frac{(1 + z)z \sum_{y \in \Omega} |y|^2}{[1 - (\Omega - 1)z]^2 (1 - z)}. \tag{5.3.9}$$

The integral on the right side of (5.3.8) can be evaluated exactly using the residue theorem. We deform the contour of integration past the singularities of the integrand at $(\Omega - 1)^{-1}$ and 1 to a large circle of radius R, and then let R go to infinity. The integral over the large circle vanishes in the limit, leaving the contributions from the residues. Denoting the residue of $f(z)$ at z_0 by $\text{Res}(f(z), z_0)$, we then have

$$\frac{1}{2\pi i} \oint_C \frac{-\nabla_k^2 \hat{G}_z(0; 2)}{z^{T+1}} dz \tag{5.3.10}$$

$$= -\text{Res}\left(\frac{-\nabla_k^2 \hat{G}_z(0; 2)}{z^{T+1}}, \frac{1}{\Omega - 1}\right) - \text{Res}\left(\frac{-\nabla_k^2 \hat{G}_z(0; 2)}{z^{T+1}}, 1\right).$$

To abbreviate the notation, let $b^2 = \Omega^{-1} \sum_{y \in \Omega} |y|^2$ and $\delta = (\Omega - 1)^{-1}$. Computing the residues and using (5.3.6) and (5.3.8) then gives

$$\langle |\omega(T)|^2 \rangle_{\tau=2} = b^2 \left[\left(\frac{1 + \delta}{1 - \delta}\right) T - \frac{2\delta(1 - \delta^T)}{(1 - \delta)^2} \right]$$

$$\sim \; b^2 \left(\frac{1+\delta}{1-\delta}\right) T. \tag{5.3.11}$$

Thus the mean-square displacement is as expected asymptotically linear in the number of steps.

5.4   Bounds on the lace expansion

Equation (5.2.15) can be used to provide a diagrammatic representation for $\Pi_z(0, x; \tau)$ like that obtained using the inclusion-exclusion relation in Section 5.1. In this section we describe this diagrammatic representation, and then use it to obtain upper bounds on $\hat{\Pi}_z(k; \tau)$ and some of its derivatives. The bounds we obtain will be in terms of norms of the two-point function. Although we drop the memory τ from the notation, the results of this section are valid for any memory $2 \le \tau \le \infty$. The activity z is a complex parameter.

By (5.2.15), (5.2.14) and (5.2.10),

$$\Pi_z(0, x) \;=\; \sum_{N=1}^{\infty}(-1)^N \Pi_z^{(N)}(0, x) \tag{5.4.1}$$

$$=\; \sum_{N=1}^{\infty} \sum_{\substack{\omega \,:\, 0 \,\to\, x \\ |\omega| \ge 1}} z^{|\omega|} \sum_{L \in \mathcal{L}_N[0,|\omega|]} \prod_{st \in L} \mathcal{U}_{st} \prod_{s't' \in \mathcal{C}(L)} (1+\mathcal{U}_{s't'}).$$

In this section we do not concern ourselves with the convergence or divergence of the above series, or of other related series which will occur. Rather, we treat these series as formal power series in z, and obtain upper bounds which are valid in this context. The question of whether or not these bounds suffice to determine convergence is postponed until Section 6.2.

A lace $L \in \mathcal{L}_N[0, |\omega|]$ consists of exactly N edges, and hence the factor $\prod_{st \in L} \mathcal{U}_{st}$ occurring in the sum over ω takes on either the value $(-1)^N$ or 0. The nonzero value will be attained for those walks ω such that $\omega(s) = \omega(t)$ for each $st \in L$.

Consider the case $N = 1$. There is a unique lace in $\mathcal{L}_1[0, |\omega|]$, namely $L_1 = \{0, |\omega|\}$. The first product in (5.4.1) is nonzero, in fact -1, for precisely those walks which end at the origin. Thus $\Pi_z^{(1)}(0, x) = 0$ if $x \ne 0$. All edges other than $\{0, |\omega|\}$ are compatible with L_1, and hence the second product in (5.4.1) is nonzero only for those walks which have no other self-intersections. The sum over walks in $\hat{\Pi}_z^{(1)}(0, x)$ is thus the sum over all self-avoiding loops (walks which begin and end at the origin and otherwise are self-avoiding, consisting of at least two steps). Let $\mathcal{U}$ denote the set of all self-avoiding loops. Then, introducing a diagrammatic notation, we can write

$$\Pi_z^{(1)}(0, x) = \delta_{0,x} \sum_{\omega \in \mathcal{U}} z^{|\omega|} \equiv \delta_{0,x} \; 0 \bigcirc .$$

We can turn the above into a bound on $\hat{\Pi}_z^{(1)}(k)$ by simply noting that

$$\hat{\Pi}_z^{(1)}(k) = \sum_{\omega \in \mathcal{U}} z^{|\omega|} = \sum_{x \in \Omega} z G_z(0, x) \tag{5.4.2}$$

and hence

$$|\hat{\Pi}_z^{(1)}(k)| \leq |z| \, \Omega \sup_{x \neq 0} G_{|z|}(0, x). \tag{5.4.3}$$

Writing $\|\cdot\|_\infty$ for the x-space supremum norm

$$\|f\|_\infty = \sup_{x \in Z^d} |f(x)|, \tag{5.4.4}$$

and introducing

$$
\begin{aligned}
H_z(x, y) &= G_z(x, y) - \delta_{x,y} \\
&= \begin{cases} G_z(x, y) & x \neq y \\ 0 & x = y, \end{cases}
\end{aligned} \tag{5.4.5}
$$

(5.4.2) can be rewritten as

$$|\hat{\Pi}_z^{(1)}(k)| \leq |z| \, \Omega \, \|H_{|z|}\|_\infty. \tag{5.4.6}$$

(Actually H_z is a function of two variables, and in writing norms of H_z we mean norms of the function $H_z(0, \cdot)$ of a single variable.)

For $N = 2$, we proceed as follows. Laces in $\mathcal{L}_2[0, |\omega|]$ are in a one-one correspondence with pairs of times s_2, t_1 with $0 < s_2 < t_1 < |\omega|$, as can be seen from Figure 5.5. For such a lace L, a walk with $\prod_{st \in L} \mathcal{U}_{st}(\omega) \neq 0$ breaks up in a natural way into three subwalks. Letting x denote the endpoint of ω, these three subwalks are of the form $\omega^{(1)} : 0 \to x$, $\omega^{(2)} : x \to 0$, $\omega^{(3)} : 0 \to x$, of respective lengths s_2, $t_1 - s_2$, $|\omega| - t_1$. We can split the factor $z^{|\omega|}$ into three corresponding factors

$$z^{|\omega|} = \prod_{i=1}^{3} z^{|\omega^{(i)}|}. \tag{5.4.7}$$

Each of the subwalks consists of at least one step. The factor $\prod_{s't' \in C(L)}(1 + \mathcal{U}_{s't'})$ ensures that each of these three subwalks is self-avoiding, since any

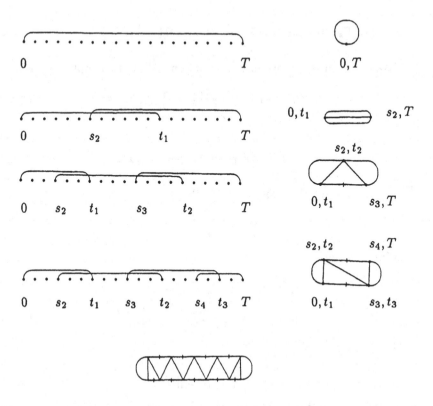

Figure 5.5: The left column shows the general form of laces consisting of 1, 2, 3 or 4 edges. The lace edges are denoted s_i, t_i, $1 \leq i \leq N$, with $s_1 = 0$ and $t_N = T$. The right column shows the self-intersections required for a walk ω with $\prod_{st \in L} \mathcal{U}_{st}(\omega) \neq 0$. For $\prod_{s't' \in \mathcal{C}(L)}(1 + \mathcal{U}_{s't'}) \neq 0$, each of the $2N - 1$ subwalks must be a self-avoiding walk, and in addition there must be mutual avoidance between some (but not all) of the subwalks. The number of loops in a diagram is equal to the number of edges in the corresponding lace. The lines which are slashed correspond to subwalks which may consist of zero steps, but the others correspond to subwalks consisting of at least one step. The eleven-loop diagram is depicted at the bottom.

bond lying entirely in one of the intervals $[0, s_2]$, $[s_2, t_1]$, $[t_1, |\omega|]$ is compatible with L. (In particular, $x \neq 0$.) This same factor also provides an avoidance interaction between the three subwalks, since there are bonds in $C(L)$ which link distinct subwalks. This interaction between distinct subwalks would be awkward to take into account exactly, but in an upper bound it can be disregarded since $1 + \mathcal{U}_{s't'} \leq 1$. Therefore

$$|\Pi_z^{(2)}(0, x)| \leq H_{|z|}(0, x)^3. \tag{5.4.8}$$

We now show how (5.4.8) can be used to estimate $\hat{\Pi}_z^{(2)}(k)$ and its second derivative with respect to a component k_μ of k. Writing ∂_μ^u for the u-th order partial derivative with respect to k_μ, and using the definition of the Fourier transform,

$$\left| \partial_\mu^u \hat{\Pi}_z^{(2)}(k) \right| \leq \sum_x |x_\mu|^u \Pi_{|z|}^{(2)}(0, x) \leq \sum_x |x_\mu|^u H_{|z|}(0, x)^3$$

$$\leq \| |x_\mu|^u H_{|z|}(0, x) \|_\infty \| H_{|z|} \|_2^2. \tag{5.4.9}$$

Here $\| \cdot \|_2$ denotes the x-space L^2-norm

$$\|f\|_2 = \left[\sum_{x \in \mathbf{Z}^d} |f(x)|^2 \right]^{1/2} \tag{5.4.10}$$

In terms of the bubble diagram (1.5.4), the right side of (5.4.9) is equal to

$$\| |x_\mu|^u H_{|z|}(0, x) \|_\infty [\mathsf{B}(|z|) - 1]. \tag{5.4.11}$$

This bound provides an indication of the critical nature of $d = 4$, in the following way. Assuming that the infrared bound $\eta \geq 0$ is indeed valid, then $\mathsf{B}(z_c)$ is finite for $d > 4$. The infrared bound also implies that the critical two-point function decays at least as fast as $|x|^{2-d}$, so that for $d > 4$ (5.4.11) will be finite at the critical point for $u \leq 2$, and hence so will $\partial_\mu^u \hat{\Pi}_{z_c}^{(2)}(k)$. For models with a suitable weak interaction, such as the nearest-neighbour model in sufficiently high dimensions, or a sufficiently "spread-out" model above four dimensions (see Chapter 6), the quantity $\mathsf{B}(z_c) - 1$ will be not only finite, but arbitrarily small, and will be the small parameter responsible for convergence of the expansion. For this reason the distinction between H_z and G_z will be crucial.

We wish to obtain bounds on $\hat{\Pi}_z^{(N)}(k)$ and some of its derivatives in an analogous way for higher values of N. To state these bounds we first need to introduce the following definition.

Definition 5.4.1 *We define a multiplication operator $\mathcal{M}$ and convolution operators $\mathcal{H}^{(0)}$ and $\mathcal{H}^{(1)}$ by*

$$(\mathcal{M}f)(x) \;=\; H_{|z|}(0,x)f(x) \qquad (5.4.12)$$

$$(\mathcal{H}^{(0)}f)(x) \;=\; \sum_y G_{|z|}(x,y)f(y) \qquad (5.4.13)$$

$$(\mathcal{H}^{(1)}f)(x) \;=\; \sum_y H_{|z|}(x,y)f(y). \qquad (5.4.14)$$

These operators depend on $|z|$, but to simplify the notation we do not make this dependence explicit.

Then we have the following bounds on $\hat{\Pi}_z^{(N)}(k)$, for $N \geq 2$. (The case $N = 1$ is special and the bound is given in (5.4.3).)

Theorem 5.4.2 *For $N \geq 2$,*

$$|\hat{\Pi}_z^{(N)}(k)| \leq \left((\mathcal{H}^{(1)}\mathcal{M})(\mathcal{H}^{(0)}\mathcal{M})^{N-2} H_{|z|} \right)(0). \qquad (5.4.15)$$

Proof. The Fourier transform on the left side of (5.4.15) is obtained by multiplying the N-th term of (5.4.1) by $e^{ik\cdot x}$ and summing over x. For an upper bound, we take absolute values inside all sums in the resulting expression. This removes the k dependence, and replaces z by $|z|$. Then we bound by 1 all factors $1 + \mathcal{U}_{st}$ in the product over compatible bonds in (5.4.1) for which s and t are times corresponding to distinct subwalks. The remaining factors ensure that all subwalks remain self-avoiding, but there is no longer any interaction between distinct subwalks. We claim that the resulting expression is then exactly equal to the right side of (5.4.15).

To see this, we note that for $N = 2, 3, 4$ we have the following diagrammatic representations, in which slashed lines correspond to $G_{|z|}$ and unslashed lines to $H_{|z|}$ (i.e. slashed lines correspond to subwalks which may consist of no steps, while unslashed lines correspond to subwalks consisting of at least one step). To begin,

where symmetry was used in the last step. Proceeding from the analogue of the right side of the above equation, and then again using symmetry, we obtain

$$\mathcal{H}^{(1)}\mathcal{M}\mathcal{H}^{(0)}\mathcal{M}H_{|z|}(x) = \boxed{\diagdown}\,{}^{0}_{x}$$

$$\mathcal{H}^{(1)}\mathcal{M}(\mathcal{H}^{(0)}\mathcal{M})^2 H_{|z|}(x) = \boxed{\diagdown}\,{}^{x}_{0}\,.$$

The pattern continues for larger N, and reproduces the diagrams representing $\hat{\Pi}_z^{(N)}$ when $x = 0$ (with no mutual avoidance between distinct subwalks). $\square$

Remark. We will require modifications of Theorem 5.4.2 of two types:
1. The first, and most common, arises when we wish to bound various derivatives of $\hat{\Pi}_z^{(N)}$, with respect to z and/or a component of k. For example, in bounding the derivative of $\hat{\Pi}_z^{(N)}$ with respect to k_μ, the differentiation of the exponential brings down a factor ix_μ. The factor x_μ can then be written as a sum of displacements along subwalks, and when absolute values are taken and interactions between subwalks neglected, the result is a sum of diagrams of the same topology as those representing $\hat{\Pi}_z^{(N)}$ itself, but with one of the subwalks in each diagram weighted by the absolute value of the μ-th component of its displacement. (Taken together, the weighted lines give a path from 0 to x.) Such a diagram is equal to the analogue of the right side of (5.4.15) in which the factor corresponding to the weighted subwalk is replaced by the corresponding multiplication or convolution by $|x_\mu|H_{|z|}(0, x)$. [Note that by (5.4.5), $x_\mu H_z(0, x) = x_\mu G_z(0, x)$.]
2. The second modification concerns an improvement to (5.4.15) for finite memory. For finite memory, the laces occurring in the lace expansion consist of bonds of length τ or less. Consequently all subwalks in diagrams representing $\hat{\Pi}_z$ consist of at most τ steps, so we may replace the operators on the right side of (5.4.15) by multiplication and convolution by the corresponding generating functions truncated at order z^τ. This improvement will be relevant in this book only in Section 6.8, which is the one place where we will not work directly with the fully self-avoiding walk.

The right side of (5.4.15) will be bounded using the following lemma.

Lemma 5.4.3 *Given functions $f_0, f_1, \ldots, f_{2M}$ on $\mathbf{Z}^d$, define $\mathcal{H}_{2j}$ and $\mathcal{M}_{2j}$ to be respectively the operations of convolution with f_{2j} and multiplication by f_{2j-1}, for $j = 1, \ldots, M$. Then for any k,*

$$\left\| \prod_{j=1}^{M} (\mathcal{H}_{2j}\mathcal{M}_{2j}) f_0 \right\|_\infty \leq \|f_k\|_\infty \prod_{\substack{0 \leq j \leq 2M \\ j \neq k}} \|f_j\|_2. \tag{5.4.16}$$

In the product over j on the left side, factors are to be taken with decreasing index from left to right.

Proof. Fix $k \in \{0, 1, 2, \ldots, 2M\}$. Using $\|\mathcal{A}\|_{p,r}$ to denote the norm of an operator $\mathcal{A} : \ell^p \to \ell^r$, the left side of (5.4.16) can be bounded above by

$$\prod_{j>k} \|\mathcal{H}_{2j}\mathcal{M}_{2j}\|_{\infty,\infty} \|\mathcal{H}_{2k}\mathcal{M}_{2k}\|_{2,\infty} \prod_{i<k} \|\mathcal{H}_{2i}\mathcal{M}_{2i}\|_{2,2} \|f_0\|_p \qquad (5.4.17)$$

where $p = 2$ if $k > 0$ and $p = \infty$ if $k = 0$. (Also the norm of $\mathcal{H}_0 \mathcal{M}_0$ should be omitted if $k = 0$.) The desired result then follows from the inequalities

$$
\begin{aligned}
\|\mathcal{H}_{2j}\mathcal{M}_{2j}\|_{\infty,\infty} &\leq \|\mathcal{H}_{2j}\|_{2,\infty}\|\mathcal{M}_{2j}\|_{\infty,2} &\leq \|f_{2j}\|_2\|f_{2j-1}\|_2 \\
\|\mathcal{H}_{2j}\mathcal{M}_{2j}\|_{2,2} &\leq \|\mathcal{H}_{2j}\|_{1,2}\|\mathcal{M}_{2j}\|_{2,1} &\leq \|f_{2j}\|_2\|f_{2j-1}\|_2 \\
\|\mathcal{H}_{2k}\mathcal{M}_{2k}\|_{2,\infty} &\leq \|\mathcal{H}_{2k}\|_{2,\infty}\|\mathcal{M}_{2j}\|_{2,2} &\leq \|f_{2k}\|_2\|f_{2k-1}\|_\infty \\
\|\mathcal{H}_{2k}\mathcal{M}_{2k}\|_{2,\infty} &\leq \|\mathcal{H}_{2k}\|_{1,\infty}\|\mathcal{M}_{2j}\|_{2,1} &\leq \|f_{2k}\|_\infty\|f_{2k-1}\|_2,
\end{aligned}
$$

where the right hand inequalities follow from the Hölder and Young inequalities. (The latter states that $\|g * h\|_s \leq \|g\|_r \|h\|_p$ for $1 \leq p, r, s \leq \infty$ satisfying $p^{-1} + r^{-1} = 1 + s^{-1}$.) $\square$

In the next theorem, Lemma 5.4.3 (in conjunction with Remark 1 below Theorem 5.4.2) is used to bound various derivatives of $\hat{\Pi}_z(k)$. For finite memory, the bounds can be improved as in Remark 2 below Theorem 5.4.2.

Theorem 5.4.4 *For any $v \geq 0$,*

$$|\partial_z^v \hat{\Pi}_z^{(1)}(k)| \leq \Omega \, \||\partial_z^v|_{z=|z|}[zH_z]\|_\infty. \qquad (5.4.18)$$

For any integer $N \geq 2$,

$$|\partial_z \hat{\Pi}_z^{(N)}(k)| \leq (2N-1)\||\partial_z|_{z=|z|}H_z\|_\infty \|H_{|z|}\|_2^N \|G_{|z|}\|_2^{N-2}. \qquad (5.4.19)$$

For any integers $v = 0, 1$, $u \geq 0$ and $N \geq 2$,

$$
\begin{aligned}
|\partial_z^v \partial_\mu^u \hat{\Pi}_z^{(N)}(k)| &\leq (2N-1)^v \left[\frac{N+1}{2}\right]^u \|x_\mu^u H_{|z|}\|_\infty \||\partial_z^v|_{z=|z|}H_z\|_2 \\
&\times \|H_{|z|}\|_2^{N-1} \|G_{|z|}\|_2^{N-2}. \qquad (5.4.20)
\end{aligned}
$$

For any $N \geq 2$ and for any positive p,

$$
\begin{aligned}
0 &\leq \hat{\Pi}_p^{(N)}(0) - \hat{\Pi}_p^{(N)}(k) \qquad (5.4.21) \\
&\leq \frac{1}{d}\sum_{\mu=1}^d (1 - \cos k_\mu)\left[\frac{N+1}{2}\right]^2 \||x|^2 H_p\|_\infty \|H_p\|_2^N \|G_p\|_2^{N-2}.
\end{aligned}
$$

For $N = 1$, $\hat{\Pi}_z^{(1)}(0) - \hat{\Pi}_z^{(1)}(k) = 0$, and hence $\partial_\mu^u \hat{\Pi}_z^{(1)}(k) = 0$ for all $u \geq 1$.

Proof. The bound (5.4.18) follows immediately from (5.4.2). For (5.4.19), applying ∂_z to $z^{|\omega|}$ brings down a factor $|\omega|$. Considering ω to consist of $2N - 1$ subwalks ω_j, we have

$$|\omega| = \sum_{j=0}^{2N-1} |\omega_j|.$$

Now we bound the diagram in which the j-th subwalk is weighted by the factor $|\omega_j|$ using Lemma 5.4.3, with the infinity norm on the j-th subwalk. This gives (5.4.19).

For (5.4.20) the k-derivative gives rise to an additional factor $|x_\mu^u|$. For the factor $|\omega|$ we proceed as for (5.4.19), obtaining a sum of terms in which one subwalk is weighted by a factor $|\omega_j|$. Then for the factor $|x_\mu^u|$ we choose a sequence of subwalks (depending on j) connecting 0 and x spatially, without using the already weighted j-th subwalk. In general no more than $(N+1)/2$ subwalks will be needed. We then write x_μ as the sum of displacements along the subwalks, and use the inequality

$$|a_1 + \ldots + a_m|^u \le m^{u-1}(|a_1|^u + \ldots + |a_m|^u)$$

to obtain a sum of diagrams where in each term one subwalk is weighted by $|\omega_j|$ and another by $|x_\mu^u|$. Then Lemma 5.4.3 is applied.

To prove (5.4.21), we note that the first inequality follows from the fact that by symmetry and (5.2.16),

$$\hat{\Pi}_p^{(N)}(0) - \hat{\Pi}_p^{(N)}(k) = \sum_x \Pi_p^{(N)}(0, x)[1 - \cos k \cdot x] \ge 0.$$

For the upper bound we write

$$\hat{\Pi}_p^{(N)}(0) - \hat{\Pi}_p^{(N)}(k) = \sum_x \Pi_p^{(N)}(0, x)(1 - e^{ik \cdot x}), \tag{5.4.22}$$

and write the last factor on the right as a telescoping sum

$$1 - e^{ik \cdot x} = \sum_{\mu=1}^{d}(1 - e^{ik_\mu x_\mu}) \prod_{\nu < \mu} e^{ik_\nu x_\nu}. \tag{5.4.23}$$

Inserting (5.4.23) into (5.4.22) and using symmetry gives

$$\hat{\Pi}_p^{(N)}(0) - \hat{\Pi}_p^{(N)}(k) = \sum_x \Pi_p^{(N)}(0, x) \sum_{\mu=1}^{d}(1 - \cos k_\mu x_\mu) \prod_{\nu < \mu} \cos k_\nu x_\nu. \tag{5.4.24}$$

Using the fact that for any real y and positive integer m,

$$0 \leq 1 - \cos my \leq m^2(1 - \cos y), \tag{5.4.25}$$

and again using symmetry, the right side of (5.4.24) can be bounded above by

$$\frac{1}{d}\sum_{\mu=1}^{d}(1 - \cos k_\mu)\, d\sum_{x} x_\mu^2 \Pi_p^{(N)}(0, x). \tag{5.4.26}$$

The bound (5.4.21) then follows, since it was shown in the proof of (5.4.20) (with $v = 0$) that the sum over x on the right side of the above equation is bounded above by the right side of (5.4.20).

Finally, the last statement of the theorem follows immediately from the first equality of (5.4.2). $\qquad\qquad\square$

We end this section with a lemma which will be used in Section 6.8 to bound $c_n(0, x)$. In the statement of the lemma, $c_{n,\sigma}(0, x)$ denotes the number of n-step memory-σ walks from 0 to x.

Lemma 5.4.5 *For memories* $\sigma < \tau \leq \infty$,

$$|\hat{\Pi}_z(k;\sigma) - \hat{\Pi}_z(k;\tau)| \leq 2\|\sum_{n=\sigma/6}^{\infty} c_{n,\sigma}(0, x)|z|^n\|_\infty$$

$$\tag{5.4.27}$$

$$\times\left[|z|\,\Omega + \sum_{N=2}^{\infty}(2N-1)\|H_{|z|}(0, \cdot; \sigma)\|_2^N\|G_{|z|}(0, \cdot; \sigma)\|_2^{N-2}\right].$$

Proof.   By Definition 5.2.1, $\mathcal{L}_\sigma[a, b] \subset \mathcal{L}_\tau[a, b]$, and for $L \in \mathcal{L}_\sigma[a, b]$, we have $\mathcal{C}_\sigma(L) \subset \mathcal{C}_\tau(L)$. The latter inclusion is often strict, but if L has no bond whose length exceeds $\sigma/2$ then $\mathcal{C}_\tau(L)$ can contain no bond whose length exceeds σ, and hence $\mathcal{C}_\sigma(L) = \mathcal{C}_\tau(L)$. Therefore in the difference on the left side of (5.4.27) there will be an exact cancellation of terms arising from laces having all bonds of length less than or equal to $\sigma/2$. Temporarily writing $\tilde{\Pi}_m$ for the contribution to $\hat{\Pi}_z(k; m)$ due to laces having at least one bond of length exceeding $\sigma/2$, we have

$$|\hat{\Pi}_z(k;\sigma) - \hat{\Pi}_z(k;\tau)| \leq |\tilde{\Pi}_\sigma| + |\tilde{\Pi}_\tau|. \tag{5.4.28}$$

A typical bond in a lace spans either two or three subwalks; see Figure 5.5. For a lace bond of length at least $\sigma/2$, at least one of the subwalks must consist of at least $\sigma/6$ steps. Now we bound the right side of (5.4.28) as in (5.4.20) (with $u = v = 0$) and (5.4.3), using the L^∞ norm on a subwalk consisting of at least $\sigma/6$ steps. The factor $2N - 1$ arises from the fact that any one of the $2N - 1$ subwalks may be the only subwalk of at least $\sigma/6$ steps. Also, the bound on $\tilde{\Pi}_\tau$ can be written in terms of memory σ rather than τ, since $c_{n,\tau}(0, x) \leq c_{n,\sigma}(0, x)$.

Finally we observe that in the L^2 norms on the right side, the generating functions H and G can be truncated at the term of order z^σ by the second Remark under Theorem 5.4.2. □

5.5 Other models

The lace expansion has been adapted to study lattice trees and animals, and percolation, and provides a way of proving mean-field critical behaviour above eight and six dimensions respectively. For these models it is not the finiteness of the bubble diagram which leads to mean-field behaviour, but rather the finiteness of the square diagram for lattice trees and animals, and of the triangle diagram for percolation. The lace expansion can be used to prove that these diagrams are finite.

In this section we discuss the lace expansion for lattice trees and animals and for percolation. The material developed here is not required elsewhere in the book. The proof of convergence of the expansion in these contexts is very similar to the proof of convergence of the expansion for the self-avoiding walk (see Section 6.2), and will not be discussed here. Instead we just derive the expansions, and briefly indicate how they can be bounded. In the bounds the square or triangle diagrams play a key role.

5.5.1 Lattice trees and animals

Let Ω be a finite set of sites in $\mathbf{Z}^d$, not containing the origin, which is symmetric with respect to the symmetries of $\mathbf{Z}^d$. We refer to a pair $\{x, y\}$ of sites with $y - x \in \Omega$ as a bond. A *lattice tree* is a connected set of bonds which has no closed loops. Although a tree T is defined as a set of bonds, we write $x \in T$ if x is an endpoint of some bond of T. We denote the number of bonds in a tree T by $|T|$. A *lattice animal* is a connected set of bonds which may contain closed loops. We denote a typical lattice animal by A and the number of bonds in A by $|A|$.

Let t_n denote the number of n-bond trees containing the origin, and a_n the number of n-bond animals containing the origin. It can be shown using subadditivity arguments that both $t_n^{1/n}$ and $a_n^{1/n}$ converge to finite positive limits λ and λ_a as n goes to infinity. The asymptotic behaviour of these quantities is believed to be of the form

$$t_n \sim \text{const.}\lambda^n n^{-\theta+1}, \quad a_n \sim \text{const.}\lambda_a^n n^{-\theta+1} \tag{5.5.1}$$

for some critical exponent θ which is the same for both trees and animals.

Another quantity of interest is the radius of gyration $R(n)$ (for trees) and $R_a(n)$ (for animals). This is defined by

$$R(n)^2 = \frac{\sum_{T \ni 0, |T|=n} \sum_{x \in T} |x - \bar{x}_T|^2}{\sum_{T \ni 0, |T|=n} \sum_{x \in T} 1} \tag{5.5.2}$$

where $\bar{x}_T = (n+1)^{-1} \sum_{x \in T} x$ is the centre of mass of T and $|x|$ is the Euclidean length of x, and similarly for $R_a(n)$. This is believed to behave as

$$R(n) \sim \text{const.} n^\nu, \quad R_a(n) \sim \text{const.} n^\nu \tag{5.5.3}$$

for some critical exponent ν which again is the same for both trees and animals.

The *two-point function* for trees or animals is defined, for sites $x, y \in \mathbf{Z}^d$, by

$$G_z(x,y) = \sum_{T \ni x,y} z^{|T|}, \quad G_z^a(x,y) = \sum_{A \ni x,y} z^{|A|}, \tag{5.5.4}$$

where z is a complex parameter known as the *activity*. We wish to derive an expansion for these two-point functions analogous to the lace expansion for the self-avoiding walk. This can be done using either the inclusion-exclusion or the resummation approach, and here we shall follow the latter.

Using the expansion, the following theorem is proven in Hara and Slade (1992c). Related results for animals appear in Hara and Slade (1990b).

Theorem 5.5.1 *For Ω the set of nearest neighbours of the origin and d sufficiently large, or for $d > 8$ and $\Omega = \{x \neq 0 : \|x\|_\infty \leq L\}$ with L sufficiently large, there are positive constants A and D such that $\theta = 5/2$ and $\nu = 1/4$ in the sense that for any $\epsilon < min\{1/2, (d-8)/4\}$,*

$$t_n = A\lambda^n n^{-3/2}[1 + O(n^{-\epsilon})]$$

and

$$R_n = Dn^{1/4}[1 + O(n^{-\epsilon})].$$

The expansion for trees

We begin with lattice trees. Given two distinct sites x, y and a tree $T \ni x, y$, the *backbone* $\beta_T(x, y)$ of T is defined to be the unique path, consisting of bonds of T, which joins x to y. Usually x and y are understood and we write simply β_T for $\beta_T(x, y)$. Sites in the backbone are labelled consecutively

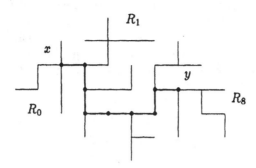

Figure 5.6: Decomposition of a tree T containing sites x and y into its backbone $\beta_T(x,y)$ and ribs $R_0, ..., R_8$. The vertices of the backbone are indicated by heavy dots.

from x to y, beginning with $\beta_T(0) = x$ and ending at (say) $\beta_T(n) = y$. Removal of the bonds in the backbone disconnects the tree into $n+1$ mutually nonintersecting trees $R_0, ..., R_n$, which we refer to as *ribs*. This decomposition is shown in Figure 5.6.

Given a set $\mathbf{R} = \{R_0, ..., R_n\}$ of $n+1$ trees R_j, we define

$$\mathcal{U}_{st}(\mathbf{R}) = \begin{cases} -1 & \text{if } R_s \text{ and } R_t \text{ share a common site} \\ 0 & \text{if } R_s \text{ and } R_t \text{ share no common site}. \end{cases} \tag{5.5.5}$$

Then the two point function can be written

$$G_z(0,x) = \sum_{\omega:0\to x} z^{|\omega|} \left[\prod_{i=0}^{|\omega|} \sum_{R_i \ni \omega(i)} z^{|R_i|} \right] \prod_{0 \le s < t \le |\omega|} (1 + \mathcal{U}_{st}(\mathbf{R})) \tag{5.5.6}$$

where each sum over R_i is a sum over trees containing $\omega(i)$, and $\mathbf{R} = (R_0, ..., R_{|\omega|})$. The sum over ω is the sum over all ordinary random walks from 0 to x (taking steps in Ω), although walks that are not self-avoiding give zero contribution to (5.5.6).

We use the terminology of Definition 5.2.1 with a change in the definition of connected graph. A graph Γ on $[a, b]$ will now be said to be *connected* if both a and b are endpoints of edges in Γ, and if in addition for each $c \in (a, b)$ there are $s, t \in [a, b]$ such that $s < c < t$ with either (i) $\{s, t\} \in \Gamma$, or (ii) $\{c, t\} \in \Gamma$ and $\{s, c\} \in \Gamma$. This notion of connectedness is less restrictive than that used for the self-avoiding walk, and is better suited for dealing with the interaction between ribs. This new definition of connected graph leads to a larger set of laces than before, where we still define a *lace* to be a minimally connected graph, i.e., a connected graph for which the removal of any edge would result in a disconnected graph. We modify the prescription associating to each connected graph Γ a unique lace $\mathcal{L}_\Gamma$, to conform with the new notion of connectedness, and now define $\mathcal{L}_\Gamma$ to consist of edges $s_1 t_1, s_2 t_2, \dots$ where

$$s_1 = a, \quad t_1 = \max\{t : at \in \Gamma\}$$

$$t_{i+1} = \max\{t : st \in \Gamma, s \le t_i\}$$

$$s_i = \min\{s : st_i \in \Gamma\}.$$

Let

$$K[a,b] = \prod_{a \le s < t \le b} (1 + \mathcal{U}_{st}) \tag{5.5.7}$$

and

$$J[0,a] = \sum_{L \in \mathcal{L}[0,a]} \prod_{st \in L} \mathcal{U}_{st} \prod_{s't' \in \mathcal{C}(L)} (1 + \mathcal{U}_{s't'}). \tag{5.5.8}$$

Then following the proof of Lemma 5.2.2 with the modified definition of connectedness gives for $b \ge 1$,

$$K[0,b] = K[1,b] + \sum_{a=1}^{b-1} J[0,a]K[a+1,b] + J[0,b]. \tag{5.5.9}$$

(The middle term on the right side is taken to be 0 if $b = 1$.) Substitution of (5.5.9) into (5.5.6) results in

$$
\begin{aligned}
G_z(0,x) = {} & \sum_{R_0 \ni 0} z^{|R_0|} \delta_{0,x} + \sum_{\substack{\omega\,:\,0\to x \\ |\omega| \ge 1}} z^{|\omega|} \left[\prod_{i=0}^{|\omega|} \sum_{R_i \ni \omega(i)} z^{|R_i|} \right] K[1,|\omega|] \\
& + \sum_{\substack{\omega\,:\,0\to x \\ |\omega| \ge 2}} z^{|\omega|} \left[\prod_{i=0}^{|\omega|} \sum_{R_i \ni \omega(i)} z^{|R_i|} \right] \sum_{a=1}^{|\omega|-1} J[0,a]K[a+1,|\omega|] \\
& + \sum_{\substack{\omega\,:\,0\to x \\ |\omega| \ge 1}} z^{|\omega|} \left[\prod_{i=0}^{|\omega|} \sum_{R_i \ni \omega(i)} z^{|R_i|} \right] J[0,|\omega|]. \tag{5.5.10}
\end{aligned}
$$

The first term on the right hand side is due to the contribution to (5.5.6) from the trivial zero-step walk, and the other terms are due to the walks ω with $|\omega| \ge 1$.

Denoting $G_z(0,0)$ by

$$g_z = G_z(0,0) = \sum_{T \ni 0} z^{|T|} \tag{5.5.11}$$

and writing

$$\Pi_z(0,x) = \sum_{\substack{\omega \,:\, 0 \to x \\ |\omega| \geq 1}} z^{|\omega|} \left[\prod_{i=0}^{|\omega|} \sum_{R_i \ni \omega(i)} z^{|R_i|} \right] J[0, |\omega|], \tag{5.5.12}$$

the first and last terms on the right side of (5.5.10) can be written as $g_z \delta_{0,x}$ and $\Pi_z(0,x)$ respectively. The second term on the right side of (5.5.10) is equal to

$$\sum_{R_0 \ni 0} z^{|R_0|} z \sum_{u \in \Omega} G_z(u,x).$$

For the third term on the right side of (5.5.10), we consider ω to be composed of an initial a-step walk ω_1 from 0 to (say) u, followed by a single step to (say) v, and then a final portion (possibly consisting of 0 steps) ω_2 from v to x. The term in question is then equal to

$$z \sum_{(u,v)} \sum_{\substack{\omega_1 \,:\, 0 \to u \\ |\omega_1| \geq 1}} z^{|\omega_1|} \left[\prod_{i=0}^{|\omega_1|} \sum_{R_i \ni \omega_1(i)} z^{|R_i|} \right] J[0, |\omega_1|]$$

$$\times \sum_{\substack{\omega_2 \,:\, v \to x \\ |\omega_2| \geq 0}} z^{|\omega_2|} \left[\prod_{i=0}^{|\omega_2|} \sum_{R_i \ni \omega_2(i)} z^{|R_i|} \right] K[0, |\omega_2|]$$

$$= z \sum_{(u,v)} \Pi_z(0,u) G_z(v,x),$$

where the sum over (u,v) denotes the sum over all directed steps with $v - u \in \Omega$. Summarizing, (5.5.10) can be rewritten as

$$G_z(0,x) = \delta_{0,x}\, g_z + \Pi_z(0,x) + z g_z \sum_{(0,u)} G_z(u,x) + z \sum_{(u,v)} \Pi_z(0,u) G_z(v,x). \tag{5.5.13}$$

This is the expansion for trees.

We can write this more compactly by writing $D(x)$ for Ω^{-1} times the indicator function of the set Ω, and defining

$$h_z(x) = \delta_{0,x} g_z + \Pi_z(0,x). \tag{5.5.14}$$

Then (5.5.13) becomes

$$G_z(0, x) = h_z(x) + h_z * z\Omega D * G_z(x). \qquad (5.5.15)$$

The expansion for animals

For lattice animals the derivation of the expansion requires some modification due to the fact that for an animal containing sites x and y there is in general not a unique path in the animal from x to y. To describe this modification, some definitions are needed.

A lattice animal A containing x and y is said to have a *double connection* from x to y if there are two *disjoint* (i.e. sharing no common bond) self-avoiding walks in A between x and y or if $x = y$. A bond $\{u, v\}$ in A is called *pivotal* for the connection from x to y if its removal would disconnect the animal into two connected components with x in one connected component and y in the other. There is a natural order to the set of pivotal bonds for the connection from x to y, and each pivotal bond is ordered in a natural way, as follows. The *first* pivotal bond for the connection from x to y (assuming there is at least one) is the pivotal bond for which there is a double connection between one endpoint of the pivotal bond and x. The endpoint for which there is a double connection to x is then the *first* endpoint of the first pivotal bond. To determine the second pivotal bond, the role of x is then played by the second endpoint of the first pivotal bond, and so on.

Given two sites x, y and an animal A containing x and y, the *backbone* of A is now defined to be the set of pivotal bonds for the connection from x to y. In general this backbone is not connected. The *ribs* of A are the connected components which remain after the removal of the backbone from A. An example is depicted in Figure 5.7. The set of all animals having a double connection between x and y is denoted $\mathcal{D}_{x,y}$, and we write

$$g_z^a(x, y) = \sum_{D \in \mathcal{D}_{x,y}} z^{|D|}. \qquad (5.5.16)$$

In particular $\mathcal{D}_{x,x}$ is the set of all animals containing x. Let B be an arbitrary finite ordered set of directed bonds:

$$B = \big((u_1, v_1), ..., (u_{|B|}, v_{|B|})\big).$$

Let $v_0 = 0$ and $u_{|B|+1} = x$. Then

$$G_z^a(0, x) = \sum_{B: |B| \geq 0} z^{|B|} \left[\prod_{i=0}^{|B|} \sum_{D_i \in \mathcal{D}_{v_i, u_{i+1}}} z^{|D_i|} \right] K[0, |B|],$$

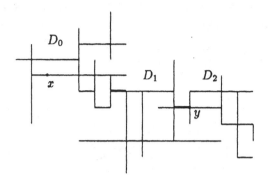

Figure 5.7: Decomposition of a lattice animal A containing x and y into backbone and ribs. The backbone, consisting of two bonds, is drawn in bold lines.

where now in the definition of $K[0, |B|]$ in (5.5.7)

$$U_{st} = \begin{cases} -1 & \text{if } D_s \text{ and } D_t \text{ share a common site} \\ 0 & \text{if } D_s \text{ and } D_t \text{ have no common site.} \end{cases}$$

Define

$$\Pi_z^a(0, y) = \sum_{B:|B|\geq 1} z^{|B|} \left[\prod_{i=0}^{|B|} \sum_{D_i \in \mathcal{D}_{v_i, u_{i+1}}} z^{|D_i|} \right] J[0, |B|] \qquad (5.5.17)$$

and

$$h_z^a(x) = g_z^a(0, x) + \Pi_z^a(0, x). \qquad (5.5.18)$$

A calculation similar to that used to derive (5.5.13), using (5.5.9), gives

$$G_z^a(0, x) = h_z^a(x) + h_z^a * z\Omega D * G_z^a(x). \qquad (5.5.19)$$

Bounds on the expansion for trees

For proving convergence of the expansions for trees, bounds are required on $\hat{\Pi}_z(k)$ and $\partial_\mu^2 \hat{\Pi}_z(k)$. In this section we indicate how appropriate bounds can be obtained on these quantities. The procedure for obtaining bounds for lattice animals is similar but more involved, and will not be discussed here.

We denote by $\mathcal{L}_N[0,a]$ the set of laces in $\mathcal{L}[0,a]$ consisting of exactly N edges, and write

$$J_N[0,a] = \sum_{L \in \mathcal{L}_N[0,a]} \prod_{st \in L} \mathcal{U}_{st} \prod_{s't' \in \mathcal{C}(L)} (1 + \mathcal{U}_{s't'}) \qquad (5.5.20)$$

and

$$\Pi_z^{(N)}(0,x) = (-1)^N \sum_{\substack{\omega\,:\,0\,\to\,x, \\ |\omega| \geq 1}} z^{|\omega|} \left[\prod_{i=0}^{|\omega|} \sum_{R_i \ni \omega(i)} z^{|R_i|} \right] J_N[0,|\omega|]. \qquad (5.5.21)$$

Then from (5.5.12) and (5.5.8) we have

$$\Pi_z(0,x) = \sum_{N=1}^{\infty} (-1)^N \Pi_z^{(N)}(0,x). \qquad (5.5.22)$$

For a nonzero contribution to $\Pi_z^{(N)}(0,x)$, the factor $\prod_{st \in L} \mathcal{U}_{st}$ in J_N enforces intersections between the ribs R_s and R_t. This leads to bounds in which the contribution to $\Pi_z^{(N)}(0,x)$ from the N-edge laces can be bounded above by an N-loop diagram. We illustrate this in detail only for the simplest cases $N = 1, 2$. We also discuss the manner of bounding diagrams only for these two simplest cases. For more details the reader is referred to Hara and Slade (1990b).

To bound the term $\Pi_z^{(1)}(0,x)$ we proceed as follows. There is a unique lace consisting of a single edge, so by definition

$$\begin{aligned}
\Pi_z^{(1)}(0,x) = &- \sum_{\substack{\omega\,:\,0\,\to\,x, \\ |\omega| \geq 1}} z^{|\omega|} \left[\prod_{i=0}^{|\omega|} \sum_{R_i \ni \omega(i)} z^{|R_i|} \right] \mathcal{U}_{0,|\omega|} \\
&\times \prod_{\substack{0 \leq s < t \leq |\omega| \\ (s,t) \neq (0,|\omega|)}} (1 + \mathcal{U}_{st}).
\end{aligned} \qquad (5.5.23)$$

The factor $\mathcal{U}_{0,|\omega|}$ gives a nonzero contribution only if R_0 and $R_{|\omega|}$ intersect, and the final product in (5.5.23) disallows any further rib intersections. We first consider the case $x \neq 0$. Relaxing the latter restriction somewhat and overcounting an enforcement of the former gives the upper bound

$$|\Pi_z^{(1)}(0,x)| \leq \sum_v \sum_{\substack{\omega\,:\,0\,\to\,x, \\ |\omega| \geq 1}} |z|^{|\omega|} \sum_{R_0 \ni 0, v} |z|^{|R_0|} \sum_{R_{|\omega|} \ni x, v} |z|^{|R_{|\omega|}|}$$

$$\times \left[\prod_{i=1}^{|\omega|-1} \sum_{R_i \ni \omega(i),\not\ni 0,x} |z|^{|R_i|} \right] \prod_{1 \le s < t \le |\omega|-1} (1+\mathcal{U}_{st}).$$

Now

$$\sum_{R_o \ni 0,v} |z|^{|R_o|} = G_{|z|}(0,v)$$

and

$$\sum_{R_{|\omega|} \ni x,v} |z|^{|R_{|\omega|}|} = G_{|z|}(x,v).$$

Since

$$\sum_{\substack{\omega\,:\,0\to x,\\ |\omega| \ge 1}} |z|^{|\omega|} \left[\prod_{i=1}^{|\omega|-1} \sum_{R_i \ni \omega(i),\not\ni 0,x} |z|^{|R_i|} \right] \prod_{1 \le s < t \le |\omega|-1} (1+\mathcal{U}_{st}) \le G_{|z|}(0,x),$$

for $x \ne 0$ we have

$$|\Pi_z^{(1)}(0,x)| \le \sum_v G_{|z|}(0,x) G_{|z|}(x,v) G_{|z|}(v,0). \tag{5.5.24}$$

When $x = 0$ in (5.5.23), we can argue similarly that

$$|\Pi_z^{(1)}(0,0)| \le \sum_{|e|=1} G_{|z|}(0,e) G_{|z|}(e,0), \tag{5.5.25}$$

by decoupling the last step of ω from its preceeding steps. Since $G_{|z|}(0,0) \ge 1$, we can now combine (5.5.24) and (5.5.25) to obtain

$$|\Pi_z^{(1)}(0,x)| \le \sum_v \left[G_{|z|}(0,v) G_{|z|}(v,x) G_{|z|}(x,0) - \delta_{0,v}\delta_{v,x} g_{|z|}^3 \right]. \tag{5.5.26}$$

Thus $\hat{\Pi}_z^{(1)}(k)$ is bounded by the triangle diagram minus the trivial contribution arising when all three vertices are the origin. For $\partial_\mu^2 \hat{\Pi}_z^{(1)}(k)$, the corresponding bound is the triangle diagram with one line weighted by x_μ^2.

A similar strategy can be used to bound $\Pi_z^{(N)}(0,x)$ for $N \ge 2$. The situation for $N = 2$ is shown in Figure 5.8. For a lace $L = (0t_1, s_2|\omega|)$ consisting of exactly two edges, there are two generic configurations possible with $\prod_{st \in L} \mathcal{U}_{st} \ne 0$, one for the case $s_2 < t_1$ and the other for $s_2 = t_1$. We illustrate the basic idea for the latter case. The contribution to $\Pi_z^{(2)}$ due to

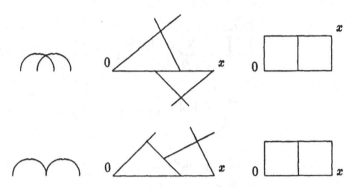

Figure 5.8: The two generic laces consisting of two bonds, schematic diagrams showing the corresponding rib intersections for a nonzero contribution to $\Pi_z^{(2)}(0,x)$, and Feynman diagrams bounding the corresponding contributions to $\Pi_z^{(2)}(0,x)$. In evaluating the diagrams, sums over vertices are constrained to disallow the coincidence of all vertices on any loop.

laces with $s_2 = t_1$ can be written

$$\sum_{\substack{\omega \,:\, 0 \to x \\ |\omega| \geq 2}} z^{|\omega|} \left[\prod_{i=0}^{|\omega|} \sum_{R_i \ni \omega(i)} z^{|R_i|} \right] \sum_{s=1}^{|\omega|-1} \mathcal{U}_{0s} \mathcal{U}_{s|\omega|} \prod_{s't' \in \mathcal{C}(\{0,s\},\{s,|\omega|\})} (1 + \mathcal{U}_{s't'}).$$

$$(5.5.27)$$

We bound this by replacing z by $|z|$, and using $1 + \mathcal{U}_{s't'} \leq 1$ for $s' < s < t'$ to decouple the interaction. For $\mathcal{U}_{0s} \neq 0$, R_0 and R_s must intersect. Let y be a site where R_0 and R_s intersect, and let $\beta_{R_s}(\omega(s), y)$ be the backbone of R_s. For $\mathcal{U}_{s|\omega|} \neq 0$, $R_{|\omega|}$ must intersect R_s, and hence there must be a rib emanating from a site on the backbone $\beta_{R_s}(\omega(s), y)$ which intersects $R_{|\omega|}$ — see Figure 5.8. Now by arguing in a similar fashion to the case $N = 1$, (5.5.27) can be bounded above by the second Feynman diagram depicted in Figure 5.8. The contribution from the other type of lace is bounded by the first Feynman diagram of Figure 5.8.

In general, $|\hat{\Pi}_z^{(N)}(k)|$ is bounded above by a sum of 2^{N-1} "ladder" diagrams, each containing exactly N non-trivial loops. The basic operation used in bounding these diagrams is repeatedly to apply the simple inequality

$$\left| \sum_x f(x) g(x) \right| \leq \sup_x |f(x)| \sum_y |g(y)|. \qquad (5.5.28)$$

For the case $N = 2$, the argument goes as follows. Let

$$\bar{T}_z = \sup_{w \in Z^d} \left[\sum_{x,y} G_z(0,x) G_z(x,y) G_z(y,w) - \delta_{0,w}(g_z)^3 \right]$$

$$\bar{W}_z = \sup_{w \in Z^d} \sum_{x,y} |x|^2 G_z(0,x) G_z(x,y) G_z(y,w)$$

$$S_z = \sum_{w,x,y} G_z(0,w) G_z(w,x) G_z(x,y) G_z(y,0) - (g_z)^4.$$

Then we have

$$|\hat{\Pi}_z^{(2)}(k)| \leq 2\bar{T}_{|z|} S_{|z|}.$$

Also, using the inequality $|\sum_{i=1}^n a_i|^2 \leq n \sum_{i=1}^n |a_i|^2$ to replace $|x|^2$ leads to a sum of diagrams in which a single line in each diagram is weighted by the square of the difference of the line endpoints, and we obtain

$$|\nabla_k^2 \hat{\Pi}_z^{(2)}(k)| \leq 13 \bar{W}_{|z|} S_{|z|}.$$

Here the factor $13 = 3^2 + 2^2$ arises from the fact that in the first diagram of Figure 5.8 the $|x|^2$ is distributed over $n = 3$ lines, while in the second $n = 2$.

Higher values of N can be handled in a similar fashion.

5.5.2 Percolation

Grimmett (1989) provides a good introduction to percolation. Here we mention only those aspects of percolation which are relevant for the development of the lace expansion. The expansion can be applied to both bond and site percolation, but for simplicity we restrict attention to bond percolation.

Let Ω be a finite set of sites in Z^d, not containing the origin, which is symmetric with respect to the symmetries of Z^d. We consider independent bond percolation on Z^d, where the bonds are the pairs $\{x,y\}$ of sites with $y - x \in \Omega$. This means that to each bond we associate an independent Bernoulli random variable $n_{\{x,y\}}$ which takes the value one with probability p and the value zero with probability $1 - p$, where p is a parameter in the closed interval $[0,1]$. If $n_{\{x,y\}} = 1$ then we say that the bond $\{x,y\}$ is *occupied*, and otherwise we say that it is vacant. A *configuration* is a realization of the random variables for all bonds. Given a configuration and any two sites x and y, we say that x and y are *connected* if there is a self-avoiding walk from x to y consisting of occupied bonds, or if $x = y$. For $d \geq 2$ this model undergoes a phase transition in the following sense:

There is a critical value $p_c \in (0,1)$ such that the probability $\theta(p)$ that the origin is connected to infinitely many sites is zero for $p < p_c$ and strictly positive for $p > p_c$.

We denote the indicator function of an event E by $I[E]$ and expectation with respect to the joint distribution of the Bernoulli random variables $n_{\{x,y\}}$ by $\langle \cdot \rangle_p$. Then the two-point function $\tau_p(x,y)$ is defined to be the probability that x and y are connected:

$$\tau_p(x,y) = \langle I[x \text{ and } y \text{ are connected}]\rangle_p. \qquad (5.5.29)$$

For $p < p_c$ the two-point function is known to decay exponentially as $|x - y| \to \infty$, so that the correlation length

$$\xi(p) = -\left[\lim_{n \to \infty} \frac{1}{n} \log \tau_p(0, ne_1)\right]^{-1} \qquad (5.5.30)$$

is finite and strictly positive. The susceptibility is defined by

$$\chi(p) = \sum_{x \in Z^d} \tau_p(0, x). \qquad (5.5.31)$$

The susceptibility is known to be finite for $p < p_c$ and to diverge as $p \nearrow p_c$.

The following power laws are believed to hold:

$$\chi(p) \sim \text{const.}(p_c - p)^{-\gamma} \text{ as } p \nearrow p_c \qquad (5.5.32)$$

$$\theta(p) \sim \text{const.}(p - p_c)^{\beta} \text{ as } p \searrow p_c \qquad (5.5.33)$$

$$\xi(p) \sim \text{const.}(p_c - p)^{-\nu} \text{ as } p \nearrow p_c \qquad (5.5.34)$$

for some dimension-dependent critical exponents γ, β, ν. Currently the only rigorous result proving existence of these critical exponents is the following theorem. In the statement of the theorem we as usual write $f(x) \simeq g(x)$ to mean that there are positive constants c_1, c_2 such that $c_1 g(x) \le f(x) \le c_2 g(x)$ for x sufficiently close to its limiting value. References to further results in this direction are given in the Notes at the end of the chapter.

Theorem 5.5.2 *For Ω the set of nearest neighbours of the origin and d sufficiently large, or for $d > 6$ and $\Omega = \{x \ne 0 : \|x\|_\infty \le L\}$ with L sufficiently large,*

$$\begin{aligned} \chi(p) &\simeq (p_c - p)^{-1} &\text{as } p \nearrow p_c \\ \theta(p) &\simeq (p - p_c)^1 &\text{as } p \searrow p_c \\ \xi(p) &\simeq (p_c - p)^{-1/2} &\text{as } p \nearrow p_c. \end{aligned}$$

It is a short step from the behaviour of $\theta(p)$ given in Theorem 5.5.2 to the conclusion that the percolation probability is zero at the critical point, i.e. that $\theta(p_c) = 0$. This has otherwise been proven only for the nearest-neighbour model in two dimensions; see the Notes for more details.

An important ingredient in the proof of Theorem 5.5.2 is an expansion for the two-point function $\tau_p(0, x)$, valid for $p < p_c$, which is analogous to the expansions for the self-avoiding walk and lattice trees and animals. Our purpose in the remainder of this section is derive the expansion, and to indicate how it can be bounded.

The expansion

For the expansion we need to define several concepts.

Definition 5.5.3 *(a) A* bond *is an unordered pair of distinct sites $\{x, y\}$ with $y - x \in \Omega$. A* directed bond *is an ordered pair (x, y) of distinct sites with $y - x \in \Omega$. A* path *from x to y is a self-avoiding walk from x to y, considered to be a set of bonds. Two paths are* disjoint *if they have no bonds in common (they may have common sites). Given a bond configuration, an* occupied path *is a path consisting of occupied bonds.*

(b) Given a bond configuration, two sites x and y are connected *if there is an occupied path from x to y or if $x = y$. We denote by $C(x)$ the random set of sites which are connected to x. Two sites x and y are* doubly-connected *if there are at least two disjoint occupied paths from x to y or if $x = y$. We denote by $D_c(x)$ the random set of sites which are doubly-connected to x. Given a bond $b = \{u, v\}$ and a bond configuration, we define $C_b(x)$ to be the set of sites which remain connected to x in the new configuration obtained by setting $n_b = 0$.*

(c) Given a set of sites $A \subset \mathbf{Z}^d$ and a bond configuration, two sites x and y are connected in A *if there is an occupied path from x to y having all of its sites in A (so in particular it is required that $x, y \in A$), or if $x = y \in A$. Two sites x and y are* connected through A *if they are connected in such a way that every occupied path from x to y has at least one bond with an endpoint in A, or if $x = y \in A$.*

(d) The restricted two-point function *is defined by*

$$\tau_p^A(x, y) = \langle I[x \text{ and } y \text{ are connected in } \mathbf{Z}^d \backslash A] \rangle_p.$$

(e) Given a bond configuration, a bond $\{u, v\}$ (occupied or not) is called pivotal *for the connection from x to y if (i) either $x \in C(u)$ and $y \in C(v)$, or $x \in C(v)$ and $y \in C(u)$, and (ii) $y \notin C_{\{u,v\}}(x)$. Similarly a directed bond (u, v) is pivotal for the connection from x to y if $x \in C_{\{u,v\}}(u)$, $y \in$*

$C_{\{u,v\}}(v)$ and $y \notin C_{\{u,v\}}(x)$. If x and y are connected then there is a natural order to the set of occupied pivotal bonds for the connection from x to y (assuming there is at least one occupied pivotal bond), and each of these pivotal bonds is directed in a natural way, as follows. The first pivotal bond from x to y is the directed occupied pivotal bond (u, v) such that u is doubly-connected to x. If (u, v) is the first pivotal bond for the connection from x to y, then the second pivotal bond is the first pivotal bond for the connection from v to y, and so on.

The basic idea behind the expansion is similar to that underlying the expansion for lattice animals. Given a configuration in which 0 and x are connected, the connected bond cluster of the origin is a lattice animal containing the sites 0 and x. The occupied pivotal bonds divide the cluster into doubly-connected pieces, as in Figure 5.7. No two of these pieces can share a common site, so there is a kind of "repulsive interaction" between these pieces. However the situation is not as simple as it was for lattice animals, because the pieces interact also when they share a common boundary bond. Rather than try to formalize this interaction into a quantity $\mathcal{U}$ as we did for lattice animals, we shall proceed instead to derive the expansion along the lines of the inclusion-exclusion approach outlined in Section 5.1.

In this approach, we think of the pivotal bonds as corresponding to the steps of a walk. The first thing to do is to extract the contribution due to the zero-step walk, which in the percolation context corresponds to the event that 0 and x are doubly-connected. Thus we have

$$\tau_p(0, x) = \langle I[x \in D_c(0)]\rangle_p + \langle I[x \in C(0), x \notin D_c(0)]\rangle_p. \qquad (5.5.35)$$

If 0 is connected to x, but not doubly, then there is a pivotal bond for the connection from 0 to x and hence a first pivotal bond, so that

$$\tau_p(0, x) = \langle I[x \in D_c(0)]\rangle_p \qquad (5.5.36)$$
$$+ \sum_{(u,v)} \langle I[x \in C(0), (u, v) \text{ is the first pivotal bond}]\rangle_p.$$

To proceed further, we need a way of writing the last term on the right side as a convolution with τ_p. This is achieved using the next lemma.

For the statement of the lemma, given sites x, y and a set of sites A we introduce two events. Let $E_1(x, y)$ be the event that $y \in D_c(x)$, and let $E_2(x, y; A)$ be the event that x is connected to y through A and there is no pivotal bond for the connection from x to y whose first endpoint is connected to x through A. In particular, $E_2(x, y; A)$ includes the event that x and y are doubly-connected and connected through A. Observe that $E_1(x, y) = E_2(x, y; \mathbf{Z}^d)$.

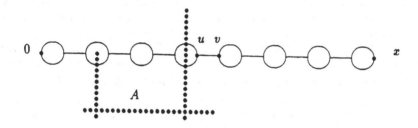

Figure 5.9: The event of Lemma 5.5.4, that $E_2(0, u; A)$ occurs and (u, v) is occupied and pivotal for the connection from 0 to x. The dotted lines represent the sites in A. There is no restriction on intersections between A and $C_{\{u,v\}}(x)$.

Lemma 5.5.4 *Given a nonempty set of sites A and a site u, let $E = E_2(0, u; A)$. Let $p < p_c$. Then*

$$\langle I[E]I[(u, v) \text{ is occupied and pivotal for the connection from } 0 \text{ to } x]\rangle_p$$
$$= p \langle I[E] \tau_p^{C_{\{u,v\}}(0)}(v, x)\rangle_p. \tag{5.5.37}$$

Proof. The event appearing in the left side of (5.5.37) is depicted in Figure 5.9. The proof is by conditioning on $C_{\{u,v\}}(0)$, which is the connected cluster of the origin which remains after setting $n_{\{u,v\}} = 0$. This cluster is finite with probability one, since $p < p_c$.

We first observe that for the event E under consideration, the event that E occurs and (u, v) is pivotal (for the connection from 0 to x) is independent of the occupation status of the bond (u, v). Therefore the left side of the identity in the statement of the lemma is equal to

$$p \langle I[E] I[(u, v) \text{ is pivotal for the connection from } 0 \text{ to } x]\rangle_p. \tag{5.5.38}$$

By conditioning on $C_{\{u,v\}}(0)$, (5.5.38) is equal to

$$p \sum_{S: S \ni 0} \langle I[E \text{ occurs}, (u, v) \text{ is pivotal}, C_{\{u,v\}}(0) = S]\rangle_p, \tag{5.5.39}$$

where the sum is over all finite sets of sites S containing 0.

In (5.5.39), the statement that (u, v) is pivotal can be replaced by the statement that v is connected to x in $\mathbf{Z}^d \backslash S$. This event depends only on the occupation status of the bonds which do not have an endpoint in S. On the other hand, the event E under consideration is determined by the occupation status of bonds which have an endpoint in $C_{\{u,v\}}(0)$. Similarly, the set $C_{\{u,v\}}(0)$ is *self-determined* in the sense that for a deterministic (finite) set of sites S, the event that $C_{\{u,v\}}(0) = S$ depends on the values of n_b only for bonds b which have one or both endpoints in S. Hence the event that both E occurs and $C_{\{u,v\}}(0) = S$ is independent of the event that v is connected to x in $\mathbf{Z}^d \backslash S$, and therefore (5.5.39) is equal to

$$p \sum_{S: S \ni 0} \langle I[E \text{ occurs and } C_{\{u,v\}}(0) = S] \rangle_p \, \tau_p^S(v, x). \qquad (5.5.40)$$

Bringing the restricted two-point function inside the expectation, replacing the superscript S by $C_{\{u,v\}}(0)$, and performing the sum over S, (5.5.40) is equal to

$$p \langle I[E] \, \tau_p^{C_{\{u,v\}}(0)}(v, x) \rangle_p. \qquad (5.5.41)$$

This completes the proof.   □

We now apply the lemma to the second term on the right side of (5.5.36), with $E = E_1(0, u) = E_2(0, u; \mathbf{Z}^d)$. The summand in this term is equal to the probability that 0 is doubly-connected to u and (u, v) is occupied and pivotal for the connection from 0 to x. Hence by the lemma it is equal to

$$p \langle I[u \in D_c(0)] \, \tau_p^{C_{\{u,v\}}(0)}(v, x) \rangle_p. \qquad (5.5.42)$$

To extract a term involving a convolution with τ_p from this quantity, we write

$$\tau_p^{C_{\{u,v\}}(0)}(v, x) = \tau_p(v, x) - [\tau_p(v, x) - \tau_p^{C_{\{u,v\}}(0)}(v, x)]. \qquad (5.5.43)$$

Using (5.5.42) and (5.5.43) in (5.5.36), we obtain

$$\begin{aligned}
\tau_p(0, x) &= \langle I[x \in D_c(0)] \rangle_p + p \sum_{(u,v)} \langle I[u \in D_c(0)] \rangle_p \tau_p(v, x) \\
&\quad - p \sum_{(u,v)} \langle I[u \in D_c(0)] \{ \tau_p(v, x) - \tau_p^{C_{\{u,v\}}(0)}(v, x) \} \rangle_p. \quad (5.5.44)
\end{aligned}$$

The above equation gives the lowest order expansion with remainder. We abbreviate the notation by writing

$$g_p(0, x) = \langle I[x \in D_c(0)] \rangle_p \qquad (5.5.45)$$

and

$$R_p^{(0)}(0, x) = p \sum_{(u,v)} \langle I[u \in D_c(0)] \{ \tau_p(v, x) - \tau_p^{C_{\{u,v\}}(0)}(v, x) \} \rangle_p. \qquad (5.5.46)$$

We denote by D the function on $\mathbf{Z}^d$ which takes the value Ω^{-1} at sites in Ω and otherwise is zero. Then (5.5.44) can be rewritten as

$$\tau_p(0, x) = g_p(0, x) + g_p * p\Omega D * \tau_p(x) - R_p^{(0)}(0, x). \tag{5.5.47}$$

To proceed further, we will use the following lemma to expand the remainder term $R_p^{(0)}(0, x)$. For the statement of the lemma, we write $I_2(v, x; A)$ for the indicator function of the event $E_2(v, x; A)$.

Lemma 5.5.5 *Given a set of sites A and two sites v and x,*

$$\tau_p(v, x) - \tau_p^A(v, x) = \langle I_2(v, x; A) \rangle_p + p \sum_{(y, y')} \langle I_2(v, y; A) \tau_p^{C_{(v, v')}(v)}(y', x) \rangle_p. \tag{5.5.48}$$

Proof. The left side is the probability of the event that v and x are connected but are not connected in $\mathbf{Z}^d \backslash A$. By definition, this is the probability that v is connected to x through A. If v is connected to x through A then either (i) there is no pivotal bond for the connection from v to x whose first endpoint is connected to v through A, or (ii) there is such a pivotal bond. Case (i) is exactly the event $E_2(v, x; A)$, and gives the first term on the right side of (5.5.48). In case (ii), let (y, y') denote the first pivotal bond for the connection from v to x such that y is connected to v through A. The contribution to the left side of (5.5.48) due to this case is

$$\sum_{(y, y')} \langle I[E_2(v, y; A) \text{ occurs and } (y, y') \text{ is occupied and pivotal}$$

$$\text{for the connection from } v \text{ to } x] \rangle_p.$$

Then by Lemma 5.5.4, with (v, y) playing the role of $(0, u)$, the contribution due to this case gives the second term on the right side of (5.5.48). $\square$

Using Lemma 5.5.5 and (5.5.46), and replacing the summation index (u, v) by (y_1, y_1'), we have

$$\begin{aligned}
R_p^{(0)}(0, x) = {} & p \sum_{(y_1, y_1')} \langle I[y_1 \in D_c(0)] \langle I_2(y_1', x; C^0_{\{y_1, y_1'\}}(0)) \rangle^{(1)} \rangle^{(0)} \\
& + p^2 \sum_{(y_1, y_1')} \sum_{(y_2, y_2')} \langle I[y_1 \in D_c(0)] \langle I_2(y_1', y_2; C^0_{\{y_1, y_1'\}}(0)) \\
& \qquad\qquad\qquad \times \tau_p^{C^1_{\{y_2, y_2'\}}(y_1')}(y_2', x) \rangle^{(1)} \rangle^{(0)}.
\end{aligned}$$

Here and in the following we simplify the notation by dropping the subscript p from the angular brackets denoting expectation. In addition we have introduced a superscript to coordinate random sets with the appropriate expectation, in nested expectations. Thus for example in the second term in the right side of the above equation, the set $C^0_{\{y_1,y_1'\}}(0)$ is random with respect to the outer expectation, but may be treated as deterministic in the evaluation of the inner expectation. Using the analogue of (5.5.43) to replace the restricted two-point function on the right side by an unrestricted two-point function plus a correction, and defining

$$\Pi_p^{(1)}(0,x) = p \sum_{(y_1,y_1')} \langle I[y_1 \in D_c(0)] \langle I_2(y_1', x; C^0_{\{y_1,y_1'\}}(0)) \rangle^{(1)} \rangle^{(0)}$$

and

$$R_p^{(1)}(0,x) = p^2 \sum_{(y_1,y_1')} \sum_{(y_2,y_2')} \langle I[y_1 \in D_c(0)] \langle I_2(y_1', y_2; C^0_{\{y_1,y_1'\}}(0))$$
$$\times \{\tau_p(y_2', x) - \tau_p^{C^1_{\{y_2,y_2'\}}(y_1')}(y_2', x)\} \rangle^{(1)} \rangle^{(0)},$$

we now have from (5.5.47) that

$$\tau_p(0,x) = g_p(0,x) - \Pi_p^{(1)}(0,x) + [g_p - \Pi_p^{(1)}] * p\Omega D * \tau_p(x) + R_p^{(1)}(0,x), \quad (5.5.49)$$

where D denotes Ω^{-1} times the indicator function of the set Ω.

The above procedure can be iterated as many times as desired. The result is the lace expansion for percolation, which is stated in the next theorem. We refrain from giving the simple but tedious details of the proof. For the statement of the theorem we write $y_0' = 0$ and introduce for $n \geq 1$,

$$C^{n-1} = C^{n-1}_{\{y_n,y_n'\}}(y_{n-1}'),$$

$$I^n = I_2(y_n', y_{n+1}; C^{n-1}),$$

$$\Pi_p^{(n)}(0,x) = p^n \sum_{(y_1,y_1')} \cdots \sum_{(y_n,y_n')} \langle I[y_1 \in D_c(0)] \langle I^1 \langle I^2 \langle I^3 \ldots \langle I^{n-1}$$
$$\times \langle I_2(y_n', x; C^{n-1}) \rangle^{(n)} \rangle^{(n-1)} \ldots \rangle^{(3)} \rangle^{(2)} \rangle^{(1)} \rangle^{(0)},$$

$$h_p^{(n)}(0,x) = g_p(0,x) + \sum_{j=1}^{n} (-1)^j \Pi_p^{(j)}(0,x)$$

and

$$R_p^{(n)}(0,x) = p^{n+1} \sum_{(y_1,y_1')} \cdots \sum_{(y_{n+1},y_{n+1}')} \langle I[y_1 \in D_c(0)] \langle I^1 \langle I^2 \ldots \langle I^n$$
$$\times \{\tau_p(y_{n+1}', x) - \tau_p^{C^n}(y_{n+1}', x)\} \rangle^{(n)} \ldots \rangle^{(3)} \rangle^{(2)} \rangle^{(1)} \rangle^{(0)}.$$

Finally defining $h_p^{(0)}(0, x) = g_p(0, x)$, we have the following theorem.

Theorem 5.5.6 *For $p < p_c$ and $N \geq 0$,*

$$\tau_p(0, x) = h_p^{(N)}(0, x) + h_p^{(N)} * p\Omega D * \tau_p(x) + (-1)^{N+1} R_p^{(N)}(0, x). \quad (5.5.50)$$

Bounds on the expansion

For effective bounds on the expansion the main requirement is to bound $\hat{g}_p(k) - 1$ and $\hat{\Pi}_p^{(n)}(k)$, as well as their second derivatives with respect to k. Bounds on the remainder term then follow readily. The method for obtaining bounds is to first obtain upper bounds in terms of Feynman diagrams, and then bound the diagrams in terms of the triangle diagram:

$$\mathsf{T}_p = \sum_{x,y \in Z^d} \tau_p(0, x)\tau_p(x, y)\tau_p(y, 0) - 1. \quad (5.5.51)$$

For second derivatives with respect to k_μ, the bubble diagram with one line weighted with x_μ^2 will also occur in upper bounds. Here we will discuss the main idea used in obtaining bounds in terms of Feynman diagrams.

To obtain a diagrammatic bound on $\hat{\Pi}_z^{(n)}(k)$ for the self-avoiding walk (or for lattice trees), an important role was played by the repulsive character of the interaction: we were able to ignore interactions between distinct subwalks in upper bounds. For percolation this step is performed using the inequality of van den Berg and Kesten (1985). This is a rather general inequality, but for our purposes the following special case will suffice.

Lemma 5.5.7 *Let $V_1, V_2, \ldots, V_n$ be sets of paths in the lattice, and let E_i $(i = 1, \ldots, n)$ be the event that at least one of the paths in V_i is occupied. Let F be the event that there are pairwise disjoint occupied paths from each of the sets $V_1, V_2, \ldots, V_n$. Then*

$$\langle I[F] \rangle_p \leq \prod_{i=1}^{n} \langle I[E_i] \rangle_p.$$

It follows immediately from Lemma 5.5.7 and the definition of g_p in (5.5.45) that

$$|\hat{g}_p(k) - 1| \leq \sum_{x \neq 0} \tau_p(0, x)^2. \quad (5.5.52)$$

The right side is the percolation bubble diagram, with the trivial unit term omitted. An additional factor of $|x|^2$ appears on the right side in the analogous bound on $\nabla_k^2 \hat{g}_p(k)$.

To bound $\hat{\Pi}_p^{(n)}(k)$, the following lemma is used. Its proof relies heavily on Lemma 5.5.7.

Lemma 5.5.8 *The following inequalities are satisfied:*
(a)

$$\langle I_2(y', x; A) \rangle_p \leq \sum_u I[u \in A] \sum_{z'} \tau_p(y', z') \tau_p(z', u) \tau_p(u, x) \tau_p(z', x)$$

(b)

$$\langle I_2(y_i', y_{i+1}; A) I[v \in C_{\{y_{i+1}, y_{i+1}'\}}(y_i')] \rangle_p \tag{5.5.53}$$

$$\leq \sum_u I[u \in A] \left(\sideset{}{'}\sum_{w,z'} \tau_p(y_i', z') \tau_p(z', u) \tau_p(u, y_{i+1}) \tau_p(y_{i+1}, w) \tau_p(w, z') \tau_p(w, v) \right.$$

$$\left. + \sum_{w,z'} \tau_p(y_i', w) \tau_p(w, v) \tau_p(w, z') \tau_p(z', u) \tau_p(u, y_{i+1}) \tau_p(y_{i+1}, z') \right),$$

where the primed sum is restricted to disallow a term with $w = z' = u = y_{i+1}$.

Proof. (a) There are two distinct ways that the event $E_2(y', x; A)$ can occur: either (i) y' and x are doubly-connected and connected through A, or (ii) y' and x are not doubly-connected and there is a last pivotal bond (z, z') for the connection from y' to x, with z' and x doubly-connected and connected through A, and with z connected to y' in $\mathbf{Z}^d \backslash A$. In case (i), either $y' \in A$, or $x \in A$, or we can find a site $u \in A \backslash \{y', x\}$ and disjoint occupied paths connecting y' to x, y' to u, and u to x. The probability of the latter of these three events is bounded above by

$$\sum_{u \in A \backslash \{y', x\}} \tau_p(y', x) \tau_p(y', u) \tau_p(u, x), \tag{5.5.54}$$

by the van den Berg–Kesten inequality. The first two of the events in case (i) can be similarly bounded, with the overall result that the contribution from case (i) is bounded above by (5.5.54) with the summation expanded to all $u \in A$. Subsequently in the proof we shall leave implicit any discussion of configurations like those in case (i) with $y' \in A$ or $x \in A$.

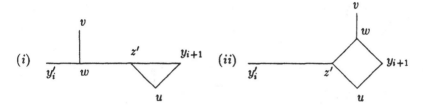

Figure 5.10: The configurations of Lemma 5.5.8(b).

For case (ii), $z' \neq y'$, and we can similarly find a site $u \in A$ and four distinct occupied paths from y' to z', z' to u, u to x, and z' to x. Hence by the van den Berg–Kesten inequality the contribution from case (ii) is bounded above by

$$\sum_{u \in A} \sum_{z' \neq y'} \tau_p(y', z') \tau_p(z', x) \tau_p(z', u) \tau_p(u, x). \qquad (5.5.55)$$

The desired result then follows by combining the contributions due to (i) and (ii).

(b) For a configuration in which $E_2(y'_i, y_{i+1}; A)$ occurs and in addition $v \in C_{\{y_{i+1}, y'_{i+1}\}}(y'_i)$, either ($i$) we can find a site $u \in A$, sites $w, z' \in \mathbf{Z}^d$ and distinct paths connecting y'_i and w, w and v, w and z', z' and u, u and y_{i+1}, and y_{i+1} and z', or (ii) we can find sites $u \in A$, $w, z' \in \mathbf{Z}^d$ and distinct paths connecting y'_i and z', z' and w, w and v, w and y_{i+1}, z' and u, and u and y_{i+1}. (Since z' and y_{i+1} are doubly-connected and connected through A, each path connecting z' to y_{i+1} passes through A, and hence $u \in A$ can be chosen such that w and u lie on distinct paths from z' to y_{i+1}). Diagrams illustrating these two possibilities are given in Figure 5.10. It is possible that for example $y'_i = z'$ in (i) and/or (ii), but we refrain from giving a detailed discussion of this or other such special cases. However we do note that in case (ii) a configuration having $z' = w = u = y_{i+1}$ is already accounted for in case (i), and we therefore need not include such configurations also in case (ii). The desired inequality then follows from the van den Berg–Kesten inequality. $\qquad \square$

This lemma can then be used in an iterative fashion to estimate the nested expectations occurring in the definition of $\hat{\Pi}_p^{(n)}$ from the inside out. We now describe this in detail for $\hat{\Pi}_p^{(1)}$, which was defined above as

$$\Pi_p^{(1)}(0, x) = p \sum_{(y_1, y'_1)} \langle I[y_1 \in D_c(0)] \langle I_2(y'_1, x; C^0_{\{y_1, y'_1\}}(0)) \rangle^{(1)} \rangle^{(0)}. \qquad (5.5.56)$$

(a) 0

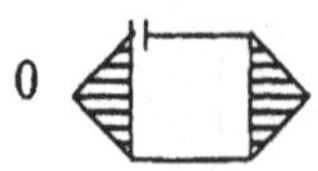

(b) 0   + 0

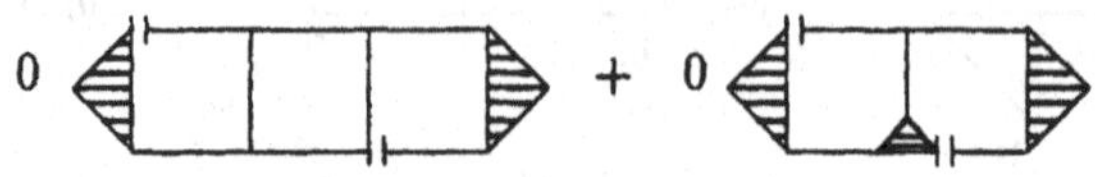

Figure 5.11: The diagrams bounding (a) $\hat{\Pi}_p^{(1)}(k)$ and (b) $\hat{\Pi}_p^{(2)}(k)$. The seven shaded triangles can shrink to a single point, but the six unshaded loops cannot.

Using Lemma 5.5.8(a) to estimate the inner expectation gives

$$\Pi_p^{(1)}(0, x) \leq p \sum_{(y_1, y_1')} \sum_{u, z'} \langle I[y_1 \in D_c(0)] I[u \in C^0_{\{y_1, y_1'\}}(0))]\rangle^{(0)}$$
$$\times \ \tau_p(y_1', z')\tau_p(z', u)\tau_p(u, x)\tau_p(z', x).$$

To estimate the remaining expectation, we observe that it is bounded above by the probability that there are two disjoint occupied paths from 0 to y_1, with one of the paths containing a site w which is connected to u by an occupied path which is disjoint from the other two. Applying the van den Berg–Kesten inequality to the expectation then gives

$$\Pi_p^{(1)}(0, x) \leq p \sum_{(y_1, y_1')} \sum_{u, w, z'} \tau_p(0, y_1)\tau_p(0, w)\tau_p(w, y_1)\tau_p(w, u)$$
$$\times \ \tau_p(y_1', z')\tau_p(z', u)\tau_p(u, x)\tau_p(z', x). \qquad (5.5.57)$$

Then $\hat{\Pi}_p^{(1)}(k)$ is bounded by the sum over x of the right side of (5.5.57). This bound is illustrated in Figure 5.11, together with the analogous bound for $\hat{\Pi}_p^{(2)}(k)$ (which uses Lemma 5.5.8(b)). In the diagrams the unshaded loops are restricted to disallow the coincidence of all vertices on the loop, and the shaded loops are unrestricted. A pair of vertical bars implies a sum over directed bonds. Any loop containing a pair of vertical bars automatically disallows the coincidence of all vertices on the loop (since the endpoints of the directed bond are distinct) and hence is unshaded.

The diagrams which bound $\hat{\Pi}_p^{(n)}$ can then be estimated in terms of

$$\sup_{w \in Z^d} \left[\sum_{x, y} \tau_p(0, x)\tau_p(x, y)\tau_p(y, w) - \delta_{0, w} \right],$$

which is closely related to the triangle diagram with the trivial unit contribution omitted. For further details the reader is referred to Hara and Slade (1990a).

5.6 Notes

Sections 5.1 and 5.2. The idea of expanding the self-avoidance interaction is a natural one which has occurred from time to time in the literature. However the lace expansion is the first such expansion which has been controlled rigorously. The lace expansion was first introduced in Brydges and Spencer (1985), who applied it to the weakly self-avoiding walk [defined below (5.2.2)], with interaction strength λ sufficiently small and dimension $d > 4$. In this context, they proved that the mean-square displacement is asymptotically linear in the number of steps and the scaling limit of the endpoint is Gaussian.

The connection between the lace expansion and the cluster expansions of statistical mechanics and constructive quantum field theory is discussed in Brydges (1986). The derivation of the lace expansion via the inclusion-exclusion relation, described in Section 5.1, was first given in Slade (1991). Lemma 5.2.6 was first proved in Lawler (1989).

Section 5.3. The memory-two walk has been studied by many authors. The expression (5.3.11) for the mean-square displacement was first obtained in Domb and Fisher (1958), in a more general setting. See also Barber and Ninham (1970), and Ernst (1988). For finite memory greater than two it is unclear how to evaluate $\hat{\Pi}_z(k; \tau)$ explicitly.

Section 5.4. The bounds given on the lace expansion generally follow the method of Brydges and Spencer (1985). Hara and Slade (1992b) use considerably more elaborate bounds to obtain convergence of the expansion for the strictly self-avoiding walk in five or more dimensions.

Section 5.5.1. For lattice trees and lattice animals subadditivity arguments showing that $t_n \leq \lambda^n$ and $a_n \leq \lambda_a^n$ are given in Klein (1981) and Klarner (1967) respectively. A lower bound $t_n \geq \text{const.}\lambda^n n^{-\text{const.}\log n}$ is given in Janse van Rensburg (1992b). A field theory representation was given in Lubensky and Isaacson (1979), where arguments were put forth that the upper critical dimension is eight. Skeleton inequalities were used in Bovier, Fröhlich and Glaus (1986) and Tasaki and Hara (1987) to illustrate the importance of the square diagram and the relevance of its finiteness at the critical point for mean-field behaviour.

Lace expansion methods were first applied to trees and animals in Hara and Slade (1990b). They considered the nearest-neighbour model in sufficiently high dimensions, and spread-out models above eight dimensions, and proved that the susceptibility (defined by summing the two-point function over the lattice) is bounded above and below by constant multiples of $(z_c - z)^{-1/2}$, and the correlation length of order two is bounded above and below by constant multiples of $(z_c - z)^{-1/4}$. These results were improved in Hara and Slade (1992c) to give the control of fixed-n quantities for trees stated in Theorem 5.5.1. The proof uses the fractional derivative methods of Section 6.3.

Section 5.5.2. Introductions to percolation, with varying perspectives, are given in the books by Grimmett (1989), Durrett (1988), Stauffer (1985) and Kesten (1982); see also the review article Kesten (1987). The lace expansion for percolation was introduced in Hara and Slade (1990a), where it was proven that the triangle condition is satisfied for the nearest-neighbour model above 48 dimensions (now improved to 42 — still not optimal!) and for spread-out models (of greater generality than stated in Theorem 5.5.2) above six dimensions. The triangle condition states that the triangle diagram $\sum_{x,y} \tau_p(0, x)\tau_p(x, y)\tau_p(y, 0)$ is finite at the critical point $p = p_c$. It is not obvious that the triangle condition holds, since $\sum_x \tau_p(0, x)$ diverges as $p \nearrow p_c$. The triangle condition was first introduced by Aizenman and Newman (1984) as a sufficient condition for mean-field behaviour of the susceptibility (expected cluster size). Later it was shown in Nguyen (1987) that the gap exponents also take their mean-field values if the triangle condition is satisfied. Then Barsky and Aizenman (1991) showed that if the triangle condition is satisfied then the critical exponents for the percolation probability $\theta(p)$ and the magnetization (the critical exponent δ) exist and take their mean-field values. Thus the portion of Theorem 5.5.2 corresponding to the susceptibility and the percolation probability is a combination of results of Aizenman and Newman (1984), Barsky and Aizenman (1991) and Hara and Slade (1990a). The control of the correlation length stated in Theorem 5.5.2 was later obtained in Hara (1990), using a modification of the lace expansion method. Section 6.5.1 describes an analogous argument for the self-avoiding walk correlation length. A further result for the critical behaviour of high dimensional percolation was obtained in Yang and Zhang (1992), who showed that given a nonnegative integer n the cluster density function $\kappa(p)$ has a finite n-th left derivative at p_c, if d is sufficiently large.

Combining the result of Hara (1990) with the upper and lower bounds in terms of the percolation correlation length for the time constant for first passage percolation [due respectively to Chayes (1991) and Chayes and Chayes (1986a)], the time constant for the nearest-neighbour model in high dimensions is seen to behave like $(p_c - p)^{1/2}$.

It follows from Theorem 5.5.2 that under the hypotheses of the theorem the percolation probability is continuous at the critical point in the sense that $\theta(p_c) = 0$. Although this is very strongly believed to be true in all dimensions, the only other instance of a rigorous proof that the critical percolation probability vanishes is for the nearest-neighbour model in two dimensions [see Grimmett (1989) for a proof and references to original work of Harris and Kesten]. Continuity at all points other than p_c is proven in Aizenman, Kesten and Newman (1987).

The lace expansion also leads to a bound on the nearest-neighbour percolation critical point of the form $(2d - 1)^{-1} \leq p_c \leq (2d)^{-1} + O(d^{-2})$; see Hara and Slade (1990a). This improves an estimate of Kesten (1990). For the spread-out model in more than six dimensions it is implicit in Hara and Slade (1990a) that the critical point satisfies $\Omega^{-1} \leq p_c \leq \Omega^{-1} + o(L^{-d-2})$. For spread-out percolation in general dimensions $d \geq 2$, Penrose (1992) has shown that $p_c \sim \Omega^{-1}$ as $L \to \infty$.

As stated, Lemma 2.1 of Hara and Slade (1990a) is incorrect: the class of events in the statement is too large. However the conclusion of the lemma is true for the events to which the lemma is applied. The error, which was pointed out by Y. Higuchi, is corrected in Lemma 5.5.4. Lemma 5.5.4 considers a slightly different event than that used in Hara and Slade (1990a), to simplify the presentation.

The lace expansion has also been applied to oriented percolation models, in which bonds are oriented in one direction. The upper critical dimension for these models is believed to be five [Obukhov (1980)]. Strong evidence in favour of this belief is given by the results of Nguyen and Yang (1991), who proved that the triangle condition holds for the nearest-neighbour model in very high dimensions or for spread-out models above five dimensions. This implies mean-field behaviour for various critical exponents as for the unoriented case. It appears that the lace expansion can also be applied to prove Gaussian behaviour of the two-point function at the critical point, at least for the nearest-neighbour model in high dimensions [Nguyen and Yang (in preparation)].

Chapter 6

Above four dimensions

6.1 Overview of the results

The lace expansion has been used to resolve many of the issues concerning the self-avoiding walk in five or more dimensions. Proving convergence of the lace expansion for $d = 5$ involves a myriad of major technical difficulties, due to the fact that the best bound on the small parameter responsible for convergence of the expansion, namely $\|H_{z_c}\|_2^2 = \mathsf{B}(z_c) - 1$, is 0.493. However many of these technical difficulties are not present if the small parameter can be taken to be arbitrarily small, and it is in the context of an arbitrarily small parameter that the proof becomes most transparent. For this reason, in this chapter we give the proof of convergence of the lace expansion and its consequences for the critical behaviour in two contexts: for the nearest-neighbour model with large d, and for the "spread-out" self-avoiding walk with steps (x, y) satisfying $0 < \|x - y\|_\infty \leq L$, for $d > 4$ and large L.

For each of these two models we will use Ω to denote the coordination number, i.e. $\Omega = 2d$ for the nearest-neighbour model and $\Omega = (2L + 1)^d - 1$ for the spread-out model. It will be shown that in either case the behaviour of $\|H_{z_c}\|_2^2$ is governed by the contribution to the corresponding ordinary random walk critical ($z = \Omega^{-1}$) bubble diagram due to the Ω terms in which two single step walks end at the same site, i.e. $\Omega/\Omega^2 = \Omega^{-1}$. Hence the small parameter can be made arbitrarily small by increasing Ω.

In the remainder of this section we summarize the results that will be obtained in this chapter. We discuss both the nearest-neighbour model and the spread-out model simultaneously, combining the statements that d is sufficiently large for the nearest-neighbour model, and L is sufficiently large for the spread-out model, into the single statement that Ω is sufficiently

171

large. We emphasize that all of the results stated in this section, with the exception of Theorem 6.1.3, have been proven in Hara and Slade (1992a,b) for the nearest-neighbour model for $d \geq 5$.

Asymptotic formulas for c_n and the mean-square displacement are given in the following theorem, whose proof can be found in Section 6.4.2.

Theorem 6.1.1 *There is an Ω_0 such that for $\Omega \geq \Omega_0$ there are positive A, D such that the following hold (assuming $d > 4$ for the spread-out model).*

(a) $c_n = A\mu^n[1 + O(n^{-\epsilon})]$ *as* $n \to \infty$, *for any* $\epsilon < \min\{(d-4)/2, 1\}$.

(b) $\langle|\omega(n)|^2\rangle = Dn[1 + O(n^{-\epsilon})]$ *as* $n \to \infty$, *for any* $\epsilon < \min\{(d-4)/4, 1\}$.

Remark. Bounds on the constants A and D will be given in Section 6.2.3. In particular, for the nearest-neighbour model in high dimensions D is strictly greater than one, indicating that the self-avoiding walk does move away from the origin more quickly than ordinary random walk, although only at the level of the diffusion constant. For the nearest-neighbour model in five dimensions the current best bounds are given in Hara and Slade (1992b) to be $1 \leq A \leq 1.493$ and $1.098 \leq D \leq 1.803$.

A corollary of (a) is that $\lim_{n\to\infty} c_{n+1}/c_n = \mu$ [cf. Equation (7.1.4)]. This is believed to be true in all dimensions, but remains unproved for $d = 2, 3, 4$. Theorem 6.1.1 is proven via a Tauberian-type theorem, after first controlling the susceptibility and correlation length of order two. The results for χ and ξ_2 are stated in the next theorem, which is proved in Section 6.2.3. [The notation $f(z) \sim g(z)$ means $\lim_{z \nearrow z_c} f(z)/g(z) = 1$.]

Theorem 6.1.2 *There is an Ω_0 such that for $\Omega \geq \Omega_0$ (assuming $d > 4$ for the spread-out model)*

$$\chi(z) \sim \frac{Az_c}{z_c - z}$$

and

$$\xi_2(z) \sim \left(\frac{Dz_c}{z_c - z}\right)^{1/2},$$

where the constants A, D are the same as in Theorem 6.1.1.

For $c_n(0, x)$ we will prove the following theorem, which gives the hyper-scaling inequality $\alpha_{sing} - 2 \leq -d/2$. In fact this inequality is believed to be an equality; see Section 2.1. Theorem 6.1.3 is the only result stated in this section which has not been proved for the nearest-neighbour model for all $d \geq 5$.

Theorem 6.1.3 *There is an Ω_0 such that for $\Omega \geq \Omega_0$ (assuming $d > 4$ for the spread-out model) there is a constant B such that*

$$\sup_{x \in \mathbb{Z}^d} c_n(0, x) \leq B\mu^n n^{-d/2}.$$

This theorem is proved in Section 6.8. An immediate consequence of Theorem 6.1.3 is the following result, which is a weaker version of the statement that $\alpha_{sing} - 2 \leq -d/2$. This weaker statement has been proven for the nearest-neighbour model for all $d \geq 5$; we comment briefly on the method of proof in the Notes for this chapter.

Corollary 6.1.4 *There is an Ω_0 such that for $\Omega \geq \Omega_0$ (assuming $d > 4$ for the spread-out model)*

$$\sup_{x \in \mathbb{Z}^d} \sum_{n=0}^{\infty} n^a c_n(0, x)\mu^{-n} < \infty$$

for any $a < (d-2)/2$.

For the correlation length $\xi(z) = 1/m(z)$ [see (1.3.15)] we have the following result, which is proved in Section 6.5.1.

Theorem 6.1.5 *There is an Ω_0 such that for $\Omega \geq \Omega_0$ (assuming $d > 4$ for the spread-out model)*

$$\xi(z) \sim \sqrt{\frac{D}{2d}} \left(\frac{z_c}{z_c - z}\right)^{1/2},$$

with the same constant D as in Theorem 6.1.1.

By Theorems 6.1.1, 6.1.2 and 6.1.5, the length scales defined by the mean square displacement, the correlation length of order two, and the correlation length are as expected all governed by the same critical exponent $\nu = 1/2$.

Using Theorem 6.1.5 it can be shown that the renormalized coupling constant $g(z)$ of (1.4.22) obeys

$$g(z) \simeq (z_c - z)^{(d-4)/2} \quad \text{as } z \nearrow z_c, \tag{6.1.1}$$

for the spread-out model with Ω sufficiently large and for the nearest-neighbour model for $d \geq 6$. Unfortunately (6.1.1) remains unproven for $d = 5$. Further details are given in the Remark under Theorem 1.5.5.

The results for the critical two-point function are stronger in k-space than in x-space, and are summarized in the following theorem, whose proof can be found in Section 6.5.2. The upper bound on $G_{z_c}(0, x)$ in the theorem, for $p < (d-2)/2$, follows immediately from Corollary 6.1.4 and the fact that $|x|^p c_n(0, x) \leq n^p c_n(0, x)$. The k-space result provides a strong infrared bound.

Theorem 6.1.6 There is an Ω_0 such that for $\Omega \geq \Omega_0$ (assuming $d > 4$ for the spread-out model) the following hold. For any p satisfying $p < (d-2)/2$ or $p \leq 2$, there is a constant $C(p)$ such that for all x, $G_{z_c}(0, x) \leq C(p)|x|^{-p}$. There is a positive constant such that the Fourier transform satisfies $\hat{G}_{z_c}(k) = const.[k^2 + O(k^{2+\epsilon})]^{-1}$ as $k \to 0$, for any $\epsilon < \min\{(d-4)/2, 1\}$. In addition, there is a positive constant such that $0 \leq \hat{G}_{z_c}(k) \leq const.k^{-2}$ for all $k \in [-\pi, \pi]^d$.

Corollary 6.1.7 There is an Ω_0 such that for $\Omega \geq \Omega_0$ (assuming $d > 4$ for the spread-out model)

$$m(z_c) = 0.$$

Proof. The bound on $\hat{G}_{z_c}(k)$ of Theorem 6.1.6 implies that the critical bubble diagram $B(z_c) = (2\pi)^{-d} \int_{[-\pi, \pi]^d} \hat{G}_{z_c}(k)^2 d^d k$ is finite (see Section 1.5). It then follows from Theorem 4.1.6 that $m(z_c) = 0$. $\Box$

To discuss the scaling limit, we first introduce some notation. Let $C_d[0, 1]$ denote the continuous $\mathbf{R}^d$-valued functions on $[0, 1]$, equipped with the supremum norm. Given an n-step self-avoiding walk ω, we define $X_n \in C_d[0, 1]$ by setting $X_n(k/n) = (Dn)^{-1/2}\omega(k)$ for $k = 0, 1, 2, \ldots, n$, and taking $X_n(t)$ to be the linear interpolation of this. We denote by dW the Wiener measure on $C_d[0, 1]$. Expectation with respect to the uniform measure on the n-step self-avoiding walks is denoted by $\langle \cdot \rangle_n$. The following theorem is proved in Section 6.6.

Theorem 6.1.8 There is an Ω_0 such that for $\Omega \geq \Omega_0$ (assuming $d > 4$ for the spread-out model), the scaled self-avoiding walk converges in distribution to Brownian motion. In other words for any bounded continuous function f on $C_d[0, 1]$,

$$\lim_{n \to \infty} \langle f(X_n) \rangle_n = \int f dW.$$

The next result concerns the existence of a measure on infinitely long self-avoiding walks. We defer the precise definition of this measure until Section 6.7, where the following theorem will be proved.

Theorem 6.1.9 *There is an Ω_0 such that for $\Omega \geq \Omega_0$ (assuming $d > 4$ for the spread-out model) the infinite self-avoiding walk exists.*

The key ingredient in the proofs of the above theorems is the convergence of the lace expansion, which is proved in the next section.

6.2 Convergence of the lace expansion

This section is divided into three parts. The first part proves a lemma which encapsulates the basic structure of the proof of convergence of the lace expansion, and also gives a number of properties of simple random walk which will be needed in the convergence proof. The second part gives the proof of convergence of the lace expansion, and states a number of consequences. The last part gives the proof of Theorem 6.1.2, i.e. existence of and mean-field values for the critical exponents for the susceptibility and the correlation length of order two.

6.2.1 Preliminaries

The following elementary lemma will be used to prove convergence of the lace expansion. It states that under an appropriate continuity assumption, if a set of inequalities implies a stronger set of inequalities, then in fact the stronger inequalities must hold.

Lemma 6.2.1 *Let $f_1, \ldots, f_n$ be nonnegative functions defined on the interval $[0, p_1)$, and let $p_0 \in [0, p_1)$ and $a < 1$ be given. Suppose that*

1. *f_i is continuous on the interval $[0, p_1)$, for $i = 1, \ldots, n$,*

2. *$f_i(p) \leq a$ for $0 \leq p \leq p_0$, for $i = 1, \ldots, n$,*

3. *for each $p \in [p_0, p_1)$, if $f_i(p) \leq 1$ for all $i = 1, \ldots, n$, then in fact $f_i(p) \leq a$ for all $i = 1, \ldots, n$. (In other words a set of inequalities implies a stronger set of inequalities.)*

Then $f_i(p) \leq a$ for all $p \in [0, p_1)$ and all $i = 1, \ldots, n$.

Proof. Define $f_{max}(p) = \max_{1 \leq i \leq n} f_i(p)$. By the second assumption, it suffices to show that $f_{max}(p) \leq a$ for $p \in [p_0, p_1)$. By the third assumption $f_{max}(p) \notin (a, 1]$ for all $p \in [p_0, p_1)$. By the first assumption $f_{max}(p)$ is continuous in $p \in [0, p_1)$. Since $f_{max}(p_0) \leq a$ by the second assumption, the above two facts imply that $f_{max}(p)$ cannot enter the forbidden interval $(a, 1]$ when $p \in [p_0, p_1)$ and hence $f_{max}(p) \leq a$ for all $p \in [0, p_1)$. $\qquad\square$

Before defining the functions f_i that we will use, we need to introduce two models of ordinary random walk corresponding to the two models of self-avoiding walk discussed in the previous section. For the usual nearest-neighbour simple random walk we denote the coordination number by $\Omega = 2d$, and also use Ω to denote the set of sites which are nearest neighbours of the origin. The critical ($z = \Omega^{-1}$) two-point function for this model is shown in (A.8) to be given by

$$C^{(0)}(0, x) = \int_{[-\pi, \pi]^d} \frac{e^{-ik \cdot x}}{1 - \hat{D}_0(k)} \frac{d^d k}{(2\pi)^d}, \tag{6.2.1}$$

where

$$\hat{D}_0(k) = \frac{1}{\Omega} \sum_{x \in \Omega} e^{ik \cdot x} = d^{-1} \sum_{\mu=1}^{d} \cos k_\mu. \tag{6.2.2}$$

Let $L \geq 1$ be an integer. For the ordinary "spread-out" random walk in $\mathbf{Z}^d$ whose steps (x, y) satisfy $0 < \|x - y\|_\infty \leq L$, we will use Ω to denote the set of $x \in \mathbf{Z}^d$ with $0 < \|x\|_\infty \leq L$, and also write Ω for the cardinality of this set, i.e. $\Omega = (2L + 1)^d - 1$. For $x \in \mathbf{Z}^d$, let $C^{(L)}(0, x)$ denote the critical spread-out ordinary random walk two-point function. This is given in (A.8) by

$$C^{(L)}(0, x) = \int_{[-\pi, \pi]^d} \frac{e^{-ik \cdot x}}{1 - \hat{D}_L(k)} \frac{d^d k}{(2\pi)^d}, \tag{6.2.3}$$

where

$$\hat{D}_L(k) = \frac{1}{\Omega} \sum_{x \in \Omega} e^{ik \cdot x} = \frac{1}{\Omega} \sum_{x \in \Omega} \cos(k \cdot x). \tag{6.2.4}$$

We write simply $C(0, x)$ and $\hat{D}(k)$ when we wish to discuss both the spread-out and nearest-neighbour models simultaneously. The following lemma is a combination of the statements of Lemmas A.3 and A.5, in which some bounds have been degraded for a unified statement.

Lemma 6.2.2 *For any $d \geq 1$ there is an Ω_0 such that for any $k \in [-\pi, \pi]^d$ and $\Omega \geq \Omega_0$,*

$$1 - \hat{D}(k) \geq \frac{k^2}{2\pi^2 d}. \tag{6.2.5}$$

For any $d \geq 1$ there is an Ω_0 such that for all $\Omega \geq \Omega_0$

$$\sup_{n \geq 0} n^{d/2} \|\hat{D}^n\|_1 < \infty. \tag{6.2.6}$$

Let s denote a fixed small positive number for the spread-out model, and let $s = 0$ for the nearest-neighbour model. There is a K such that for all

Ω *(assuming $d > 4$ for the spread-out model and $d \geq 5$ for the nearest-neighbour model)*

$$\|\hat{C}\|_2^2 - 1 = \left\|\frac{1}{1-\hat{D}}\right\|_2^2 - 1 \leq K\Omega^{-1+s}, \qquad (6.2.7)$$

and

$$\left\|\frac{\partial_\mu^2 \hat{D}}{[1-\hat{D}]^2}\right\|_1 + 2\left\|\frac{(\partial_\mu \hat{D})^2}{[1-\hat{D}]^3}\right\|_1 \leq K\Omega^{-1+s+2/d} \qquad (6.2.8)$$

(the $2/d$ in the exponent can be omitted for the nearest-neighbour model). The above norms are all L^p norms on $[-\pi, \pi]^d$ with measure $(2\pi)^{-d}d^dk$. The constant K depends on the dimension (but not on L) for the spread-out model, and is a universal constant for the nearest-neighbour model.

In the following we will maintain the convention that K and Ω_0 depend on the dimension when a statement is applied to the spread-out model, but are universal constants when the same statement is applied to the nearest-neighbour model.

6.2.2 The convergence proof

To prove convergence of the lace expansion, we will use Lemma 6.2.1 with $n = 2$, $p_0 = \Omega^{-1}$, $p_1 = z_c$, $a = 2/3$,

$$f_1(p) = \frac{\|H_p\|_2^2}{3K\Omega^{-1+s}} \quad \text{and} \quad f_2(p) = \frac{\|x_\mu^2 G_p\|_\infty}{3K\Omega^{-1+s+2/d}}, \qquad (6.2.9)$$

with K the constant of Lemma 6.2.2. Here s is as in the statement of Lemma 6.2.2, and the $2/d$ can be omitted from the exponent in the definition of f_2 for the nearest-neighbour model.

The following three results confirm that the hypotheses of Lemma 6.2.1 are satisfied, either for the nearest-neighbour model in sufficiently high dimensions, or for the spread-out model in more than four dimensions with Ω sufficiently large. It will then follow from the lemma that $\|H_p\|_2^2$ and $\|x_\mu^2 G_p\|_\infty$ are both small (for large Ω) uniformly in $p \in [0, z_c)$. This will give good bounds on the lace expansion, when combined with Theorem 5.4.4.

For simplicity we deal explicitly only with the strictly self-avoiding walk, although the results of this section also hold for all finite memories $2 \leq \tau < \infty$, subject to the replacement of z_c by the finite memory critical point $z_c(0; \tau)$. In particular, the constants of Corollaries 6.2.6 and 6.2.7 and Theorem 6.2.9 are independent of τ. Finite memory is used only to prove the bound on $c_n(0, x)$ of Theorem 6.1.3.

Lemma 6.2.3 *The above functions f_1 and f_2 are continuous on the interval $[0, z_c)$.*

Proof. We begin with f_1. Since the subcritical two-point function decays exponentially by (1.3.14), $\|H_p\|_2^2$ is finite for $p < z_c$. This norm can be rewritten as a power series in p with positive coefficients, which therefore must have radius of convergence at least z_c. Hence it is continuous in $p \in [0, z_c)$.

For f_2, we fix $r \in [0, z_c)$. Arguing as in the derivation of (1.3.14), there is a constant M, depending on r but not on x, such that for any $p \in [0, r]$ and any x,

$$\frac{d}{dp} x_\mu^2 G_p(0, x) \leq M. \tag{6.2.10}$$

Hence for $p_1 < p_2 \leq r$ we have

$$
\begin{aligned}
0 &\leq f_2(p_2) - f_2(p_1) \\
&\leq (3K)^{-1} \Omega^{1-s-2/d} \sup_x x_\mu^2 [G_{p_2}(0, x) - G_{p_1}(0, x)] \\
&\leq (3K)^{-1} \Omega^{1-s-2/d} M(p_2 - p_1).
\end{aligned}
$$

This implies continuity of f_2 for $p < r$, and hence for $p < z_c$ since r is arbitrary. $\square$

Lemma 6.2.4 *For $p \in [0, \Omega^{-1}]$, $f_i(p) \leq 1/3$ for $i = 1, 2$.*

Proof. For $p \in [0, \Omega^{-1}]$, $G_p(0, x) \leq G_{1/\Omega}(0, x)$. Since in general the self-avoiding walk two-point function is bounded above by the ordinary random walk two-point function having the same activity, $G_p(0, x) \leq G_{1/\Omega}(0, x) \leq C(0, x)$. Now $H_p(0, x) = G_p(0, x) - \delta_{0,x}$, so $\|H_p\|_2^2 = \|G_p\|_2^2 - 1 \leq \|C\|_2^2 - 1$. Hence by the Parseval relation $\|H_p\|_2^2 \leq \|\hat{C}\|_2^2 - 1$, and the desired bound on f_1 follows from (6.2.7). For f_2 we use the Fourier transform to write

$$
\begin{aligned}
x_\mu^2 G_p(0, x) &\leq x_\mu^2 C(0, x) = -\int_{[-\pi, \pi]^d} \partial_\mu^2 \hat{C}(k) e^{-ik \cdot x} \frac{d^d k}{(2\pi)^d} \\
&= -\int_{[-\pi, \pi]^d} \left[\frac{\partial_\mu^2 \hat{D}}{(1 - \hat{D})^2} + \frac{2(\partial_\mu \hat{D})^2}{(1 - \hat{D})^3} \right] e^{-ik \cdot x} \frac{d^d k}{(2\pi)^d}. \tag{6.2.11}
\end{aligned}
$$

The desired bound then follows from (6.2.8). $\square$

This leaves the last and most substantial assumption of Lemma 6.2.1 to be shown. The following result confirms that the final hypothesis of

Lemma 6.2.1 is satisfied for f_1 and f_2 of (6.2.9), i.e. that for each $p \in [\Omega^{-1}, z_c)$, if $f_i(p) \leq 1$ ($i = 1, 2$) then in fact $f_i(p) \leq 2/3$ ($i = 1, 2$).

Remark. The next theorem states that a pair of inequalities implies a stronger pair. In conjunction with Lemmas 6.2.1, 6.2.3 and 6.2.4, this means that in fact the stronger pair of inequalities holds. Hence the weaker inequalities also hold, and any consequences of the weaker inequalities used in the course of the proof [such as the infrared bound (6.2.19)] will have been shown to hold, once the theorem is proved.

Theorem 6.2.5 *There is an Ω_0 such that for $\Omega \geq \Omega_0$ (with $d > 4$ for the spread-out model) the following implication holds. For any $p \in [\Omega^{-1}, z_c)$, if*

$$\|H_p\|_2^2 \leq 3K\Omega^{-1+s} \quad \text{and} \quad \|x_\mu^2 G_p\|_\infty \leq 3K\Omega^{-1+s+2/d} \qquad (6.2.12)$$

then in fact

$$\|H_p\|_2^2 \leq 2K\Omega^{-1+s} \quad \text{and} \quad \|x_\mu^2 G_p\|_\infty \leq 2K\Omega^{-1+s+2/d}. \qquad (6.2.13)$$

Here s is as in the statement of Lemma 6.2.2, and the $2/d$ in the exponent in the bound on $\|x_\mu^2 G_p\|_\infty$ can be omitted for the nearest-neighbour model.

Proof. We assume the weaker pair of bounds, and prove the stronger pair. For the proof we will work with Fourier transforms. As will be described in more detail below [in the paragraph containing (6.2.27)], the assumed bounds (6.2.12), together with Theorem 5.4.4, imply (absolute) convergence of the lace expansion. Hence by (5.2.18),

$$\hat{F}_p(k) \equiv \hat{G}_p(k)^{-1} = 1 - p\Omega\hat{D}(k) - \hat{\Pi}_p(k). \qquad (6.2.14)$$

Since $F_p(0) = \chi(p)^{-1} > 0$ for $p < z_c$, it follows by adding and subtracting $\hat{F}_p(0)$ to $\hat{F}_p(k)$ that for $p \geq \Omega^{-1}$

$$\begin{aligned} \hat{F}_p(k) &= \hat{F}_p(0) + p\Omega[1 - \hat{D}(k)] + \hat{\Pi}_p(0) - \hat{\Pi}_p(k) \\ &\geq [1 - \hat{D}(k)] + [\hat{\Pi}_p(0) - \hat{\Pi}_p(k)]. \end{aligned} \qquad (6.2.15)$$

The basic idea of the proof is that the assumed bounds imply that the second term on the right side is a small perturbation of the first, which in turn implies that $\hat{G}_p = 1/\hat{F}_p(k)$ is bounded above by a small perturbation of its ordinary random walk counterpart, and hence by Lemma 6.2.2 the improved bounds hold.

We now bound the difference $\hat{\Pi}_p(0) - \hat{\Pi}_p(k)$, using Theorem 5.4.4. It follows from (5.4.1), (5.2.16), symmetry, and (5.4.21) that

$$
\begin{aligned}
\hat{\Pi}_p(0) - \hat{\Pi}_p(k) \;\geq\;& -\sum_{j=1}^{\infty}[\hat{\Pi}_p^{(2j+1)}(0) - \hat{\Pi}_p^{(2j+1)}(k)] \\
\geq\;& -\sum_{\mu=1}^{d}(1 - \cos k_\mu)\|x_1^2 H_p\|_\infty \\
& \times \sum_{j=1}^{\infty}(j+1)^2\|H_p\|_2^{2j+1}\|G_p\|_2^{2j-1}.
\end{aligned}
\tag{6.2.16}
$$

For the norm $\|G_p\|_2$, we note that by definition $H_p(0,x) = G_p(0,x) - \delta_{0,x}$, and hence using (6.2.12) we have

$$\|G_p\|_2^2 = \|H_p\|_2^2 + 1 \leq 2 \tag{6.2.17}$$

for sufficiently large Ω. The right side of (6.2.16) is dominated for large Ω by the $j = 1$ term, and hence by (6.2.5) and (6.2.12) we have

$$\hat{\Pi}_p(0) - \hat{\Pi}_p(k) \geq -K_1\Omega^{-u}[1 - \hat{D}(k)], \tag{6.2.18}$$

where $u = 3/2$ for the nearest-neighbour model and $u = 5/2 - 5s/2 - 2/d$ for the spread-out model, and K_1 is a constant which is independent of L for the spread-out model and independent of d for the nearest-neighbour model. We will use K_1 as a "variable constant" in what follows, to denote various constants which are independent of L or d as in (6.2.18) and whose precise values are irrelevant. Substituting (6.2.18) into (6.2.15) gives the infrared bound

$$\hat{F}_p(k) \geq [1 - K_1\Omega^{-u}][1 - \hat{D}(k)]. \tag{6.2.19}$$

We are now in a position to obtain the improved bound on $\|H_p\|_2^2$. By the Parseval relation and (6.2.17),

$$\|H_p\|_2^2 = \|\hat{G}_p\|_2^2 - 1,$$

where the norm on the right side denotes the L^2 norm on $[-\pi, \pi]^d$ with measure $(2\pi)^{-d}d^dk$. Hence by (6.2.19) we have

$$
\begin{aligned}
\|H_p\|_2^2 \;=\;& \left\|\frac{1}{\hat{F}_p}\right\|_2^2 - 1 \\
\leq\;& (1 + K_1\Omega^{-u})\left\|\frac{1}{1 - \hat{D}}\right\|_2^2 - 1.
\end{aligned}
\tag{6.2.20}
$$

Applying (6.2.7), this gives

$$\|H_p\|_2^2 \leq (1 + K_1\Omega^{-u+1-s})K\Omega^{-1+s}. \tag{6.2.21}$$

For the nearest-neighbour model $-u + 1 - s = -1/2$, while for the spread-out model with $d > 4$, $-u + 1 - s = -3/2 + 3s/2 + 2/d < 0$. This gives the desired result that for Ω sufficiently large

$$\|H_p\|_2^2 \leq 2K\Omega^{-1+s}. \tag{6.2.22}$$

We turn now to the bound on $\|x_\mu^2 G_p\|_\infty$. We give the proof with the $2/d$ present in the exponent, but for the nearest-neighbour model this can be omitted by following the same proof. [The significant difference between the two models occurs in (6.2.31).]

In terms of the Fourier transform we can write

$$x_\mu^2 G_p(0, x) = -\int \partial_\mu^2 \hat{G}_p(k)e^{-ik\cdot x} \frac{d^d k}{(2\pi)^d}. \tag{6.2.23}$$

Explicit computation of the derivative on the right side gives the following expression, in which we have simplified the notation by dropping arguments and denoting partial differentiation with respect to k_μ by the subscript μ.

$$\hat{G}_{\mu,\mu} = p\Omega\frac{\hat{D}_{\mu,\mu}}{\hat{F}^2} + 2(p\Omega)^2\frac{\hat{D}_\mu^2}{\hat{F}^3} + \frac{\hat{\Pi}_{\mu,\mu}}{\hat{F}^2} + 4p\Omega\frac{\hat{D}_\mu\hat{\Pi}_\mu}{\hat{F}^3} + 2\frac{\hat{\Pi}_\mu^2}{\hat{F}^3}. \tag{6.2.24}$$

We insert (6.2.24) into (6.2.23), and take absolute values inside the integral and the sum of five terms. Applying (6.2.19) to bound $\hat{F}$ from below gives $\hat{F}^{-j} \leq (1 + K_1\Omega^{-u})(1 - \hat{D})^{-j}$ for $j \geq 1$. Applying (6.2.8) and using $1 \leq p\Omega$ then yields

$$
\begin{aligned}
x_\mu^2 G_p(0, x) \quad \leq \quad & (1 + K_1\Omega^{-u})(p\Omega)^2 \\
& \times \left[K\Omega^{-1+s+2/d} + \|\hat{\Pi}_{\mu,\mu}\|_\infty \left\| \frac{1}{1 - \hat{D}} \right\|_2^2 \right. \\
& \left. + 4\left\| \frac{\hat{D}_\mu\hat{\Pi}_\mu}{(1 - \hat{D})^3} \right\|_1 + 2\left\| \frac{\hat{\Pi}_\mu^2}{(1 - \hat{D})^3} \right\|_1 \right]. \tag{6.2.25}
\end{aligned}
$$

The last three terms on the right side are error terms. Before bounding these, we first bound the factor $p\Omega$. By (6.2.12),

$$\|H_p\|_\infty \leq \|x_\mu^2 G_p\|_\infty \leq 3K\Omega^{-1+s+2/d}; \tag{6.2.26}$$

this follows from the facts that $H_p(0, 0) = 0$, and for $x \neq 0$, $H_p(0, x) = G_p(0, x)$ and $1 \leq x_\mu^2$ for some μ. (This bound on $\|H_p\|_\infty$ is inefficient for the spread-out model, for which the factor $\Omega^{2/d}$ on the right side should not be necessary, but it is adequate for our needs.) Applying (6.2.26) and (6.2.12) to (5.4.18) and (5.4.20), we see that for sufficiently large Ω the lace expansion converges and

$$|\hat{\Pi}_p(k)| \leq p\Omega K_1 \Omega^{-1+s+2/d}. \qquad (6.2.27)$$

Since $\chi(p)^{-1} = 1 - p\Omega - \hat{\Pi}_p(0) > 0$,

$$p\Omega \leq 1 - \hat{\Pi}_p(0) \leq 1 + p\Omega K_1 \Omega^{-1+s+2/d},$$

so that for Ω sufficiently large

$$p \leq \Omega^{-1}[1 + K_1 \Omega^{-1+s+2/d}]. \qquad (6.2.28)$$

Since $-u \leq -1 + s + 2/d$, the factor $(1 + K_1 \Omega^{-u})(p\Omega)^2$ in (6.2.25) can be replaced by $1 + K_1 \Omega^{-1+s+2/d}$, for Ω large.

We next consider bounds on the derivatives of $\hat{\Pi}_p$ appearing in (6.2.25). It follows from (6.2.12) and (5.4.20) that

$$|\partial_\mu^2 \hat{\Pi}_p(k)| \leq K_1 \Omega^{-2+2s+2/d} \qquad (6.2.29)$$

(the $N = 2$ loop term dominates). We also will need a bound on $\partial_\mu \hat{\Pi}_p(k)$. Since by symmetry this derivative is zero whenever $k_\mu = 0$, it follows from Taylor's Theorem and the above bound on the second derivative that

$$|\partial_\mu \hat{\Pi}_p(k)| \leq K_1 \Omega^{-2+2s+2/d}|k_\mu|. \qquad (6.2.30)$$

Similarly,

$$|\partial_\mu D(k)| \leq K_1 \Omega^{2/d}|k_\mu|. \qquad (6.2.31)$$

Turning now to the three error terms in (6.2.25), for the first we use (6.2.29) and (6.2.7) to bound it above by $K_1 \Omega^{-2+2s+2/d}$. For the other two terms we first note that by symmetry, (6.2.5) and (6.2.7),

$$\left\| \frac{k_\mu^2}{(1-\hat{D})^3} \right\|_1 \leq K_1. \qquad (6.2.32)$$

Hence by (6.2.30) and (6.2.31) the second error term is bounded above by

$$K_1 \Omega^{2/d} \Omega^{-2+2s+2/d} \left\| \frac{k_\mu^2}{(1-\hat{D})^3} \right\|_1 \leq K_1 \Omega^{-2+2s+4/d}. \qquad (6.2.33)$$

Finally, the last error term can be bounded above by $K_1 \Omega^{-4+4s+4/d}$, using (6.2.30) and then (6.2.32). Taking Ω sufficiently large then gives the desired result

$$x_\mu^2 G_p(0, x) \leq 2K\Omega^{-1+s+2/d}. \tag{6.2.34}$$

$\square$

The following results, which follow relatively easily from Theorem 6.2.5, will be fundamental in the rest of the chapter.

Corollary 6.2.6 *For $\Omega \geq \Omega_0$ (with $d > 4$ for the spread-out model),*

$$\begin{aligned}
\|H_z\|_\infty &\leq 2K\Omega^{-1+s+2/d}, \\
\|x_\mu^2 G_z\|_\infty &\leq 2K\Omega^{-1+s+2/d},
\end{aligned}$$

and

$$\|H_z\|_2^2 \leq 2K\Omega^{-1+s}$$

for all complex z in the closed disk $|z| \leq z_c$. Here s is as in the statement of Lemma 6.2.2, and for the nearest-neighbour model the $2/d$ can be omitted from the exponent in the first two inequalities.

Proof. Since the left sides are largest at $z = z_c$, we can restrict attention to this case. The left sides are monotone increasing in real positive z, and satisfy the above bounds uniformly in $z < z_c$ by Theorem 6.2.5 (see (6.2.26) and the Remark preceding Theorem 6.2.5). Therefore the same bounds hold at $z = z_c$ by the monotone convergence theorem. $\square$

Corollary 6.2.7 *For $\Omega \geq \Omega_0$ (with $d > 4$ for the spread-out model), there is a constant K_1 such that the following bounds hold uniformly in $k \in [-\pi, \pi]^d$ and $|z| \leq z_c$:*

$$\begin{aligned}
|\hat{\Pi}_z(k)| &\leq K_1 \Omega^{-1+s+2/d} \\
|\partial_\mu \hat{\Pi}_z(k)| &\leq K_1 \Omega^{-2+2s+2/d}|k_\mu| \\
|\partial_\mu^2 \hat{\Pi}_z(k)| &\leq K_1 \Omega^{-2+2s+2/d}.
\end{aligned}$$

In fact the series representations of these quantities are bounded absolutely (absolute values inside sums over x, N) and uniformly by the right sides. The critical point obeys

$$\Omega^{-1} \leq z_c \leq \Omega^{-1}[1 + K_1 \Omega^{-1+s+2/d}].$$

Also,

$$1 - z_c\Omega - \hat{\Pi}_{z_c}(0) = 0.$$

For any $p \in [0, z_c]$

$$\hat{\Pi}_p(0) - \hat{\Pi}_p(k) \geq -K_1 \Omega^{-u}[1 - \hat{D}(k)]$$

and for any $p \in [\Omega^{-1}, z_c]$

$$\hat{F}_p(k) \geq [1 - K_1 \Omega^{-u}][1 - \hat{D}(k)].$$

Here s is as in the statement of Lemma 6.2.2, and for the nearest-neighbour model the $2/d$ can be omitted from the exponents in the first four inequalities. The exponent u is equal to $3/2$ for the nearest-neighbour model, while for the spread-out model $u = 5/2 - 5s/2 - 2/d$.

Proof.   Given Corollary 6.2.6, the first four inequalities follow exactly as in the proof of Theorem 6.2.5. It then follows from the dominated convergence theorem that for $u \in \{0, 1, 2\}$, $\partial_\mu^u \hat{\Pi}_z(k)$ is continuous on the closed disk $|z| \leq z_c$. Since $\chi(p) \to \infty$ as $p \nearrow z_c$ by (1.3.6), we have

$$\chi(p)^{-1} = \hat{F}_p(0) \to 1 - z_c \Omega - \hat{\Pi}_{z_c}(0) = 0.$$

The last two bounds of the corollary follow from (6.2.18) and (6.2.19) for $p < z_c$, and then follow at z_c by taking the limit. $\square$

By Corollary 6.2.7 and the fact that $-\nabla_k^2 \hat{D}(0) \geq 1$, there is a constant C_4 such that for Ω sufficiently large and $p \in [\Omega^{-1}, z_c]$,

$$\begin{aligned}
\nabla_k^2 \hat{F}_p(0) &= -p\Omega\nabla_k^2 \hat{D}(0) - \nabla_k^2 \hat{\Pi}_p(0) \\
&\geq C_4 > 0.
\end{aligned} \qquad (6.2.35)$$

The following lemma will allow for bounds on $\partial_z \hat{\Pi}_z(k)$ in the closed disk $|z| \leq z_c$.

Lemma 6.2.8 *For any $p \in (0, z_c]$ and $m = 1, 2, 3, \ldots$,*

$$\partial_p^m G_p(0, x) \leq m!\, p^{-m}\, H_p * \cdots * H_p * G_p(x), \qquad (6.2.36)$$

where there are m factors of H_p in the convolution.

Proof.   By definition,

$$\partial_p^m G_p(0, x) = m!\, p^{-m} \sum_{\substack{\omega\, :\, 0\, \to\, x \\ |\omega| \geq m}} \binom{|\omega|}{m} p^{|\omega|},$$

where the sum is over all self-avoiding walks from 0 to x. The binomial coefficient on the right side counts the number of ways to choose $0 < i_1 <$

$i_2 < \cdots < i_m \leq |\omega|$, so it is also the number of ways to break ω into $m + 1$ pieces such that the first m pieces each consist of at least one step. The upper bound then follows by neglecting the mutual avoidance between these pieces. $\qquad\square$

Theorem 6.2.9 *For Ω sufficiently large (with $d > 4$ for the spread-out model),*

$$|\partial_z \hat{\Pi}_z(k)| \leq 7K\Omega^{s+2/d} \qquad (6.2.37)$$

uniformly in $k \in [-\pi, \pi]^d$ and $|z| \leq z_c$. In fact the series representation of the left side is bounded absolutely (absolute values inside sums over x and N) and uniformly by the right side. Here K is the constant of Lemma 6.2.2, s is as in the statement of Lemma 6.2.2, and the $2/d$ can be omitted from the exponent for the nearest-neighbour model. Hence for Ω sufficiently large there is a positive constant C_3 such that for any $p \in (0, z_c]$

$$-\partial_z \hat{F}_p(0) = \Omega + \partial_z \hat{\Pi}_p(0) \geq C_3 > 0 \qquad (6.2.38)$$

Proof. The bound (6.2.38) clearly follows from (6.2.37), so it suffices to obtain (6.2.37). But by (5.4.18), (5.4.19), Corollary 6.2.6 and the upper bound on $z_c\Omega$ of Corollary 6.2.7, to prove (6.2.37) it suffices to show that

$$\||\partial_z|_{z=|z|} H_z\|_\infty \leq 4K\Omega^{s+2/d}. \qquad (6.2.39)$$

Since $H_p(0, x)$ is a power series with nonnegative coefficients, it suffices to obtain (6.2.39) at $z = z_c$. By Lemma 6.2.8 and the fact that $G_z(0, x) = H_z(0, x) + \delta_{0,x}$,

$$\begin{aligned}
\partial_z H_{z_c}(0, x) = \partial_z G_{z_c}(0, x) &\leq z_c^{-1} H_{z_c} * H_{z_c}(x) + z_c^{-1} H_{z_c}(0, x) \\
&\leq z_c^{-1} \|H_{z_c}\|_2^2 + z_c^{-1} H_{z_c}(0, x).
\end{aligned}$$

The desired result now follows from Corollary 6.2.6 and the fact that z_c is bounded below by Ω^{-1}. $\qquad\square$

We conclude this section with an upper bound on the susceptibility, which in particular implies that it is finite in the closed disk $|z| \leq z_c$ everywhere except at the critical point itself.

Theorem 6.2.10 *For Ω sufficiently large (with $d > 4$ for the spread-out model), the inverse susceptibility $\hat{F}_z(0) = 1 - z\Omega - \hat{\Pi}_z(0)$ satisfies*

$$|\hat{F}_z(0)| \geq \frac{\Omega}{2}|z_c - z| \qquad (6.2.40)$$

for all z with $|z| \leq z_c$.

Proof. Let $|z| \le z_c$. By Corollary 6.2.7 $\hat{F}_{z_c}(0) = 0$ and hence

$$
\begin{aligned}
|\hat{F}_z(0)| &= \left| \int_{z_c}^{z} \partial_z \hat{F}_z(0) dz \right| \\
&= |z_c - z| \left| \Omega + \int_0^1 \partial_z \hat{\Pi}_{(1-t)z_c + tz}(0) dt \right|.
\end{aligned}
\tag{6.2.41}
$$

The lemma then follows, using Theorem 6.2.9.   □

6.2.3   Proof of Theorem 6.1.2

The critical bubble diagram $\mathsf{B}(z_c) = \|G_{z_c}\|_2^2 = 1 + \|H_{z_c}\|_2^2$ is finite by Corollary 6.2.6. It follows from Theorem 1.5.3 that $\bar{\gamma} = 1$, in the sense that there are positive constants c_1 and c_2 such that for all $p < z_c$,

$$
c_1(z_c - p)^{-1} \le \chi(p) \le c_2(z_c - p)^{-1}.
\tag{6.2.42}
$$

To obtain the stronger *asymptotic* behaviour stated in Theorem 6.1.2, we observe that since $\hat{F}_{z_c}(0) = 1 - z_c\Omega - \hat{\Pi}_{z_c}(0) = 0$ by Corollary 6.2.7,

$$
\begin{aligned}
\chi(z) &= \frac{1}{\hat{F}_z(0) - \hat{F}_{z_c}(0)} \\
&= \left(\frac{1}{z_c - z} \right) \left(\Omega + \frac{\hat{\Pi}_{z_c}(0) - \hat{\Pi}_z(0)}{z_c - z} \right)^{-1}.
\end{aligned}
\tag{6.2.43}
$$

It then follows from Theorem 6.2.9 that as $z \nearrow z_c$

$$
\chi(z) \sim [\Omega + \partial_z \hat{\Pi}_{z_c}(0)]^{-1} (z_c - z)^{-1}.
\tag{6.2.44}
$$

Defining

$$
A = z_c^{-1} [\Omega + \partial_z \hat{\Pi}_{z_c}(0)]^{-1}
\tag{6.2.45}
$$

gives the statement of Theorem 6.1.2 for the susceptibility.

For the correlation length of order 2, we note that by symmetry and direct calculation,

$$
\xi_2(z)^2 = \frac{-\nabla_k^2 \hat{G}_z(0)}{\hat{G}_z(0)} = [-z\Omega \nabla_k^2 \hat{D}(0) - \nabla_k^2 \hat{\Pi}_z(0)]\chi(z).
\tag{6.2.46}
$$

The desired asymptotic behaviour of $\xi_2(z)$ now follows from the asymptotic behaviour of $\chi(z)$ and (6.2.35), if we define

$$
D = A[-z_c\Omega \nabla_k^2 \hat{D}(0) - \nabla_k^2 \hat{\Pi}_{z_c}(0)].
\tag{6.2.47}
$$

[Continuity at z_c of $\nabla_k^2 \hat{\Pi}_z(0)$ is discussed in the proof of Corollary 6.2.7.]

□

We end this section with bounds on the constants A and D, for simplicity restricting the discussion to the nearest-neighbour model in high dimensions.

Proposition 6.2.11 *For the nearest-neighbour model with d sufficiently large, there are positive universal constants c_1, c_2, c_3 such that*

$$1 \le A \le 1 + c_1 d^{-1} \text{ and } 1 + c_2 d^{-1} \le D \le 1 + c_3 d^{-1}.$$

In particular D is strictly greater than 1.

Proof. For the first bound we conclude from Theorem 1.5.3 that $1 \le A \le B(z_c)$. But by Corollary 6.2.6, $B(z_c) \le 1 + c_1 d^{-1}$ for some constant c_1.

For the bound on the diffusion constant D, we have from (6.2.47) and (6.2.45) that

$$D = \frac{1 - (2dz_c)^{-1} \nabla_k^2 \hat{\Pi}_{z_c}(0)}{1 + (2d)^{-1} \partial_z \hat{\Pi}_{z_c}(0)}. \tag{6.2.48}$$

It suffices to show that there are positive constants a_i such that

$$-a_1 d^{-3/2} \le -(2dz_c)^{-1} \nabla_k^2 \hat{\Pi}_{z_c}(0) \le a_2 d^{-1} \tag{6.2.49}$$

and

$$-a_3 d^{-1} \le (2d)^{-1} \partial_z \hat{\Pi}_{z_c}(0) \le -a_4 d^{-1}. \tag{6.2.50}$$

Beginning with (6.2.49), it follows from Corollary 6.2.7 and the fact that $2dz_c \ge 1$ that

$$|(2dz_c)^{-1} \nabla_k^2 \hat{\Pi}_{z_c}(0)| \le a_2 d^{-1}. \tag{6.2.51}$$

This gives the upper bound of (6.2.49). For the lower bound, by symmetry it can be concluded that for fixed μ

$$-\nabla_k^2 \hat{\Pi}_{z_c}(0) \ge -d \sum_{j=1}^{\infty} \sum_x x_\mu^2 \Pi_{z_c}^{(2j+1)}(0, x). \tag{6.2.52}$$

By (5.4.20) and Corollary 6.2.6 the right side is bounded below by a multiple of $-d^{-3/2}$.

Turning now to (6.2.50), the lower bound follows immediately from (6.2.37). For the upper bound, we write

$$\partial_z \hat{\Pi}_{z_c}(0) = -\partial_z \hat{\Pi}_{z_c}^{(1)}(0) + \sum_{N=2}^{\infty} (-1)^N \partial_z \hat{\Pi}_{z_c}^{(N)}(0). \tag{6.2.53}$$

The first term on the right side (with its minus sign) is bounded above by the contribution due to the walk which steps to a neighbour of the origin and then back to the origin, which is $-\partial_z(2dz^2) = -4dz_c \leq -2$. Thus it suffices to show that the second term on the right side is bounded in absolute value by a multiple of d^{-1}. This follows from Corollary 6.2.6 and the bound $\|\partial_z H_{z_c}\|_\infty \leq K_1$ of (6.2.39), together with (5.4.19). $\quad\square$

6.3 Fractional derivatives

In this section we describe some elementary properties of what we term fractional derivatives. This terminology is somewhat inaccurate, but is useful in a suggestive sense in the analysis of the large-n asymptotics of power series coefficients. Given a power series $f(z) = \sum_{n=0}^{\infty} a_n z^n$ and $\epsilon \geq 0$, we define the fractional derivative

$$\delta_z^\epsilon f(z) = \sum_{n=0}^{\infty} n^\epsilon a_n z^n. \tag{6.3.1}$$

Note that for ϵ equal to a positive integer, δ_z^ϵ does not give the usual derivative. We will use (6.3.1) with $\epsilon \in (0,1)$. Allowing ϵ to take on arbitrary negative values defines a relative of the antiderivative, as follows. For $\alpha > 0$ we define

$$\delta_z^{-\alpha} f(z) = \sum_{n=1}^{\infty} n^{-\alpha} a_n z^n. \tag{6.3.2}$$

Both of the above quantities will be finite at least strictly within the circle of convergence of $f(z)$.

The following lemma provides formulas which are convenient for estimating fractional derivatives.

Lemma 6.3.1 *Let $f(z) = \sum_{n=0}^{\infty} a_n z^n$ have radius of convergence R. Then for any z with $|z| < R$, and for any $\alpha > 0$,*

$$\delta_z^{-\alpha} f(z) = C_\alpha \int_0^\infty [f(ze^{-\lambda^{1/\alpha}}) - f(0)] d\lambda, \tag{6.3.3}$$

where $C_\alpha = [\alpha\Gamma(\alpha)]^{-1}$. In addition, for any z with $|z| < R$ and for any $\epsilon \in (0,1)$,

$$\delta_z^\epsilon f(z) = C_{1-\epsilon} z \int_0^\infty f'(ze^{-\lambda^{1/(1-\epsilon)}}) e^{-\lambda^{1/(1-\epsilon)}} d\lambda. \tag{6.3.4}$$

The identities (6.3.3) and (6.3.4) also hold for $z = R$, if $a_n \geq 0$.

Proof. Let $|z| < R$. We first note that for any $\alpha > 0$,

$$n^{-\alpha} = \frac{1}{\alpha \Gamma(\alpha)} \int_0^\infty e^{-n\lambda^{1/\alpha}} d\lambda, \qquad (6.3.5)$$

as can be seen by making the substitution $y = n\lambda^{1/\alpha}$ in the integral on the right side. Therefore

$$\sum_{n=1}^\infty n^{-\alpha} a_n z^n = C_\alpha \sum_{n=1}^\infty a_n \int_0^\infty (ze^{-\lambda^{1/\alpha}})^n d\lambda. \qquad (6.3.6)$$

Since the right side converges absolutely the order of integration and summation can be interchanged to yield (6.3.3).

For (6.3.4), we write $n^\epsilon = n^{-(1-\epsilon)} n$ and use (6.3.5) with $\alpha = 1 - \epsilon$ to obtain

$$\sum_{n=0}^\infty n^\epsilon a_n z^n = C_{1-\epsilon} z \sum_{n=1}^\infty n a_n \int_0^\infty (ze^{-\lambda^{1/(1-\epsilon)}})^{n-1} e^{-\lambda^{1/(1-\epsilon)}} d\lambda. \qquad (6.3.7)$$

Since the right side converges absolutely we can interchange the order of summation and integration to obtain

$$\sum_{n=0}^\infty n^\epsilon a_n z^n = C_{1-\epsilon} z \int_0^\infty f'(ze^{-\lambda^{1/(1-\epsilon)}}) e^{-\lambda^{1/(1-\epsilon)}} d\lambda. \qquad (6.3.8)$$

Now suppose that $a_n \geq 0$ and take $z = R$. Then the above interchanges of sum and integral are justified by Fubini's Theorem. $\qquad\square$

The following lemma provides an error estimate analogous to the error estimate in Taylor's theorem. In applications of the lemma, R will be the radius of convergence of f.

Lemma 6.3.2 *Let $\epsilon \in (0,1)$ and let $f(z) = \sum_{n=0}^\infty a_n z^n$. Let $R > 0$ and suppose that $A_\epsilon \equiv \sum_{n=0}^\infty n^\epsilon |a_n| R^{n-\epsilon} < \infty$, so in particular $f(z)$ converges for $|z| \leq R$. Then for any z with $|z| \leq R$,*

$$|f(z) - f(R)| \leq 2^{1-\epsilon} A_\epsilon |R - z|^\epsilon. \qquad (6.3.9)$$

Suppose that $B_\epsilon \equiv \sum_{n=1}^\infty n^{1+\epsilon} |a_n| R^{n-1-\epsilon} < \infty$, so in particular $f'(z) = \sum_{n=1}^\infty n a_n z^{n-1}$ converges for $|z| \leq R$. Then for any z with $|z| \leq R$,

$$|f(z) - f(R) - f'(R)(z - R)| \leq \frac{2^{1-\epsilon}}{1+\epsilon} B_\epsilon |R - z|^{1+\epsilon}. \qquad (6.3.10)$$

Proof. We just give the proof of (6.3.10). The proof of (6.3.9) is similar and simpler. By definition,

$$f(z) - f(R) - f'(R)(z-R) = (z-R) \sum_{n=1}^{\infty} a_n R^{n-1} \sum_{j=0}^{n-1} \left[\left(\frac{z}{R} \right)^j - 1 \right]. \quad (6.3.11)$$

But in general

$$w^j - 1 = (w-1)^\epsilon \left[\frac{w^j - 1}{w-1} \right]^\epsilon (w^j - 1)^{1-\epsilon}$$

$$= (w-1)^\epsilon \left[\sum_{m=0}^{j-1} w^m \right]^\epsilon (w^j - 1)^{1-\epsilon}. \quad (6.3.12)$$

Taking absolute values in (6.3.12) and using $w = z/R$ and $|w| \leq 1$ gives

$$\left| \left(\frac{z}{R} \right)^j - 1 \right| \leq |z - R|^\epsilon j^\epsilon 2^{1-\epsilon} R^{-\epsilon}. \quad (6.3.13)$$

Since $\sum_{j=0}^{n-1} j^\epsilon \leq (1+\epsilon)^{-1} n^{1+\epsilon}$, (6.3.10) follows from (6.3.11) and (6.3.13).
□

The intuition behind the following lemma is that if a power series with radius of convergence R behaves like $|R - z|^{-b}$ near $z = R$, for some $b \geq 1$, then roughly speaking it should have coefficient of z^n not much worse than order $R^{-n} n^{b-1}$.

Lemma 6.3.3 *Let $f(z) = \sum_{n=0}^{\infty} a_n z^n$ have radius of convergence greater than or equal to $R > 0$.*
(i) Suppose that for $|z| < R$, $|f(z)| \leq const.|R - z|^{-b}$ for some $b \geq 1$. Then $|a_n| \leq O(R^{-n} n^\alpha)$, for any $\alpha > b - 1$.
(ii) If for some $b \geq 1$ a bound on the derivative of the form $|f'(z)| \leq const.|R - z|^{-b}$ holds for every $|z| < R$, then $|a_n| \leq O(R^{-n} n^{-\alpha})$ for any $\alpha < 2 - b$.

Proof. (i) Fix $b \geq 1$ and let $\alpha > b - 1$. Since $n^{-\alpha} a_n$ is the coefficient of z^n in the fractional antiderivative $\delta_z^{-\alpha} f(z)$,

$$n^{-\alpha} a_n = \frac{1}{2\pi i} \oint \delta_z^{-\alpha} f(z) \frac{dz}{z^{n+1}}, \quad (6.3.14)$$

where the integral is around a circle of radius $r < R$ centred at the origin. By Lemma 6.3.1,

$$n^{-\alpha} |a_n| \leq const. r^{-n} \int_{-\pi}^{\pi} d\theta \int_0^{\infty} d\lambda |f(re^{i\theta} e^{-\lambda^{1/\alpha}}) - f(0)|. \quad (6.3.15)$$

Since $f(z) - f(0) = O(|z|)$ for z near zero, the contribution to the integral with respect to λ due to $\lambda \in [1, \infty)$ is finite. Using the assumed bound on $f(z)$, we thus have

$$n^{-\alpha}|a_n| \leq \text{const.} r^{-n} \left[1 + \int_{-\pi}^{\pi} d\theta \int_0^1 d\lambda |R - re^{i\theta}e^{-\lambda^{1/\alpha}}|^{-b} \right].$$

Replacing the R on the right side by r gives an upper bound. Taking the limit $r \to R$ in the upper bound leads to

$$n^{-\alpha}|a_n| \leq \text{const.} R^{-n-b} \left[1 + \int_{-\pi}^{\pi} d\theta \int_0^1 d\lambda |1 - e^{i\theta}e^{-\lambda^{1/\alpha}}|^{-b} \right]. \quad (6.3.16)$$

To check that the integral on the right side is finite, it suffices to show that the corresponding quantity with limits of integration $\theta = \pm 1$ is finite (or any other small finite interval containing $\theta = 0$). Thus it suffices to verify that

$$\int_0^1 d\theta \int_0^1 d\lambda |1 - e^{i\theta}e^{-\lambda^{1/\alpha}}|^{-b} < \infty. \quad (6.3.17)$$

As we now show, it is an exercise in calculus to see that the left side is bounded for $\alpha > b - 1 \geq 0$.

Making the substitution $u = \lambda^{1/\alpha}$ and writing the absolute value on the right side as the square root of the sum of the squares of its real and imaginary parts leads to an upper bound for (6.3.17) of the form

$$\int_0^1 d\theta \int_0^1 du\, u^{\alpha-1}[(1 - e^{-u})^2 + e^{-2u}\theta^2]^{-b/2}. \quad (6.3.18)$$

The change of variables $\theta_1 = \theta e^{-u}/(1 - e^{-u})$ in (6.3.18) gives

$$\int_0^1 du\, u^{\alpha-1} \frac{1 - e^{-u}}{e^{-u}} (1 - e^{-u})^{-b} \int_0^{e^{-u}/(1-e^{-u})} d\theta_1 [1 + \theta_1^2]^{-b/2}. \quad (6.3.19)$$

The θ_1-integral is bounded uniformly in u if $b > 1$, while if $b = 1$ it is finite for u near 1 and $O(|\log u|)$ for u near 0. Hence for $b \geq 1$, (6.3.19) is bounded above by a multiple of

$$\int_0^1 du\, u^{\alpha-b} |\log u|, \quad (6.3.20)$$

which is finite for $\alpha > b - 1$.

(ii) Given the bound on the derivative, it follows from (i) that $|na_n| \leq O(R^{-n}n^p)$ for any $p > b - 1$. Therefore $|a_n| \leq O(R^{-n}n^{p-1})$ for any $\alpha \equiv 1 - p < 2 - b$. $\qquad \square$

Remark. The hypothesis $b \geq 1$ in Lemma 6.3.3(i) is not artificial. For example, let $f(z) = \sum_{n=1}^{\infty} n^{-2} z^{2^n}$. Then $f(z)$ is finite for $|z| \leq 1$ so in particular $|f(z)| \leq \text{const.} |1 - z|^{-b}$ for any $b \in [0, 1)$. However $a_N = [\log_2 N]^{-2}$ for $N = 2^n$, so $a_n \neq O(n^{b-1+\epsilon})$ for $\epsilon \in (0, 1 - b)$.

The following lemma is a kind of Tauberian theorem, in which information more detailed than merely the asymptotic form of a power series near its singularity provides information about the large-n asymptotics of the coefficients of the power series.

Lemma 6.3.4 *Let*

$$f(z) = \frac{1}{\varphi(z)} = \sum_{n=0}^{\infty} b_n z^n,$$

where $\varphi(z) = \sum_{n=0}^{\infty} a_n z^n$. Suppose that for some $\epsilon \in (0, 1)$

$$\sum_{n=0}^{\infty} n^{1+\epsilon} |a_n| R^n < \infty,$$

so in particular $\varphi(z)$ and $\varphi'(z)$ are finite when $|z| = R$. Assume in addition that $\varphi'(R) \neq 0$. Suppose that $\varphi(R) = 0$ and that $\varphi(z) \neq 0$ for $|z| \leq R, z \neq R$. Then

$$f(z) = \frac{1}{-\varphi'(R)} \frac{1}{R - z} + O(|R - z|^{\epsilon-1}) \qquad (6.3.21)$$

uniformly in $|z| \leq R$, and

$$b_n = R^{-n-1} \left[\frac{1}{-\varphi'(R)} + O(n^{-\alpha}) \right] \quad \text{as } n \to \infty, \qquad (6.3.22)$$

for every $\alpha < \epsilon$.

Proof. Since $\varphi(R) = 0$,

$$f(z) = -\frac{1}{\varphi(R) - \varphi(z)} \qquad (6.3.23)$$

$$= -\frac{1}{\varphi'(R)(R - z) + [\varphi(R) - \varphi(z) - \varphi'(R)(R - z)]}.$$

Let

$$h(z) = \frac{\varphi(R) - \varphi(z) - \varphi'(R)(R - z)}{R - z} \qquad (6.3.24)$$

and

$$\psi(z) = -\frac{h(z)}{\varphi'(R) + h(z)} = -1 - \frac{\varphi'(R)}{\varphi(z)}(R - z). \qquad (6.3.25)$$

Then ψ is analytic in $|z| < R$. Also,

$$f(z) = -\frac{1}{\varphi'(R)} \frac{1}{R - z}[1 + \psi(z)]. \qquad (6.3.26)$$

Since $h(z) = O(|R - z|^\epsilon)$ uniformly in $|z| \leq R$ by Lemma 6.3.2, it is also the case that $\psi(z) = O(|R - z|^\epsilon)$ uniformly in $|z| \leq R$. This proves (6.3.21).

Let C_r be the circle of radius r centred at the origin and oriented counterclockwise. The coefficient b_n is given by the contour integral

$$b_n = \frac{1}{2\pi i} \int_{C_{R/2}} \frac{f(z)}{z^{n+1}} dz, \qquad (6.3.27)$$

so by (6.3.26)

$$b_n = -\frac{1}{\varphi'(R)} \left[\frac{1}{R^{n+1}} + \frac{1}{2\pi i} \int_{C_{R/2}} \frac{\psi(z)}{(R - z)z^{n+1}} dz \right]. \qquad (6.3.28)$$

It remains to show that the second term in (6.3.28) gives a correction of the desired size.

We use statement (ii) of Lemma 6.3.3 for the correction term, as follows. A straightforward calculation using the bound on the $(1 + \epsilon)$-derivative of φ assumed in the statement of the lemma, together with Lemma 6.3.2, gives

$$\left| \frac{d}{dz} \frac{\psi(z)}{R - z} \right| \leq O(|R - z|^{\epsilon - 2}) \qquad (6.3.29)$$

uniformly in $|z| \leq R$. Hence the coefficient of z^n of $(R - z)^{-1}\psi(z)$ is bounded above by $O(R^{-n}n^{-\alpha})$, for every $\alpha < \epsilon$, by Lemma 6.3.3(ii). This gives the required bound on the second term of (6.3.28). $\qquad \square$

6.4 c_n and the mean-square displacement

This section consists of two parts. In the first part we obtain bounds on fractional derivatives involving $\hat{\Pi}_z(k)$, and then in the second part these bounds are used in conjunction with the results of Section 6.3 to prove Theorem 6.1.1.

6.4.1 Fractional derivatives of the two-point function

We begin by obtaining bounds on norms of fractional derivatives of the two-point function. Bounds on fractional derivatives of $\hat{\Pi}_z(k)$ are then obtained, using a generalization of Theorem 5.4.4 involving fractional z-derivatives.

The results of this section hold for finite or infinite memory, subject to the replacement of z_c by the finite memory critical point $z_c(0; \tau)$ in all occurrences. We use K_2 and c in this section to denote constants which may depend on Ω, and which may change from one occurrence to the next. They are however independent of the memory.

For $\lambda \geq 0$ we define

$$p_\lambda = z_c e^{-\lambda^{1/(1-\epsilon)}}, \tag{6.4.1}$$

and as usual we write

$$\hat{F}_z(k) = \frac{1}{\hat{G}_z(k)} = 1 - z\Omega\hat{D}(k) - \hat{\Pi}_z(k). \tag{6.4.2}$$

The following lemma will be used to bound norms of fractional derivatives of the two-point function.

Lemma 6.4.1 *For Ω sufficiently large (with $d > 4$ for the spread-out model), there is a positive constant c such that for any k or λ*

$$\hat{F}_{p_\lambda}(k) \geq c[1 - e^{-\lambda^{1/(1-\epsilon)}}\hat{D}(k)]. \tag{6.4.3}$$

Proof. Since $\hat{F}_{z_c}(0) = 0$,

$$
\begin{aligned}
\hat{F}_{p_\lambda}(k) &= [\hat{F}_{p_\lambda}(k) - \hat{F}_{p_\lambda}(0)] + [\hat{F}_{p_\lambda}(0) - \hat{F}_{z_c}(0)] \tag{6.4.4}\\
&= p_\lambda\Omega[1 - \hat{D}(k)] + [\hat{\Pi}_{p_\lambda}(0) - \hat{\Pi}_{p_\lambda}(k)] + \int_{p_\lambda}^{z_c}[-\partial_p\hat{F}_p(0)]dp.
\end{aligned}
$$

By Theorem 6.2.9,

$$\int_{p_\lambda}^{z_c}[-\partial_p\hat{F}_p(0)]dp \geq C_3(z_c - p_\lambda). \tag{6.4.5}$$

Also, by Corollary 6.2.7,

$$\hat{\Pi}_{p_\lambda}(0) - \hat{\Pi}_{p_\lambda}(k) \geq -K_1\Omega^{-1/2}[1 - \hat{D}(k)]. \tag{6.4.6}$$

Take $\Omega \geq \max\{4C_3/3, 64K_1^2\}$, and consider first the case of λ bounded away from infinity in such a way that $p_\lambda \geq 4K_1\Omega^{-3/2}$. Then by (6.4.4)–(6.4.6)

$$
\begin{aligned}
\hat{F}_{p_\lambda}(k) &\geq \frac{3p_\lambda\Omega}{4}[1 - \hat{D}(k)] + C_3(z_c - p_\lambda) \\
&\geq C_3 z_c[1 - e^{-\lambda^{1/(1-\epsilon)}}\hat{D}(k)], \tag{6.4.7}
\end{aligned}
$$

which gives (6.4.3) for this range of λ. For λ such that $p_\lambda \leq 4K_1\Omega^{-3/2}$, we have $p_\lambda \leq (2\Omega)^{-1}$, and so we use

$$\hat{G}_{p_\lambda}(k) \leq \hat{G}_{p_\lambda}(0)$$

and bound the right side by the ordinary random walk susceptibility at $p = (2\Omega)^{-1}$, which is finite. Therefore $\hat{F}_{p_\lambda}(k)$ is bounded below by a constant, and so (6.4.3) holds (decreasing c if necessary).   □

We are now able to obtain bounds on fractional derivatives of the two-point function.

Theorem 6.4.2 *For Ω sufficiently large (with $d > 4$ for the spread-out model) there is a positive constant K_2 (which may depend on ϵ and Ω) such that for any $p \in [0, z_c]$,*

$$\|\delta_p^\epsilon \partial_p G_p\|_\infty \leq K_2 \quad \text{if} \quad 0 < \epsilon < \min\{(d-4)/2, 1\}, \tag{6.4.8}$$

$$\|\delta_p^\epsilon G_p\|_2 \leq K_2 \quad \text{if} \quad 0 < \epsilon < \min\{(d-4)/4, 1\}, \tag{6.4.9}$$

and

$$\|x_\mu^2 \delta_p^\epsilon G_p\|_\infty \leq K_2 \quad \text{if} \quad 0 < \epsilon < \min\{(d-4)/2, 1\}. \tag{6.4.10}$$

Proof. Let $\epsilon \in (0, 1)$. For an upper bound, we take $p = z_c$. We define p_λ as in (6.4.1). The proof of each of these three inequalities is similar, and we focus mainly on the first one. By Lemma 6.3.1 [using the fact that $G_p(0, x)$ has nonnegative coefficients $c_n(0, x)$], we have

$$\delta_p^\epsilon \partial_p G_p(0, x) = C_{1-\epsilon} z_c \int_0^\infty \partial_p^2 G_p(0, x)\big|_{p=p_\lambda} e^{-\lambda^{1/(1-\epsilon)}} d\lambda. \tag{6.4.11}$$

Using Lemma 6.2.8 to bound the derivative of G_p in the integrand, and then going to the Fourier transform, we can bound the right side as

$$\delta_p^\epsilon \partial_p G_p(0, x) \leq 2 \int \frac{d^d k}{(2\pi)^d} C_{1-\epsilon} z_c \int_0^\infty d\lambda e^{-\lambda^{1/(1-\epsilon)}} p_\lambda^{-2} \left| \hat{H}_{p_\lambda}(k)^2 \hat{G}_{p_\lambda}(k) \right|. \tag{6.4.12}$$

Using the fact that $\hat{H}_z(k) = [z\Omega\hat{D}(k) + \hat{\Pi}_z(k)]\hat{F}_z(k)^{-1}$, and then using (6.2.37) to bound $z^{-1}\hat{\Pi}_z(k)$, it can be shown that there is a constant K_4 (depending on Ω but not on λ) such that

$$p_\lambda^{-1}|\hat{H}_{p_\lambda}(k)| \leq K_4 \hat{F}_{p_\lambda}(k)^{-1}. \tag{6.4.13}$$

We bound the right side of (6.4.12), using (6.4.13) and (6.4.3), by

$$2K_4^2 \int \frac{d^d k}{(2\pi)^d} C_{1-\epsilon} z_c \int_0^\infty d\lambda e^{-\lambda^{1/(1-\epsilon)}} \hat{F}_{p_\lambda}(k)^{-3}$$

$$\leq \text{const.} \int \frac{d^d k}{(2\pi)^d} \int_0^\infty d\lambda e^{-\lambda^{1/(1-\epsilon)}} [1 - e^{-\lambda^{1/(1-\epsilon)}} \hat{D}(k)]^{-3}.$$

Now by (6.3.4), the right side of the above inequality is equal to

$$\text{const.} \int \frac{d^d k}{(2\pi)^d} \frac{1}{\hat{D}(k)} \delta_p^\epsilon \left[(1 - p\hat{D}(k))^{-2} \right] \Big|_{p=1}$$

$$= \text{const.} \int \frac{d^d k}{(2\pi)^d} \sum_{n=2}^{\infty} n(n-1)^\epsilon \hat{D}(k)^{n-2}.$$

By (6.2.6), the right side is finite for $1 + \epsilon - (d/2) < -1$. This proves (6.4.8).

For (6.4.9), we proceed in a similar fashion. Using Lemmas 6.3.1 and 6.2.8 and the Parseval relation gives

$$\left\| \delta_p^\epsilon G_p \right\|_2 \leq \left\| C_{1-\epsilon} z_c \int_0^\infty d\lambda e^{-\lambda^{1/(1-\epsilon)}} p_\lambda^{-1} \hat{H}_{p_\lambda}(k) \hat{G}_{p_\lambda}(k) \right\|_2 ,$$

where the norm on the left is with respect to normalized Lebesgue measure on $[-\pi, \pi]^d$. Arguing as above and using the triangle inequality for $\| \cdot \|_2$ gives

$$\| \delta_p^\epsilon G_p \|_2 \leq \text{const.} \sum_{n=1}^{\infty} n^\epsilon \| \hat{D}(k)^{n-1} \|_2.$$

The desired bound now follows from the fact that $\| \hat{D}(k)^n \|_2 \leq O(n^{-d/4})$, by (6.2.6).

For (6.4.10), by Lemma 6.3.1 we have

$$x_\mu^2 \delta_p^\epsilon G_p(0, x) = C_{1-\epsilon} z_c \int_0^\infty d\lambda e^{-\lambda^{1/(1-\epsilon)}} x_\mu^2 \partial_p G_{p_\lambda}(0, x). \qquad (6.4.14)$$

It follows from (6.2.10) that $x_\mu^2 \partial_p G_{p_\lambda}(0, x)$ is bounded uniformly in x and $\lambda \geq \lambda_0$, for any fixed positive λ_0. Taking for simplicity $\lambda_0 = 1$, it suffices to bound

$$\int_0^1 d\lambda e^{-\lambda^{1/(1-\epsilon)}} x_\mu^2 \partial_p G_{p_\lambda}(0, x). \qquad (6.4.15)$$

Applying Lemma 6.2.8 and the Fourier transform, and noting that p_λ is bounded below by p_1 for $\lambda \in [0, 1]$, the above integral is bounded above by

$$\text{const.} \int_0^1 d\lambda e^{-\lambda^{1/(1-\epsilon)}} \int \frac{d^d k}{(2\pi)^d} \left| \partial_\mu^2 [\hat{H}_{p_\lambda}(k) \hat{G}_{p_\lambda}(k)] \right|. \qquad (6.4.16)$$

It follows from Corollary 6.2.7 that $|\partial_\mu^2 \hat{F}_{p_\lambda}(k)|$ is bounded (uniformly in λ). Also, it follows from Taylor's theorem and the bound on $|\partial_\mu \hat{\Pi}_z(k)|$ of Corollary 6.2.7, together with (6.2.5), that

$$\sum_{\mu=1}^{d} [\partial_\mu \hat{F}_{p_\lambda}(k)]^2 \leq \text{const.} k^2 \leq \text{const.} [1 - e^{-\lambda^{1/(1-\epsilon)}} \hat{D}(k)].$$

It then follows from direct computation of the second derivative of

$$[\hat{F}_{p\lambda}(k)^{-1} - 1]\hat{F}_{p\lambda}(k)^{-1}$$

occurring in (6.4.16), together with (6.4.3) and symmetry, that

$$x_\mu^2 \delta_p^\epsilon G_p(0, x) \leq \text{const.}[1 + \int_0^\infty d\lambda e^{-\lambda^{1/(1-\epsilon)}} \int \frac{d^d k}{(2\pi)^d} \hat{F}_{p\lambda}(k)^{-3}].$$

Now the discussion below (6.4.13) can be applied. □

The following corollary of Theorem 6.4.2 will be used to prove Theorem 6.1.1.

Corollary 6.4.3 *For Ω sufficiently large (with $d > 4$ for the spread-out model), there is a K_2 (which may depend on ϵ and Ω) such that for any $k \in [-\pi, \pi]^d$ and $|z| \leq z_c$,*

$$|\delta_z^\epsilon \partial_z \hat{\Pi}_z(k)|, \quad |\delta_z^\epsilon \partial_\mu^u \hat{\Pi}_z(k)| \leq K_2, \tag{6.4.17}$$

for $u = 0, 1, 2$, where the first bound holds for any nonnegative $\epsilon < \min\{(d-4)/2, 1\}$ and the second for any nonnegative $\epsilon < \min\{(d-4)/4, 1\}$. In fact the series representations of the left side are bounded absolutely by K_2.

Proof. We write the left sides as sums over sites x and number of loops N. For upper bounds, we take absolute values inside sums over both x and N, and consider $z = z_c$. For the first bound, the derivatives bring down a factor $|\omega|^{1+\epsilon}$. This can be distributed among the subwalks of the N loop diagram, using Hölder's inequality in the form

$$|\omega|^{1+\epsilon} = \left[\sum_{j=0}^{2N-1} |\omega_j|\right]^{1+\epsilon} \leq (2N-1)^\epsilon \sum_{j=0}^{2N-1} |\omega_j|^{1+\epsilon}.$$

The resulting diagrams can then be bounded using Lemma 5.4.3, with the subwalk weighted by $|\omega|^{1+\epsilon}$ bounded with the L^∞ norm. Convergence then follows using an extension of Theorem 5.4.4 for fractional derivatives, together with Corollary 6.2.6 and (6.4.8).

Similarly, for the second bound, the derivatives bring down factors $|\omega|^\epsilon$ and $|x_\mu^u|$. As in the proof of Theorem 5.4.4, these can be distributed among the subwalks in a diagram, with each factor on a distinct subwalk (the one-loop diagram does not contribute). The subwalk weighted with $|x_\mu^u|$ is bounded using the L^∞ norm, and all other subwalks are bounded using the L^2 norm, using (6.4.9) for the subwalk weighted with $|\omega|^\epsilon$. Convergence follows using Corollary 6.2.6. □

6.4.2   Proof of Theorem 6.1.1

In this section we give the proof of Theorem 6.1.1. We begin with c_n.

Proof of Theorem 6.1.1(a). The susceptibility is given by

$$\chi(z) = \frac{1}{\hat{F}_z(0)} = \frac{1}{1 - z\Omega - \hat{\Pi}_z(0)}. \tag{6.4.18}$$

Fix $\epsilon < \min\{(d-4)/2, 1\}$. By Corollary 6.4.3, for any $\epsilon' < \min\{(d-4)/2, 1\}$, $\sum_n n^{1+\epsilon'}|\pi_n|z_c^n < \infty$, where π_n is the coefficient of z^n in the power series representation of $\hat{\Pi}_z(0)$. Moreover by Theorem 6.2.9 $\partial_z \hat{F}_{z_c}(0) \neq 0$, and by Theorem 6.2.10 the only singularity of $\chi(z)$ on the circle $|z| = z_c$ is at $z = z_c$. It then follows immediately from Lemma 6.3.4 that

$$\begin{aligned}
c_n &= z_c^{-n-1}\left[\frac{1}{-\partial_z \hat{F}_{z_c}(0)} + O(n^{-\epsilon})\right] \\
&= A\mu^n[1 + O(n^{-\epsilon})],
\end{aligned} \tag{6.4.19}$$

where in agreement with (6.2.45)

$$A = \frac{1}{-z_c \partial_z \hat{F}_{z_c}(0)}. \tag{6.4.20}$$

$\square$

We now turn to the mean-square displacement.

Proof of Theorem 6.1.1(b). By definition of the Fourier transform,

$$\langle |\omega(n)|^2 \rangle_n = \frac{-\nabla_k^2 \hat{c}_n(0)}{c_n}. \tag{6.4.21}$$

The asymptotic behaviour of the denominator on the right side was obtained in (6.4.19), and we now proceed to analyze the numerator.

Since $\hat{c}_n(k)$ is the coefficient of z^n in $\hat{G}_z(k)$,

$$-\nabla_k^2 \hat{c}_n(0) = -\frac{1}{2\pi i}\oint \nabla_k^2 \hat{G}_z(0)\frac{dz}{z^{n+1}} = \frac{1}{2\pi i}\oint \frac{\nabla_k^2 \hat{F}_z(0)}{\hat{F}_z(0)^2}\frac{dz}{z^{n+1}}, \tag{6.4.22}$$

where the integrals are performed around a small circle centred at the origin. We define an error term $E(z)$ by

$$\frac{\nabla_k^2 \hat{F}_z(0)}{\hat{F}_z(0)^2} = \frac{\nabla_k^2 \hat{F}_{z_c}(0)}{[\partial_z \hat{F}_{z_c}(0)]^2 (z_c - z)^2} + E(z). \tag{6.4.23}$$

Inserting the right side of (6.4.23) into the right side of (6.4.22), the integral corresponding to the first term can be performed exactly to give

$$-\nabla_k^2 \hat{c}_n(0) = \frac{\nabla_k^2 \hat{F}_{z_c}(0)}{[\partial_z \hat{F}_{z_c}(0)]^2}(n+1)z_c^{-(n+2)} + \frac{1}{2\pi i} \oint E(z)\frac{dz}{z^{n+1}}. \qquad (6.4.24)$$

The remaining task is to bound the last term in (6.4.24). This is done using Lemma 6.3.3. Let $\epsilon < \min\{(d-4)/4, 1\}$. It follows from Lemma 6.3.3(i) that if it can be shown that $|E(z)| \leq \text{const.}|z_c - z|^{-2+\epsilon}$ for all $|z| \leq z_c$, then the second term on the right side of (6.4.24) is $O(z_c^{-n}n^\alpha)$ for every $\alpha > 1-\epsilon$. Assuming for the moment this bound on the error term and using (6.4.19), we then have the desired result

$$\langle |\omega(n)|^2 \rangle_n = Dn + O(n^\alpha), \qquad (6.4.25)$$

with

$$D = \frac{\nabla_k^2 \hat{F}_{z_c}(0)}{z_c[-\partial_z \hat{F}_{z_c}(0)]}. \qquad (6.4.26)$$

We now establish the upper bound on $|E(z)|$ used in the previous paragraph. We first use (6.4.23) to write $E(z)$ as a difference of two fractions, and then write this difference over a common denominator and add and subtract $\nabla_k^2 \hat{F}_z(0)\hat{F}_z(0)^2$ in the numerator. This leads to

$$E(z) = T_1 + T_2 \qquad (6.4.27)$$

with

$$T_1 = [\partial_z \hat{F}_{z_c}(0)]^{-2} \frac{\nabla_k^2 \hat{F}_z(0) - \nabla_k^2 \hat{F}_{z_c}(0)}{(z_c - z)^2} \qquad (6.4.28)$$

and

$$T_2 = \frac{-\nabla_k^2 \hat{F}_z(0)[\hat{F}_z(0)^2 - [\partial_z \hat{F}_{z_c}(0)]^2(z_c - z)^2]}{[\partial_z \hat{F}_{z_c}(0)]^2 \hat{F}_z(0)^2(z_c - z)^2}. \qquad (6.4.29)$$

For T_1, we use existence of an ϵ-derivative in the numerator by Corollary 6.4.3, together with the Taylor theorem type bound of (6.3.9) to conclude that

$$|T_1| \leq O(|z_c - z|^{\epsilon-2}). \qquad (6.4.30)$$

For T_2 we factor the difference of squares in the numerator and bound the denominator using Theorem 6.2.10, obtaining

$$|T_2| \leq \text{const.}(z_c - z)^{-4}[\hat{F}_z(0) + \partial_z \hat{F}_{z_c}(0)(z_c - z)][\hat{F}_z(0) - \partial_z \hat{F}_{z_c}(0)(z_c - z)]. \qquad (6.4.31)$$

The middle factor on the right side is $O(|z_c - z|^{1+\epsilon})$, by Corollary 6.4.3 and Lemma 6.3.2. For the last factor, the second term is clearly $O(|z_c - z|)$, as is the first, by virtue of the bound on the middle factor. Therefore $|T_2| \leq O(|z_c - z|^{\epsilon-2})$. $\qquad \square$

6.5   Correlation length and infrared bound

6.5.1   The correlation length

In this section we prove Theorem 6.1.5, which states that for sufficiently large Ω,

$$\xi(z) \sim \sqrt{\frac{D}{2d}} \left(\frac{z_c}{z_c - z}\right)^{1/2} \quad \text{as } z \nearrow z_c. \tag{6.5.1}$$

We work with the fully self-avoiding walk, with positive activity $p < z_c$, and as usual write $m(p) = \xi(p)^{-1}$.

By Proposition 4.1.1(b) and Theorem 4.1.6 [and the fact that $B(z_c) < \infty$ by Corollary 6.2.6], $m(p)$ is strictly positive and finite for $p < z_c$, and $m(p) \searrow 0$ as $p \nearrow z_c$. For any function f defined on $\mathbf{Z}^d$, and $m \in \mathbf{R}$, we define

$$f^{(m)}(x) = f(x)e^{mx_1}. \tag{6.5.2}$$

The following lemma, whose proof is deferred to the end of this section, is a key ingredient in the proof of (6.5.1).

Lemma 6.5.1 *For Ω sufficiently large (with $d > 4$ for the spread-out model), there is a $\delta > 0$ (which may depend on Ω) such that for $p \in [z_c - \delta, z_c)$ and $m < m(p)$,*

$$\|H_p^{(m)}\|_2^2 \equiv \sum_x [H_p^{(m)}(0,x)e^{mx_1}]^2 \leq 2K\Omega^{-1+s},$$

where K and s are as in the statement of Lemma 6.2.2.

Lemma 6.5.1 leads to the following result.

Corollary 6.5.2 *Let Ω be sufficiently large (with $d > 4$ for the spread-out model). There is a positive constant K_3 which is independent of p and m (but may depend on ϵ and Ω) such that*

$$\sum_x |x_1|^{2+\epsilon} |\Pi_p^{(m)}(0,x)| \leq K_3$$

for all $p \in [z_c - \delta, z_c)$, $m < m(p)$ and $\epsilon < \min\{(d-4)/2, 1\}$.

Proof.   The sum on the left side involves diagrams having two or more loops, weighted with both $|x_1|^{2+\epsilon}$ and e^{mx_1}. We split the former among subwalks along one side of the diagram using Hölder's inequality, and factor the latter along subwalks on the other side of the diagram. We then bound the resulting diagrams using Lemma 5.4.3. The subwalk weighted with $|x_1|^{2+\epsilon}$ is bounded using the infinity norm, as follows:

$$\sup_x |x_1|^{2+\epsilon} \sum_n c_n(0,x)p^n \leq \sup_x |x_1|^2 \sum_n n^\epsilon c_n(0,x)p^n$$

$$= \|x_1^2 \delta_p^\epsilon G_p\|_\infty. \tag{6.5.3}$$

The right side is finite for ϵ as in the statement of the corollary, by Theorem 6.4.2. All other subwalks are bounded as in Lemma 5.4.3 using the L^2 norm, yielding factors of $\|H_p\|_2$, $\|G_p\|_2 = 1 + \|H_p\|_2$, $\|H_p^{(m)}\|_2$ and $\|G_p^{(m)}\|_2 = 1 + \|H_p^{(m)}\|_2$. The sum of all diagrams is then bounded above by a geometric series with an m-dependent ratio. The geometric series converges for Ω sufficiently large by Lemma 6.5.1 and Corollary 6.2.6, uniformly for p and m as in the statement of the lemma. $\qquad\square$

Proof of Theorem 6.1.5. The proof is modelled on the corresponding random walk result in Theorem A.2(b). Let $p \in [z_c - \delta, z_c)$. For $m < m(p)$, let $\chi^{(m)}(p) = \sum_x G_p^{(m)}(0,x)$. Because $G_p(0,x)$ decays exponentially with decay rate $m(p)$ by Lemma 4.1.5, $\chi^{(m)}(p)$ is finite if $m < m(p)$. By multiplying (5.2.17) by e^{mx_1} and then taking the Fourier transform, we obtain

$$\hat{G}_p^{(m)}(k) = \frac{1}{1 - p\Omega\hat{D}^{(m)}(k) - \hat{\Pi}_p^{(m)}(k)}. \tag{6.5.4}$$

[The Fourier transform $\hat{\Pi}_p^{(m)}$ exists for $p \in [z_c - \delta, z_c)$ and $m < m(p)$, by Corollary 6.5.2.] The function $\hat{D}^{(m)}(k)$ is defined by

$$\hat{D}^{(m)}(k) = \frac{1}{\Omega} \sum_{y \in \Omega} e^{my_1} e^{ik \cdot y}. \tag{6.5.5}$$

Using (6.5.4) and (6.5.5),

$$\chi(p)^{-1} - \chi^{(m)}(p)^{-1} = p \sum_{y \in \Omega} [\cosh my_1 - 1] + \hat{\Pi}_p^{(m)}(0) - \hat{\Pi}_p(0). \tag{6.5.6}$$

We intend to take the limit as $m \nearrow m(p)$ in (6.5.6). As a first observation, we show that for any $p < z_c$,

$$\lim_{m \nearrow m(p)} \chi^{(m)}(p) = \infty. \tag{6.5.7}$$

This can be seen as follows. For simplicity we deal in the remainder of this paragraph only with the nearest-neighbour model; a modified argument applies to the spread-out model. Let $B_R = \{y \in \mathbf{Z}^d : ||y||_\infty \leq R\}$ and $\partial B_R = \{y \in \mathbf{Z}^d : ||y||_\infty = R\}$. For $y \in \partial B_R$ let $G_x^R(0, y) = \sum_x z^{|\omega|}$ where the sum is over all nearest-neighbour self-avoiding walks from 0 to y which hit ∂B_R for the first and only time at y. Then for $x \notin B_R$, we have the following Lieb-Simon type inequality:

$$G_p(0, x) \leq \sum_{y \in \partial B_R} G_p^R(0, y)G_p(y, x) \leq \sum_{y \in \partial B_R} G_p(0, y)G_p(y, x). \qquad (6.5.8)$$

Multiplying this inequality by $e^{m(p)x_1}$, it follows from Lemma A.1 that if $\chi^{(m(p))}(p)$ were finite then $G^{(m(p))}(0, x)$ would decay exponentially, contradicting the definition of $m(p)$. It then follows from the monotone convergence theorem, and the fact that $\chi^{(m)}(p)$ is finite if $m < m(p)$, that (6.5.7) holds.

The remainder of the proof is concerned with showing that the limit of the right side of (6.5.6), as $m \nearrow m(p)$, is a multiple of $m(p)^2$ plus a higher order correction. Together with Theorem 6.1.2, which states that $\chi(p) \sim \text{const.}(z_c - p)^{-1}$, this will show that $m(p)^2$ is asymptotic to a multiple of $z_c - p$ as $p \nearrow z_c$.

By definition and symmetry,

$$\hat{\Pi}_p^{(m)}(0) - \hat{\Pi}_p(0) = \sum_x [\cosh mx_1 - 1]\Pi_p(0, x)$$

$$= \frac{m^2}{2d} \sum_x |x|^2 \Pi_p(0, x)$$

$$+ \sum_x \left[\cosh mx_1 - 1 - \frac{m^2 x_1^2}{2}\right] \Pi_p(0, x). \qquad (6.5.9)$$

Fix $\epsilon < \min\{(d-4)/2, 1\}$. There is a positive constant C such that

$$0 \leq \cosh mx_1 - 1 - \frac{m^2 x_1^2}{2} \leq Cm^{2+\epsilon}|x_1|^{2+\epsilon} \cosh mx_1. \qquad (6.5.10)$$

Hence by Corollary 6.5.2,

$$\left|\sum_x \left[\cosh mx_1 - 1 - \frac{m^2 x_1^2}{2}\right] \Pi_p(0, x)\right| \leq Cm^{2+\epsilon} \sum_x |x_1|^{2+\epsilon}|\Pi_p^{(m)}(0, x)|$$

$$\leq CK_3 m^{2+\epsilon},$$

uniformly in $m < m(p)$ and $p \in [z_c - \delta, z_c)$. Hence the limit of the left side as $m \nearrow m(p)$ is $O[m(p)^{2+\epsilon}]$.

From this fact, together with (6.5.7) and (6.5.9), we conclude that taking the limit as $m \nearrow m(p)$ in (6.5.6) gives

$$
\begin{aligned}
\chi(p)^{-1} &= p \sum_{y \in \Omega} [\cosh(m(p)y_1) - 1] \\
&\quad + \frac{m(p)^2}{2d} \sum_{x} |x|^2 \Pi_p(0, x) + O(m(p)^{2+\epsilon}).
\end{aligned}
\tag{6.5.11}
$$

Since as noted at the beginning of this section $m(p) \to 0$ as $p \nearrow z_c$, the right side of (6.5.11) is asymptotic to

$$
m(p)^2 \left[\frac{z_c}{2} \sum_{y \in \Omega} y_1^2 + \frac{1}{2d} \sum_{x} |x|^2 \Pi_{z_c}(0, x) \right] = \frac{m(p)^2}{2d} \nabla_k^2 \hat{F}_{z_c}(0)
$$

as $p \nearrow z_c$. The right side is positive, by (6.2.35). Also, by Theorem 6.1.2 the left side of (6.5.11) is asymptotic to $(Az_c)^{-1}(z_c - p)$. Thus by (6.4.20) and (6.4.26) we have

$$
m(p)^2 \sim \frac{2d}{A\nabla_k^2 \hat{F}_{z_c}(0)} \frac{z_c - p}{z_c} = \frac{2d}{D} \frac{z_c - p}{z_c},
\tag{6.5.12}
$$

which proves Theorem 6.1.5.   □

We now complete the remaining step of the proof.

Proof of Lemma 6.5.1. The proof uses Lemma 6.2.1 with m playing the role of p, $n = 1$, $a = 2/3$, $p_0 = 0$, $p_1 = m(p)$ and

$$
f_1(m) = \frac{\|H_p^{(m)}\|_2^2}{3K\Omega^{-1+s}}.
$$

We begin by considering the hypotheses of Lemma 6.2.1 in this context. First, for any $p < z_c$, $f_1(m)$ is continuous in $m \in [0, m(p))$. This follows from the fact that if $p < z_c$ and $m < m(p)$ then $\|H_p^{(m)}\|_2^2 < \infty$ (by Lemma 4.1.5), together with the monotone convergence theorem. Next, by Corollary 6.2.6, $f_1(0) \leq 2/3$ for all $p < z_c$, if Ω is sufficiently large.

It thus suffices to show that the remaining, and substantial, hypothesis of Lemma 6.2.1 is satisfied, namely that there is a $\delta > 0$ such that given $p \in [z_c - \delta, z_c)$, $m < m(p)$ and Ω sufficiently large (independently of p, m), if $\|H_p^{(m)}\|_2^2 \leq 3K\Omega^{-1+s}$ then in fact $\|H_p^{(m)}\|_2^2 \leq 2K\Omega^{-1+s}$. The remainder of the proof is concerned with showing the existence of such a δ.

Denoting the reciprocal of $\hat{G}_p^{(m)}(k)$ by $\hat{F}_p^{(m)}(k)$, we have

$$|\hat{G}_p^{(m)}(k)| \leq \frac{1}{|\text{Re}\,\hat{F}_p^{(m)}(k)|}. \tag{6.5.13}$$

Now

$$\begin{aligned}
\text{Re}\,\hat{F}_p^{(m)}(k) &\geq \text{Re}\,\hat{F}_p^{(m)}(k) - \hat{F}_p^{(m)}(0) \\
&= p\Omega\text{Re}[\hat{D}^{(m)}(0) - \hat{D}^{(m)}(k)] + \text{Re}[\hat{\Pi}_p^{(m)}(0) - \hat{\Pi}_p^{(m)}(k)].
\end{aligned}$$

But for $p \geq \Omega^{-1}$,

$$\begin{aligned}
p\Omega\text{Re}[\hat{D}^{(m)}(0) - \hat{D}^{(m)}(k)] &= p\sum_{y\in\Omega} e^{my_1}[1 - \cos k\cdot y] \\
&= p\sum_{y\in\Omega} \cosh my_1 [1 - \cos k\cdot y] \\
&\geq 1 - \hat{D}(k). \tag{6.5.14}
\end{aligned}$$

Also, by Corollary 6.2.7 we have

$$\begin{aligned}
&\text{Re}[\hat{\Pi}_p^{(m)}(0) - \hat{\Pi}_p^{(m)}(k)] \\
&= [\hat{\Pi}_p(0) - \hat{\Pi}_p(k)] + \text{Re}\left[\left(\hat{\Pi}_p^{(m)}(0) - \hat{\Pi}_p^{(m)}(k)\right) - \left(\hat{\Pi}_p(0) - \hat{\Pi}_p(k)\right)\right] \\
&\geq -K_1\Omega^{-u}[1 - \hat{D}(k)] + \sum_x [\cosh mx_1 - 1]\Pi_p(0, x)[1 - \cos k\cdot x]
\end{aligned}$$

(with $u = 3/2$ for the nearest-neighbour model and $u = 5/2 - 5s/2 - 2/d$ for the spread-out model). Therefore

$$\begin{aligned}
\text{Re}\,\hat{F}_p^{(m)}(k) &\geq [1 - K_1\Omega^{-u}][1 - \hat{D}(k)] \\
&\quad + \sum_x [\cosh mx_1 - 1]\Pi_p(0, x)[1 - \cos k\cdot x].
\end{aligned}$$

Since

$$0 \leq \cosh t - 1 \leq \text{const.}|t|^\epsilon \cosh t$$

for $0 \leq \epsilon \leq 2$, and since [by (6.2.5)]

$$0 \leq 1 - \cos k\cdot x \leq \frac{(k\cdot x)^2}{2} \leq \pi^2 d|x|^2[1 - \hat{D}(k)],$$

we have

$$\begin{aligned}
&\left|\sum_x [\cosh mx_1 - 1]\Pi_p(0, x)[1 - \cos k\cdot x]\right| \\
&\leq c_1 dm^\epsilon[1 - \hat{D}(k)]\sum_x |x|^2 |x_1|^\epsilon |\Pi_p^{(m)}(0, x)|,
\end{aligned}$$

where c_1 is a universal constant. Now the proof of Corollary 6.5.2 goes through equally well assuming $||H_p^{(m)}||_2^2 < 3K\Omega^{-1+s}$ rather than $||H_p^{(m)}||_2^2 < 2K\Omega^{-1+s}$, so under this assumption we have a bound on the sum over x in the right side of the above inequality by a constant independent of p and $m < m(p)$ (but possibly depending on ϵ and Ω). Changing the value of K_1, we then have

$$|\hat{G}_p^{(m)}(k)| \leq [1 + K_1\Omega^{-u} + Cm^\epsilon]\frac{1}{[1 - \hat{D}(k)]}, \qquad (6.5.15)$$

where C is independent of p but may depend on Ω.

Using (6.5.15), the Parseval relation, the fact that $||H_p^{(m)}||_2^2 = ||G_p^{(m)}||_2^2 - 1$, and the ordinary random walk bound of (6.2.7) gives

$$\begin{aligned} ||H_p^{(m)}||_2^2 &\leq [1 + K_1\Omega^{-u} + Cm^\epsilon]\left\|\frac{1}{1 - \hat{D}}\right\|_2^2 - 1 \\ &\leq K\Omega^{-1+s}[1 + K_1\Omega^{-u} + Cm^\epsilon] + K_1\Omega^{-u} + Cm^\epsilon. \end{aligned}$$

For $d > 4$, $-u < -1 + s$, so if Ω is sufficiently large, say $\Omega \geq \Omega_1$ (not depending on m, p), then

$$K_1\Omega^{-u} \leq \min\{1/4, K\Omega^{-1+s}/4\}.$$

For fixed $\Omega \geq \Omega_1$ we then choose δ sufficiently small that

$$Cm(p)^\epsilon \leq \min\{1/4, K\Omega^{-1+s}/4\}$$

for $p \in [z_c - \delta, z_c)$; this is possible because $m(p) \to 0$ as $p \nearrow z_c$. Then for $p \in [z_c - \delta, z_c)$ and $m < m(p)$ we have

$$\begin{aligned} ||H_p^{(m)}||_2^2 &\leq K\Omega^{-1+s}\left[1 + \frac{1}{4} + \frac{1}{4}\right] + \frac{1}{4}K\Omega^{-1+s} + \frac{1}{4}K\Omega^{-1+s} \\ &= 2K\Omega^{-1+s}. \end{aligned}$$

$\square$

6.5.2 The infrared bound

In this section we give the proof of Theorem 6.1.6, apart from the bound $G_{z_c}(0, x) \leq C(p)|x|^{-p}$ for $p < (d-2)/2$, which is a consequence of Corollary 6.1.4. The bound $G_{z_c}(0, x) \leq \text{const.}|x|^{-2}$ has already been established in Corollary 6.2.6, and the upper bound on $\hat{G}_{z_c}(k)$ follows from the last inequality of Corollary 6.2.7. Let $\epsilon < \min\{(d-4)/2, 1\}$. Here we show that as $k \to 0$,

$$\hat{G}_{z_c}(k) = \frac{2d}{\nabla_k^2 \hat{F}_{z_c}(0)} \frac{1}{k^2 + O(k^{2+\epsilon})}. \qquad (6.5.16)$$

Since $\hat{G}_{z_c}(0)^{-1} = 0$ by Corollary 6.2.7,

$$\begin{aligned}
\hat{G}_{z_c}(k)^{-1} &= \hat{G}_{z_c}(k)^{-1} - \hat{G}_{z_c}(0)^{-1} \\
&= z_c \Omega [1 - \hat{D}(k)] + \hat{\Pi}_{z_c}(0) - \hat{\Pi}_{z_c}(k). \qquad (6.5.17)
\end{aligned}$$

The last two terms on the right side can be written as

$$\frac{k^2}{2d} \sum_x |x|^2 \Pi_{z_c}(0, x) + \sum_x \left[1 - \cos k \cdot x - \frac{(k \cdot x)^2}{2} \right] \Pi_{z_c}(0, x).$$

The quantity in square brackets can be bounded above by $c_2 k^{2+\epsilon} |x|^{2+\epsilon}$, for some universal constant c_2. The sum over x can then be bounded as in Corollary 6.5.2. The result is

$$\hat{G}_{z_c}(k)^{-1} = k^2 [C^{-1} + O(k^\epsilon)], \qquad (6.5.18)$$

where

$$C^{-1} = \frac{1}{2d} \left[-z_c \Omega \nabla_k^2 \hat{D}(0) + \sum_x |x|^2 \Pi_{z_c}(0, x) \right] = \frac{1}{2d} \nabla_k^2 \hat{F}_{z_c}(0). \qquad (6.5.19)$$

6.6   Convergence to Brownian motion

Given an n-step self-avoiding walk ω, we define

$$X_n(k/n) = (Dn)^{-1/2} \omega(k), \quad k = 0, 1, \ldots, n \qquad (6.6.1)$$

where D is the diffusion constant given in (6.4.26). We then obtain a continuous function X_n on the interval $[0, 1]$, taking values in $\mathbf{R}^d$, by defining $X_n(t)$ to be the linear interpolation of $X_n(k/n)$. In this section we prove Theorem 6.1.8, which states that if Ω is sufficiently large (with $d > 4$ for the spread-out model) then $X_n(t)$ converges in distribution to Brownian motion, or in other words that

$$\lim_{n \to \infty} \langle f(X_n) \rangle_n = \int f \, dW \qquad (6.6.2)$$

for every bounded continuous function f on $C_d[0,1]$ (the latter denotes the $\mathbf{R}^d$-valued continuous functions on $[0,1]$, equipped with the supremum norm). Here W is the Wiener measure, normalized such that

$$\int e^{i\mathbf{k}\cdot B_t}dW = e^{-k^2 t/2d}, \qquad (6.6.3)$$

and the angular brackets on the left side of (6.6.2) denote expectation with respect to the uniform measure on the set of n-step self-avoiding walks beginning at the origin.

To prove this result it suffices to show both convergence of the finite-dimensional distributions to Gaussian distributions and tightness [see e.g. Billingsley (1968)]. Tightness follows readily from Theorem 6.1.1 and will be discussed at the end of this section.

For convergence of the finite-dimensional distributions, we need to prove that for any positive integer N, any $0 < t_1 < t_2 < \ldots < t_N \leq 1$ and any bounded, continuous function g on $\mathbf{R}^{dN}$,

$$\lim_{n\to\infty} \langle g(X_n(t_1), \ldots, X_n(t_N)) \rangle_n = \int g(B_{t_1}, \ldots, B_{t_N}) dW. \qquad (6.6.4)$$

Since weak convergence of probability measures on $\mathbf{R}^m$ is implied by convergence of the corresponding characteristic functions, it suffices to consider only

$$g(\mathbf{x}) = e^{i\mathbf{k}\cdot\mathbf{x}}, \qquad (6.6.5)$$

where $\mathbf{x} = (x^{(1)}, \ldots, x^{(N)})$ with each $x^{(i)} \in \mathbf{R}^d$, and similarly for $\mathbf{k}$, and $\mathbf{k}\cdot\mathbf{x} = \sum_{i=1}^{N} k^{(i)} \cdot x^{(i)}$. Equivalently, we can replace this g by

$$g(\mathbf{x}) = \exp\left[i\sum_{j=1}^{N} k^{(j)} \cdot (x^{(j)} - x^{(j-1)})\right], \qquad (6.6.6)$$

which will be better suited to take into account the "effective independence" of the self-avoiding walk on distinct intervals $[t_i, t_{i+1}]$.

Let $\mathbf{a} = (a_1, \ldots, a_N)$, with each a_i a nonnegative integer, and let

$$\Delta\omega(\mathbf{a}) = (\omega(a_1), \omega(a_2) - \omega(a_1), \ldots, \omega(a_N) - \omega(a_{N-1})). \qquad (6.6.7)$$

We define

$$M(\mathbf{k}, \mathbf{a}) = \sum_{\omega:|\omega|=a_N} e^{i\mathbf{k}\cdot\Delta\omega(\mathbf{a})} K[0, a_N], \qquad (6.6.8)$$

where the sum over ω is a sum over simple random walks, and $K[a, b]$ was defined in (5.2.6). (We work in this section with infinite memory, i.e., with the fully self-avoiding walk.) Inserting (6.6.6) into (6.6.4) and using the above notation, we see that for convergence of the finite-dimensional distributions it suffices to show that for $N = 1, 2, 3, \ldots$,

$$\lim_{n \to \infty} c_{ntN}^{-1} M(k/\sqrt{Dn}, nt) = \exp\left[-\frac{1}{2d} \sum_{i=1}^{N} \left(k^{(i)}\right)^2 (t_i - t_{i-1})\right]. \quad (6.6.9)$$

[The nt and nt_N on the left are to be interpreted as $(\lfloor nt_1 \rfloor, \ldots, \lfloor nt_N \rfloor)$ and $\lfloor nt_n \rfloor$ respectively; similar shorthand is used throughout this section. Also, $(k^{(i)})^2$ denotes the square of the Euclidean norm of the vector $k^{(i)}$. Finally, there is no difficulty in the replacement of $X_n(t)$ by $(Dn)^{-1/2}\omega(\lfloor nt \rfloor)$ that has been made in the left side of (6.6.9).]

We obtain (6.6.9) for $N = 1$ in Section 6.6.1, and prove (6.6.9) for $N \geq 2$ by induction on N in Section 6.6.2.

6.6.1 The scaling limit of the endpoint

In this section we prove (6.6.9) for $N = 1$. In fact, a minor generalization of (6.6.9) will be needed to take the induction step, and we prove the generalization here.

Theorem 6.6.1 *Let Ω be sufficiently large (with $d > 4$ for the spread-out model). Let h_n be any fixed nonnegative sequence with $\lim_{n \to \infty} h_n = 0$, and let $g = \{g_n\}$ be any real sequence with $|g_n| \leq h_n$ for all n. Fix $t > 0$ and let $T = t(1 - g_n)$. Then for any $k \in \mathbf{R}^d$,*

$$\lim_{n \to \infty} \frac{\hat{c}_{nT}(k/\sqrt{Dn})}{c_{nT}} = \exp\left[-k^2 t/2d\right], \quad (6.6.10)$$

uniformly in g.

Proof. Fix any $\epsilon < \min\{(d-4)/4, 1\}$. By (6.4.19), the denominator of (6.6.10) can be written

$$c_{nT} = z_c^{-nT-1}\{[-\partial_z \hat{F}_{z_c}(0)]^{-1} + O(n^{-\epsilon})\}, \quad (6.6.11)$$

uniformly in g. The numerator of (6.6.10) is the coefficient of z^{nT} in $\hat{G}_z(k/\sqrt{Dn})$, and hence is given by

$$\hat{c}_{nT}(k/\sqrt{Dn}) = \frac{1}{2\pi i} \oint \frac{1}{\hat{F}_z(k/\sqrt{Dn})} \frac{dz}{z^{nT+1}}, \quad (6.6.12)$$

where the integration contour is a small circle centred at the origin. The task now is to obtain the asymptotic form of the integral on the right side.

We extract the leading contribution to the right side of (6.6.12) as follows. We subtract $\hat{F}_{z_c}(0) = 0$ from $\hat{F}_z(k/\sqrt{Dn})$, and then add and subtract

$$\partial_z \hat{\Pi}_{z_c}(0)(z_c - z) + \frac{1}{2d} \nabla_k^2 \hat{\Pi}_{z_c}(0) \frac{k^2}{Dn}.$$

The result can be written

$$\hat{F}_z(k/\sqrt{Dn}) = \alpha - \beta z + E, \tag{6.6.13}$$

where

$$\alpha = \alpha(k/\sqrt{Dn}) = -z_c \partial_z \hat{F}_{z_c}(0) - \frac{1}{2d} \nabla_k^2 \hat{\Pi}_{z_c}(0) \frac{k^2}{Dn}, \tag{6.6.14}$$

$$\beta = \beta(k/\sqrt{Dn}) = -\partial_z \hat{F}_{z_c}(0) - \Omega[1 - \hat{D}(k/\sqrt{Dn})], \tag{6.6.15}$$

and

$$E = -\hat{\Pi}_z(k/\sqrt{Dn}) - \partial_z \hat{\Pi}_{z_c}(0)(z_c - z) + \hat{\Pi}_{z_c}(0) + \frac{k^2}{2dDn} \nabla_k^2 \hat{\Pi}_{z_c}(0). \tag{6.6.16}$$

The error term can be written

$$E = E_1 + E_2 + E_3, \tag{6.6.17}$$

where

$$E_1 = \hat{\Pi}_{z_c}(k/\sqrt{Dn}) - \hat{\Pi}_z(k/\sqrt{Dn}) - \partial_z \hat{\Pi}_{z_c}(k/\sqrt{Dn})(z_c - z), \tag{6.6.18}$$

$$E_2 = [\partial_z \hat{\Pi}_{z_c}(k/\sqrt{Dn}) - \partial_z \hat{\Pi}_{z_c}(0)](z_c - z), \tag{6.6.19}$$

and

$$E_3 = \hat{\Pi}_{z_c}(0) - \hat{\Pi}_{z_c}(k/\sqrt{Dn}) + \frac{k^2}{2dDn} \nabla_k^2 \hat{\Pi}_{z_c}(0). \tag{6.6.20}$$

By (6.6.13),

$$\frac{1}{\hat{F}_z(k/\sqrt{Dn})} = \frac{1}{\alpha - \beta z} - \frac{E}{(\alpha - \beta z)\hat{F}_z(k/\sqrt{Dn})}. \tag{6.6.21}$$

We now insert (6.6.21) into (6.6.12). The integral corresponding to the first term on the right side of (6.6.21) is $\beta^{nT} \alpha^{-(nT+1)}$. A straightforward calculation using the definition of D in (6.4.26) and the fact that $1 - \hat{D}(u) \sim -(1/2d)u^2 \nabla^2 \hat{D}(0)$ shows that

$$\frac{\beta^{nT}}{\alpha^{nT+1}} \sim \frac{1}{-\partial_z \hat{F}_{z_c}(0) z_c^{nT+1}} \left[1 - \frac{k^2}{2dn}\right]^{nT}$$

$$\sim \frac{1}{-\partial_z \hat{F}_{z_c}(0) z_c^{nT+1}} \exp[-k^2 t/2d], \tag{6.6.22}$$

uniformly in g. Comparing (6.6.11), the theorem follows from (6.6.22) if it can be shown that

$$\frac{1}{2\pi i} \oint \frac{E}{(\alpha - \beta z)\hat{F}_z(k/\sqrt{Dn})} \frac{dz}{z^{nT+1}} = o(z_c^{-nT}) \qquad (6.6.23)$$

uniformly in g. We show that (6.6.23) holds by using Lemma 6.3.3.

The first step is to obtain lower bounds on the two factors in the denominator of the integrand of (6.6.23). We begin with $|\alpha - \beta z| = \beta|\alpha/\beta - z|$. For large n, β is bounded away from zero by Theorem 6.2.9. Also, it can be seen from (6.6.22) that $\alpha/\beta \geq z_c$ for n large. Hence there is a positive constant such that for large n and $|z| \leq z_c$,

$$|\alpha - \beta z| \geq \text{const.}|z_c - z|. \qquad (6.6.24)$$

For a lower bound on $|\hat{F}_z(k/\sqrt{Dn})|$, we write

$$\hat{F}_z(k/\sqrt{Dn}) = \hat{F}_z(k/\sqrt{Dn}) - \hat{F}_{z_c}(k/\sqrt{Dn}) + \hat{F}_{z_c}(k/\sqrt{Dn}). \qquad (6.6.25)$$

By Corollary 6.4.3 and Lemma 6.3.2, the first two terms on the right side combine to give $-\partial_z \hat{F}_{z_c}(k/\sqrt{Dn})(z_c - z) + O(|z_c - z|^{1+\epsilon})$. By the dominated convergence theorem the derivative appearing here is continuous in k, and hence differs from its value at $k = 0$ by $o(1)$. Thus we have

$$\hat{F}_z(k/\sqrt{Dn}) = [-\partial_z \hat{F}_{z_c}(0) + O(|z_c - z|^\epsilon) + o(1)](z_c - z) + \hat{F}_{z_c}(k/\sqrt{Dn}). \qquad (6.6.26)$$

By Corollary 6.2.7, the last term on the right side is nonnegative. Since the first term in square brackets on the right side is also nonnegative by Theorem 6.2.9, it follows that for z in a small neighbourhood of z_c (inside the closed disk of radius z_c), the right side of (6.6.26) is bounded below by $\text{const.}|z_c - z|$ (for large n). Outside of this neighbourhood, by Lemma 6.2.10 there is a constant $c > 0$ such that $|\hat{F}_z(0)| \geq c$. Hence $|\hat{F}_z(k/\sqrt{Dn})| \geq c/2$ if n is sufficiently large, since by Corollary 6.2.7 there is a bound on $|\nabla_k \hat{F}_z(k)|$ which is uniform in k and $|z| \leq z_c$. Therefore for $|z| \leq z_c$ we have

$$|\hat{F}_z(k/\sqrt{Dn})| \geq \text{const.}|z_c - z|. \qquad (6.6.27)$$

We now turn to upper bounds on E_i for $i = 1, 2, 3$. Beginning with E_1, it follows from (6.6.24), (6.6.27), and a straightforward calculation using Corollary 6.4.3 and Lemma 6.3.2 that

$$\left| \frac{d}{dz} \frac{E_1}{(\alpha - \beta z)(\hat{F}_z(k/\sqrt{Dn}))} \right| \leq O(|z_c - z|^{\epsilon-2}), \qquad (6.6.28)$$

and hence by Lemma 6.3.3(*ii*) (6.6.23) is satisfied if E is replaced by E_1.
For E_2, we show

$$|E_2| \leq n^{-\epsilon/2}|z_c - z|, \tag{6.6.29}$$

which suffices by Lemma 6.3.3(*i*). To do so we write $\pi_m(x)$ for the coefficient of z^m in $\Pi_z(0, x)$, so that

$$\partial_z \hat{\Pi}_{z_c}(k/\sqrt{Dn}) - \partial_z \hat{\Pi}_{z_c}(0) = -\sum_{x,m} m\pi_m(x)z_c^{m-1}[1 - \cos(k \cdot x/\sqrt{Dn})]. \tag{6.6.30}$$

Since $|1 - \cos t| \leq O(t^\epsilon)$ for small $\epsilon \leq 2$, and since $|x|^\epsilon|\pi_m(x)| \leq m^\epsilon|\pi_m(x)|$, the right side of (6.6.30) is $O(n^{-\epsilon/2})$ by Corollary 6.4.3, which gives (6.6.29).

Finally, for E_3 we use symmetry to write

$$E_3 = \sum_{x,m} \pi_m(x)z_c^m \left[1 - \cos\frac{k \cdot x}{\sqrt{Dn}} - \frac{(k \cdot x)^2}{2Dn}\right]. \tag{6.6.31}$$

For small positive ϵ, $|1 - \cos t - t^2/2| \leq O(|t|^{2+\epsilon})$. Since $|x|^{2+\epsilon}|\pi_m(x)| \leq m^\epsilon|x|^2|\pi_m(x)|$, it follows from Corollary 6.4.3 that $|E_3| \leq O(n^{-1-\epsilon/2})$. Then Lemma 6.3.3(*i*) gives (6.6.23) for E replaced by E_3. □

6.6.2 The finite-dimensional distributions

In this section we complete the proof of Theorem 6.1.8 by showing that (6.6.9) holds for $N \geq 2$. The proof of (6.6.9) is by induction on N, with the case $N = 1$ having been treated in the previous section. Lemma 5.2.5 is a basic element of the induction argument.

To perform the induction step, some flexibility is needed in the number of steps in the walk. Let $g = \{g_n\}$ be any sequence satisfying $0 \leq g_n \leq n^{-1/2}$, and let $\mathbf{T} = (t_1, t_2, \ldots, t_{N-1}, T)$, where $T = t_N(1 - g_n)$. It suffices to prove the following theorem.

Theorem 6.6.2 *Let Ω be sufficiently large (with $d > 4$ for the spread-out model) and let $N \geq 2$. Suppose that*

$$\lim_{n \to \infty} c_{nT}^{-1} M(\mathbf{k}/\sqrt{Dn}, n\mathbf{T}) = \exp\left[-\frac{1}{2d}\sum_{i=1}^{N}\left(k^{(i)}\right)^2 (t_i - t_{i-1})\right] \tag{6.6.32}$$

holds uniformly in g, when N is replaced by $N - 1$. Then in fact (6.6.32) holds as stated, uniformly in g.

Proof.   To simplify the notation we write $\kappa = (\kappa_1, \ldots, \kappa_N) = \mathbf{k}/\sqrt{Dn}$. By (6.6.8), and Lemma 5.2.5 with $m = nt_{N-1}$,

$$M(\kappa, n\mathbf{T}) = \sum_{I \ni nt_{N-1}} \sum_{\omega:|\omega|=nT} e^{i\kappa \cdot \Delta\omega(n\mathbf{T})} K[0, I_1] J[I_1, I_2] K[I_2, nT].$$

(6.6.33)

The sum over I is the sum over intervals $[I_1, I_2]$ of integers with either $0 \le I_1 < nt_{N-1} < I_2 \le nT$ or $I_1 = I_2 = nt_{N-1}$.

In (6.6.33) we factor the walk ω into three independent subwalks on the subintervals $[0, I_1], I = [I_1, I_2]$ and $[I_2, nT]$. We fix a sequence b_n with $\lim_{n\to\infty} b_n = \infty$ and $b_n = o(n^{1/2})$, for example $b_n = n^{1/4}$. It will become apparent that the significant contribution to the right side of (6.6.33) is due to intervals I with $|I| \le b_n$. We take n sufficiently large that for such I, $nt_{N-2} < I_1 \le nt_{N-1} \le I_2 < nT$.

Denote by $M_{\mathbf{k}}^{\le}$ and $M_{\mathbf{k}}^{>}$ respectively the contributions to the right side of (6.6.33) due to $|I| \le b_n$ and $|I| > b_n$. By factoring the exponential we can resum to obtain

$$M_{\mathbf{k}}^{\le} \equiv \sum_{\substack{I \ni nt_{N-1} \\ |I| \le b_n}} M(\kappa_1, \ldots, \kappa_{N-1}; nt_1, \ldots, nt_{N-2}, I_1)$$

$$\times \sum_{\omega:|\omega|=|I|} E_1(\omega, I) J[0, |I|] \hat{c}_{nT-I_2}(\kappa_N),$$

(6.6.34)

where

$$\begin{aligned}
E_1(\omega, I) &= \exp[i\kappa_{N-1} \cdot \omega(nt_{N-1} - I_1) + i\kappa_N \cdot \omega(I_2 - nt_{N-1})] \\
&= 1 + O(b_n n^{-1/2})
\end{aligned}$$

(6.6.35)

uniformly in ω and $|I| \le b_n$. For $|I| \le b_n$ and n sufficiently large, $I_1 \in [nt_{N-1}(1 - n^{-1/2}), nt_{N-1}]$. Hence by the induction hypothesis,

$$M(\kappa_1, \ldots, \kappa_{N-1}; nt_1, \ldots, nt_{N-2}, I_1)$$

$$= c_{I_1} \left[\exp\left[-\frac{1}{2d} \sum_{i=1}^{N-1} \left(k^{(i)} \right)^2 (t_i - t_{i-1}) \right] + E_2(I) \right],$$

(6.6.36)

where $|E_2(I)| = o(1)$ uniformly in $|I| \le b_n$. Similarly it follows from Theorem 6.6.1 that for $|I| \le b_n$,

$$\hat{c}_{nT-I_2}(\kappa_n) = c_{nT-I_2} \left[\exp\left[-\frac{1}{2d} \left(k^{(N)} \right)^2 (t_N - t_{N-1}) \right] + E_3(I) \right],$$

(6.6.37)

where $|E_3(I)| = o(1)$ uniformly in $|I| \leq b_n$.

Substituting (6.6.35)-(6.6.37) into (6.6.34) leads to

$$M_{\mathbf{k}}^{\leq} = \exp\left[-\frac{1}{2d}\sum_{i=1}^{N}\left(k^{(i)}\right)^2 (t_i - t_{i-1})\right] M_0^{\leq} + A, \qquad (6.6.38)$$

where

$$|A| \leq o(1) \sum_{\substack{I \ni nt_{N-1} \\ |I| \leq b_n}} c_{I_1} \sum_{|\omega|=|I|} |J[0,|I|]| c_{nT-I_2}. \qquad (6.6.39)$$

Since $M(0, nT) = c_{nT}$, we have $M_0^{\leq} = c_{nT} - M_0^{>}$. Hence

$$
\begin{aligned}
c_{nT}^{-1} M(\kappa, nT) &= \exp\left[-\frac{1}{2d}\sum_{i=1}^{N}\left(k^{(i)}\right)^2 (t_i - t_{i-1})\right]\left[1 - c_{nT}^{-1} M_0^{>}\right] \\
&\quad + c_{nT}^{-1} A + c_{nT}^{-1} M_{\mathbf{k}}^{>}.
\end{aligned} \qquad (6.6.40)
$$

Now by Theorem 6.1.1(a) and (6.6.39),

$$c_{nT}^{-1}|A| \leq o(1) \sum_{|I|=1}^{b_n} |I| \sum_{|\omega|=|I|} |J[0,|I|]| |z_c|^{|I|} \qquad (6.6.41)$$

(here $|I|$ is merely a summation index and the sum is no longer a sum over intervals). In (6.6.41) the factor $|I|$ counts the number of possibilities for $nt_{N-1} \in I$. Extending the summation over $|I|$ on the right side to infinity, it follows from the (absolute) bound on $\partial_z \hat{\Pi}$ of Theorem 6.2.9 that

$$c_{nT}^{-1}|A| \leq o(1). \qquad (6.6.42)$$

It suffices now to show that $c_{nT}^{-1} M_{\mathbf{k}}^{>} = o(1)$ as $n \to \infty$. Arguing as for $M_{\mathbf{k}}^{\leq}$,

$$c_{nT}^{-1}|M_{\mathbf{k}}^{>}| \leq O(1) \sum_{|I|=b_n+1}^{\infty} |I| \sum_{|\omega|=|I|} |J[0,|I|]| |z_c|^{|I|}. \qquad (6.6.43)$$

The right side goes to zero as $n \to \infty$ since by Theorem 6.2.9

$$\sum_{|I|=1}^{\infty} |I| \sum_{\omega:|\omega|=|I|} |J[0,|I|]| |z_c|^{|I|} < \infty.$$

$\square$

6.6.3   Tightness

Tightness is proved via the following lemma. Although not stated explicitly in Billingsley (1968), the lemma follows in a straightforward manner from Theorem 8.4 and results on pages 87-89 [both references to Billingsley (1968)]. For the statement of the lemma, we define a process closely related to $X_n(t)$ by

$$Y_n(t) = (Dn)^{-1/2}\omega(\lfloor nt \rfloor). \qquad (6.6.44)$$

Lemma 6.6.3 *The sequence $\{X_n\}$ is tight if there exist constants $K \geq 0$ and $a > 1/2$ such that for $0 \leq t_1 < t_2 < t_3 \leq 1$ and for all n,*

$$\langle |Y_n(t_2) - Y_n(t_1)|^{2a}|Y_n(t_3) - Y_n(t_2)|^{2a}\rangle_n \leq K|t_2 - t_1|^a|t_3 - t_2|^a, \quad (6.6.45)$$

where the angular brackets denote expectation with respect to the uniform measure on the set of n-step self-avoiding walks.

We will use Theorem 6.1.1 to show that (6.6.45) holds with $a = 1$. With $a = 1$ the left side of (6.6.45) is equal to

$$\frac{1}{D^2 n^2 c_n} \sum_{|\omega|=n} |\omega(nt_2) - \omega(nt_1)|^2 |\omega(nt_3) - \omega(nt_2)|^2 K[0,n], \qquad (6.6.46)$$

where the sum on the right side is over all n-step ordinary random walks, and brackets indicating integer part have been omitted to simplify the notation. The inequality

$$K[0,n] \leq K[0,nt_1]K[nt_1,nt_2]K[nt_2,nt_3]K[nt_3,n] \qquad (6.6.47)$$

allows for the replacement of the sum over ω by sums over independent subwalks on the intervals $[0,nt_1]$, $[nt_1,nt_2]$, $[nt_2,nt_3]$, $[nt_3,n]$, for an upper bound on (6.6.46). Also, by Theorem 6.1.1(a),

$$c_n^{-1} \leq \text{const.} c_{nt_1}^{-1} c_{nt_2-nt_1}^{-1} c_{nt_3-nt_2}^{-1} c_{n-nt_3}^{-1}. \qquad (6.6.48)$$

Using the above two inequalities in (6.6.46) gives

$$\langle |Y_n(t_2) - Y_n(t_1)|^2 |Y_n(t_3) - Y_n(t_2)|^2 \rangle_n$$
$$\leq \text{const.} n^{-2} c_{nt_2-nt_1}^{-1} \sum_{|\omega|=nt_2-nt_1} |\omega(nt_2 - nt_1)|^2 K[0, nt_2 - nt_1]$$
$$\times c_{nt_3-nt_2}^{-1} \sum_{|\omega|=nt_3-nt_2} |\omega(nt_3 - nt_2)|^2 K[0, nt_3 - nt_2]$$
$$= \text{const.} n^{-2} \langle |\omega(nt_2 - nt_1)|^2 \rangle_{nt_2-nt_1} \langle |\omega(nt_3 - nt_2)|^2 \rangle_{nt_3-nt_2}.$$

But by Theorem 6.1.1(b), the expectations on the right side are bounded above by a multiple of $(nt_2 - nt_1)(nt_3 - nt_2)$, which gives (6.6.45).

6.7 The infinite self-avoiding walk

In this section we give the proof of Theorem 6.1.9. Throughout the section we work only with the fully self-avoiding walk, i.e. $\tau = \infty$.

We begin by defining the infinite self-avoiding walk. Given $n \geq m$ and an m-step self-avoiding walk ω, we let $P_{m,n}(\omega)$ denote the fraction of n-step walks whose first m steps are given by ω. In other words, $P_{m,n}(\omega)$ is the fraction of n-step self-avoiding walks which *extend* ω. Then we define

$$P_m(\omega) = \lim_{n \to \infty} P_{m,n}(\omega) \qquad (6.7.1)$$

if the limit exists. If the limit does exist, then the probability measures P_m on m-step walks will be *consistent* in the sense that for each $n \geq m$ and each m-step self-avoiding walk ω,

$$P_m(\omega) = \sum_{\rho > \omega} P_n(\rho), \qquad (6.7.2)$$

where the sum is over all n-step self-avoiding walks ρ which extend ω. This consistency property allows for the definition via cylinder sets of a measure P_∞ on the set of all infinite self-avoiding walks. The measure P_∞ is the *infinite self-avoiding walk*.

Although the limit (6.7.1) is believed to exist in all dimensions, the closest results to existence of the limit in general dimensions are Theorems 7.4.2 and 7.4.5(a). These state that for any m-step self-avoiding walk ω which can be extended to an infinite self-avoiding walk, $\liminf_{n \to \infty} P_{m,n}(\omega) > 0$ and $\lim_{n \to \infty} P_{m,n+2}(\omega)/P_{m,n}(\omega) = 1$. (For bridges the situation is easier, and existence of the infinite bridge in all dimensions is proven in Section 8.3.) The remainder of this section is devoted to a proof that the limit in (6.7.1) exists if Ω is sufficiently large (with $d > 4$ for the spread-out model).

Given a nonnegative integer m, let $\mathbf{k} = (k^{(1)}, \ldots, k^{(m)})$, where $k^{(i)} \in [-\pi, \pi]^d$. Given $n \geq m$ and an n-step self-avoiding walk ω, let ω_m be the first m steps of ω, and

$$\mathbf{k} \cdot \omega_m = \sum_{i=1}^{m} k^{(i)} \cdot \omega(i).$$

Let

$$\bar{\varphi}_{m,n}(\mathbf{k}) = \sum_{|\omega|=n} e^{i\mathbf{k} \cdot \omega_m} K[0, n],$$

where the sum is over all n-step ordinary random walks and K was introduced in (5.2.6). We also define

$$\varphi_{m,n}(\mathbf{k}) = \sum_{|\omega|=m} e^{i\mathbf{k} \cdot \omega} P_{m,n}(\omega) = \frac{1}{c_n} \bar{\varphi}_{m,n}(\mathbf{k}),$$

where the sum is over all m-step self-avoiding walks (walks which do have self-intersections would make no contribution so the sum can also be considered to be over ordinary random walks). Since $\{P_{m,n}\}_n$ is clearly tight, a standard convergence theorem [see Billingsley (1968), p. 46] guarantees that existence of the limit (6.7.1) follows from existence of the limit

$$\varphi_m(\mathbf{k}) = \lim_{n \to \infty} \varphi_{m,n}(\mathbf{k}), \qquad (6.7.3)$$

for all $\mathbf{k} \in [-\pi, \pi]^{md}$.

We now recall some notation and a lemma from Section 5.2. For $m \geq 0$, we defined a quantity similar to the Fourier transform $\hat{G}_z(k)$ of the two-point function by

$$\Gamma_z(\mathbf{k}, m) = \sum_{n=m}^{\infty} \bar{\varphi}_{m,n}(\mathbf{k}) z^n.$$

Since $|\Gamma_z(\mathbf{k}, m)| \leq \chi(|z|)$, this power series converges for $|z| < z_c$. We also defined a quantity similar to $\hat{\Pi}_z(k)$, again for $m \geq 0$, by

$$\Psi_z(\mathbf{k}, m) = \sum_{s=m}^{\infty} z^s \sum_{|\omega|=s} e^{i\mathbf{k}\cdot\omega_m} J[0, s], \qquad (6.7.4)$$

where the sum is over ordinary random walks. It follows from the absolute bound on the lace expansion of Theorem 6.2.9 that for $v = 0, 1$, $\partial_z^v \Psi_z(\mathbf{k}, m)$ is bounded by a finite constant uniformly in $\mathbf{k}$ and $|z| \leq z_c$. For $j < m$ we define $\bar{\mathbf{k}}_j = (k^{(j+1)}, \ldots, k^{(m)})$. In Lemma 5.2.6, it was shown that for $m \geq 1$

$$\Gamma_z(\mathbf{k}, m) = z\Omega\hat{D}(\sum_{j=1}^{m} k^{(j)})\Gamma_z(\bar{\mathbf{k}}_1, m-1)$$

$$+ \sum_{s=2}^{m-1} z^s \sum_{|\omega|=s} \exp[i\sum_{j=1}^{m} k^{(j)} \cdot \omega(\min\{j, s\})] J[0, s]\Gamma_z(\bar{\mathbf{k}}_s, m-s)$$

$$+ \Psi_z(\mathbf{k}, m)\chi(z).$$

Let $N_z(\mathbf{k}, m) = \chi(z)^{-1}\Gamma_z(\mathbf{k}, m)$. The above identity and induction on m then can be used to argue that for $v = 0, 1$, $\partial_z^v N_z(\mathbf{k}, m)$ is uniformly bounded in $\mathbf{k}$ and $|z| \leq z_c$.

To prove existence of the limit (6.7.3), we proceed as follows. By definition of N_z,

$$\bar{\varphi}_{m,n}(\mathbf{k}) = \frac{1}{2\pi i} \oint N_z(\mathbf{k}, m)\chi(z)\frac{dz}{z^{n+1}} \qquad (6.7.5)$$

$$= N_{z_c}(\mathbf{k}, m)c_n + \frac{1}{2\pi i} \oint [N_z(\mathbf{k}, m) - N_{z_c}(\mathbf{k}, m)]\chi(z)\frac{dz}{z^{n+1}},$$

where the contour is a small circle centred at the origin. It suffices to show that the second term on the right side is $o(c_n)$, which by Theorem 6.1.1(a) is equivalent to $o(z_c^{-n})$. Hence by Lemma 6.3.3(*ii*) it suffices to show that for $|z| \leq z_c$,

$$\left| \frac{d}{dz}[N_z(\mathbf{k}, m) - N_{z_c}(\mathbf{k}, m)]\chi(z) \right| \leq O(|z_c - z|^{-1}), \tag{6.7.6}$$

for this would imply that the second term on the right side of (6.7.5) is $O(z_c^{-n} n^{-\alpha})$, for every $\alpha < 1$.

Now since $|\partial_z N_z|$ is uniformly bounded for $|z| \leq z_c$,

$$\frac{d}{dz}[N_z(\mathbf{k}, m) - N_{z_c}(\mathbf{k}, m)]\chi(z) = O(1)\chi(z) + O(|z_c - z|)\frac{d}{dz}\chi(z).$$

The first term on the right side is $O(|z_c - z|^{-1})$ by Theorem 6.2.10. It follows easily from Theorem 6.2.9 and Theorem 6.2.10 that the second term on the right side is also $O(|z_c - z|^{-1})$. Thus we have (6.7.6), and the proof of Theorem 6.1.9 is complete.

6.8 The bound on $c_n(0, x)$

In this section we prove Theorem 6.1.3, which states that for the nearest-neighbour model in sufficiently high dimensions or for the spread-out model with $d > 4$ and L sufficiently large, there is a positive constant B such that

$$\sup_{x \in \mathbb{Z}^d} c_n(0, x) \leq B\mu^n n^{-d/2}. \tag{6.8.1}$$

This shows that if as believed $c_n(0, x) \sim \text{const.}\mu^n n^{\alpha_{sing}-2}$ then $\alpha_{sing} - 2 \leq -d/2$, i.e. we have proved an inequality corresponding to the conjectured hyperscaling relation $\alpha_{sing} - 2 = -d\nu$. This is the only result stated in Section 6.1 which has not yet been proved for the nearest-neighbour model for all $d \geq 5$. We assume in this section that n and $\|x\|_1$ have the same parity; otherwise $c_n(0, x) = 0$.

For reasons to be discussed momentarily, with some reluctance we reintroduce a finite memory as in Section 5.2; recall that the estimates of Section 6.2 are uniform in the memory. Let $c_{n,\tau}(0, x)$ denote the number of n-step walks ending at x which are self-avoiding with memory τ. For any $\tau \in [0, \infty]$ and for any x,

$$c_n(0, x) \leq c_{n,\tau}(0, x) = \int_{[-\pi,\pi]^d} \frac{d^d k}{(2\pi)^d} \hat{c}_{n,\tau}(k)e^{-ik \cdot x}. \tag{6.8.2}$$

Denote by $\vec{\pi}$ the d-dimensional vector whose components are all π. Then since n and $\|x\|_1$ have the same parity,

$$\hat{c}_{n,\tau}(k - \vec{\pi}) = (-1)^n \hat{c}_{n,\tau}(k). \tag{6.8.3}$$

Using this fact, together with periodicity, we can then conclude that the integral on the right side of (6.8.2) is twice the integral over $[-\pi/2, \pi/2] \times [-\pi, \pi]^{d-1}$, and hence

$$c_n(0, x) \leq \frac{2}{(2\pi)^d} \int_{-\pi/2}^{\pi/2} dk_1 \int_{[-\pi,\pi]^{d-1}} dk_2 \ldots dk_d \, \hat{c}_{n,\tau}(k) e^{-ik \cdot x}. \tag{6.8.4}$$

To estimate the integral on the right side, we will first use contour integration to estimate the integrand. In Theorem 6.6.1 we have already obtained good control of $\hat{c}_n(k)$ for k of order $n^{-1/2}$, but now we require estimates valid for all k.

In (5.2.5), we introduced $z_c(k; \tau)$ as the radius of convergence of $\hat{G}_z(k; \tau)$. Thus $z_c(k; \tau)$ is the zero of $\hat{F}_z(k; \tau) = \hat{G}_z(k; \tau)^{-1}$ which is closest to the origin. Our contour integration method requires us to track this zero as a function of k, and because it will occur frequently we abbreviate the notation by writing $z_\tau(k) \equiv z_c(k; \tau)$ and $z_\tau \equiv z_c(0; \tau)$. In general zeroes of $\hat{F}_z(k; \tau)$ occur in complex conjugate pairs since $\hat{G}_z(k; \tau) = \hat{G}_{\bar{z}}(k; \tau)$, but it will be shown that for small k there is a unique, and hence real, zero near z_τ. Clearly $z_\tau(k) \geq z_\tau$, since $|\hat{G}_z(k; \tau)| \leq \hat{G}_z(0; \tau)$. Similarly, $z_\tau \leq z_c$. Without introducing a finite memory we are unable to control $\hat{\Pi}_z$ beyond z_c and therefore are unable to analyze $z_\infty(k)$ for k bounded away from 0. However with a memory we will see that there is an analytic continuation of $\hat{G}_z(k; \tau)$ beyond z_τ, which permits us to control the integral in (6.8.4).

For $\epsilon > 0$ and $\tau \geq 4$ we define

$$D_\tau(\epsilon) \equiv \{z : |z| \leq z_\tau[1 + \epsilon\tau^{-1} \log \tau]\}. \tag{6.8.5}$$

We wish to show that there is an analytic continuation of $\hat{G}_z(k; \tau)$ to the disk $D_\tau(\epsilon)$, for some positive ϵ. As a first step we have the following consequence of Theorem 6.4.2.

Theorem 6.8.1 *Let Ω be sufficiently large (with $d > 4$ for the spread-out model). There is a positive constant ϵ_0 (which may depend on d, Ω but not on τ) such that for any $\tau \leq \infty$, $\epsilon < \epsilon_0$ and $p \in (0, z_\tau]$,*

$$\|\delta_p^\epsilon \partial_p G_p(0, \cdot; \tau)\|_\infty \leq 5K\Omega^{s+2/d}, \tag{6.8.6}$$

$$\|\delta_p^\epsilon G_p(0, \cdot; \tau)\|_2^2 \leq 3K\Omega^{-1+s}, \tag{6.8.7}$$

and

$$||\,|x|^2\delta_p^\epsilon G_p(0,\cdot;\tau)||_\infty \le 3K\Omega^{-1+s+2/d}. \tag{6.8.8}$$

Here K is the constant of Lemma 6.2.2, $s = 0$ for the nearest-neighbour model, $s = 1/20$ for the spread-out model, and the $2/d$ in the exponents can be omitted for the nearest-neighbour model.

Proof. Theorem 6.4.2 (with finite memory) immediately gives finite bounds (uniform in τ) for the left sides, for any $\epsilon < \min\{(d-4)/4, 1\}$. In fact, if we take $\epsilon < \min\{(d-4)/8, 1/2\}$ (say), then Theorem 6.4.2 can be used to give finite bounds (again uniform in τ) on the derivatives with respect to ϵ of the norms on the left sides. Hence the left sides can be made as close as desired to their $\epsilon = 0$ values by taking ϵ sufficiently small, independent of τ. The theorem then follows from the $\epsilon = 0$ bounds given in Corollary 6.2.6 and (6.2.39). □

This leads to the following corollary, which extends Corollary 6.2.7 and Theorem 6.2.9. The corollary in particular provides an analytic continuation of $\hat{\Pi}_z(k;\tau)$ to $D_\tau(\epsilon_0)$, and hence a meromorphic continuation of $\hat{G}_z(k;\tau)$ to the same disk.

Corollary 6.8.2 Let Ω be sufficiently large (with $d > 4$ for the spread-out model). There is a positive constant K_1 which is independent of Ω such that for any $\epsilon \le \epsilon_0$, any τ, k, and any $z \in D_\tau(\epsilon_0)$,

$$|\hat{\Pi}_z(k;\tau)| \le K_1\Omega^{-1+s+2/d}, \tag{6.8.9}$$

$$|\partial_z\hat{\Pi}_z(k;\tau)| \le K_1\Omega^{s+2/d}, \tag{6.8.10}$$

$$|\partial_\mu\hat{\Pi}_z(k;\tau)| \le K_1\Omega^{-2+2s+2/d}|k_\mu|, \tag{6.8.11}$$

$$|\partial_\mu^2\hat{\Pi}_z(k;\tau)| \le K_1\Omega^{-2+2s+2/d}, \tag{6.8.12}$$

$$|\hat{\Pi}_z(0;\tau) - \hat{\Pi}_z(k;\tau)| \le K_1\Omega^{-2+2s+2/d}k^2. \tag{6.8.13}$$

Proof. By Theorem 5.4.4 and the second Remark below Theorem 5.4.2, the quantities on the left sides of (6.8.10)–(6.8.13) can be bounded in terms of various norms involving

$$H_z(0,x;\tau) = \sum_{n=1}^\tau c_{n,\tau}(0,x)z^n. \tag{6.8.14}$$

Here we have used the fact that for finite memory, all diagram subwalks have length at most τ, and hence the sum in the above equation can be truncated at $n = \tau$.

For example, the left side of the second inequality can be bounded in terms of the norms

$$\|\partial_z H_{|z|}(0,\cdot\,;\tau)\|_\infty \quad \text{and} \quad \|H_{|z|}(0,\cdot\,;\tau)\|_2.$$

Each of these norms can be bounded using Theorem 6.8.1. To see this, we note that for $z \in D_\tau(\epsilon)$ and $v \in \{0,1\}$,

$$\partial_z^v H_{|z|}(0,x;\tau) \leq \sum_{n=1}^{\tau} n^v c_{n,\tau}(0,x) z_\tau^{n-v}(1+\epsilon\tau^{-1}\log\tau)^{n-v}. \tag{6.8.15}$$

For $1 \leq n \leq \tau$,

$$(1+\epsilon\tau^{-1}\log\tau)^n \leq (1+\epsilon\tau^{-1}\log\tau)n^\epsilon. \tag{6.8.16}$$

Taking $v = 0$, using (6.8.16) in (6.8.15), and extending the summation to all $n \geq 1$ gives

$$\|H_{|z|}(0,\cdot\,;\tau)\|_\infty \leq (1+\epsilon\tau^{-1}\log\tau)\|\delta_z^\epsilon G_{z_\tau}(0,x;\tau)\|_\infty. \tag{6.8.17}$$

Similarly,

$$\|H_{|z|}(0,\cdot\,;\tau)\|_2 \leq (1+\epsilon\tau^{-1}\log\tau)\|\delta_z^\epsilon G_{z_\tau}(0,\cdot\,;\tau)\|_2. \tag{6.8.18}$$

The case of $v = 1$ is slightly more involved:

$$\partial_z H_{|z|}(0,x;\tau) \leq (1+\epsilon\tau^{-1}\log\tau)\sum_{n=1}^{\tau} n^{1+\epsilon} c_{n,\tau}(0,x) z_\tau^{n-1}. \tag{6.8.19}$$

Since

$$\delta_z^\epsilon \partial_z G_z(0,x;\tau) = \sum_{n=2}^\infty (n-1)^\epsilon n c_{n,\tau}(0,x) z^{n-1}, \tag{6.8.20}$$

it follows from (6.8.19) that

$$\|\partial_z H_{|z|}(0,\cdot\,;\tau)\|_\infty \leq (1+\epsilon\tau^{-1}\log\tau)\left(2^\epsilon\|\delta_z^\epsilon \partial_z G_{z_\tau}(0,\cdot\,;\tau)\|_\infty + 1\right). \tag{6.8.21}$$

The second bound of the corollary then follows just as in Section 6.2, using Theorem 6.8.1.

The other bounds are similar, apart from the last one, which follows by integration of the third bound. $\qquad\qquad\qquad\qquad\qquad\qquad\square$

By (5.2.18),
$$\hat{F}_z(k;\tau) = 1 - z\Omega\hat{D}(k) - \hat{\Pi}_z(k;\tau).$$

Also, by Corollary 6.2.7, $\hat{F}_{z_\tau}(0;\tau) = 0$. The next lemma provides an extension of Theorem 6.2.10 for finite memory.

Lemma 6.8.3 *Let* Ω *be sufficiently large (with* $d > 4$ *for the spread-out model). Then*

$$|\hat{F}_z(0; \tau)| \geq \frac{\Omega}{2}|z_\tau - z| \qquad (6.8.22)$$

for all τ, *and for all* $z \in D_\tau(\epsilon_0)$. *In particular,* z_τ *is the unique zero of* $\hat{F}_z(0; \tau)$ *in* $D_\tau(\epsilon_0)$.

Proof. The proof is identical to that of Theorem 6.2.10, using Corollary 6.8.2. $\qquad\square$

The next lemma gives bounds which allow for the estimation of $\hat{c}_{n,\tau}(k)$ by contour integration. There are separate estimates for two distinct sets of k values.

Lemma 6.8.4 *Let* Ω *be sufficiently large (with* $d > 4$ *for the spread-out model). There is a positive constant* c_1 *(depending on* d, Ω *but not on* τ*) such that the following hold.*
(a) For $k \in [-\pi/2, \pi/2] \times [-\pi, \pi]^{d-1}$ *with* $k^2 \leq c_1\tau^{-1}\log\tau$, $\hat{F}_z(k; \tau)$ *has a unique zero* $z_\tau(k)$ *in* $D_\tau(\epsilon_0)$, *which is in fact located inside* $D_\tau(\epsilon_0/2)$. *Moreover,* $z_\tau(k)$ *is a simple zero and*

$$|\hat{F}_z(k; \tau)| \geq \frac{\Omega}{2}|z - z_\tau(k)| \qquad (6.8.23)$$

for all τ *and all* $z \in D_\tau(\epsilon_0)$.
(b) There is a positive $\epsilon_1 < \epsilon_0$ *(depending on* d, Ω *but not on* τ*) such that for* $k \in [-\pi/2, \pi/2] \times [-\pi, \pi]^{d-1}$ *with* $k^2 \geq c_1\tau^{-1}\log\tau$, $\hat{F}_z(k; \tau)$ *has no zero in* $D_\tau(\epsilon_1)$. *Moreover there is a constant* c_2 *(depending on* d, Ω *but not on* τ*) such that for these values of* k

$$|\hat{F}_z(k; \tau)| \geq c_2[\tau^{-1}\log\tau + |z\arg z|] \qquad (6.8.24)$$

for all τ *and all* $z \in D_\tau(\epsilon_1)$. *(Here* $\arg z \in [-\pi, \pi]$.)*

Proof. (a) Consider k near the origin. By Lemma 6.8.3 and Rouché's Theorem, to see that there is a unique zero in $D_\tau(\epsilon_0)$ it suffices to show that $|\hat{F}_z(k; \tau) - \hat{F}_z(0; \tau)| < |\hat{F}_z(0; \tau)|$ on the boundary of $D_\tau(\epsilon_0)$. But the left side is equal to

$$\left|z\Omega[1 - \hat{D}(k)] + \hat{\Pi}_z(0; \tau) - \hat{\Pi}_z(k; \tau)\right| \leq c'k^2 \qquad (6.8.25)$$

by (6.8.13) and the fact that $1 - \hat{D}(k)$ is order k^2, where c' is independent of τ. By Lemma 6.8.3, on the boundary of $D_\tau(\epsilon_0)$

$$|\hat{F}_z(0; \tau)| \geq \frac{z_\tau\Omega\epsilon_0}{2}\tau^{-1}\log\tau. \qquad (6.8.26)$$

Since $z_\tau \geq z_0 = 1/\Omega$, uniqueness of the zero follows if

$$k^2 < \frac{\epsilon_0}{2c'}\tau^{-1}\log\tau. \tag{6.8.27}$$

Thus we take $c_1 < \epsilon_0/(2c')$.

To see that the zero is located in $D_\tau(\epsilon_0/2)$, we proceed as follows. Differentiation of the equation $\hat{F}_{z_\tau(k)}(k;\tau) = 0$ with respect to k_μ, together with Corollary 6.8.2, gives

$$|\partial_\mu z_\tau(k)| = \left|\frac{\partial_\mu \hat{F}}{\partial_z \hat{F}}\right| \leq \text{const.}|k_\mu| \tag{6.8.28}$$

for small k. Therefore

$$z_\tau(k) = z_\tau(0) + \int_0^1 \frac{d}{dt} z_\tau(tk)dt \leq z_\tau(0) + \text{const.}k^2, \tag{6.8.29}$$

with the constant independent of τ. This gives the desired result, if we take c_1 sufficiently small.

For the lower bound on $\hat{F}$, we have

$$
\begin{aligned}
|\hat{F}_z(k;\tau)| &= |\hat{F}_z(k;\tau) - \hat{F}_{z_\tau(k)}(k;\tau)| \\
&= |z_\tau(k) - z|\,|\Omega D(k) + \int_0^1 \partial_z \hat{\Pi}_{z_\tau + t(z_\tau(k)-z)}(k;\tau)dt| \\
&\geq \frac{\Omega}{2}|z - z_\tau(k)|,
\end{aligned}
\tag{6.8.30}
$$

for k near zero and Ω sufficiently large, by Corollary 6.8.2.

(b) It suffices to prove the lower bound (6.8.24). For this we add and subtract $\hat{\Pi}_z(0;\tau)$ to $\hat{F}_z(k;\tau)$, and then subtract $\hat{F}_{z_\tau}(0;\tau) = 0$, obtaining

$$\hat{F}_z(k;\tau) = \Omega(z_\tau - z\hat{D}(k)) + \hat{\Pi}_{z_\tau}(0;\tau) - \hat{\Pi}_z(0;\tau) + \hat{\Pi}_z(0;\tau) - \hat{\Pi}_z(k;\tau). \tag{6.8.31}$$

Using Corollary 6.8.2 then gives

$$|\hat{F}_z(k;\tau)| \geq \Omega|z_\tau - z\hat{D}(k)| - K_1\Omega^{s+2/d}|z_\tau - z| - K_1 k^2 \Omega^{-2+2s+2/d}. \tag{6.8.32}$$

For the middle term on the right side, we use

$$|z_\tau - z| \leq |z_\tau - z\hat{D}(k)| + |z|\,|1 - \hat{D}(k)|. \tag{6.8.33}$$

It follows from the fact that $|1 - \cos t| \leq t^2/2$ for all $t \in \mathbb{R}$ that $|1 - \hat{D}(k)|$ is bounded above by $k^2/(2d)$ for the nearest-neighbour model and by $(k^2\Omega^{2/d})/2$ for the spread-out model. We write this upper bound as $a(\Omega)k^2$. Then using $|z| \leq 2z_\tau$ we have

$$
\begin{aligned}
|\hat{F}_z(k;\tau)| \geq \ & (\Omega - K_1\Omega^{s+2/d})|z_\tau - z\hat{D}(k)| \\
& - [2z_\tau a(\Omega)K_1\Omega^{s+2/d} + K_1\Omega^{-2+2s+2/d}]k^2.
\end{aligned}
$$

Next we write $z = |z|e^{i\theta}$, and use the inequality $|a| + |b| \leq \sqrt{2}|a + ib|$ to obtain

$$
|z_\tau - z\hat{D}(k)| \geq \frac{1}{\sqrt{2}}\left[|z_\tau - |z|\hat{D}(k)\cos\theta| + |z\hat{D}(k)\sin\theta|\right]. \tag{6.8.34}
$$

Now for $z \in D_\tau(\epsilon_1)$, and for k^2 large as stated in the lemma but within a small sphere of radius $O(1)$ centred at the origin [so that $\hat{D}(k)$ is bounded away from zero], it follows from (6.2.5) that

$$
\begin{aligned}
|z\hat{D}(k)\cos\theta| \ & \leq \ z_\tau(1 + \epsilon_1\tau^{-1}\log\tau)\left(1 - \frac{k^2}{2\pi^2 d}\right) \\
& \leq \ z_\tau\left(1 - \frac{k^2}{4\pi^2 d}\right),
\end{aligned}
$$

if we choose ϵ_1 sufficiently small (independent of τ). For this range of k and for Ω sufficiently large we then have

$$
|\hat{F}_z(k;\tau)| \geq c_2[k^2 + |z||\arg z|] \tag{6.8.35}
$$

for some constant c_2 depending on d, Ω, but not on τ.

For k^2 at least $O(1)$ from the origin, $\hat{D}(k)$ is bounded away from 1. Hence by (6.8.9) and the fact that $z_\tau\Omega = 1 - \hat{\Pi}_{z_\tau}(0;\tau)$, for Ω sufficiently large there are $\beta, \beta' \in (0, 1)$ such that

$$
\begin{aligned}
|\hat{F}_z(k;\tau)| \ & \geq \ 1 - |z\Omega\hat{D}(k)| - |\hat{\Pi}_z(k;\tau)| \\
& \geq \ 1 - \beta(1 + o(1))(1 + \epsilon_1\tau^{-1}\log\tau) - o(1) \\
& > \ \beta'.
\end{aligned}
$$

[Here $o(1)$ denotes a quantity which goes to zero as Ω increases, uniformly in τ.]

This completes the proof. $\square$

We are now in a position to estimate $\hat{c}_{n,\tau}(k)$.

Lemma 6.8.5 *Let* Ω *be sufficiently large (with* $d > 4$ *for the spread-out model).*

(a) For $k \in [-\pi/2, \pi, 2] \times [-\pi, \pi]^{d-1}$ with $k^2 \leq c_1 \tau^{-1} \log \tau$,

$$\hat{c}_{n,\tau}(k) = z_\tau(k)^{-(n+1)} \left[-\partial_z \hat{F}_{z_\tau(k)}(k; \tau)^{-1} \right.$$
$$\left. + O[(1 + (\epsilon_0/3)\tau^{-1} \log \tau)^{-(n+1)} \log \tau] \right]. \quad (6.8.36)$$

(b) For $k \in [-\pi/2, \pi, 2] \times [-\pi, \pi]^{d-1}$ with $k^2 \geq c_1 \tau^{-1} \log \tau$,

$$\hat{c}_{n,\tau}(k) = z_\tau^{-(n+1)}(1 + \epsilon_1 \tau^{-1} \log \tau)^{-(n+1)} O[\log \tau]. \quad (6.8.37)$$

Here $O[f(n, \tau)]$ denotes a quantity which is bounded in absolute value by $const.|f(n, \tau)|$, with the constant independent of n, τ.

Proof. (a) Let C be the circle of radius $z_\tau/2$ centred at the origin, oriented clockwise. Then

$$\hat{c}_{n,\tau}(k) = \frac{1}{2\pi i} \oint_C \hat{G}_z(k; \tau) \frac{dz}{z^{n+1}}. \quad (6.8.38)$$

Since $|z_\tau(k)| \geq z_\tau$, $z_\tau(k)$ is not inside C. By Lemma 6.8.4 and the Residue Theorem, deforming the contour from C to the boundary $\partial D_\tau(\epsilon_0)$ of $D_\tau(\epsilon_0)$ gives

$$\hat{c}_{n,\tau}(k) = z_\tau(k)^{-(n+1)} [-\partial_z \hat{F}_{z_\tau(k)}(k; \tau)^{-1}$$
$$+ \frac{1}{2\pi i} \oint_{\partial D_\tau(\epsilon_0)} \hat{G}_z(k; \tau) \left(\frac{z_\tau(k)}{z} \right)^{n+1} dz].$$

To bound the integral on the right side we first note that since $z_\tau(k) \in D_\tau(\epsilon_0/2)$, for $z \in \partial D_\tau(\epsilon_0)$ we have

$$\left| \frac{z_\tau(k)}{z} \right|^{n+1} \leq O[(1 + (\epsilon_0/3)\tau^{-1} \log \tau)^{-(n+1)}]. \quad (6.8.39)$$

Also, by (6.8.23)

$$\oint_{\partial D_\tau(\epsilon_0)} |\hat{G}_z(k; \tau)||dz| \leq O[|\log|z_\tau(1 + \epsilon_0 \tau^{-1} \log \tau) - z_\tau(k)|, |]$$
$$\leq O[\log \tau]. \quad (6.8.40)$$

This proves (6.8.36).

(b) We perform contour integration as in part (a), deforming the contour to $\partial D_\tau(\epsilon_1)$. Here there is no singularity inside $\partial D_\tau(\epsilon_1)$, and only the error term contributes. Explicitly,

$$\hat{c}_{n,\tau}(k) = \frac{1}{2\pi i} \oint_{\partial D_\tau(\epsilon_1)} \frac{1}{\hat{F}_z(k; \tau) z^{n+1}} dz. \quad (6.8.41)$$

The factor $z^{-(n+1)}$ in the integrand is responsible for everything in the upper bound stated in the lemma except for the $\log \tau$, which comes from integrating $1/\hat{F}$ using the lower bound of Lemma 6.8.4(b). $\qquad\square$

We need one more lemma before proving Theorem 6.1.3. Recall from Lemma 1.2.3 that as the memory τ goes to infinity, z_τ converges to z_c. The next lemma gives a bound on the rate of this convergence.

Lemma 6.8.6 *Let Ω be sufficiently large (with $d > 4$ for the spread-out model) and $\epsilon < \min\{(d-4)/2, 1\}$. There is a K_1 (possibly depending on d, ϵ, Ω but not on τ) such that for all τ,*

$$0 \le z_c - z_\tau \le K_1 \tau^{-(1+\epsilon)}. \qquad (6.8.42)$$

Proof. Consider two memories $\sigma < \tau$. Then $z_\sigma \le z_\tau$. Since $\hat{F}_{z_\tau}(0; \tau) = \hat{F}_{z_\sigma}(0; \sigma) = 0$,

$$(z_\tau - z_\sigma)\Omega = [\hat{\Pi}_{z_\sigma}(0; \sigma) - \hat{\Pi}_{z_\sigma}(0; \tau)] + [\hat{\Pi}_{z_\sigma}(0; \tau) - \hat{\Pi}_{z_\tau}(0; \tau)]. \qquad (6.8.43)$$

Using (6.8.10) to estimate the second term on the right side, for some constant C we have

$$0 \le C\Omega(z_\tau - z_\sigma) \le \hat{\Pi}_{z_\sigma}(0; \sigma) - \hat{\Pi}_{z_\sigma}(0; \tau). \qquad (6.8.44)$$

By Lemma 5.4.5 and Corollary 6.2.6, the right side is bounded by a multiple of

$$\| \sum_{n=\sigma/6}^{\infty} c_{n,\sigma}(0, \cdot) z_\sigma^n \|_\infty, \qquad (6.8.45)$$

Inserting $1 \le [6n/\sigma]^{1+\epsilon}$ into the summation, the above norm is bounded above by a multiple of

$$\sigma^{-(1+\epsilon)} \delta_z^\epsilon \partial_z G_z(0, x; \sigma)|_{z=z_\sigma}. \qquad (6.8.46)$$

The bound on the fractional derivative of (6.8.46) given in Theorem 6.4.2, together with (6.8.44), then gives

$$0 \le z_\tau - z_\sigma \le K_1 \sigma^{-(1+\epsilon)}, \qquad (6.8.47)$$

for some constant K_1. Letting $\tau \to \infty$, we obtain (6.8.42). $\qquad\square$

Proof of Theorem 6.1.3. We now take $\tau = n^{1/b}$ with $b \in (1, 1 + \epsilon)$, and use (6.8.4). As a first observation, by Lemma 6.8.6 we have

$$\lim_{n \to \infty} \left(\frac{z_{n^{1/b}}}{z_c} \right)^n = \lim_{n \to \infty} [1 - O(n^{-(1+\epsilon)/b})]^n = 1. \qquad (6.8.48)$$

To estimate the integral of $|\hat{c}_{n,n^{1/b}}(k)|$, we divide the integral over k into two parts as in Lemma 6.8.5. For k as in part (b) of the lemma, we have from (6.8.48) and the lemma that

$$|\hat{c}_{n,n^{1/b}}(k)| \leq C\mu^n n^{-C'n^{1-1/b}} \log n \leq \mu^n O(n^{-d/2}). \qquad (6.8.49)$$

Integrating this over k as in part (b) of the lemma gives a bound of the desired form (in fact the decay is much better than required).

We now bound $\hat{c}_{n,n^{1/b}}$ for k as in part (a) of the lemma. The quantity in square brackets on the right side of (6.8.36) is bounded above uniformly in k and $\tau = n^{1/b}$, so we just concentrate on the factor $z_\tau(k)^{-n}$. By (6.8.48), for $\tau = n^{1/b}$ we have

$$z_\tau(k)^{-n} = [1 + o(1)]\mu^n \left(\frac{z_\tau}{z_\tau(k)} \right)^n. \qquad (6.8.50)$$

By (6.8.29),

$$z_\tau(k) = z_\tau \left[1 + z_\tau^{-1} \sum_{\mu=1}^d \int_0^1 \partial_\mu z_\tau(tk) k_\mu dt \right]. \qquad (6.8.51)$$

It suffices to show that the right side is bounded below by $z_\tau[1 + C_1 k^2]$, for some constant $C_1 > 0$ which is independent of both τ and small k. In fact, given such a bound we would have

$$|\hat{c}_{n,n^{1/b}}(k)| \leq C\mu^n[1 - C_2 k^2]^n \leq C\mu^n e^{-nC_2 k^2}, \qquad (6.8.52)$$

and extending the integration domain to $\mathbf{R}^d$ then gives an upper bound of the required form $\mu^n n^{-d/2}$.

We now complete the proof by obtaining such a lower bound on the right side of (6.8.51). As in (6.8.28), we have

$$\partial_\mu z_\tau(l) = \frac{\partial_\mu \hat{F}_{z_\tau(l)}(l; \tau)}{-\partial_z \hat{F}_{z_\tau(l)}(l; \tau)}. \qquad (6.8.53)$$

The denominator on the right side is positive and bounded above uniformly in τ and small l, by Corollary 6.8.2. For the numerator, we use $z_\tau(l) \geq \Omega^{-1}$ and Corollary 6.8.2 to obtain

$$\partial_\mu \hat{F}_{z_\tau(l)}(l; \tau) \geq \Omega^{-1} \sum_{x \in \Omega} x_\mu \sin l \cdot x - K_1 \Omega^{-2+2s+2/d}|l_\mu|. \qquad (6.8.54)$$

Expanding $\sin l \cdot x$ and using symmetry, the first term on the right side is given by $\Omega^{-1} \sum_{x \in \Omega} x_\mu^2 l_\mu$ plus a term of order l^3 which is negligible for l going to zero depending on n. Thus the first term dominates the second, and we have the desired lower bound. This completes the proof. $\qquad \square$

6.9 Notes

Section 6.1. The results stated in Section 6.1 for sufficiently large Ω have all been proven in Hara and Slade (1992a,1992b) for the nearest-neighbour model with $d \geq 5$, except for Theorem 6.1.4. Their proof relies on the fortuitous fact that the critical bubble diagram is not too large in five dimensions. Since the critical bubble diagram can be expected to diverge as $d \to 4^+$ (with any reasonable definition for noninteger dimensions), the proof is not entirely natural. A more natural proof (which we hope will one day be forthcoming) would rely on the fact that the bubble is finite rather than small. The methods used for the nearest-neighbour model with $d \geq 5$ are quite similar to those of Chapter 6, except for the proof of convergence of the lace expansion. The latter, while similar in spirit to the proof given in Section 6.2, is enormously more complex (and in fact is computer assisted) due to the fact that the small parameter cannot be taken to be arbitrarily small.

For the nearest-neighbour model with $d \geq 5$, Corollary 6.1.4 cannot be deduced from Theorem 6.1.3 since the latter has not yet been proven in this context. Instead, in Hara and Slade (1992a) (6.4.8) is extended to show that the supremum norm of the a-th derivative of the critical two-point function is finite [rather than just the $(1 + \epsilon)$-th derivative].

Section 6.2. The first proof of convergence of the lace expansion was for weakly self-avoiding walk with $d > 4$, in Brydges and Spencer (1985). The proof began by considering walks which are self-avoiding with finite memory τ, and then used an intricate induction on the memory. Later, in Yang and Klein (1988), the same methods were applied to a weakly self-avoiding walk taking steps of arbitrary length m parallel to the coordinate axes in $\mathbf{Z}^d$ with probability proportional to m^{-2}, and it was shown that if the self-avoidance is sufficiently weak then the scaling limit of the endpoint has a Cauchy distribution if $d > 2$ (which is the distribution of the scaling limit in any dimension for the corresponding random walk without the self-avoidance constraint). An alternate proof of Brydges and Spencer's results, which uses the lace expansion and an induction on finite memory but avoids the use of generating functions, is given in Golowich and Imbrie (1992).

In Slade (1987) and Slade (1989) it was proven that $\nu = 1/2$ and $\gamma = 1$ for the strictly self-avoiding walk in sufficiently high dimensions. Convergence of the expansion was proven using Lemma 6.2.1. No estimate was given of how high the dimension had to be.

The convergence proof used in Section 6.2 is the prototype for the proofs of convergence of the lace expansions for lattice trees and animals and for percolation, used to prove Theorems 5.5.1 and 5.5.2.

Section 6.3. The fractional derivative analysis is taken from Hara and Slade (1992a).

Section 6.4. This section follows the methods of Hara and Slade (1992a).

Section 6.5. The proof of mean-field behaviour for the correlation length in Theorem 6.1.5 is modelled on the proof of the analogous result for percolation in Hara (1990), and follows Hara and Slade (1992a).

Section 6.6. It was proven in Slade (1988, 1989) that the scaling limit of the self-avoiding walk is Gaussian in sufficiently high dimensions, using a finite-memory cut-off. Hara and Slade (1992a) used the fractional derivative argument presented here.

Section 6.7. The definition used here for the infinite self-avoiding walk was introduced in Lawler (1980). Lawler (1989) constructed the infinite self-avoiding walk in sufficiently high dimensions.

Section 6.8. The bound on $c_n(0, x)$ obtained in Section 6.8 is new. The method of proof takes its inspiration from Brydges and Spencer (1985). The proof is unsatisfactory in its use of finite memory; it should be possible to improve Theorem 6.1.4 to prove that $c_n(0, x)$ is asymptotic to a multiple of $n^{-d/2}$ (for fixed x, as $n \to \infty$), without making use of finite memory. To do so remains an open problem. No estimate has been made of how high the dimension need be for the proof to work for the nearest-neighbour model, but we would guess something on the order of $d \geq 10$.

Chapter 7

Pattern theorems

7.1 Patterns

In this chapter we shall prove a useful theorem due to Kesten (1963) about the occurrence of patterns on self-avoiding walks, and investigate a number of its applications. Briefly, a *pattern* is a (short) self-avoiding walk that occurs as part of a longer self-avoiding walk. Kesten's Pattern Theorem says that if a given pattern can possibly occur several times on a self-avoiding walk, then it must occur at least aN times on almost all N-step self-avoiding walks, for some $a > 0$ (in this context, "almost all" means "except for an exponentially small fraction"). This can be viewed as a weak analogue of classical "large deviations" estimates for the strong law of large numbers, which say that certain events have exponentially small probabilities [see for example Chapter 1 of Ellis (1985)].

Another statistic of interest regarding patterns is the frequency of occurrence of a particular pattern at the beginning of self-avoiding walks. In general dimension d, it is an open problem to prove that the fraction of N-step self-avoiding walks that begin with a given pattern converges as N tends to infinity. This has been done in certain special cases: for $d \geq 5$ (see Section 6.7), and for bridges in every dimension (see Section 8.3). The existence of such a limit would provide a natural definition of a probability measure for infinite self-avoiding walks. We can only prove the following weaker results in the general case: if P is a pattern that can occur at the beginning of an arbitrarily long self-avoiding walk, then the fraction of N-step self-avoiding walks beginning with this pattern is bounded away from zero as N tends to infinity; also, the ratio of these fractions for N and $N+2$ converges to one. These results and some extensions will be discussed in

Section 7.4. The proofs of these results rely heavily upon Kesten's original pattern theorem.

Kesten originally applied his pattern theorem to prove the following *ratio limit theorems*:

$$\lim_{N \to \infty} \frac{c_{N+2}}{c_N} = \mu^2, \tag{7.1.1}$$

$$\lim_{N \to \infty} \frac{q_{2N+2}}{q_{2N}} = \mu^2, \tag{7.1.2}$$

$$\lim_{N \to \infty} \frac{b_{N+1}}{b_N} = \mu. \tag{7.1.3}$$

We shall prove these results in Section 7.3. Unfortunately, the same methods do not allow us to prove

$$\lim_{N \to \infty} \frac{c_{N+1}}{c_N} = \mu. \tag{7.1.4}$$

Equation (7.1.4) in $\mathbf{Z}^d$ has only been proven for $d \geq 5$ (see Theorem 6.1.1); finding a proof for $d = 2, 3, 4$ remains an open problem. To get a feeling for why (7.1.1) is easier to prove than (7.1.4), consider the following easy exercise: prove that

$$c_{N+2} \geq c_N \qquad \text{for every } N. \tag{7.1.5}$$

The idea of the proof is given in Figure 7.1. (In detail: Given an N-step

$$x_1 = M$$

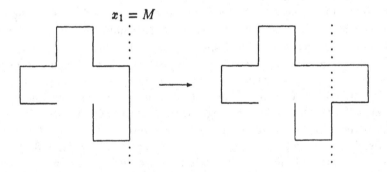

Figure 7.1: The idea behind the proof that $c_{N+2} \geq c_N$: increasing the length of a self-avoiding walk by 2.

self-avoiding walk ω, let $M = \max\{\omega_1(i) : 0 \leq i \leq N\}$. On the one hand, if a step of ω joins two points u and v in the hyperplane $x_1 = M$, then replace that step by three steps: u to $u + (1, 0, \ldots, 0)$ to $v + (1, 0, \ldots, 0)$ to v. On the other hand, if this hyperplane does not contain a step of ω, then it must contain an endpoint $[\omega(0)$ or $\omega(N)]$; in this case, add two steps to the end of the walk in the $+x_1$ direction. In either case we get an $(N + 2)$-step self-avoiding walk, from which ω can be determined unambiguously.) Now try the following exercise: prove that

$$c_{N+1} \geq c_N \qquad \text{for every } N. \qquad (7.1.6)$$

It is much harder to construct a one-to-one mapping from the set of N-step walks to the set of $(N + 1)$-step walks, but it can in fact be done; for the lengthy details, see O'Brien (1990). Finally, we observe that (7.1.6) is easy to prove on the triangular lattice and other lattices which are not bipartite (i.e. that contain self-avoiding polygons with an odd number of steps). On such lattices, it turns out that (7.1.4) can be proven by the methods of this chapter; see the Remark preceding Theorem 7.3.2.

Pattern theorems have found several other applications, including: evaluating the ergodicity properties of certain Monte Carlo algorithms (see Sections 9.4.1 and 9.4.2), investigating self-avoiding walks restricted to subsets of $\mathbf{Z}^d$ (see Section 8.2), and establishing the frequency of knots in three-dimensional self-avoiding polygons (see Section 8.4).

It is now time to make precise definitions about patterns and their occurrence. To begin with, we can take the word "pattern" to be a synonym for "self-avoiding walk".

Definition 7.1.1 *A pattern* $P = (p(0), \ldots, p(n))$ *is said to occur at the j-th step of the self-avoiding walk* $\omega = (\omega(0), \ldots, \omega(N))$ *if there exists a vector v in* $\mathbf{Z}^d$ *such that* $\omega(j + k) = p(k) + v$ *for every* $k = 0, \ldots, n$. *(Evidently, v must be* $\omega(j) - p(0)$.)

Definition 7.1.2 *Let* $\mathcal{S}_N$ *denote the set of N-step self-avoiding walks ω such that* $\omega(0) = 0$. *For $k \geq 0$ and P a pattern, let $c_N[k, P]$ denote the number of walks in $\mathcal{S}_N$ for which P occurs at no more than k different steps. Let $\mathcal{F}_N[P]$ denote the subset of walks in $\mathcal{S}_N$ for which P occurs at the 0-th step. We say that P is a* proper front pattern *if $\mathcal{F}_N[P]$ is non-empty for all sufficiently large N. We say that P is a* proper internal pattern *if for every k there is a self-avoiding walk on which P occurs at k or more different steps.*

Kesten's Pattern Theorem tells us that if P is a proper internal pattern, then there exists an $a > 0$ such that

$$\limsup_{N \to \infty} \left(c_N[aN, P] \right)^{1/N} < \mu. \qquad (7.1.7)$$

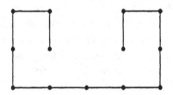

Figure 7.2: This pattern can occur twice on a self-avoiding walk in Z^2, but not three times.

The theorem actually tells us a bit more; see Theorem 7.2.3.

The basic results about "front patterns" say that if P is a proper front pattern, then

$$\liminf_{N \to \infty} \frac{|\mathcal{F}_N[P]|}{c_N} > 0 \qquad (7.1.8)$$

(where $|\cdot|$ denotes cardinality) and

$$\lim_{n \to \infty} \frac{|\mathcal{F}_{N+2}[P]|}{|\mathcal{F}_N[P]|} = \mu^2. \qquad (7.1.9)$$

Further results about front patterns appear in Section 7.4.

We take this opportunity to note some equivalent characterizations of proper internal patterns.

Proposition 7.1.3 *Let P be a pattern. The following are equivalent:*
(a) P is a proper internal pattern;
(b) There exists a cube $Q = \{x : 0 \le x_i \le b\}$ and a self-avoiding walk ϕ such that: P occurs at some step of ϕ, ϕ is contained in Q, and the two endpoints of ϕ are corners of Q;
(c) There exists a self-avoiding walk ω such that P occurs at three or more steps of ω.

We remark that if (b) above holds for P, then it is always possible to take

$$b = 2 + \max\{\|u - v\|_\infty : u \text{ and } v \text{ are sites of } P\}.$$

The proof of this proposition is straightforward, except for showing that (c) implies the other assertions. This implication is proven in Hammersley and Whittington (1985). Although we shall not require part (c) in this book, it is worth noting that the proposition is false if we change "three" to "two" in part (c), since there exist patterns which can occur at the beginning and end of a self-avoiding walk but never in the middle; an example in Z^2 is the pattern $NWS^2E^4N^2WS$ shown in Figure 7.2. (In this notation, N denotes a step in the direction $(0, 1)$ ["North"], etc.)

7.2 Kesten's Pattern Theorem

In this section, we shall formulate and prove Kesten's Pattern Theorem in its full generality, following the structure of Kesten's original proof. The general version of the theorem is a bit stronger than (7.1.7): in addition to specifying a pattern, one may also require that a certain amount of space around the pattern be unoccupied. The precise generalization is as follows.

Definition 7.2.1 *A cube is any set of the form*

$$Q = \{x \in \mathbf{Z}^d : a_i \leq x_i \leq a_i + b \text{ for all } i = 1, \ldots, d\},$$

where $a_1, \ldots, a_d$, and b are integers, and $b > 0$. Each cube has 2^d corners (extreme points of the convex hull). If Q is a cube as above, then let $\overline{Q}$ denote the cube which is two units larger in all directions:

$$\overline{Q} = \{x \in \mathbf{Z}^d : a_i - 2 \leq x_i \leq a_i + b + 2 \text{ for all } i = 1, \ldots, d\};$$

and let ∂Q denote the set of points in $\overline{Q}$ but not in Q (a kind of "external boundary" of Q),

$$\partial Q = \overline{Q} \setminus Q.$$

An outer point *of ∂Q is a point of ∂Q which has at least one nearest neighbour that is not in $\overline{Q}$.*

Definition 7.2.2 *Suppose that Q is a cube and P is an n-step pattern such that $p(0)$ and $p(n)$ are corners of Q, and $p(i) \in Q$ for every $i = 0, \ldots, n$ (in particular, P is a proper internal pattern; see Proposition 7.1.3). We say that (P, Q)* occurs at the j-th step *of the self-avoiding walk ω if there exists a v in $\mathbf{Z}^d$ such that $\omega(j + k) = p(k) + v$ for every $k = 0, \ldots, n$, and $\omega(i)$ is not in $Q + v$ for every $i < j$ and every $i > j + n$. For every $k \geq 0$, let $c_N[k, (P, Q)]$ denote the number of self-avoiding walks in S_N for which (P, Q) occurs at no more than k different steps.*

Theorem 7.2.3 *(a) Let Q be a cube and P be a pattern as in Definition 7.2.2. Then there exists an $a > 0$ such that*

$$\limsup_{N \to \infty} (c_N[aN, (P, Q)])^{1/N} < \mu. \tag{7.2.1}$$

(b) For any proper internal pattern P, there exists an $a > 0$ such that

$$\limsup_{N \to \infty} (c_N[aN, P])^{1/N} < \mu. \tag{7.2.2}$$

Before proceeding, we shall show that part (b) of the theorem [which is (7.1.7)] follows from part (a). Let P be a proper internal pattern, and choose ϕ and Q as in Proposition 7.1.3(b). Since P occurs on ϕ, any walk on which (ϕ, Q) occurs at m different steps must have P occurring at m or more different steps. Therefore

$$c_N[k, P] \leq c_N[k, (\phi, Q)] \quad \text{for every } k \geq 0,$$

from which we see that part (a) of Theorem 7.2.3 is indeed stronger than part (a). Thus it suffices to prove part (a).

The first ingredient in the proof of Theorem 7.2.3(a) is the following basic geometrical lemma. Part (a) of the lemma will construct a pattern that exactly fills a cube. Part (b) will show that we can splice a proper internal pattern onto a self-avoiding walk if we erase the part of the walk that occupies the corresponding enlarged cube $\overline{Q}$.

Lemma 7.2.4 *(a) Let Q be a cube in $\mathbf{Z}^d$. Then there exists a self-avoiding walk ω, whose endpoints are corners of Q, which is entirely contained in Q and visits every point of Q. (In particular, the number of steps in ω is one less than the number of points in Q.)*
(b) Let $P = (p(0), \ldots, p(k))$ be a pattern contained in the cube Q, whose endpoints are corners of Q. Let x and y be two distinct outer points of ∂Q. Then there exists a self-avoiding walk ω' with the following properties: its initial point is x and its last point is y; it is entirely contained in $\overline{Q}$; there exists a j such that $\omega'(j + i) = p(i)$ for every $i = 0, \ldots, k$; and $\omega'(i) \in \partial Q$ whenever $i < j$ or $i > j + k$. In particular, (P, Q) occurs at the j-th step of ω'.

Proof. (a) This is proven by induction on the dimension. It is obvious in one dimension. Assume that it has been proven for dimension $d - 1$. For simplicity, assume

$$Q = \{x \in \mathbf{Z}^d : 0 \leq x_i \leq b, i = 1, \ldots, d\}.$$

The intersection of Q with each of the hyperplanes $x_d = l$ ($l = 0, \ldots, b$) is a $(d - 1)$-dimensional cube embedded in $\mathbf{Z}^d$; call it Q^l. By the inductive hypothesis, there is a self-avoiding walk that starts at the origin and fills up Q^0 while staying inside Q^0, and whose last point is a corner of Q^0. Since every corner of Q^l is a nearest neighbour of a corner of Q^{l+1}, it is clear that we can find the desired walk for Q by filling up each of the $(d - 1)$-dimensional cubes $Q^0, \ldots, Q^d$ in turn.

(b) First choose a self-avoiding walk $\omega^{[1]}$ from x to $p(0)$ which does not touch y and contains only outer points of ∂Q (except necessarily for the

last two points of the walk). Then one can find a self-avoiding walk $\omega^{[2]}$
from $p(k)$ to y which stays in ∂Q and never touches $\omega^{[1]}$ (to do this, simply
avoid outer points until the very end). Then the desired walk ω' is the
concatenation of $\omega^{[1]}$, P, and $\omega^{[2]}$.   □

Now we must add to our stockpile of notation. We fix a positive integer
r which will be the "radius" of the cube Q of interest that occurs in the
statement of the Pattern Theorem. For a given N-step self-avoiding walk
ω, we extend Definition 7.2.1 by specifying cubes centred at points of ω:
for $j = 0, \ldots, N$, let

$$
\begin{aligned}
Q(j) &= \{x \in \mathbf{Z}^d : |x_i - \omega_i(j)| \le r \text{ for every } i = 1, \ldots, d\}, \\
\overline{Q}(j) &= \{x \in \mathbf{Z}^d : |x_i - \omega_i(j)| \le r + 2 \text{ for every } i = 1, \ldots, d\}, \\
\partial Q(j) &= \overline{Q}(j) \setminus Q(j).
\end{aligned}
$$

We say that E^* *occurs at the j-th step of ω* if $Q(j)$ is completely covered
by ω [i.e. for every v in $Q(j)$ there exists an i such that $v = \omega(i)$]. See
Figure 7.3. For every $k \ge 1$, we say that E_k *occurs at the j-th step of ω* if
at least k points of $\overline{Q}(j)$ are covered by ω; and we say that $\tilde{E}_k$ *occurs at
the j-th step of ω* if E^* or E_k (or both) occur there.

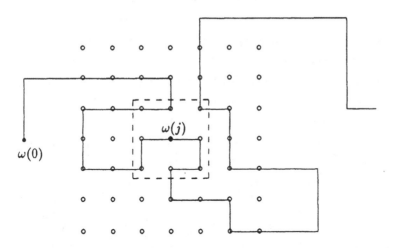

Figure 7.3: An example in $\mathbf{Z}^2$ with $r = 1$: The nine sites inside the dashed
box comprise $Q(j)$. The sites of $\overline{Q}(j)$ are marked with o. Both E^* and E_{29}
occur at the j-th step of this walk.

In the following, we will use E to denote any of E^*, E_k, or $\tilde{E}_k$. If m
is a positive integer, we say that $E(m)$ *occurs at the j-th step of ω* if E

occurs at the m-th step of the $2m$-step walk $(\omega(j - m), \ldots, \omega(j + m))$. [If $j - m < 0$ or $j + m > N$, then an obvious modification must be made in this definition: for $j - m < 0$, it means that E occurs at the j-th step of $(\omega(0), \ldots, \omega(j + m))$; for $j + m > N$, it means that E occurs at the m-th step of $(\omega(j - m), \ldots, \omega(N))$.] In particular, if $E(m)$ occurs at the j-th step of ω, then E occurs at the j-th step of ω; this would not necessarily be true if we replaced E by (P, Q). For every $k \geq 0$, let $c_N[k, E]$ (respectively, $c_N[k, E(m)]$) denote the number of self-avoiding walks in $\mathcal{S}_N$ for which E (respectively, $E(m)$) occurs at no more than k different steps. Observe that $c_N[k, E(m)]$ is non-increasing in m for fixed N and k because occurrences of $E(m)$ are more frequent as m increases.

The next lemma says that if E occurs on almost all walks, then (for some m) $E(m)$ must occur on almost all walks (in fact, it must occur *often* on almost all walks). Thus, if a self-avoiding walk is likely to fill a cube, then it is also likely to fill a cube within some bounded number of steps.

Lemma 7.2.5 *If*

$$\liminf_{N \to \infty} (c_N[0, E])^{1/N} < \mu, \tag{7.2.3}$$

then there exists an $a_1 > 0$ and an integer m such that

$$\limsup_{N \to \infty} (c_N[a_1 N, E(m)])^{1/N} < \mu. \tag{7.2.4}$$

Proof. Since $c_N[0, E] = c_N[0, E(N)]$, it follows that there exist $\epsilon > 0$ and an integer m such that

$$c_m[0, E(m)] < (\mu(1 - \epsilon))^m$$

and

$$c_m < (\mu(1 + \epsilon))^m.$$

Consider an N-step self-avoiding walk ω, and let $M = \lfloor N/m \rfloor$. If $E(m)$ occurs at most k times in ω, then $E(m)$ occurs in at most k of the M m-step subwalks

$$(\omega((i - 1)m), \omega((i - 1)m + 1), \ldots, \omega(im)) \qquad (i = 1, \ldots, M).$$

Counting the number of ways in which k or fewer of these subwalks can contain an occurrence of $E(m)$ (and remembering to count the last $N - Mm$ steps of ω), we are led to the bound

$$c_N[k, E(m)] \leq \sum_{j=0}^{k} \binom{M}{j} (c_m)^j (c_m[0, E(m)])^{M-j} c_{N-Mm} \tag{7.2.5}$$

$$\leq \mu^{mM} c_{N-Mm} \sum_{j=0}^{k} \binom{M}{j} (1 + \epsilon)^{jm} (1 - \epsilon)^{Mm-jm}.$$

It suffices to show that there is a $\rho > 0$ and a $t < 1$ such that

$$c_N[\rho M, E(m)]^{1/M} < t\mu^m \tag{7.2.6}$$

for all sufficiently large M, since this gives (7.2.4) whenever $0 < a_1 < \rho/m$. But if ρ is a small positive number, then

$$\sum_{j=0}^{\rho M} \binom{M}{j} (1+\epsilon)^{jm} (1-\epsilon)^{Mm-jm} \tag{7.2.7}$$

$$\leq (\rho M + 1) \binom{M}{\rho M} \left(\frac{1+\epsilon}{1-\epsilon}\right)^{\rho M m} (1-\epsilon)^{Mm}.$$

(For readability, we often write ρM instead of $\lfloor \rho M \rfloor$.) As $M \to \infty$, the M-th root of the right-hand side of (7.2.7) converges to

$$\frac{1}{\rho^\rho (1-\rho)^{1-\rho}} \left(\frac{1+\epsilon}{1-\epsilon}\right)^{\rho m} (1-\epsilon)^m,$$

which is less than 1 whenever $0 < \rho < \rho_0$, for some sufficiently small ρ_0. Combining this with (7.2.5), we see that (7.2.6) holds if $0 < \rho < \rho_0$ and M is sufficiently large. $\qquad \square$

Remark. Although we will not need this fact, it is worth pointing out that the lim inf in (7.2.3) is in fact a limit. This follows from $c_{N+M}[0, E] \leq c_N[0, E]c_M[0, E]$ and Lemma 1.2.2.

The next lemma is the heart of the proof of the Pattern Theorem. It says that almost all walks fill some cube (of the fixed radius r). The starting point of the proof is the observation that all walks cover at least $r+3$ points of the cube of radius $r+2$ centred at the origin; so if the lemma were false, then there would exist a K such that almost all walks cover K points of some cube (and in fact many cubes), but almost never cover $K+1$ points of any cube. This is used to obtain a contradiction.

Lemma 7.2.6 $\liminf_{N\to\infty} c_N[0, E^*]^{1/N} < \mu$.

Proof. Assume that the lemma is false, i.e. assume that

$$\lim_{N\to\infty} c_N[0, E^*]^{1/N} = \mu. \tag{7.2.8}$$

We make three observations: First, $c_N[0, \tilde{E}_k]$ is a nondecreasing function of k. Secondly, if E^* does not occur on a given walk then $E_{(2r+5)^d}$ cannot occur; therefore

$$c_N[0, E^*] \leq c_N[0, \tilde{E}_{(2r+5)^d}] \leq c_N, \tag{7.2.9}$$

and hence (7.2.8) implies that

$$\lim_{N\to\infty} c_N[0,\tilde{E}_{(2r+5)^d}]^{1/N} = \mu. \tag{7.2.10}$$

Thirdly, $c_N[0,\tilde{E}_{r+3}] = 0$ for all $N \geq r+2$ [since the first $r+3$ points of any walk ω must be in $\overline{Q}(0)$]. We conclude from these observations that there exists a K [with $r+3 \leq K < (2r+5)^d$] such that

$$\liminf_{N\to\infty} c_N[0,\tilde{E}_K]^{1/N} < \mu \tag{7.2.11}$$

and

$$\liminf_{N\to\infty} c_N[0,\tilde{E}_{K+1}]^{1/N} = \mu. \tag{7.2.12}$$

By (7.2.11) and Lemma 7.2.5, there exist an $a_1 > 0$ and an integer m such that

$$\limsup_{N\to\infty} c_N[a_1 N, \tilde{E}_K(m)]^{1/N} < \mu. \tag{7.2.13}$$

Define the set of self-avoiding walks

$$T_N = \{\omega \in S_N : \tilde{E}_{K+1} \text{ never occurs}; E_K(m) \text{ occurs at least } a_1 N \text{ times}\}. \tag{7.2.14}$$

Observe that replacing $E_K(m)$ by $\tilde{E}_K(m)$ in (7.2.14) does not change anything, since the condition that $\tilde{E}_{K+1}$ never occurs ensures that $E^*(m)$ never occurs. The cardinality of T_N satisfies

$$|T_N| \geq c_N[0,\tilde{E}_{K+1}] - c_N[a_1 N, \tilde{E}_K(m)], \tag{7.2.15}$$

and therefore, by (7.2.12) and (7.2.13),

$$\lim_{N\to\infty} |T_N|^{1/N} = \mu. \tag{7.2.16}$$

Thus, there is a number K such that it is not unusual to find lots of cubes with exactly K points occupied and *no* cubes with more than K points occupied. The rest of the proof is simply a matter of counting. The main idea is the following. Given such a walk $\omega \in T_N$, consider the collection of all cubes that have exactly K points covered. Remove the pieces of ω that cover a particular (small) subcollection of these cubes, and consider all possible ways of replacing them with pieces that entirely fill the same cubes. This is not a one-to-one transformation, and the length of the resulting walk is a bit different, but we can still arrange it so that the number of resulting walks is larger than $|T_N|$ by an exponential factor, and this will contradict (7.2.16).

Suppose that ω is an N-step self-avoiding walk such that $\tilde{E}_{K+1}$ never occurs on ω and $E_K(m)$ occurs at the j_1-th, j_2-th,..., j_s-th steps of ω (and perhaps at other steps as well). Suppose in addition that

$$0 < j_1 - m, \ j_s + m < N, \ \text{and} \ j_l + m < j_{l+1} - m \ \text{for all} \ l = 1, \ldots, s-1 \tag{7.2.17}$$

and

$$\overline{Q}(j_1), \ldots, \overline{Q}(j_s) \ \text{are pairwise disjoint.} \tag{7.2.18}$$

For $l = 1, \ldots, s$, let

$$\sigma_l = \min\{i : \omega(i) \in \overline{Q}(j_l)\} \quad \text{and} \quad \eta = \max\{i : \omega(i) \in \overline{Q}(j_l)\}.$$

Since $E_K(m)$ occurs at the j_l-th step and E_{K+1} does not occur at the j_l-th step, there must be exactly K points of $\overline{Q}(j_l)$ that are occupied by points of ω, and those points must lie between $\omega(j_l - m)$ and $\omega(j_l + m)$ on the walk. Therefore $j_l - m \leq \sigma_l < j_l < \eta \leq j_l + m$ for every l. Consider all possible ways of replacing each subwalk $(\omega(\sigma_l), \ldots, \omega(\eta))$ by a subwalk that stays inside $\overline{Q}(j_l)$ and completely covers $Q(j_l)$ [such subwalks exist by Lemma 7.2.4; we can do this operation simultaneously for all subwalks because we have ensured that there is no overlap amongst the subwalks nor amongst the cubes $\overline{Q}(j_l)$]. The result is always a self-avoiding walk ψ on which E^* occurs at least s times, and whose length N' satisfies

$$N' < N + s(2r + 5)^d. \tag{7.2.19}$$

Now consider all triples (ω, ψ, J) where: ω is a self-avoiding walk in T_N; $J = \{j_1, \ldots, j_s\}$ is a subset of $\{1, \ldots, N\}$ such that (7.2.17) and (7.2.18) hold, $E_K(m)$ occurs at each j_l in J, and $s = \lfloor \delta N \rfloor$ (here δ is a small positive number that will be specified at the end of the proof); and ψ is a self-avoiding walk that can be obtained from ω and J by the procedure of the preceding paragraph. We shall estimate the number of such triples both from above and below to obtain a contradiction. For both estimates, we shall use the observation that each cube $\overline{Q}(j)$ intersects exactly $V \equiv (4r + 9)^d$ cubes of "radius" $r + 2$ [this is because $\overline{Q}(j)$ intersects the cube of radius $r + 2$ centred at x if and only if $\|\omega(j) - x\|_\infty \leq 2(r + 2)$].

First, the number of such triples is at least the cardinality of T_N times the minimum number of possible choices of J for walks ω in T_N. Each ω in T_N contains at least $a_1 N$ occurrences of $E_K(m)$, and so we can find $h_1 < \ldots < h_u$, where $u = \lfloor a_1 N/((2m + 2)V) \rfloor - 2$, such that (i) $E_K(m)$ occurs at the h_l-th step of ω for every $l = 1, \ldots, u$, (ii) $0 < h_1 - m$, $h_u + m < N$, and $h_l + m < h_{l+1} - m$ for every $l = 1, \ldots, u - 1$, and (iii) the cubes $\overline{Q}(h_1), \ldots, \overline{Q}(h_u)$ are pairwise disjoint. Clearly, any subset of

$\{h_1, \ldots, h_u\}$ that has cardinality $\lfloor \delta N \rfloor$ is a possible choice for J. So if we set $\rho = a_1/((2m+2)V)$, then (dropping $\lfloor \cdot \rfloor$ from the notation)

$$\text{number of triples} \geq |T_N| \binom{\rho N - 2}{\delta N}. \tag{7.2.20}$$

For an upper bound, consider a triple (ω, ψ, J). Observe that E^* occurs at least $|J| = \lfloor \delta N \rfloor$ times on ψ; it may occur more than $|J|$ times because making a change in a cube $\overline{Q}(j_l)$ can produce occurrences of E^* in some of the cubes of radius $r+2$ that intersect $\overline{Q}(j_l)$. However, since E^* never occurs on ω, we infer that E^* occurs no more than $V|J|$ times on ω. Therefore, given ψ, there are at most $\binom{V\delta N}{\delta N}$ possibilities for the locations of the cubes $\overline{Q}(j_l)$, $l = 1, \ldots, |J|$. Given ψ and the locations of these $|J|$ cubes, each cube $\overline{Q}(j_l)$ determines a subwalk of ψ that replaced some subwalk of ω. Since each of the replaced subwalks of ω had length $2m$ or less, there are at most $(\sum_{i=0}^{2m} c_i)^{\delta N}$ possibilities for ω if we know both ψ and the locations of the $|J|$ cubes. Finally, if we know ω and the locations of the cubes, then J is uniquely determined. So if we define $Z = \sum_{i=0}^{2m} c_i$, then using $\binom{V\delta N}{\delta N} \leq 2^{V\delta N}$ and (7.2.19) we see that

$$\text{number of triples} \leq 2^{V\delta N} Z^{\delta N} \sum_{i=0}^{N+(2r+5)^d \delta N} c_i. \tag{7.2.21}$$

We now combine (7.2.20) and (7.2.21), take N-th roots, and let $N \to \infty$; by (7.2.16), we obtain

$$\mu \frac{\rho^\rho}{\delta^\delta (\rho - \delta)^{\rho - \delta}} \leq 2^{V\delta} \mu^{1+(2r+5)^d \delta} Z^\delta.$$

Setting $Y = 2^V \mu^{(2r+5)^d} Z$ and $t = \delta/\rho$ gives

$$1 \leq (t^t (1-t)^{1-t} Y^t)^\rho.$$

To obtain a contradiction, then, it suffices to show that the function $f(t) = t^t(1-t)^{1-t}Y^t$ is less than 1 for sufficiently small $t > 0$; this is true because $\lim_{t \searrow 0} f(t) = 1$ and $\lim_{t \searrow 0} f'(t) = -\infty$. This contradiction proves the lemma. $\square$

We are now ready to prove Kesten's Pattern Theorem. The ideas for this proof are really the same as those already used in the proof of Lemma 7.2.6.

Proof of Theorem 7.2.3. First, assume without loss of generality that the cube in the statement of the theorem is

$$Q = \{x \in \mathbf{Z}^d : |x_i| \leq r, i = 1, \ldots, d\}.$$

Assume that the theorem is false; then for every $a > 0$,

$$\limsup_{N \to \infty} c_N[aN, (P, Q)]^{1/N} = \mu. \tag{7.2.22}$$

We shall say that E^{**} *occurs at the j-th step of ω* if the cube $\overline{Q}(j)$ is completely covered by ω. By Lemmas 7.2.6 and 7.2.5, there exist $a' > 0$ and m' such that

$$\limsup_{N \to \infty} c_N[a'N, E^{**}(m')]^{1/N} < \mu. \tag{7.2.23}$$

Let $a > 0$ be a small unspecified number, and let H_N denote the following set of walks:

$$H_N = \{\omega \in \mathcal{S}_N : (P, Q) \text{ occurs at most } aN \text{ times on } \omega;$$
$$E^{**}(m') \text{ occurs at least } a'N \text{ times } \}.$$

The cardinality of H_N satisfies

$$|H_N| \geq c_N[aN, (P, Q)] - c_N[a'N, E^{**}(m)],$$

and therefore, by (7.2.22) and (7.2.23),

$$\lim_{N \to \infty} |H_N|^{1/N} = \mu. \tag{7.2.24}$$

Let δ be a small positive number, to be specified at the end of the proof. Consider all triples (ω, v, J) such that: ω is in H_N; $J = \{j_1, \ldots, j_s\}$ is a subset of $\{1, \ldots, N\}$ such that $E^{**}(m')$ occurs at each j_l, (7.2.17) holds with m replaced by m', and $s = \lfloor \delta N \rfloor$; and v is a self-avoiding walk obtained by replacing the occurrence of $E^{**}(m')$ at each j_l by an occurrence of (P, Q), analogously to the method described in the proof of Lemma 7.2.6 [σ_l and τ_l are defined in the same way, and we use part (b) of Lemma 7.2.4]. We remark that the occurrences of $E^{**}(m')$ guarantee that (7.2.18) holds. Arguing as we did for (7.2.20), we see that

$$\text{number of triples} \geq |H_N| \binom{\rho N - 2}{\delta N}, \tag{7.2.25}$$

where now $\rho = a'/(2m' + 2)$. For the upper bound, we use the fact that v has at most $aN + 2m'V\delta N$ occurrences of (P, Q). [This allows for (i) at most aN occurrences of (P, Q) on ω, and (ii) the possibility that changing a single occurrence of $E^{**}(m')$ to a (P, Q) may create several other occurrences of (P, Q) either by creating additional occurrences of P or by vacating sites of other cubes.] Also, note that v has at most N steps. Therefore the analogue of (7.2.21) here is

$$\text{number of triples} \le 2^{aN+2m'V\delta N} Z'^{\delta N} \sum_{i=0}^{N} c_i, \qquad (7.2.26)$$

where $Z' = \sum_{i=0}^{2m'} c_i$. We now combine (7.2.25) and (7.2.26), put $a = \delta$, take N-th roots, and let $N \to \infty$; by (7.2.24), we obtain

$$\mu \frac{\rho^\rho}{\delta^\delta (\rho - \delta)^{\rho-\delta}} \le 2^{\delta+2m'V\delta} Z'^{\delta} \mu.$$

As in the proof of Lemma 7.2.6, this leads to a contradiction for sufficiently small δ, and so the theorem is proven.   □

7.3   The main ratio limit theorem

The principal task of this section is to prove Equations (7.1.1), (7.1.2), and (7.1.3). The proof of each will be based on Lemma 7.3.1 and Theorem 7.3.2, which will also be used in the next section as the basis for analogous results for walks with specified end patterns.

   Lemma 7.3.1 gives three conditions which together are sufficient for the ratio limit theorems to hold. The first two conditions will be relatively easy to verify in our cases of interest; the third will follow from Theorem 7.3.2 below.

Lemma 7.3.1 *Let $\{a_N\}$ be a sequence of positive numbers and let $\phi_N = a_{N+2}/a_N$. Assume that*
(i) $\lim_{N \to \infty} a_N^{1/N} = \mu$,
(ii) $\liminf_{N \to \infty} \phi_N > 0$, and
(iii) there exists a constant $D > 0$ such that

$$\phi_N \phi_{N+2} \ge (\phi_N)^2 - \frac{D}{N} \qquad (7.3.1)$$

for all sufficiently large N. Then

$$\lim_{N \to \infty} \phi_N = \mu^2. \qquad (7.3.2)$$

Proof.   First observe that (ii) and (iii) imply that there exists a constant $B > 0$ such that

$$\phi_{N+2} \ge \phi_N - \frac{B}{N} \qquad \text{for all sufficiently large } N. \qquad (7.3.3)$$

Let $\sigma_N = \phi_N - \mu^2$. To prove the lemma, we shall show (by contradiction) that the lim sup of σ_N cannot be strictly positive, nor can the lim inf be strictly negative.

First assume that $\limsup_{N \to \infty} \sigma_N > 0$. Then there exists an $\epsilon > 0$ (possibly $\epsilon = +\infty$) and a sequence $N(1) < N(2) < \ldots$ such that $\lim_{j \to \infty} \sigma_{N(j)} = \epsilon$. For each $j \geq 1$, define

$$M(j) = \left\lfloor \frac{N(j)\sigma_{N(j)}}{2B} \right\rfloor ;$$

note that $M(j) \to \infty$ as $j \to \infty$. For sufficiently large j and every $0 \leq k < M(j)$, (7.3.3) implies that

$$\phi_{N(j)+2k} \geq \phi_{N(j)} - \frac{kB}{N(j)}$$

$$\geq \mu^2 + \sigma_{N(j)} - \frac{M(j)B}{N(j)}$$

$$\geq \mu^2 + \frac{\sigma_{N(j)}}{2}.$$

Therefore

$$\frac{a_{N(j)+2M(j)}}{a_{N(j)}} = \prod_{k=0}^{M(j)-1} \phi_{N(j)+2k} \geq \left(\mu^2 + \frac{\sigma_{N(j)}}{2}\right)^{M(j)}$$

Take $M(j)$-th roots of this inequality, and let $j \to \infty$, obtaining $\mu^2 \geq \mu^2 + \epsilon/2$, which is a contradiction. Therefore $\limsup_{N \to \infty} \sigma_N \leq 0$.

Next, assume that $\liminf_{N \to \infty} \sigma_N < 0$. The procedure is similar to that of the preceding paragraph. Since σ_N is bounded below, there exists an $\epsilon > 0$ and a sequence $N(1) < N(2) < \ldots$ such that $\lim_{j \to \infty} \sigma_{N(j)} = -\epsilon$, and such that $\sigma_{N(j)} < 0$ for every j. Without loss of generality, we can assume that the constant B of (7.3.3) satisfies $B \geq \mu^2$. For each $j \geq 1$, define

$$M(j) = \left\lfloor \frac{N(j)|\sigma_{N(j)}|}{4B} \right\rfloor ;$$

since $-\mu^2 < \sigma_{N(j)} < 0$, it follows that

$$M(j) \leq \frac{N(j)\mu^2}{4B} \leq \frac{N(j)}{4}.$$

For sufficiently large j and every $0 < k \leq M(j)$, (7.3.3) implies that

$$\phi_{N(j)-2k} \leq \phi_{N(j)} + \frac{kB}{N(j) - 2k}$$

$$\leq \quad \mu^2 - |\sigma_{N(j)}| + \frac{N(j)|\sigma_{N(j)}|}{4(N(j) - N(j)/2)}$$

$$= \quad \mu^2 - \frac{|\sigma_{N(j)}|}{2}.$$

As before, we obtain

$$\frac{a_{N(j)}}{a_{N(j)-2M(j)}} = \prod_{k=1}^{M(j)} \phi_{N(j)-2k} \leq \left(\mu^2 - \frac{|\sigma_{N(j)}|}{2}\right)^{M(j)}$$

We take $M(j)$-th roots of this inequality and let $j \to \infty$, obtaining the contradiction $\mu^2 \leq \mu^2 - \epsilon/2$. Therefore $\liminf_{N\to\infty} \sigma_N \geq 0$. □

Remark. It is apparent that Lemma 7.3.2 remains true if we replace $N + 2$ by $N + 1$ everywhere. Our inability to prove that $c_{N+1}/c_N \to \mu$ in $\mathbf{Z}^d$ ($d = 2, 3, 4$) is due to the failure of our proof of the corresponding analogue of the next theorem. As will become clear during the course of the proof, the reason for this failure can be seen most simply in Figure 7.4: there does not exist a pair of patterns U and V in $\mathbf{Z}^d$ having the same endpoints whose lengths differ by 1. However on a lattice where such a pair of patterns exists, for example the triangular lattice, we can modify our argument easily to show that $c_{N+1}/c_N \to \mu$ on that lattice.

Theorem 7.3.2 *There exists a constant $D > 0$ such that*

$$\phi_N \phi_{N+2} \geq (\phi_N)^2 - \frac{D}{N} \qquad \text{for all sufficiently large } N, \qquad (7.3.4)$$

where ϕ_N is defined according to any one of the following:
(a) $\phi_N = c_{N+2}/c_N$ for every N;
(b) $\phi_N = b_{N+2}/b_N$ for every N; or
(c) $\phi_N = c_{N+2}(0, x)/c_N(0, x)$ for all N of the same parity as $\|x\|_1$, where x is a given point of $\mathbf{Z}^d$.

Proof. First we define two patterns, $U = (u(0), \dots, u(9))$ and $V = (v(0), \dots, v(11))$. Each begins at the origin and lies in the (x_1, x_2)-plane (i.e. $u_i(j) = 0 = v_i(j)$ for all $i = 3, \dots, d$ and every j). The steps in the (x_1, x_2)-plane are N³E³S³ for U, and N³ESENES³ for V (see Figure 7.4). Let Q be the cube

$$Q = \{x \in \mathbf{Z}^d : 0 \leq x_i \leq 3 \text{ for every } i = 1, \dots, d\},$$

so that U and V are both contained in Q, and their endpoints are corners of Q. The main idea is that (U, Q) and (V, Q) must both occur many times on almost all self-avoiding walks, and changing a U to a V increases the length of a walk by two; this gives us a way to transform N-step walks into $(N+2)$-step walks, and $(N+2)$-step walks into $(N+4)$-step walks. We will then do some counting based on all possibilities for these transformations.

As usual, $\mathcal{S}_N$ is the set of N-step self-avoiding walks whose initial point is the origin. If we are in case (a) of the theorem, let W_N be $\mathcal{S}_N$; if we are in case (b), let W_N be the set of all bridges in $\mathcal{S}_N$; and if we are in case (c), let W_N be the set of all walks ω in $\mathcal{S}_N$ such that $\omega(N) = x$. Let w_N denote the cardinality of W_N. Then

$$\lim_{N \to \infty} w_N^{1/N} = \mu \qquad (7.3.5)$$

[where we have used Equation (3.1.10) for part (b) and Corollary 3.2.6 for part (c)]. For integers $N > 0$, $i \geq 0$, and $j \geq 0$, let $W_N(i,j)$ be the set of all walks in W_N on which (U,Q) occurs at precisely i different steps and (V,Q) occurs at precisely j different steps. Let $w_N(i,j)$ denote the cardinality of $W_N(i,j)$. For integers $a, b \geq 0$, define

$$w_N(\geq a, \geq b) = \sum_{i \geq a, j \geq b} w_N(i,j).$$

In particular, $w_N(\geq 0, \geq 0) = w_N$.

Consider the collection of all pairs (ω, ω') such that $\omega \in W_N(i,j)$ and ω' can be obtained from ω by changing one occurrence of (U,Q) to an occurrence of (V,Q). In other words, (ω, ω') is an allowed pair if there exists a k such that (U,Q) occurs at the k-th step of ω, (V,Q) occurs at the k-th step of ω', $\omega(l) = \omega'(l)$ for all $l = 0, \ldots, k$, and $\omega(l) = \omega'(l+2)$ for all $l = k+9, \ldots, N$. In particular, $\omega' \in W_{N+2}(i-1, j+1)$. Counting the number of allowed pairs in two ways, we see that

Number of pairs $= i w_N(i,j) = (j+1) w_{N+2}(i-1, j+1)$.

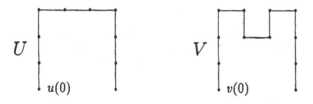

Figure 7.4: The patterns U and V in the (x_1, x_2)-plane.

Therefore

$$w_{N+2}(\geq 0, \geq 1) = \sum_{i\geq 1, j\geq 0} w_{N+2}(i-1, j+1) = \sum_{i\geq 1, j\geq 0} \frac{iw_N(i,j)}{j+1} \quad (7.3.6)$$

and

$$w_{N+4}(\geq 0, \geq 2) = \sum_{i\geq 2, j\geq 0} w_{N+4}(i-2, j+2) = \sum_{i\geq 2, j\geq 0} \frac{i(i-1)w_N(i,j)}{(j+1)(j+2)}. \quad (7.3.7)$$

The Schwarz inequality tells us that

$$\left(\sum_{i\geq 1, j\geq 0} \frac{iw_N(i,j)}{j+1} \right)^2 \leq \left(\sum_{i\geq 1, j\geq 0} w_N(i,j) \right) \left(\sum_{i\geq 1, j\geq 0} \frac{i^2 w_N(i,j)}{(j+1)^2} \right); \quad (7.3.8)$$

combining this with (7.3.6) implies that

$$[w_{N+2}(\geq 0, \geq 1)]^2 \leq w_N \left(\sum_{i\geq 1, j\geq 0} \frac{i^2 w_N(i,j)}{(j+1)^2} \right). \quad (7.3.9)$$

For $N \geq 1$, define

$$\Xi_N = \frac{w_{N+4}(\geq 0, \geq 2)}{w_N} - \left(\frac{w_{N+2}(\geq 0, \geq 1)}{w_N} \right)^2 \quad (7.3.10)$$

and

$$\hat{\Xi}_N = \phi_N \phi_{N+2} - (\phi_N)^2 - \Xi_N. \quad (7.3.11)$$

The error term $\hat{\Xi}_N$ is easy to bound:

$$|\hat{\Xi}_N| \leq \left| \frac{w_{N+4} - w_{N+4}(\geq 0, \geq 2)}{w_N} \right| + \left| \frac{w_{N+2}^2 - [w_{N+2}(\geq 0, \geq 1)]^2}{w_N^2} \right|$$

$$\leq \frac{c_{N+4}[1, (V, Q)]}{w_N} + \frac{2w_{N+2}c_{N+2}[0, (V, Q)]}{w_N^2},$$

and hence Theorem 7.2.3 and Equation (7.3.5) imply that $\hat{\Xi}_N$ decays to 0 exponentially fast. Therefore to prove the theorem it suffices to find a lower bound for Ξ_N of the form $-\text{const.}/N$. By Theorem 7.2.3 and Equation (7.3.5), there exists an $a > 0$ such that

$$\limsup_{N\to\infty} \left(1 - \frac{w_N(\geq 0, \geq aN)}{w_N} \right)^{1/N} < 1. \quad (7.3.12)$$

Using (7.3.7) and (7.3.9),

$$
\begin{aligned}
\Xi_N \;&\geq\; \left(\sum_{i \geq 0, j \geq 0} \frac{i(i-1)w_N(i,j)}{(j+1)(j+2)} \;-\; \sum_{i \geq 0, j \geq 0} \frac{i^2 w_N(i,j)}{(j+1)^2} \right) \frac{1}{w_N} \\
&=\; \left(\sum_{i \geq 0, j \geq 0} \frac{(-i^2 - ij - i)w_N(i,j)}{(j+1)^2(j+2)} \right) \frac{1}{w_N}.
\end{aligned}
$$

Since $w_N(i,j) = 0$ if $i > N$ or $j > N$, we can bound the factor $-i^2 - ij - i$ below by $-3N^2$. Splitting the sum over j into $aN \leq j \leq N$ and $0 \leq j < aN$, we then obtain

$$
\Xi_N \geq \frac{-3N^2 w_N(\geq 0, \geq aN)}{(aN)^3 w_N} + (-3N^2)\left(1 - \frac{w_N(\geq 0, \geq aN)}{w_N} \right).
$$

By (7.3.12), the second term in the last line above decays to 0 exponentially fast, and the first term is asymptotic to $-3/a^3 N$. Thus the theorem is proven. $\qquad\square$

Before we proceed with the proofs of the main ratio limit theorems, we prove a lemma that will be needed to prove assumption (*ii*) of Lemma 7.3.1 in the fixed-endpoint case.

Lemma 7.3.3 *Let x be a nonzero point of $\mathbf{Z}^d$. Then $c_{N+2}(0, x) \geq c_N(0, x)$ for all sufficiently large N having the same parity as $\|x\|_1$.*

Proof. The idea is similar to the proof that $c_{N+2} \geq c_N$ as depicted in Figure 7.1, but now we must not touch the endpoints. Fix an integer $A > \|x\|_\infty$. Suppose $N > (2A+1)^d$, and let ω be an N-step self-avoiding walk with $\omega(0) = 0$ and $\omega(N) = x$. Then at least one point of ω must lie outside the cube $\{y \in \mathbf{Z}^d : \|y\|_\infty \leq A\}$; notice that the endpoints of ω lie inside this cube. Let $M = \max\{\|\omega(i)\|_\infty : 0 \leq i \leq N\}$. Observe that $M > A$. Then there exists $j \in \{0, \dots, N\}$ and $i \in \{1, \dots, d\}$ such that $|\omega_i(j)| = M$. Choose j as small as possible; then, since $\omega(j)$ is not an endpoint of ω, we must have $\omega_i(j) = \omega_i(j+1)$. Let v be the vector whose coordinates are all 0 except the i-th, which is $+1$ if $\omega_i(j) = M$ and is -1 if $\omega_i(j) = -M$. Thus, v is the unit outer normal vector to the cube $\{y : \|y\|_\infty \leq M\}$ at the point $\omega(j)$. Define the new $(N+2)$-step walk ω^* by

$$
\omega^*(k) = \begin{cases}
\omega(k), & k = 0, \dots, j; \\
\omega(j) + v, & k = j+1; \\
\omega(j+1) + v, & k = j+2; \\
\omega(k-2), & k = j+3, \dots, N+2.
\end{cases}
$$

(Thus we replace the step from $\omega(j)$ to $\omega(j+1)$ by three steps.) Then ω^* is self-avoiding, and has the same endpoints as ω.

No two ω's can give rise to the same ω^*, because the the two added points have larger norm $\|\cdot\|_\infty$ than any other points of ω^* and hence are unambiguously determined. This proves the lemma. $\square$

We are now ready to prove the main ratio limit theorem.

Theorem 7.3.4 *(a)* $\lim_{N\to\infty} c_{N+2}/c_N = \mu^2$.
(b) For every fixed nonzero x in $\mathbf{Z}^d$, $\lim_{N\to\infty} c_{N+2}(0,x)/c_N(0,x) = \mu^2$ (here, N is restricted to having the same parity as $\|x\|_1$).
(c) $\lim_{N\to\infty} q_{2N+2}/q_{2N} = \mu^2$.
(d) $\lim_{N\to\infty} b_{N+1}/b_N = \mu$.

Proof.   Part (a) follows immediately from Lemma 7.3.1, Theorem 7.3.2(a), and (7.1.5) (which implies $\phi_N \geq 1$). Similarly, part (b) follows from Lemma 7.3.1, Corollary 3.2.6, Theorem 7.3.2(c), and Lemma 7.3.3. Part (c) is a direct consequence of part (b) and the basic relation (3.2.1).

Part (d) requires some additional work. First we apply Lemma 7.3.1 with $a_N = b_N$ [the hypotheses of the lemma follow from Corollary 3.1.6, Theorem 7.3.2(b), and the inequality $b_{N+2}/b_N \geq 1$, which is a consequence of (1.2.15)] to obtain

$$\lim_{N\to\infty} \frac{b_{N+2}}{b_N} = \mu^2. \qquad (7.3.13)$$

For every integer j, define

$$L_j = \liminf_{N\to\infty} \frac{b_{N-j}}{b_N};$$

we want to show that $L_1 = \mu^{-1}$ and that the lim inf is in fact a limit.

By (7.3.13), $L_{j+2} = \mu^{-2}L_j$ for every j. Therefore

$$L_j = \mu^{-j} \text{ for all even } j, \text{ and } L_j = \mu^{1-j}L_1 \text{ for all odd } j.$$

From (4.2.2), we see that for every j and every $N > j$,

$$\frac{b_{N-j}}{b_N} = \sum_{s=1}^{N-j} \lambda_s \frac{b_{N-j-s}}{b_N}.$$

Applying Fatou's Lemma to the above equation gives

$$L_j \geq \sum_{s=1}^{\infty} \lambda_s L_{j+s}. \qquad (7.3.14)$$

Define

$$\Sigma_o = \sum_{s \geq 1, s \text{ odd}} \lambda_s \mu^{-s} \quad \text{and} \quad \Sigma_e = \sum_{s \geq 1, s \text{ even}} \lambda_s \mu^{-s}.$$

By (4.2.4), $\Sigma_o + \Sigma_e = 1$. Applying (7.3.14) with $j = 0$ yields

$$1 \geq L_1 \mu \Sigma_o + \Sigma_e,$$

which implies that $L_1 \mu \leq 1$. Next, applying (7.3.14) with $j = 1$ gives

$$L_1 \geq \mu^{-1} \Sigma_o + L_1 \Sigma_e,$$

which implies that $L_1 \mu \geq 1$. Therefore $L_1 = \mu^{-1}$, i.e.

$$\limsup_{N \to \infty} \frac{b_{N+1}}{b_N} = \mu.$$

Combining this with (7.3.13), we finally obtain

$$\liminf_{N \to \infty} \frac{b_{N+1}}{b_N} = \liminf_{N \to \infty} \frac{b_{N+1}}{b_{N-1}} \frac{b_{N-1}}{b_N} = \mu^2 L_1 = \mu,$$

and so part (d) is proven. $\qquad\square$

7.4 End patterns

In this section, we shall prove Equations (7.1.8) and (7.1.9), as well as various extensions of these results. To begin, we extend the notion of *front patterns* from Definition 7.1.2 to the analogous notion of *tail patterns*.

Definition 7.4.1 *Let* $P = (p(0), \ldots, p(n))$ *and* $R = (r(0), \ldots, r(m))$ *be patterns. Let* $T_N[R]$ *denote the subset of walks in* S_N *for which* R *occurs at the* $(N - m)$-*th step. We say that* R *is a* proper tail pattern *if* $T_N[R]$ *is non-empty for all sufficiently large* N. *Let* $S_N[P, R]$ *denote the intersection of* $\mathcal{F}_N[P]$ *with* $T_N[R]$. *For every* x *in* $\mathbf{Z}^d$, *let* $S_N[x; P, R]$ *denote the set of all walks in* $S_N[P, R]$ *whose last point* $\omega(N)$ *is* x.

Consideration of front patterns and tail patterns together leads to results such as (7.4.7), which is used to analyze the behaviour of the "slithering-snake" Monte Carlo algorithm in Section 9.4.2, and Proposition 7.4.4, a lower bound for $c_N(0, x)$ which is stronger than the earlier bound (3.2.11).

In Section 6.7, we saw how the lace expansion is used to prove the existence of $\lim_{N \to \infty} |\mathcal{F}_N[P]|/c_N$ in high dimensions. (That section used the notation $P_{n,N}(P)$ to denote $\mathcal{F}_N[P]/c_N$ where $n = |P|$.) This limit is believed to exist in every dimension, but this remains unproven in 2, 3, or 4 dimensions, where the best results are Theorem 7.4.5 below and the following theorem.

Theorem 7.4.2 *If P is a proper front pattern and R is a proper tail pattern, then*

$$\liminf_{N \to \infty} \frac{|\mathcal{F}_N[P]|}{c_N} > 0 \tag{7.4.1}$$

and

$$\liminf_{N \to \infty} \frac{|\mathcal{T}_N[R]|}{c_N} > 0. \tag{7.4.2}$$

Proof. It suffices to prove (7.4.1), since (7.4.2) then follows by considering walks with reversed steps. Suppose $P = (p(0), \ldots, p(n))$. Since P is a proper front pattern, there must be a cube Q and a self-avoiding walk ω^P of length n' with the following properties: ω^P is entirely contained in Q; $\omega^P(n')$ is a corner of Q; and P occurs at the 0-th step of ω^P (see Figure 7.5).   By

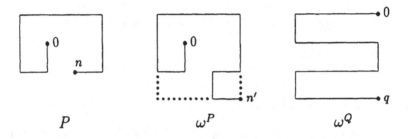

Figure 7.5: Proof of Theorem 7.4.2: the proper front patterns P, ω^P, and ω^Q. The dotted lines in the centre picture denote the boundary of Q.

Lemma 7.2.4(a), there exists a self-avoiding walk ω^Q whose last point equals $\omega^P(n')$, whose first point is another corner of Q, which is entirely contained in Q and visits every point of Q. Let q denote the number of steps in ω^Q. Evidently, $q \geq n'$.

Our first observation is that for every $N \geq n'$

$$|\mathcal{F}_N[P]| \geq |\mathcal{F}_N[\omega^P]| \geq |\mathcal{F}_{N+q-n'}[\omega^Q]|. \tag{7.4.3}$$

The first inequality is obvious, since $\mathcal{F}_N[\omega^P] \subset \mathcal{F}_N[P]$. For the second, we can define a one-to-one transformation $\omega \mapsto \omega^*$ from $\mathcal{F}_{N+q-n'}[\omega^Q]$ to $\mathcal{F}_N[\omega^P]$ as follows: for each ω in $\mathcal{F}_{N+q-n'}[\omega^Q]$, let ω^* be the (unique) N-step walk that has ω^P occurring at its 0-th step and whose last $N - n'$ steps are identical to those of ω, and translated so that $\omega^*(0) = 0$ (see Figure 7.6). Then ω^* is self-avoiding, hence it must be in $\mathcal{F}_N[\omega^P]$. Now, because of the observation (7.4.3), and because $c_N \leq c_{N+q-n'}$ [by (7.1.6)], it suffices

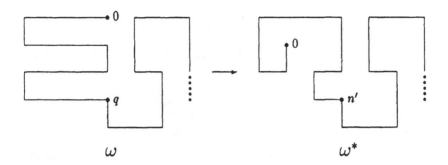

Figure 7.6: Proof of Theorem 7.4.2: the transformation of ω in $\mathcal{F}_{N+q-n'}[\omega^Q]$ to ω^* in $\mathcal{F}_N[\omega^P]$.

to show that

$$\liminf_{M\to\infty} \frac{|\mathcal{F}_M[\omega^Q]|}{c_M} > 0. \tag{7.4.4}$$

[We remark that it is not necessary to invoke (7.1.6) here; we could instead use (7.1.5) and $c_N \leq 2dc_{N-1}$ to conclude that $c_N \leq 2dc_{N+q-n'}$, which suffices for (7.4.4).]

Since ω^Q is a proper internal pattern (see Proposition 7.1.3), Theorem 7.2.3 says that there exists an $\epsilon > 0$ and an even integer k such that

$$c_k[0,\omega^Q] \leq (\mu(1-\epsilon))^k. \tag{7.4.5}$$

For all integers $l \geq j \geq 0$, let $\mathcal{G}_{l,j}$ be the set of walks ω in $\mathcal{S}_l$ such that ω^Q occurs at the j-th step of ω, and let $\mathcal{H}_{l,j} = \cup_{i=0}^{j}\mathcal{G}_{l,i}$. Thus $\mathcal{H}_{l,j}$ is the set of l-step self-avoiding walks starting at the origin on which ω^Q occurs at one of the first j steps. Then by (7.4.5),

$$
\begin{aligned}
|\mathcal{S}_{M+k} \setminus \mathcal{H}_{M+k,k}| &\leq c_k[0,\omega^Q]c_M \\
&\leq \mu^k(1-\epsilon)^k c_M.
\end{aligned}
$$

Therefore

$$
\begin{aligned}
c_{M+k} - c_M\mu^k(1-\epsilon)^k &\leq |\mathcal{H}_{M+k,k}| \\
&\leq \sum_{j=0}^{k} |\mathcal{G}_{M+k,j}| \\
&\leq \sum_{j=0}^{k} c_j|\mathcal{F}_M[\omega^Q]|c_{k-j}.
\end{aligned}
$$

We divide this by $\mu^k c_M$, obtaining

$$\frac{c_{M+k}}{\mu^k c_M} - (1-\epsilon)^k \leq \left(\sum_{j=0}^{k} \frac{c_j c_{k-j}}{\mu^k}\right) \frac{|\mathcal{F}_M[\omega^Q]|}{c_M}. \qquad (7.4.6)$$

Since k is a fixed even number, Theorem 7.3.4(a) implies that c_{M+k}/c_M converges to μ^k, and hence the left side of (7.4.6) has a strictly positive limit as $M \to \infty$. This proves (7.4.4) and the theorem. $\square$

An extension of the preceding proof allows one to prove the stronger statement

$$\liminf_{N \to \infty} \frac{|\mathcal{S}_N[P, R]|}{c_N} > 0 \qquad (7.4.7)$$

whenever P and R are proper front and tail patterns, respectively. For details, see Madras (1988).

We shall now consider the occurrence of end patterns on walks with specified endpoints.

Proposition 7.4.3 *Let e be a point in $\mathbf{Z}^d$ with $|e| = 1$. Let P and R be patterns such that $\mathcal{S}_N[e; P, R]$ is non-empty for all sufficiently large odd N. Then*

$$\liminf_{N \to \infty, N \text{ odd}} \frac{|\mathcal{S}_N[e; P, R]|}{c_N(0, e)} > 0.$$

Proof.   Assume $P = (p(0), \ldots, p(n))$ and $R = (r(0), \ldots, r(m))$ with $p(0) = 0$ and $r(m) = e$. Let $P' = (r(0), \ldots, r(m), p(0), \ldots, p(n))$. Since the pattern P' can occur on arbitrarily large self-avoiding polygons, P' must be a proper internal pattern. Therefore, by Theorem 7.2.3 there exists an $\epsilon > 0$ such that

$$\limsup_{N \to \infty} (c_N[\epsilon N, P'])^{1/N} < \mu. \qquad (7.4.8)$$

Let $\mathcal{S}_N(e)$ be the set of walks in $\mathcal{S}_N$ having $\omega(N) = e$, and let $\mathcal{S}'_N(e)$ be the set of walks in $\mathcal{S}_N(e)$ on which P' occurs at more than ϵN different steps. By (7.4.8) and Corollary 3.2.6,

$$|\mathcal{S}'_N(e)| \geq c_N(0, e) - c_N[\epsilon N, P'] \geq \frac{1}{2} c_N(0, e) \qquad (7.4.9)$$

for all sufficiently large (odd) N. If ω is in $\mathcal{S}_N(e)$ and P' occurs at the j-th step of ω, then $(\omega(j+m+1), \ldots, \omega(N), \omega(0), \ldots, \omega(j+m))$ is a translation of a self-avoiding walk ψ in $\mathcal{S}_N[e; P, R]$. Consider all pairs (ω, ψ) such that $\omega \in \mathcal{S}'_N(e)$ and ψ can be obtained from ω in this way. On the one hand, since each ω gives rise to at least ϵN different ψ's, the number of such pairs is bounded below by $\epsilon N |S'_N(e)|$. On the other hand, each ψ can be obtained from no more than N different ω's, and so the number of pairs is bounded above by $N |S_N[e; P, R]|$. Therefore

$$N |S_N[e; P, R]| \geq \epsilon N |S'_N(e)| \geq \frac{\epsilon N}{2} c_N(0, e)$$

for all sufficiently large N [the second inequality is given by (7.4.9)]. The theorem follows. □

One would like to prove the analogue of Proposition 7.4.3 when e is replaced by any given point x in $\mathbf{Z}^d$ (and N is restricted to having the same parity as $||x||_1$). However, the best known result is the following. Let e be a nearest neighbour of the origin, and let x be a non-zero point in $\mathbf{Z}^d$. Assume that $S_N[x; P, R]$ is non-empty for all sufficiently large N with the same parity as $||x||_1$. Then for odd $||x||_1$ we have

$$\liminf_{N \to \infty, N \text{ odd}} \frac{|S_N[x; P, R]|}{c_N(0, e)} > 0, \tag{7.4.10}$$

and for even $||x||_1$ (7.4.10) holds after we replace N by $N + 1$ in the numerator. For the proof, see Madras (1988). If we knew that $c_N(0, e)/c_{N'}(0, x)$ (where N' equals N or $N + 1$, according to whether $||x||_1$ is odd or even) had a positive lower bound for sufficiently large N, then we could immediately deduce the desired analogue of Proposition 7.4.3. Unfortunately, it remains an open problem to prove this lower bound, which is a particular case of Conjecture 1.4.1. We can however use Proposition 7.4.3 to prove a corresponding upper bound. This does not help to generalize Proposition 7.4.3, but it does prove a special case of Conjecture 1.4.1.

Proposition 7.4.4 *Let e and x be non-zero points of $\mathbf{Z}^d$, with $||e||_2 = 1$. Then there exists a positive constant A and an integer N_A (both depending on x) such that*

$$c_N(0, e) \leq A c_N(0, x) \qquad \text{for all } N \geq N_A$$

if $||x||_1$ is odd, and

$$c_N(0, e) \leq A c_{N+1}(0, x) \qquad \text{for all } N \geq N_A$$

if $||x||_1$ is even.

Proof. Let $(r(0), \ldots, r(m+1))$ be a proper internal pattern having $r(0) = x$, $r(m) = e$, and $r(m + 1) = 0$. Then m has the opposite parity to $||x||_1$. Let $R = (r(0), \ldots, r(m))$ and let P be the 0-step pattern (0). Then

Proposition 7.4.3 holds for this P and R, so there exists a $\kappa > 0$ such that $|\mathcal{S}_N[e; P, R]| \geq \kappa c_N(0, e)$ for all sufficiently large N. The first $N - m$ steps of a walk in $\mathcal{S}_N[e; P, R]$ is a self-avoiding walk from 0 to x, and so $|\mathcal{S}_N[e; P, R]| \leq c_{N-m}(0, x)$. The proposition now follows from these two inequalities and Lemma 7.3.3. $\qquad\qquad\qquad\qquad\qquad\qquad\qquad\qquad\qquad\qquad\qquad\quad\square$

We are now ready to prove ratio limit theorems for the number of walks with specified end patterns. The procedure is the same as in Section 7.3.

Theorem 7.4.5 *Let P be a proper front pattern and let R be a proper tail pattern. Then:*
(a) $\lim_{N \to \infty} |\mathcal{F}_{N+2}[P]|/|\mathcal{F}_N[P]| = \mu^2$.
(b) $\lim_{N \to \infty} |\mathcal{S}_{N+2}[P, R]|/|\mathcal{S}_N[P, R]| = \mu^2$.
(c) *Suppose in addition that x is a fixed nonzero point of $\mathbf{Z}^d$ and that $\mathcal{S}_N[x; P, R]$ is non-empty for all sufficiently large N having the same parity as $\|x\|_1$. Then $\lim_{N \to \infty} |\mathcal{S}_{N+2}[x; P, R]|/|\mathcal{S}_N[x; P, R]| = \mu^2$ (where N is restricted to having the same parity as $\|x\|_1$).*

Proof. We apply Lemma 7.3.1 in each case. Beginning with the most substantial hypothesis of the lemma, we observe that the analogue of Theorem 7.3.2 holds in each of the three present cases. In fact, the same proof works, with the following modifications:

1. Let W_N be $\mathcal{F}_N[P]$ in case (a), $\mathcal{S}_N[P, R]$ in case (b), and $\mathcal{S}_N[x; P, R]$ in case (c).

2. In the definition of $W_N(i, j)$, count only those occurrences of (U, Q) and (V, Q) which do not touch the end patterns; i.e. only count occurrences after the $|P|$-th step, and no later than the $(N - |R| - 9)$-th step for (U, Q) and the $(N - |R| - 11)$-th step for (V, Q).

3. $\lim_{N \to \infty} w_N^{1/N} = \mu$ by Theorem 7.4.2 for case (a), Equation (7.4.7) for case (b), and Equation (7.4.10) and Corollary 3.2.6 for case (c).

Now we verify that the hypotheses of Lemma 7.3.1 all hold. The previous paragraph shows that assumption (iii) of Lemma 7.3.1 holds in each of the present three cases. Also, assumption (i) holds in each case by point 3 in the preceding paragraph. So it only remains to check the second assumption of Lemma 7.3.1 in each case.

For case (a), Theorem 7.3.4(a) and Theorem 7.4.2 imply that

$$\liminf_{N \to \infty} \frac{|\mathcal{F}_{N+2}[P]|}{|\mathcal{F}_N[P]|} \geq \liminf_{N \to \infty} \frac{|\mathcal{F}_{N+2}[P]|}{c_N} = \liminf_{N \to \infty} \frac{|\mathcal{F}_{N+2}[P]|}{c_{N+2}} \mu^2 > 0.$$

Case (b) is similar, using (7.4.7). For case (c), we use the inequality $|S_{N+2}[x; P, R]| \geq |S_N[x; P, R]|$ for all sufficiently large N (the proof is exactly as the same as for Lemma 7.3.3, except that A must be taken large enough so that $\omega(j)$ is not on either end pattern; $A = \max\{|P|, ||x||_1 + |R|\}$ suffices). □

7.5 Notes

Sections 7.2 and 7.3. The results of these sections are due to Kesten (1963). In that paper Kesten also proved the following bounds on the convergence rates in the ratio limit theorems: for all sufficiently large N,

$$\left| \frac{c_{N+2}}{c_N} - \mu^2 \right| \leq K N^{-1/3}, \qquad (7.5.1)$$

$$- K N^{-1/3} \leq \frac{c_{N+2}(0, x)}{c_N(0, x)} - \mu^2 \leq K N^{-1/4}, \qquad (7.5.2)$$

where K is a constant [and N has the same parity as $||x||_1$ in (7.5.2)].

We conjecture that the following strengthening of the Pattern Theorem is true: for every proper internal pattern P, there exists a $t = t(P) > 0$ such that for any $\epsilon > 0$ only exponentially few N-step walks have fewer than $(t - \epsilon)N$ or more than $(t + \epsilon)N$ occurrences of P. A more modest open problem is to prove that the expected number of occurrences of a proper internal pattern P on an N-step walk is asymptotic to tN as $N \to \infty$, for some $t = t(P) > 0$ (where expectation is with respect to the uniform probability measure on S_N).

The proof of part (d) of Theorem 7.3.4 is essentially a special case of a ratio limit theorem in renewal theory; see Proposition 1.2 in Chapter 3 of Orey (1971).

Section 7.4. The results of this section are due to Madras (1988). That paper also showed that the convergence rate of (7.5.1) holds in Theorem 7.4.5(a,b), and the rate of (7.5.2) holds in Theorem 7.4.5(c).

Chapter 8

Polygons, slabs, bridges and knots

8.1 Bounds for the critical exponent α_{sing}

In this section we shall discuss rigorous bounds for the critical exponent α_{sing}, which is defined by the scaling relation

$$c_N(0, x) \sim \text{const.} \mu^N N^{\alpha_{sing}-2} \qquad \text{for fixed nonzero } x. \qquad (8.1.1)$$

(Here, N is assumed to have the same parity as $\|x\|_1$, since otherwise $c_N(0, x) = 0$.) It is believed that α_{sing} is independent of x in (8.1.1) (recall in particular Conjecture 1.4.1). Since scaling behaviour of this form has not yet been proven, our bounds will actually be bounds on the behaviour of generating functions and the like.

We begin by observing that α_{sing} is intimately related to the scaling behaviour of self-avoiding polygons. Recall from Equation (3.2.1) that if e is a nearest neighbour of the origin, then for all even $N \geq 4$ the number of N-step polygons satisfies

$$q_N = \frac{dc_{N-1}(0, e)}{N}, \qquad (8.1.2)$$

and therefore is expected to scale like

$$q_N \sim \text{const.} \mu^N N^{\alpha_{sing}-3}. \qquad (8.1.3)$$

Recalling (3.2.5) (a consequence of subadditivity) and (3.2.9), we have

$$q_N \leq (d - 1)\mu^N, \qquad (8.1.4)$$

and consequently

$$c_N(0, e) \leq \left(\frac{d-1}{d}\right) N \mu^N \qquad \text{if } |e| = 1. \qquad (8.1.5)$$

These two inequalities may be summarized by the statement that if α_{sing} exists, then

$$\alpha_{sing} \leq 3. \qquad (8.1.6)$$

The main part of this section is devoted a proof of some better bounds, namely, that if (8.1.1) holds, then

$$\alpha_{sing} \begin{cases} \leq & \frac{5}{2} & (d = 2) \\ \leq & 2 & (d = 3) \\ < & 2 & (d > 3). \end{cases} \qquad (8.1.7)$$

Unfortunately, we cannot prove that the corresponding termwise upper bounds [e.g., $q_N = O(N^{-1/2}\mu^N)$ in $d = 2$] hold. Rather, we only know (8.1.7) in certain weaker senses. For example, recall that if $a_N = O(N^p)$ then $\sum_N a_N s^N = O((1-s)^{-(1+p)})$ as $s \nearrow 1$ (for $p > -1$), but that the converse need not hold [cf. (1.3.9)-(1.3.11)]. With this in mind, consider the generating function $Q(z)$ of polygons:

$$Q(z) \equiv \sum_{N=1}^{\infty} q_N z^N. \qquad (8.1.8)$$

The bound (8.1.4) says that as z increases to $z_c = \mu^{-1}$, $Q(z)$ cannot diverge faster than $O((z_c - z)^{-1})$. We will prove that there exists a constant C', depending on d, such that

$$Q(z) \leq \sum_{N=1}^{\infty} C' N^{-(d-1)/2} \left(\frac{z}{z_c}\right)^{N-1} \qquad \text{for all } 0 \leq z \leq z_c. \qquad (8.1.9)$$

Thus as z increases to z_c, $Q(z)$ has at most a square root divergence in two dimensions, at most a logarithmic divergence in three dimensions, and is bounded above three dimensions. This is one interpretation of (8.1.7); more general interpretations will be proven below.

Conjectured values for α_{sing} are given by the hyperscaling relation $\alpha_{sing} = 2 - d\nu$ (see Section 2.1) in conjunction with the values of ν given in (1.1.12), as follows:

$$\alpha_{sing} = \begin{cases} \frac{1}{2} & d = 2 \\ 0.23\ldots & d = 3 \\ 2 - \frac{d}{2} & d \geq 4 \end{cases} \qquad (8.1.10)$$

In particular, $G_{z_c}(0, x) = \sum_{N=0}^{\infty} c_N(0, x)\mu^{-N}$ is expected to be finite in all dimensions, and hence so is $Q(z_c)$. Thus the bounds of (8.1.7) are probably far from optimal. However, for $d > 4$, it can in fact be proven using quite different methods that $\alpha_{sing} \leq 2 - d/2$ (see Theorem 6.1.3 and Corollary 6.1.4).

Before we proceed with the proof of (8.1.7), let us say a few words about lower bounds. We are not aware of any rigorous lower bound of the form $q_N \geq \text{const.} N^{-p}\mu^N$. However, we do have a lower bound in terms of bridges from Theorem 3.2.4. If we assume that bridges have the scaling behaviour

$$b_N \sim \text{const.} \mu^N N^{\gamma_{Bridge}-1}, \tag{8.1.11}$$

then $0 \leq \gamma_{Bridge} \leq 1$ because $\sum b_N/\mu^N$ diverges (by Corollary 3.1.8) and because $b_N \leq \mu^N$ [by (1.2.17)]. Therefore, under this assumption Equation (3.2.6) implies that whenever e is a nearest neighbour of the origin,

$$c_{2M+1}(0, e) \geq B\mu^{2M} M^{2\gamma_{Bridge}-d-4} \qquad \text{for all } M \geq 1, \tag{8.1.12}$$

for some positive constant B. This gives

$$\alpha_{sing} \geq 2\gamma_{Bridge} - d - 2 \geq -d - 2 \tag{8.1.13}$$

if we assume that the scaling exponents exist. We remark that by Proposition 7.4.4, the assumption (8.1.11) also implies the following generalization of (8.1.12): for any nonzero $x \in \mathbf{Z}^d$, there exists a constant B (depending on x) such that

$$c_N(0, x) \geq B\mu^N N^{2\gamma_{Bridge}-d-4}$$

for all sufficiently large N of the same parity as $\|x\|_1$.

The upper bounds (8.1.7) for α will be deduced from Propositions 8.1.2 and 8.1.4 below. Proposition 8.1.2 is a geometrical result which shows that there are fewer self-avoiding polygons than there are bridges of a certain kind; thus it suffices to get upper bounds on the number of such bridges. Proposition 8.1.4 uses the renewal theory approach of Section 4.2, but applied to slightly different quantities. So before we continue, let us define these new quantities.

Definition 8.1.1 *For each point y in $\mathbf{Z}^{d-1}$ and each positive integer N, let $b_N(y)$ (respectively, $\lambda_N(y)$) denote the number of N-step bridges (respectively, irreducible bridges) with $\omega(0) = 0$ and $(\omega_2(N), \ldots, \omega_d(N)) = y$. Recalling Definitions 4.1.7 and 4.2.3, we see that*

$$b_N(y) = \sum_{L=0}^{\infty} b_{N,L}(y) \qquad and \qquad \lambda_N(y) = \sum_{L=1}^{\infty} \lambda_{N,L}(y).$$

The following renewal-type equation may be deduced directly, or by summing over L in (4.2.8):

$$b_N(y) = \sum_{k=1}^{N} \sum_{v \in \mathbf{Z}^{d-1}} \lambda_k(v) b_{N-k}(y - v) \quad (N \geq 1). \tag{8.1.14}$$

Now define the random vector (X, Y) such that X takes values in $\{1, 2, \ldots\}$ and Y takes values in $\mathbf{Z}^{d-1}$, and

$$\Pr\{(X, Y) = (k, y)\} = \frac{\lambda_k(y)}{\mu^k} \text{ for } k \geq 1, y \in \mathbf{Z}^{d-1}. \tag{8.1.15}$$

The marginal distributions of X and Y are then given by

$$\Pr\{X = k\} = \lambda_k / \mu^k, \qquad k = 1, 2, \ldots, \tag{8.1.16}$$

$$\Pr\{Y = y\} = \sum_{k=1}^{\infty} \frac{\lambda_k(y)}{\mu^k}, \qquad y \in \mathbf{Z}^{d-1}. \tag{8.1.17}$$

Let $\{(X_n, Y_n) : n \geq 1\}$ be a sequence of independent copies of (X, Y). Then

$$\frac{b_N(y)}{\mu^N} = \Pr\{X_1 + \ldots + X_k = N \text{ and } Y_1 + \ldots + Y_k = y \text{ for some } k\}. \tag{8.1.18}$$

To see this, observe that iteration of (8.1.14) gives

$$b_N(y) = \sum_{k=1}^{N} \left[\sum \prod_{i=1}^{k} \lambda_{n_i}(v_i) \right],$$

where the inner sum is over all $n_1, \ldots, n_k \geq 1$ and all $v_1, \ldots, v_k$ in $\mathbf{Z}^{d-1}$ such that $n_1 + \ldots + n_k = N$ and $v_1 + \ldots + v_k = y$. See Section 8.3 for more about the probabilistic interpretation of these equations.

The following result gives an upper bound for the number of polygons in terms of bridges.

Proposition 8.1.2 *Let w be a nearest neighbour of the origin in $\mathbf{Z}^{d-1}$. Then*

$$q_N \leq d(d - 1) b_N(w) \quad \text{for all even } N \geq 4. \tag{8.1.19}$$

Proof. There is a simple geometric picture behind this proposition (Figure 8.1): Cut the polygon at a site with minimal x_1 coordinate and at a site with maximal x_1 coordinate. Reflect one of the two pieces through the hyperplane of the maximal x_1 coordinate. If this is done properly, the result will be a bridge with certain properties.

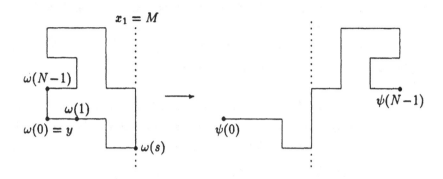

Figure 8.1: Proof of Proposition 8.1.2: Changing a polygon into a bridge by reflecting a piece through the hyperplane $x_1 = M$ (dotted line).

We now give the details. Let $e^{(1)} = (1, 0, 0, \ldots, 0) \in \mathbf{Z}^d$. For even $N \geq 4$, let Q'_N denote the set of all N-step self-avoiding polygons whose lexicographically smallest point is the origin and which do not lie entirely in the hyperplane $x_1 = 0$. (Observe that such a polygon lies in the half-space $x_1 \geq 0$.) We claim that

$$|Q'_N| \geq \frac{2}{d} q_N. \tag{8.1.20}$$

To prove the claim, consider all pairs $(\mathcal{P}, i)$, where $1 \leq i \leq d$ and $\mathcal{P}$ is an N-step polygon that does not lie entirely in a hyperplane $x_i = $ constant. The number of such pairs (up to translation) is bounded below by $2q_N$, since every polygon has at least two coordinates in which it is not constant. Also, the number of pairs is bounded above by $d|Q'_N|$ since for each i, $|Q'_N|$ is the number of N-step self-avoiding polygons (up to translation) that do not lie in a hyperplane $x_i = $ constant, by symmetry. Combining these bounds gives $2q_N \leq d|Q'_N|$, which proves (8.1.20).

Now, given a polygon $\mathcal{P}$ in Q'_N, let y be the lexicographically smallest point among all points x on $\mathcal{P}$ such that x and $x + e^{(1)}$ are joined by a bond of $\mathcal{P}$. Observe in particular that $y_1 = 0$. Let $\omega \equiv \omega[\mathcal{P}]$ be the $(N-1)$-step self-avoiding walk which has $\omega(0) = y$, $\omega(1) = y + e^{(1)}$, and whose remaining sites are given by traversing $\mathcal{P}$; thus ω is one of the $2N$ self-avoiding walks corresponding to the polygon $\mathcal{P}$ in the sense of Definition 3.2.1. See Figure 8.1. (We remark that the two-dimensional picture is deceptively simple; in particular, for $d \geq 3$, the point y need not be the lexicographically smallest point of $\mathcal{P}$.) Let $u = \omega(N-1) - y$. Then u is a unit vector, but $u \neq e^{(1)}$ [because $\omega(N-1) \neq \omega(1)$, so $u \neq \omega(1) - y = e^{(1)}$] and $u \neq -e^{(1)}$ [because

$\omega_1(j) \geq 0 = y_1$ for every j]. Therefore u is orthogonal to $e^{(1)}$, and so

$$\omega_1(N - 1) = y_1 = 0. \qquad (8.1.21)$$

We shall write $u = (0, \tilde{u})$, where $\tilde{u} \in \mathbf{Z}^{d-1}$ and $\|\tilde{u}\|_1 = 1$.
    Let
$$M = \max\{\omega_1(i) : 0 \leq i \leq N - 1\}$$

and let s be the smallest i such that $\omega_1(i) = M$. For each v in $\mathbf{Z}^d$, let $T_M[v]$ denote the reflection of v in the hyperplane $x_1 = M$:

$$T_M[(v_1, v_2, \ldots, v_d)] = (2M - v_1, v_2, \ldots, v_d).$$

Define the sequence of points

$$\psi(i) = \begin{cases} \omega(i) & \text{if } 0 \leq i < s \\ T_M[\omega(i)] & \text{if } s \leq i \leq N - 1. \end{cases}$$

Observe that $\psi = (\psi(0), \ldots, \psi(N - 1))$ is a self-avoiding walk because $\psi_1(i) < M$ for all $i < s$ and $\psi_1(i) \geq M$ for all $i \geq s$. Also observe that $\psi_1(0) = 0 \leq \psi_1(i) \leq 2M = \psi_1(N - 1)$ for all $i = 1, \ldots, N - 2$, by the definition of ψ and (8.1.21). Since the first inequality need not be strict, the walk ψ may not be bridge; however, if we concatenate it to the 1-step walk ϕ from 0 to $e^{(1)}$, then the result $\zeta \equiv \phi \circ \psi$ is an N-step bridge. Moreover $\zeta(0) = 0$ and $\zeta(N)$ is a nearest neighbour of a site on the x_1-axis; in fact

$$\zeta(N) = e^{(1)} + \psi(N - 1) - \psi(0) = e^{(1)} + T_M[y + (0, \tilde{u})] - y = (2M + 1, \tilde{u}). \qquad (8.1.22)$$

It is not hard to see that ζ uniquely determines the original polygon $\mathcal{P}$ in Q'_N. This discussion and symmetry show that

$$|Q'_N| \leq \sum_{\tilde{u} \in \mathbf{Z}^{d-1}:\|\tilde{u}\|_1=1} b_N(\tilde{u}) = 2(d - 1)b_N(w), \qquad (8.1.23)$$

where w was defined in the statement of the proposition. Combining (8.1.23) with (8.1.20) completes the proof. $\qquad \square$

    The next lemma is central to the proof of Proposition 8.1.4. It gives a bound for the probability that an ordinary (but not necessarily nearest-neighbour) random walk is at a given point x at the m-th step. It is in fact a special case of a general result about random walks, which we now describe. Suppose that $Y_1, Y_2, \ldots$ are independent, identically distributed $\mathbf{Z}^p$-valued random variables ($p \geq 1$). If the Y_i's have mean 0 and finite second moment, then $|Y_1 + \cdots + Y_m|$ grows like $m^{1/2}$ and the local central limit theorem implies that

$$\Pr\{Y_1 + \cdots + Y_m = x\} \sim \text{const.} m^{-p/2} \quad \text{for fixed } x \in \mathbf{Z}^p \qquad (8.1.24)$$

[see for example Proposition 7.9 or 7.10 of Spitzer (1976)][1]. If the Y_i's have nonzero mean, or more seriously if $E|Y_i|^2 = +\infty$, then (8.1.24) does not hold in general. However, in these cases $|Y_1 + \cdots + Y_m|$ tends to grow faster than $m^{1/2}$, and so one expects the probability of returning to a given point to be smaller than the right-hand side of (8.1.24). In fact, the following is true for any random walk $Y_1 + \cdots + Y_m$ that is truly p-dimensional, in the sense that the set of possible values of Y_1 is not contained in a hyperplane (including hyperplanes not passing through the origin): there exists a finite constant C (depending on the distribution of Y_1) such that

$$\Pr\{Y_1 + \cdots + Y_m = x\} \le Cm^{-p/2} \quad \text{for every } m \ge 1, x \in \mathbf{Z}^p. \qquad (8.1.25)$$

In Theorem A.6 we shall prove that (8.1.25) holds under somewhat stronger assumptions on the distribution of Y_1. This will be sufficient to prove the following special case, which is what we need in this section.

Lemma 8.1.3 *Let $Y_1, Y_2, \ldots$ be independent $\mathbf{Z}^{d-1}$-valued random variables each having the same distribution as Y given in (8.1.17). Then there exists a finite constant C such that*

$$\Pr\{Y_1 + \cdots + Y_m = x\} \le Cm^{-(d-1)/2} \quad \text{for every } m \ge 1, x \in \mathbf{Z}^{d-1}. \qquad (8.1.26)$$

Proof. This is an immediate consequence of Theorem A.6 with $p = d-1$. The hypotheses of Theorem A.6 are clearly satisfied because in the present case $\Pr\{Y = y\}$ is nonzero for every y in $\mathbf{Z}^{d-1}$. $\qquad \square$

With the bound (8.1.26) in hand, we are ready to prove the next proposition.

Proposition 8.1.4 *Let $h_1, h_2, \ldots$ be a nonincreasing sequence of nonnegative real numbers. Fix $w \in \mathbf{Z}^{d-1}$ with $\|w\|_1 = 1$. Then there exists a constant C (depending only on the dimension d) such that*

$$\sum_{N=1}^{\infty} h_N \left(\frac{b_N(w)}{\mu^N} \right) \le \sum_{N=1}^{\infty} C h_N N^{-(d-1)/2}. \qquad (8.1.27)$$

---

[1] If the possible values of Y_1 all lie in an r-dimensional subspace with $r < p$, or if the random walk is periodic, then a routine modification of (8.1.24) holds. This will not concern us here.

Consequently, there exists a constant C' such that

$$\sum_{N=1}^{\infty} h_N \left(\frac{q_N}{\mu^N} \right) \leq \sum_{N=1}^{\infty} C' h_N N^{-(d-1)/2}. \tag{8.1.28}$$

Proof. Let $h_1 \geq h_2 \geq \ldots \geq 0$. Let $I[A]$ denote the indicator function of the event A. Recall that the random variable X is always greater than or equal to 1, so if $X_1 + \cdots + X_k = N$ for some k, then k must be less than or equal to N. Using this observation and (8.1.18), we obtain

$$\sum_{N=1}^{\infty} h_N \left(\frac{b_N(w)}{\mu^N} \right)$$

$$= E(\sum_{N=1}^{\infty} \sum_{k=1}^{N} h_N I[X_1 + \cdots + X_k = N \text{ and } Y_1 + \cdots + Y_k = w])$$

$$\leq E(\sum_{k=1}^{\infty} \sum_{N=k}^{\infty} h_k I[X_1 + \cdots + X_k = N \text{ and } Y_1 + \cdots + Y_k = w])$$

$$= E(\sum_{k=1}^{\infty} h_k I[Y_1 + \cdots + Y_k = w])$$

$$\leq \sum_{k=1}^{\infty} C h_k k^{-(d-1)/2}$$

where the last inequality follows from Lemma 8.1.3. This proves (8.1.27). Combining this with Proposition 8.1.2 gives (8.1.28), with $C' = Cd(d-1)$.

$\square$

The following corollary is immediate.

Corollary 8.1.5 *If $h_1, h_2, \ldots$ is a nonincreasing nonnegative sequence for which the sum $\sum_{N=1}^{\infty} h_N N^{-(d-1)/2}$ converges, then $\sum_{N=1}^{\infty} h_N(q_N/\mu^N)$ also converges.*

We are now ready to prove Equation (8.1.9) as well as some more general interpretations of the bounds (8.1.7) for α_{sing}.

Corollary 8.1.6 *(a) If $0 \leq z \leq z_c$, then*

$$Q(z) \leq \sum_{N=1}^{\infty} C' N^{-(d-1)/2} \left(\frac{z}{z_c} \right)^{N-1}$$

(b) As $M \to \infty$,

$$\sum_{N=1}^{M} \frac{q_N}{\mu^N} = \begin{cases} O(M^{1/2}) & (d = 2) \\ O(\log M) & (d = 3) \\ O(1) & (d > 3). \end{cases}$$

(c) Let $d = 2$. For $0 \le z < z_c$,

$$\sum_{N=1}^{\infty} q_N N^{1/2} \left(\frac{z}{z_c}\right)^N = O\left(\frac{1}{z_c - z}\right).$$

Proof. Parts (a), (b), and (c) use (8.1.28) of Proposition 8.1.4, with the following choices for h_N:
(a) Set $h_N = (z/z_c)^N$.
(b) Given M, set $h_N = 1$ if $N \le M$ and $h_N = 0$ otherwise.
(c) Fix z. Define $f(a) = a^{1/2}(z/z_c)^a$ for $a > 0$. The function f is maximized at $a_0 = -[2\log(z/z_c)]^{-1}$, and $f'(a) < 0$ for $a > a_0$. So if we put

$$h_N = \begin{cases} f(a_0) & \text{if } N \le a_0 \\ f(N) & \text{if } N > a_0 \end{cases}$$

then $\{h_N\}$ is a nonincreasing positive sequence. The result follows from an application of (8.1.28) and some routine estimates. □

The above results are not strong enough to give a bound such as $q_N/\mu^N = O(N^{-(d-1)/2})$, although this would follow (at least for $d \le 3$) from any one of several "mild" conditions on the sequence $\{q_N/\mu^N\}$, such as monotonicity [see for example the proof of Theorem 2.4.3 in Lawler (1991)]. Unfortunately, it is not easy to check the regularity conditions of this sequence. We do know that the N-th root of the N-th term approaches 1, but this is not enough to let us say much. For example, it is consistent with the above corollaries that $\limsup q_N/\mu^N$ is strictly positive; fortunately, it is not hard to rule out this possibility.

Corollary 8.1.7

$$\lim_{N \to \infty} \frac{q_N}{\mu^N} = \lim_{N \to \infty} \frac{b_N(0)}{\mu^N} = 0. \tag{8.1.29}$$

Before proving this result, we require a simple lemma.

Lemma 8.1.8 *Let $u, u' \in \mathbf{Z}^{d-1}$, and let $j = \|u' - u\|_1$. Then*

$$b_n(u) \le b_{n+j+1}(u') \quad \text{for every } n \ge 0. \tag{8.1.30}$$

Proof. Consider a self-avoiding walk in $\mathbf{Z}^d$ that starts at the origin, takes one step in the $+x_1$ direction, and then goes to $(1, u' - u)$ in the minimum number of steps. This walk is a $(j + 1)$-step irreducible bridge. This shows that $\lambda_{j+1}(u' - u) \geq 1$. Using this fact and (8.1.14) shows that

$$b_n(u) \leq \lambda_{j+1}(u' - u)b_n(u) \leq b_{n+j+1}(u'), \qquad (8.1.31)$$

which proves the lemma.   □

Proof of Corollary 8.1.7. Fix $w \in \mathbf{Z}^{d-1}$ with $\|w\|_1 = 1$. Lemma 8.1.8 tells us that $b_{N-2}(w) \leq b_N(0)$, and combining this with Proposition 8.1.2 shows $q_{N-2} \leq d(d - 1)b_N(0)$. Therefore it suffices to prove the second equality in (8.1.29).

Let $\mathcal{B}_N(0)$ denote the set of N-step bridges ω that have $\omega(0) = 0$ and $\omega(N)$ on the x_1-axis. Let J_N denote the number of bridges in $\mathcal{B}_N(0)$ that cannot be expressed as the concatenation of a bridge from $\mathcal{B}_M(0)$ with a bridge from $\mathcal{B}_{N-M}(0)$ for some M $(0 < M < N)$. Then the following renewal equation holds:

$$b_N(0) = \sum_{k=1}^{N} J_k b_{N-k}(0) \quad (N \geq 1) . \qquad (8.1.32)$$

Put $f_n = J_n/\mu^n$ and $v_n = b_n(0)/\mu^n$. Then we have $v_0 = 1$ and $v_N = \sum_{k=1}^{N} f_k v_{N-k}$ for every $N \geq 1$, so we can use the Renewal Theorem. Since $v_n \leq 1$ for every n, Theorem B.1(c) implies that $\sum_{k=1}^{\infty} f_k \leq 1$. Therefore v_N converges as $N \to \infty$, by Theorem B.1(a, b).

Finally, suppose that v_N converges to a strictly positive limit. Then by Lemma 8.1.8 it follows that $\liminf b_N(w)/\mu^N > 0$. Now setting $h_N = N^{-1}$ in (8.1.27) results in a divergent sum being dominated by a convergent sum. This contradiction shows that v_N must converge to 0, which proves the corollary.   □

Remark. Another way to prove Corollary 8.1.7 is to use (8.1.18), which says that $b_N(y)/\mu^N$ is the probability that the random walk in $\mathbf{Z}^d$ with jumps (X_i, Y_i) (started from the origin) ever hits the point (N, y). Since this random walk is clearly transient, this probability must converge to 0 as $|(N, y)| \to \infty$, by Proposition 25.3 of Spitzer (1976). This in fact proves the stronger result that

$$\lim_{i \to \infty} \frac{b_{N_i}(y_i)}{\mu^{N_i}} = 0$$

whenever $\{N_i\}$ is a sequence of positive integers and $\{y_i\}$ is a sequence of points in $\mathbf{Z}^{d-1}$ such that $\lim_{i \to \infty}(|N_i| + |y_i|) = +\infty$.

8.2 Walks with geometrical constraints

In this section we shall consider self-avoiding walks that are restricted to lie in certain subsets of $\mathbf{Z}^d$. The emphasis will be on the connective constants associated with different subsets.

We begin by defining one class of subsets of $\mathbf{Z}^d$. For integers $k \in \{1, \ldots, d-1\}$ and $T \geq 0$, let $\mathcal{R} \equiv \mathcal{R}[k, T]$ be the subset

$$\begin{aligned}
\mathcal{R}[k, T] &= \mathbf{Z}^k \times \{0, 1, \ldots, T\}^{d-k} \\
&= \{x \in \mathbf{Z}^d : 0 \leq x_i \leq T, i = k+1, \ldots, d\}. \quad (8.2.1)
\end{aligned}$$

The set $\mathcal{R}$ is often called a *tube* if $k = 1$ or a *slab* if $k = d - 1$. We can visualize $\mathcal{R}[k, T]$ as a k-dimensional subspace that has been thickened up in the directions orthogonal to the subspace. We shall say that a translation by the vector v is *horizontal* if and only if $v \in \mathbf{Z}^k \times \{0\}^{d-k}$. The horizontal translations are precisely those translations with respect to which $\mathcal{R}$ is invariant.

For such a subset $\mathcal{R} = \mathcal{R}[k, T]$ and each integer $N \geq 0$, we denote by $\mathcal{S}_N \langle \mathcal{R} \rangle$ the set of all N-step self avoiding walks ω whose sites lie entirely in $\mathcal{R}$ and such that $\omega_i(0) = 0$ for $i = 1, \ldots, k$. Thus every N-step self-avoiding walk that lies in $\mathcal{R}$ is the horizontal translation of a unique member of $\mathcal{S}_N \langle \mathcal{R} \rangle$. We also let $c_N \langle \mathcal{R} \rangle = |\mathcal{S}_N \langle \mathcal{R} \rangle|$, which is the number of equivalence classes of self-avoiding walks in $\mathcal{R}$ up to horizontal translations.

For any finite T, one expects self-avoiding walks in $\mathcal{R}[k, T]$ to behave qualitatively like self-avoiding walks in $\mathbf{Z}^k$; in particular, they should have the same critical exponents, on the grounds of universality. As we mention in the Notes, this has to a large extent been proven rigorously for the case $k = 1$ [Klein (1980), Alm and Janson (1990)].

We shall now discuss the connective constant $\mu \langle \mathcal{R} \rangle$ for a region $\mathcal{R} = \mathcal{R}[k, T]$ for fixed k and T. For every walk ω in $\mathcal{S}_{N+M} \langle \mathcal{R} \rangle$, both the first N steps and the last M steps of ω are also self-avoiding walks in $\mathcal{R}$, and so

$$c_{N+M} \langle \mathcal{R} \rangle \leq c_N \langle \mathcal{R} \rangle c_M \langle \mathcal{R} \rangle \quad \text{for all } N, M \geq 1. \quad (8.2.2)$$

Therefore we see from Lemma 1.2.2 that

$$\mu \langle \mathcal{R} \rangle \equiv \lim_{N \to \infty} c_N \langle \mathcal{R} \rangle^{1/N} = \inf_{N \geq 1} c_N \langle \mathcal{R} \rangle^{1/N}. \quad (8.2.3)$$

Next, for each N we define $b_N \langle \mathcal{R} \rangle$ to be the number of bridges in $\mathcal{S}_N \langle \mathcal{R} \rangle$. We cannot assert $b_{N+M} \langle \mathcal{R} \rangle \geq b_N \langle \mathcal{R} \rangle b_M \langle \mathcal{R} \rangle$ because the concatenation of two bridges lying in $\mathcal{R}$ may not lie in $\mathcal{R}$. However, suppose that v and η are bridges in $\mathcal{S}_N \langle \mathcal{R} \rangle$ and $\mathcal{S}_M \langle \mathcal{R} \rangle$ respectively. Let $\tilde{\eta}$ be the translation of η by the vector $(v_1(N) + 1, v_2(N), \ldots, v_k(N), 0, \ldots, 0)$. Let ρ be a self-avoiding walk that takes one step from $v(N)$ to $v(N) + (1, 0, \ldots, 0)$ and then goes to $\tilde{\eta}(0)$ in the minimal number of steps. Then ρ is a bridge of span 1 whose length satisfies

$$|\rho| = 1 + \sum_{i=k+1}^{d} |v_i(N) - \eta_i(0)| \leq A \equiv (d - k)T + 1.$$

Let ω be the walk that begins with v, followed by ρ, followed by η, followed by $A - |\rho|$ steps in the $+x_1$ direction. Then ω is a bridge of length $N + M + A$ that lies in $\mathcal{R}$. This argument shows that

$$b_{N+M+A}\langle \mathcal{R} \rangle \geq b_N\langle \mathcal{R} \rangle b_M\langle \mathcal{R} \rangle \quad \text{for all } M, N \geq 1. \tag{8.2.4}$$

It follows from this and Lemma 1.2.2 that there exists a $\mu_{Bridge}\langle \mathcal{R} \rangle$ such that

$$\mu_{Bridge}\langle \mathcal{R} \rangle = \lim_{N \to \infty} b_N\langle \mathcal{R} \rangle^{1/N} \tag{8.2.5}$$

and

$$b_N\langle \mathcal{R} \rangle \leq \mu_{Bridge}\langle \mathcal{R} \rangle^{N+(d-k)T+1} \quad \text{for all } N \geq 1. \tag{8.2.6}$$

In detail: define

$$a_n = \begin{cases} 1 & \text{if } 1 \leq n \leq A \\ b_{n-A}\langle \mathcal{R} \rangle & \text{if } n > A; \end{cases}$$

then $a_{n+m} \geq a_n a_m$ for all $n, m \geq 1$ by (8.2.4) and the fact that $b_N\langle \mathcal{R} \rangle$ is nondecreasing in N. Now apply Lemma 1.2.2.

Next we wish to show that

$$\mu_{Bridge}\langle \mathcal{R} \rangle = \mu\langle \mathcal{R} \rangle. \tag{8.2.7}$$

We shall do this by following the Hammersley-Welsh argument of Section 3.1. For each N, let $h_N\langle \mathcal{R} \rangle$ denote the number of half-space walks in $\mathcal{S}_N\langle \mathcal{R} \rangle$. Recall that $P_D(N)$ is the number of partitions of N into distinct integers. Then for every $N \geq 1$,

$$h_N\langle \mathcal{R} \rangle \leq P_D(N) b_N\langle \mathcal{R} \rangle, \tag{8.2.8}$$

which is proven exactly as in Proposition 3.1.5 (the same argument works because $\mathcal{R}$ is invariant under reflection through a hyperplane $x_1 = $ constant). We can now imitate the proof of Theorem 3.1.1, using

$$c_N\langle \mathcal{R} \rangle \leq \sum_{m=0}^{N} h_{N-m}\langle \mathcal{R} \rangle h_{m+1}\langle \mathcal{R} \rangle \tag{8.2.9}$$

[which is proven in exactly the same way as the first inequality of (3.1.7)] as well as (8.2.8) and (8.2.6) to reach the analogue of (3.1.8)

$$c_N \langle \mathcal{R} \rangle \leq b_{N+1} \langle \mathcal{R} \rangle e^{\text{const.} N^{1/2}} \qquad (8.2.10)$$

This proves that $\mu \langle \mathcal{R} \rangle \leq \mu_{Bridge} \langle \mathcal{R} \rangle$. The reverse inequality is immediate since $c_N \langle \mathcal{R} \rangle \geq b_N \langle \mathcal{R} \rangle$, so (8.2.7) follows.

The next theorem says that for any k, $\mu \langle \mathcal{R}[k, T] \rangle$ is strictly increasing in T and converges to μ as $T \to \infty$. First we remark that the strict inequality

$$\mu \langle \mathcal{R}[k, T] \rangle < \mu \qquad (8.2.11)$$

is an immediate consequence of the Pattern Theorem (Theorem 7.2.3). In detail: if P is the walk consisting of $T + 1$ steps in the $+x_d$ direction, then P is a proper internal pattern that never occurs on a walk that lies in $\mathcal{R}[k, T]$, and so (7.1.7) implies (8.2.11). Based on this argument, one would also expect that $\mu \langle \mathcal{R}[k, T] \rangle < \mu \langle \mathcal{R}[k, T+1] \rangle$ for every T; however, the Pattern Theorem as we have stated it is not applicable in this context. An appropriate extension of the Pattern Theorem has been described by Soteros and Whittington (1989), but we shall instead give a different proof of this inequality below.

Theorem 8.2.1 *Suppose* $1 \leq k \leq d - 1$. *Then*

$$\lim_{T \to \infty} \mu \langle \mathcal{R}[k, T] \rangle = \mu \qquad (8.2.12)$$

and

$$\mu \langle \mathcal{R}[k, T] \rangle < \mu \langle \mathcal{R}[k, T+1] \rangle \quad \text{for every } T. \qquad (8.2.13)$$

Proof. We prove (8.2.12) first. Recall from Definition 8.1.1 that $b_N(0)$ denotes the number of N-step bridges in $\mathbf{Z}^d$ that begin at the origin and end on the x_1 axis. By Lemma 8.1.8 and Proposition 8.1.2 we see that $q_N \leq d(d-1)b_{N+2}(0)$; combining this with Corollary 3.2.5 and the elementary bound $b_N(0) \leq c_N$, we see that $\lim_{N \to \infty} b_N(0)^{1/N}$ exists and equals μ. Therefore for any $\epsilon > 0$ there exists an integer $s \geq 1$ such that $b_s(0)^{1/s} > \mu - \epsilon$. Let $\mathcal{R} = \mathcal{R}[k, 2s]$. Since every s-step bridge ω in $\mathbf{Z}^d$ having $\omega_i(0) = s$ for $i = k+1, \ldots, d$ must lie entirely in $\mathcal{R}$, we see that

$$b_{js} \langle \mathcal{R} \rangle \geq [b_s(0)]^j \geq (\mu - \epsilon)^{js} \qquad (8.2.14)$$

for every integer $j \geq 1$. Hence $\mu \langle \mathcal{R}[k, 2s] \rangle \geq \mu - \epsilon$. Equation (8.2.12) follows from this and (8.2.11), since ϵ is arbitrary and $\mu \langle \mathcal{R}[k, T] \rangle$ is nondecreasing in T.

The proof of (8.2.13) consists of some direct extensions of results from Sections 4.2 and 3.1. For each $N \geq 0$, let $\mathcal{B}_N \langle T \rangle$ denote the set of N-step bridges ω lying in $\mathcal{R}[k, T]$ such that $\omega(0) = 0$ and $\omega_i(N) = 0$ for $i = k+1, \ldots, d$, and let $\beta_N \langle T \rangle = |\mathcal{B}_N \langle T \rangle|$. Let $\lambda_N \langle T \rangle$ denote the number of bridges in $\mathcal{B}_N \langle T \rangle$ which cannot be expressed as the concatenation of a bridge in $\mathcal{B}_M \langle T \rangle$ to a bridge in $\mathcal{B}_{N-M} \langle T \rangle$ for some $M \in \{1, \ldots, N-1\}$. We thus obtain the following analogue of (4.2.2):

$$\beta_N \langle T \rangle = \sum_{s=1}^{N} \lambda_s \langle T \rangle \beta_{N-s} \langle T \rangle + \delta_{N,0}. \qquad (8.2.15)$$

Now consider the generating functions

$$B_z \langle T \rangle = \sum_{N=0}^{\infty} \beta_N \langle T \rangle z^N \quad \text{and} \quad \Lambda_z \langle T \rangle = \sum_{N=1}^{\infty} \lambda_N \langle T \rangle z^N,$$

and let $z \langle T \rangle = (\mu \langle \mathcal{R}[k, T] \rangle)^{-1}$. We claim that $B_z \langle T \rangle$ diverges at $z = z \langle T \rangle$, analogously to Corollary 3.1.8.

To prove this claim, we first observe that for any bridge ω in $\mathcal{S}_N \langle \mathcal{R} \rangle$ there exist two $(T(d-k)+1)$-step bridges ϕ and ψ such that the concatenation $\phi \circ \omega \circ \psi$ is in $\mathcal{B}_{N+2T(d-k)+2} \langle T \rangle$. Therefore $b_N \langle \mathcal{R} \rangle \leq \beta_{N+2T(d-k)+2} \langle T \rangle$ for every N. Hence it suffices to show that $\sum_{N=0}^{\infty} b_N \langle \mathcal{R} \rangle z^N$ diverges at $z = z \langle T \rangle$. But this can be proven by the same argument that was used to prove Corollary 3.1.8. Therefore the claim is true.

Since $B_z \langle T \rangle$ diverges at $z \langle T \rangle$, we can copy the argument leading from (4.2.2) to (4.2.4) to show that $\Lambda_{z \langle T \rangle} \langle T \rangle = 1$ for every T. Since $\lambda_N \langle T \rangle \leq \lambda_N \langle T+1 \rangle$ for every N with strict inequality for at least one N, we see that $\Lambda_z \langle T+1 \rangle > \Lambda_z \langle T \rangle$ for every $z > 0$, and in particular that

$$\Lambda_{z \langle T \rangle} \langle T+1 \rangle > \Lambda_{z \langle T \rangle} \langle T \rangle = 1. \qquad (8.2.16)$$

Hence the unique positive solution of $\Lambda_z \langle T+1 \rangle = 1$ is strictly less than $z \langle T \rangle$, i.e. $z \langle T+1 \rangle < z \langle T \rangle$. This proves (8.2.13). $\qquad \Box$

We have seen that in regions $\mathcal{R} = \mathcal{R}[k, T]$, the connective constants $\mu \langle \mathcal{R} \rangle$ and $\mu_{Bridge} \langle \mathcal{R} \rangle$ are equal. On the basis of the fact that μ equals $\mu_{Polygon}$ in $\mathbf{Z}^d$ [Corollary 3.2.5], one might also expect the analogous connective constant for polygons in $\mathcal{R}$ to equal $\mu \langle \mathcal{R} \rangle$ as well. However, as the next theorem shows, this is not always the case.

Theorem 8.2.2 *Let $q_N \langle T \rangle$ denote the number of N-step self-avoiding polygons in $\mathcal{R} = \mathcal{R}[k, T]$ up to horizontal translation.*

(a) The limit $\lim_{N \to \infty} q_N \langle \mathcal{R} \rangle^{1/N}$ *(taken through even values of N only) exists. Denote the limit by* $\mu_{Polygon} \langle \mathcal{R} \rangle$.
(b) If $k = 1$, *then* $\mu_{Polygon} \langle \mathcal{R} \rangle < \mu \langle \mathcal{R} \rangle$.
(c) If $k > 1$, *then* $\mu_{Polygon} \langle \mathcal{R} \rangle = \mu \langle \mathcal{R} \rangle$.

Proof. We limit ourselves to an outline of the proof. Part (a) can be proven by a concatenation argument as in Theorem 3.2.3, with a modification similar to what was needed for (8.2.4). The idea behind part (b) is the Pattern Theorem, as extended in Lemma 4.1 of Soteros and Whittington (1989). Let P be a self-avoiding walk contained in $\{x \in \mathcal{R}[1,T] : x_1 = 0\}$ that visits every point of this set. [Such a walk exists by Lemma 7.2.4(a).] Then P can occur many times on a long self-avoiding walk that lies in $\mathcal{R}$, but it cannot occur more than twice on a self-avoiding polygon that lies in $\mathcal{R}$. So the number of polygons should be exponentially smaller than the number of walks in $\mathcal{R}[1,T]$.

Finally, part (c) follows from a result similar to Theorem 3.2.4, with the following changes in the proof. First, the vector **v** that is orthogonal to the line containing 0 and x should also be parallel to the (x_1, x_2)-coordinate plane (observe that $\mathcal{R}$ is infinite in the x_1 and x_2 directions). Secondly, when we form $\bar{\omega}$, we allow for a few extra steps between $(\omega(i), \ldots, \omega(M))$ and $(\omega(0), \ldots, \omega(i))$ when we concatenate them, as in (8.2.4), so that $\bar{\omega}$ stays inside $\mathcal{R}$; we do the same for $\bar{v}$, and the same again when we join $\bar{\omega}$ and $\bar{v}$ to obtain ϱ. $\quad\square$

Next we look at a different class of subsets of $\mathbf{Z}^d$. Let f_i $(i = 2, \ldots, d)$ be given functions from $\{0, 1, \ldots, \}$ to $\{0, 1, \ldots, \infty\}$. Let $\mathcal{R}_f$ denote the region

$$\mathcal{R}_f = \{x \in \mathbf{Z}^d : x_1 \geq 0, 0 \leq x_i \leq f_i(x_1), i = 2, \ldots, d\},$$

which is often called a "wedge". Let $c_N \langle \mathcal{R}_f \rangle$ be the number of self-avoiding walks in $\mathcal{R}_f$ that begin at the origin.

Theorem 8.2.3 *(a) Suppose* $\limsup_{x \to \infty} f_i(x) < \infty$ *for at least one i. Then* $\limsup_{N \to \infty} c_N \langle \mathcal{R}_f \rangle^{1/N} < \mu$.
(b) Suppose $\lim_{x \to \infty} f_i(x) = +\infty$ *for every i. Then* $\lim_{N \to \infty} c_N \langle \mathcal{R}_f \rangle^{1/N}$ *exists and equals* μ.

Proof. Part (a) is an immediate consequence of the Pattern Theorem [as with (8.2.11)]. For part (b), it suffices by Theorem 8.2.1(a) to show that

$$\liminf_{N \to \infty} c_N \langle \mathcal{R}_f \rangle^{1/N} \geq \mu \langle \mathcal{R}[1,T] \rangle \tag{8.2.17}$$

for every T. By assumption, for every T there exists an $a = a(T) > 0$ such that $f_i(x) \geq T$ for every $x \geq a$ and every $i = 2, \ldots, d$. Thus $\{x \in$

$\mathcal{R}[1, T] : x_1 \geq a\}$ is a subset of $\mathcal{R}_J$. It follows that if ω is a bridge in $\mathcal{S}_N \langle \mathcal{R}[1, T] \rangle$ and $u \geq a$, then the translation of ω by the vector $(u, 0, \ldots, 0)$ is a bridge that lies entirely in $\mathcal{R}_J$. Let ρ be a walk of length $\|\omega(0)\|_1$ from the origin to $\omega(0)$. Observe that $|\rho| \leq (d-1)T$. Then the walk consisting of $a + (d-1)T - |\rho|$ steps in the $+x_1$ direction, followed by ρ, followed by ω, is an $[N + a + (d-1)T]$-step self-avoiding walk that lies in $\mathcal{R}_J$. Therefore

$$c_M \langle \mathcal{R}_J \rangle \geq b_{M-a-(d-1)T} \langle \mathcal{R}[1, T] \rangle \qquad (8.2.18)$$

for every $M \geq a + (d-1)T$. Taking M-th roots of both sides of this inequality and letting $M \to \infty$, we conclude from (8.2.5) and (8.2.7) that (8.2.17) holds. The theorem follows.   □

8.3   The infinite bridge

In this section we shall prove that there exists a measure on "infinite bridges" which may be defined in a natural manner analogous to the definition of the measure on infinite self-avoiding walks in five or more dimensions, as discussed in Section 6.7. Although the results are analogous, the proofs are completely different. The results of the present section are valid in any dimension.

The basic idea is the following. Suppose that you wish to know the probability that a long bridge (uniformly chosen from among all n-step bridges) begins with a particular (finite) sequence of steps, say four steps in the $+x_1$ direction followed by three in the $-x_2$ direction. The answer should not depend much on the length n of the bridge. In fact, we shall prove that the probability converges as $n \to \infty$. We can think of the limit as the probability that an infinitely long bridge begins with the given sequence of steps.

The basis for the results of this section is the renewal theory framework of Section 4.2. The first result also relies on the ratio limit theorem for bridges [Theorem 7.3.4(d)], which says that

$$\lim_{n \to \infty} \frac{b_{n+1}}{b_n} = \mu. \qquad (8.3.1)$$

A more constructive method of defining a measure on infinite bridges is discussed in the second half of the section. We shall see that the the two definitions are equivalent.

The following part of our discussion parallels the beginning of Section 6.7. Let $\mathcal{B}_n$ denote the set of all n-step bridges that begin at 0. Given $n \geq m$ and an m-step self-avoiding walk ω, we let $P_{m,n}^B(\omega)$ denote the fraction of n-step bridges that *extend* ω; in the terminology of Definition 7.1.1,

$$P^B_{m,n}(\omega) = \frac{|\mathcal{F}_n(\omega) \cap \mathcal{B}_n|}{b_n}. \tag{8.3.2}$$

Observe that $P^B_{n,m}(\omega)$ can be nonzero only if ω is a half-space walk (recall Definition 3.1.2). Next we define

$$P^B_m(\omega) = \lim_{n \to \infty} P^B_{m,n}(\omega) \tag{8.3.3}$$

if the limit exists. We shall prove below that the limit always exists, which implies that the probability measures P^B_m on m-step walks are *consistent* in the sense of (6.7.2).

The existence of the consistent measures P^B_m allows us to define a measure P^B_∞ on the set of all infinite self-avoiding walks ζ that satisfy $\zeta_1(0) = 0 < \zeta_1(i)$ for every $i \geq 1$, as follows. Using the notation $\zeta[0, m] \equiv (\zeta(0), \zeta(1), \ldots, \zeta(m))$ for $m \geq 0$, the measures P^B_m provide the values of P^B_∞ on *cylinder sets* of the form $\{\zeta[0, m] = \omega\}$ via

$$P^B_\infty\{\zeta[0, m] = \omega\} = P^B_m(\omega) \quad \text{for every } m\text{-step self-avoiding walk } \omega. \tag{8.3.4}$$

This is sufficient to guarantee the existence and uniqueness of the probability measure P^B_∞ on "infinite bridges" ζ.

We now prove that the measures P^B_m are indeed well-defined.

Theorem 8.3.1 *Let ω be an m-step self-avoiding walk. Then the limit (8.3.3) exists.*

Proof. First, we shall introduce some notation and use it to write an explicit expression for the limit. If β is an n-step bridge that extends ω, then let $M(\beta, \omega)$ denote the smallest value of i with the following properties: $i \geq m$; $\beta_1(j) \leq \beta_1(i)$ for every $j = 0, 1, \ldots, i$; and $\beta_1(i) < \beta_1(l)$ for every $l = i+1, \ldots, n$. Thus we see that $M(\beta, \omega)$ equals the smallest $i \geq m$ such that $\beta_1(i)$ is a break point of β (recall Definition 4.3.4) if such an i exists, and equals n otherwise. Next, for each $k \geq m$, let $\mathcal{E}_k(\omega)$ denote the set of bridges β in $\mathcal{B}_k$ that extend ω and satisfy $M(\beta, \omega) = k$. The point of this definition is that if β is an n-step bridge that extends ω, then there is a unique value of i ($m \leq i \leq n$), namely $i = M(\beta, \omega)$, with the property that $(\beta(0), \ldots, \beta(i))$ is in $\mathcal{E}_i(\omega)$ and $(\beta(i), \ldots, \beta(n))$ is an $(n-i)$-step bridge. Therefore

$$|\mathcal{F}_n(\omega) \cap \mathcal{B}_n| = \sum_{i=m}^{n} |\mathcal{E}_i(\omega)| b_{n-i}. \tag{8.3.5}$$

We shall prove that the limit (8.3.3) exists and satisfies

$$P_m^B(\omega) = \sum_{k=m}^{\infty} |\mathcal{E}_k(\omega)| \mu^{-k}. \tag{8.3.6}$$

The first step towards proving this is to divide (8.3.5) by b_n and let $n \to \infty$; then (8.3.2), (8.3.1) and Fatou's lemma imply that

$$\liminf_{n \to \infty} P_{m,n}^B(\omega) \geq \sum_{k=m}^{\infty} |\mathcal{E}_k(\omega)| \mu^{-k}. \tag{8.3.7}$$

Thus it suffices to prove the reverse inequality for the lim sup.

Suppose $k \geq m$, and consider an arbitrary β in $\mathcal{E}_k(\omega)$. If β is an irreducible bridge then let $I = 0$, and otherwise let I be the largest value of i such that $\beta_1(i)$ is a break point of β. Then $I \leq m$ [since β is in $\mathcal{E}_k(\omega)$], $(\beta(0), \dots, \beta(I)) = (\omega(0), \dots, \omega(I))$, and $(\beta(I), \dots, \beta(k))$ is an irreducible bridge, so we deduce that

$$|\mathcal{E}_k(\omega)| \leq \sum_{i=0}^{m} \lambda_{k-i} \tag{8.3.8}$$

for every $k \geq m$.

Next consider an arbitrary $J \geq m$. For any $n > J$, we use (8.3.5), (8.3.8), and finally (4.2.2) to obtain

$$
\begin{aligned}
|\mathcal{F}_n(\omega) \cap \mathcal{B}_n| - \sum_{k=m}^{J} |\mathcal{E}_k(\omega)| b_{n-k} &\leq \sum_{k=J+1}^{n} \left(\sum_{i=0}^{m} \lambda_{k-i} \right) b_{n-k} \\
&\leq \sum_{i=0}^{m} \left(\sum_{k=J+1-m+i}^{n} \lambda_{k-i} b_{n-k} \right) \\
&= \sum_{i=0}^{m} \left(b_{n-i} - \sum_{r=0}^{J-m} \lambda_r b_{n-i-r} \right). \tag{8.3.9}
\end{aligned}
$$

Now divide (8.3.9) by b_n and let $n \to \infty$. By (8.3.2) and (8.3.1), we see that

$$
\begin{aligned}
\limsup_{n \to \infty} P_{m,n}^B(\omega) - \sum_{k=m}^{J} |\mathcal{E}_k(\omega)| \mu^{-k} &\leq \sum_{i=0}^{m} \left(\mu^{-i} - \sum_{r=0}^{J-m} \lambda_r \mu^{-i-r} \right) \\
&\leq \sum_{i=0}^{m} \mu^{-i} \left(1 - \sum_{r=0}^{J-m} \lambda_r \mu^{-r} \right). \tag{8.3.10}
\end{aligned}
$$

By (4.2.4), the right side of (8.3.10) tends to 0 as $J \to \infty$, and so

$$\limsup_{n \to \infty} P^B_{m,n}(\omega) \leq \sum_{k=m}^{\infty} |\mathcal{E}_k(\omega)| \mu^{-k}. \qquad (8.3.11)$$

Together with (8.3.7), this proves the theorem. $\square$

We now give an alternate construction of a random infinite bridge in terms of a process that builds it up randomly one piece at a time. Consider the probability distribution on the set of all irreducible bridges (starting at the origin) that assigns probability μ^{-m} to each m-step irreducible bridge. This is indeed a probability distribution because $\sum_{m=1}^{\infty} \lambda_m \mu^{-m} = 1$ [by Equation (4.2.4)]. Next consider a random sequence of "pieces" $\eta^{[1]}, \eta^{[2]}, \ldots$ chosen independently from this distribution, and let $\rho = \eta^{[1]} \circ \eta^{[2]} \circ \cdots$ be the "infinite bridge" obtained by concatenating the pieces in order. Let Q^B be the probability measure governing $\eta^{[1]}, \eta^{[2]}, \ldots$ and hence ρ. Then the following result is true:

Theorem 8.3.2 *For every $m \geq 1$ and for every m-step self-avoiding walk ω,*

$$Q^B\{\rho[0, m] = \omega\} = P^B_\infty\{\zeta[0, m] = \omega\}. \qquad (8.3.12)$$

This theorem implies that Q^B and P^B_∞ yield the same probability law for the infinite bridge.

Before proving Theorem 8.3.2, we make some observations that relate the construction of ρ to the renewal theory framework of Section 8.1. Let $X_i = |\eta^{[i]}|$ be the number of steps in the i-th piece, and let Y_i be the vector in $\mathbb{Z}^{d-1}$ consisting of the second through d-th coordinates of the last point of the i-th piece. Then the joint distribution of (X_i, Y_i) is given by (8.1.15). Also, Equation (8.1.18) shows that $b_N(y)/\mu^N$ is the probability that $\rho_1(i) \leq \rho_1(N) < \rho_1(j)$ for all $i < N$ and all $j > N$ and that $(\rho_2(N), \ldots, \rho_2(N)) = y$.

Proof of Theorem 8.3.2. We shall use the random variables $X_i = |\eta^{[i]}|$ ($i \geq 1$), as defined in the preceding paragraph. For every $k \geq 1$, let A_k be the event that $\rho_1(k)$ is a break point of the infinite bridge ρ, i.e.

$$\begin{aligned} A_k &= \{\rho_1(r) \leq \rho_1(k) \text{ for every } r = 0, \ldots, k \text{ and} \\ &\qquad \rho_1(s) > \rho_1(k) \quad \text{for every } s > k \} \\ &= \{X_1 + \cdots + X_i = k \text{ for some } i \geq 1\}. \end{aligned}$$

We claim that for every $k \geq 1$

$$Q^B(\{\rho[0, k] = \beta\} \cap A_k) = \mu^{-k} \quad \text{for every bridge } \beta \in \mathcal{B}_k. \qquad (8.3.13)$$

To prove this, suppose that $\beta \in \mathcal{B}_k$. Then for some $j \geq 1$ we can write $\beta = \phi^{[1]} \circ \cdots \circ \phi^{[j]}$, where each $\phi^{[i]}$ is an irreducible bridge (in fact, $j - 1$ equals the number of break points in β). Then

$$Q^B(\{\rho[0, k] = \beta\} \cap A_k) = Q^B\{\eta^{[1]} = \phi^{[1]}, \ldots, \eta^{[j]} = \phi^{[j]}\}$$
$$= \mu^{-|\phi^{[1]}|}\mu^{-|\phi^{[2]}|} \cdots \mu^{-|\phi^{[j]}|}.$$

This proves the claim because $|\phi^{[1]}| + \cdots + |\phi^{[j]}| = |\beta| = k$.

Next fix $m \geq 1$, and fix an m-step self-avoiding walk ω. Define the random variable

$$T = \min\{i : X_1 + \cdots + X_i \geq m\}.$$

Recalling the definition of $\mathcal{E}_k(\omega)$ for $k \geq m$ from the proof of Theorem 8.3.1, we observe that the two events $\{\rho[0, k] \in \mathcal{E}_k(\omega)\}$ and A_k both occur if and only if $\{\rho[0, m] = \omega\}$ and $\{X_1 + \cdots + X_T = k\}$ both occur. Therefore

$$Q^B\{\rho[0, m] = \omega\} = \sum_{k=m}^{\infty} \Pr\{\rho[0, m] = \omega \text{ and } X_1 + \cdots + X_T = k\}$$
$$= \sum_{k=m}^{\infty} \Pr(\{\rho[0, k] \in \mathcal{E}_k(\omega)\} \cap A_k)$$
$$= \sum_{k=m}^{\infty} |\mathcal{E}_k(\omega)|\mu^{-k},$$

where the last equality is a consequence of (8.3.13). Finally, we combine the above equation with (8.3.6) to complete the proof.   □

8.4   Knots in self-avoiding polygons

Recall that every self-avoiding polygon corresponds to a simple closed curve that is determined by the set of bonds of the polygon. Through this correspondence, it makes sense to say whether or not a given self-avoiding polygon in $\mathbf{Z}^3$ is knotted. Furthermore, we can ask about the probability that a randomly chosen N-step polygon is unknotted. As we shall see below, this probability tends to 0 exponentially fast as $N \to \infty$.

The issue of knots arises in certain areas of polymer physics. For example, knots in polymers can give rise to defects during crystallization. Also, knots can occur on closed circular molecules of DNA, and they are believed to be relevant to understanding the actions of certain enzymes on these molecules. For further discussion and references, see Sumners and Whittington (1988) and Soteros, Sumners and Whittington (1992).

Let $Q[N]$ denote the set of N-step self-avoiding polygons in $\mathbf{Z}^3$ whose lexicographically smallest point is the origin, and let $R[N]$ denote the subset of unknotted polygons in $Q[N]$. The following theorem is due independently to Sumners and Whittington (1988) and Pippenger (1989).

Theorem 8.4.1 *There exists a μ_0 strictly less than μ such that*

$$\lim_{N \to \infty} |R_N|^{1/N} = \mu_0. \tag{8.4.14}$$

(As usual, this limit is taken through even values of N, since $R[N]$ is empty if N is odd.)

We shall only present an outline of the proof of this theorem. For the remaining details, which are mostly topological in nature, see Sumners and Whittington (1988), Pippenger (1989), or Soteros, Sumners and Whittington (1992).

First, the existence of the limit in (8.4.14) follows from an application of the concatenation arguments of Theorem 3.2.3, together with the fact that the concatenation of two unknotted polygons yields another unknotted polygon. So μ_0 is well-defined by (8.4.14), and it remains to explain why $\mu_0 < \mu$. The key to this is the Pattern Theorem 7.2.3. Let P be the pattern and let C be the cube that are depicted in Figure 8.2. In particular, P is entirely contained in C, and the two endpoints of P are corners of C. If

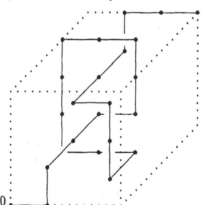

Figure 8.2: The knot-like pattern P (solid lines) and its associated cube C (dotted lines) from the proof of Theorem 8.4.1. The length of each side of C is 3.

U is a polygon and if the part of U that lies inside the cube C is precisely

the single subwalk P, then $\mathcal{U}$ must be a knotted polygon. This topological property may be rephrased in terms of Definition 7.2.2 as follows. Suppose that $\mathcal{U} \in Q[N]$ and that $\mathcal{U}$ corresponds (in the sense of Definition 3.2.1) to a self-avoiding walk ω in $\mathcal{S}_{N-1}(e)$ with $\|e\|_1 = 1$. If (P, C) occurs on ω, then $\mathcal{U}$ must be knotted. This implies that

$$|R[N]| \leq c_{N-1}[0, (P, C)]. \tag{8.4.15}$$

Combining this inequality with $c_{N-1}[0, (P, C)] \leq c_{N-1}[a(N-1), (P, C)]$ and applying Theorem 7.2.3(a), we conclude that $\limsup_{N \to \infty} |R[N]|^{1/N} < \mu$. Therefore $\mu_0 < \mu$. This completes our discussion of the proof of Theorem 8.4.1.

It is apparent from the above argument that in fact all but exponentially few N-step polygons contain aN occurrences of (P, C) for some $a > 0$, and that we can use other kinds of knot-like patterns in place of P (perhaps in larger cubes C). Therefore most large polygons are in fact rather complex knots. These ideas are explored further in Soteros *et al.* (1992).

We remark that Theorem 8.4.1 can also be proven using part (b) of the Pattern Theorem 7.2.3 instead of the stronger part (a). To do this, one uses the fact that the pattern P is a "tight knot", i.e. there is no room for another part of the polygon on which P occurs to pass back through P and "untie" the knot.

8.5   Notes

Section 8.1. The results of this section are due to Madras (1991b).

Section 8.2. Walks in restricted geometries serve as a model of polymers in the presence of physical barriers; for example, we may wish to know how much entropy is lost when a polymer is confined to a pore or capillary. Surveys and references are given in Whittington (1982) and Whittington and Soteros (1991).

Whittington (1983) proved that $\mu\langle \mathcal{R} \rangle$ and $\mu_{Bridge}\langle \mathcal{R} \rangle$ exist and are equal. Hammersley and Whittington (1985) proved (8.2.11) as well as Theorems 8.2.1 and 8.2.3. Theorem 8.2.2(b) is due to Soteros and Whittington (1988) and independently to Alm and Janson (1990). Percolation and Ising models on sets $\mathcal{R}_f$ have been studied as well; see references in Hammersley and Whittington (1985) as well as Chayes and Chayes (1986c).

Klein (1980) used "transfer matrices" to analyze self-avoiding walks in $\mathcal{R}[1, T]$ as well as in more general "one-dimensional" lattice subsets. In addition to proving that $c_N\langle \mathcal{R} \rangle \sim \text{const.}\mu\langle \mathcal{R} \rangle^N$ (i.e. $\gamma\langle \mathcal{R} \rangle = 1$), he also argued that $\langle |\omega(N)|^2 \rangle \sim \text{const.} N$ (i.e. $\nu\langle \mathcal{R} \rangle = 1$). Alm and Janson (1990)

used similar methods to perform a more detailed rigorous analysis in general
"one-dimensional" lattices. They also proved that $\gamma\langle\mathcal{R}\rangle = 1$ and $\nu\langle\mathcal{R}\rangle = 1$,
and that the "infinite self-avoiding walk" exists in the sense of Section 6.7.

Section 8.3. The results of this section are new.

Chapter 9

Analysis of Monte Carlo methods

9.1 Fundamentals and basic examples

Monte Carlo methods are useful for getting statistical estimates on the values of the connective constant, critical exponents, and other quantities related to self-avoiding walks. Essentially, a Monte Carlo simulation is a computer experiment which observes random versions of a particular system. After we obtain enough data, we can use statistical techniques to get estimates and confidence intervals for the desired quantities.

For definiteness, consider the exponent ν [defined in (1.1.5)], which measures the length scale of self-avoiding walks. There are several unresolved questions about ν, such as: Are the conjectured values (1.1.12) and (1.1.14) correct in 2, 3, and 4 dimensions? In particular, is the Flory exponent 3/5 too large in three dimensions? Do the hyperscaling relations (1.4.14) and (1.4.24) hold? In two dimensions, does the average area enclosed by an N-step self-avoiding polygon scale like $N^{2\nu}$? Good numerical estimates can give evidence in support of (or against) these and other conjectures. As we saw in Section 2.3, such evidence can also be relevant for analogous conjectures in other models; for example, if hyperscaling fails for self-avoiding walks in three dimensions, then it is likely to fail for other N-vector models as well.

To get a taste of some of the numerical values that various researchers have obtained, let us focus on the value of ν in three dimensions. An early study by Rosenbluth and Rosenbluth (1955) used *biased sampling* (see Section 9.3.1) to generate walks of up to 64 steps, obtaining an esti-

281

mate of 0.61 for ν. Stellman and Gans (1972) generated walks of up to 298 steps using a continuum version of the *pivot algorithm* (see Section 9.4.3) to obtain an estimate of 0.610 ± 0.008 for ν (this and the following are *95% confidence intervals* for ν; see Section 9.2.1). Grishman (1973) generated walks of length 500 using a combination of the *dimerization* and *enrichment* algorithms (see Sections 9.3.2 and 9.3.3), producing an estimate of 0.602 ± 0.009. However, these early results, which used relatively short walks, are biased by significant systematic errors due to unincluded correction-to-scaling terms (see Section 9.2.1). Rapaport (1985) generated walks of length up to 2400 using a combination of dimerization and enrichment, and estimated 0.592 ± 0.004. Madras and Sokal (1988) used the pivot algorithm to generate walks of up to 3000 steps, and obtained 0.592 ± 0.003. A very recent study (Li and Sokal, private communication), which uses the pivot algorithm to generate walks of up to 80,000 steps, indicates that the true value of ν is even lower: the preliminary estimate is 0.5883 ± 0.0013, which is in remarkable agreement with the field theoretic renormalization group prediction of 0.5880 ± 0.0015 obtained by Le Guillou and Zinn-Justin (1989). This brief history illustrates that correction-to-scaling terms are a serious danger, and that exponent estimates based on short walks must be interpreted with caution.

There are good reasons why Monte Carlo is "easier" for self-avoiding walks than for spin systems. First, there is only one limit to worry about, namely the length of the walk going to infinity. In a spin system, one has to take a limit going to a critical temperature as well as a thermodynamic limit of a finite lattice increasing to $\mathbf{Z}^d$. The latter is absent for self-avoiding walks, which can be simulated without any errors arising from the finite volume of the lattice. Secondly, spin system simulations typically exhibit "critical slowing-down": as the correlation length ξ diverges, you must look at finite lattices of at least ξ^d sites to learn anything, and you must look at each site before you get a new data point. This is not an inherent restriction for self-avoiding walks, since you only have to look at sites occupied by the walk. This suggests the possibility of more efficient algorithms in which critical slowing-down is much less severe.

Another frequently used numerical method is exact enumeration and extrapolation. This approach computes exact values of certain quantities for small values of N and then tries to infer an asymptotic behaviour from these numbers. We will not discuss this method in this book; the interested reader is referred to Guttmann (1989a).

To conduct a Monte Carlo experiment for the estimation of ν, one can for example proceed as follows.

(a) Select several values of N, say $N_1, \ldots, N_m$.

(b) For each N_i, generate many N_i-step self-avoiding walks at random. Use these to get an estimate $\hat{Y}_i$ of $\langle|\omega(N_i)|^2\rangle$, along with an estimate of the uncertainty in $\hat{Y}_i$.

(c) Fit a curve of the form $Y = AN^{2B}$ through the points $(N_i, \hat{Y}_i)$. The "best" value of B will be the estimate of ν.

Of course, each step raises many questions about how to proceed. In (a), how many and which values of N should be chosen? In (b), how many is "many"? What is the most efficient way to generate walks at random? How can the uncertainty best be estimated, and how does this uncertainty vary with N? In (c), how do we use the estimated uncertainties to fit data to a curve that is only believed to be *asymptotically* correct? These are the kinds of question that will be addressed in this chapter. We shall concentrate, however, on what one can say *rigorously* about the properties of these methods. The reader who wishes to pursue other aspects of Monte Carlo in more depth should consult the references listed in the Notes at the end of this chapter.

The remainder of this section will discuss some basic examples of Monte Carlo methods for generating self-avoiding walks, and will use them to illustrate various themes that appear throughout the chapter. Section 9.2 focuses on some statistical aspects—both practical and theoretical—of Monte Carlo methods. Sections 9.3 through 9.6 will treat various methods in detail. The longer proofs and calculations are deferred to Section 9.7.

We use our usual notation that $\mathcal{S}_N$ is the set of all N-step self-avoiding walks that begin at the origin. We shall restrict our attention to walks that begin at the origin, unless explicitly stated otherwise.

We begin with the basic question: How can we choose an N-step self-avoiding walk at random? (In this context, "at random" means that all walks in $\mathcal{S}_N$ are equally likely. For now, N is a given integer.) One simple method is the following:

Elementary Simple Sampling (ESS). This algorithm generates ordinary simple random walks until it obtains an N-step walk that is self-avoiding.

1. Let $w(0)$ be the origin and set $i = 0$.
2. Increase i by one. Choose one of the $2d$ neighbours of $w(i-1)$ at random, and let $w(i)$ be that point.
3. If $w(i) = w(j)$ for some $j = 0, 1, \ldots, i-1$, then go back to Step 1. Otherwise, go to Step 2 if $i < N$, and stop if $i = N$.

When this algorithm terminates, the walk $W \equiv (w(0), \ldots, w(N))$ is self-avoiding. Moreover, we claim that for any $\omega \in \mathcal{S}_N$ we have $\Pr\{W = \omega\} =$

$1/c_N$. To see this, let S_N^o be the set of all N-step (ordinary) simple walks. If we keep choosing members of S_N^o uniformly at random until one of them is in S_N, then the final result is evidently uniformly distributed on S_N. But this is essentially what the above algorithm does; Step 3 is just a short-cut to avoid generating the last $N - i$ steps of a walk that we already know intersects itself by the i-th step. Thus the ESS algorithm indeed generates a self-avoiding walk at random. However, it can be very slow when N is even moderately large: the probability that an N-step simple random walk is self-avoiding is $c_N/(2d)^N$, so the expected number of attempts (i.e. returns to Step 1) is $(2d)^N/c_N$. Therefore, using the notation T_X to represent the expected amount of computer time required for algorithm X to generate a single N-step self-avoiding walk, we have

$$T_{ESS} = \left(\frac{2d}{\mu}\right)^{N+o(N)} \tag{9.1.1}$$

We can improve on the efficiency of ESS by only generating simple random walks with no immediate reversals, as follows:

Non-Reversed Simple Sampling (NRSS). This algorithm generates simple random walks with no immediate reversals until it obtains an N-step walk that is self-avoiding.

1. Let $w(0)$ be the origin. Choose one of the $2d$ neighbours of the origin at random, and let $w(1)$ be that point. Set $i = 1$.
2. Increase i by one. Of the $2d - 1$ neighbours of $w(i - 1)$ that are different from $w(i - 2)$, choose one at random, and let $w(i)$ be that point.
3. If $w(i) = w(j)$ for some $j = 0, 1, \ldots, i - 1$, then go back to Step 1. Otherwise, go to Step 2 if $i < N$, and stop if $i = N$.

Arguing as for the ESS algorithm, we see that the NRSS algorithm generates a self-avoiding walk uniformly from S_N, and it takes an average of $2d(2d - 1)^{N-1}/c_N$ attempts to do so. Therefore

$$T_{NRSS} = \left(\frac{2d - 1}{\mu}\right)^{N+o(N)}, \tag{9.1.2}$$

which is better than (9.1.1), but still not very good.

Before continuing, we should mention the following "obvious" algorithm, which (perhaps surprisingly at first sight) *does not work*:

Myopic Self-Avoiding Walk (MSAW). Execute a random walk, at each step choosing only from those sites that have not yet been visited.

1. Let $w(0)$ be the origin, and set $i = 0$.
2. Increase i by one. Of the neighbours of $w(i-1)$ that are not in the set $\{w(0), \ldots, w(i-2)\}$, choose one at random, and let $w(i)$ be that point. (If all of the neighbours of $w(i-1)$ are in this set, then the walk is trapped, so return to Step 1.)
3. Repeat Step 2 if $i < N$, and stop if $i = N$.

This algorithm produces a walk in S_N, but with the *wrong distribution*. To see where the problem is, consider four-step walks on $\mathbf{Z}^2$: the probability of obtaining the walk NEEE on a given attempt is $\frac{1}{4} \times \frac{1}{3} \times \frac{1}{3} \times \frac{1}{3}$, but the probability of obtaining the walk NESE is $\frac{1}{4} \times \frac{1}{3} \times \frac{1}{3} \times \frac{1}{2}$. Thus, the probabilities are not uniform on S_N. In fact, the probabilities become very far from uniform for large N. The algorithm MSAW actually defines a different model, which is essentially the same as the "true self-avoiding walk" of Section 10.4.

Other algorithms for generating independent self-avoiding walks are described in Section 9.3. To varying degrees, they all suffer from the problem that they are inefficient for large walks. In fact we have the following

Open Problem: Is there an algorithm A which generates a single N-step self-avoiding walk, with distribution that is exactly uniform on S_N, such that the average time T_A is bounded by a polynomial in N?

Actually, the problem is only open in low dimensions: for $d \geq 5$, the average time of the dimerization algorithm of Section 9.3.2 is known to be bounded by a polynomial. Dimerization is also the most efficient known algorithm for generating a single walk exactly uniformly in any dimension, with an expected running time of $N^{O(\log N)}$ (if the usual scaling assumptions are true; see Section 9.3.2). However, there do exist more efficient algorithms that generate self-avoiding walks with a distribution that is arbitrarily close to uniform. These algorithms do not attempt to generate a sequence of independent self-avoiding walks, but rather they use a *Markov chain* to generate a sequence of self-avoiding walks that is not independent. Such methods are known as *dynamic*[1], as opposed to *static*, Monte Carlo methods. Roughly speaking, dynamic methods generate new walks by modifying (or "updating") walks that have been previously generated, while static methods build up walks from scratch. Static methods yield independent walks (or independent groups of walks), while dynamic methods yield correlated sequences of walks.

---

[1] This usage is distinct from the term "polymer dynamics", which refers to the (real) motion of polymers.

The basic idea of the dynamic approach is the following. Suppose that π is a probability distribution on some set S (i.e., for each $i \in S$, $\pi(i)$ is the probability of i, and $\sum_{i \in S} \pi(i) = 1$), and that we wish to generate a random object with the distribution π. If we can find a Markov chain with state space S whose unique equilibrium distribution is π, then the fundamental theory of Markov chains tells us that running this chain for a long time will produce observations whose distribution approaches π. In our case, we may take $S = S_N$ and $\pi(\omega) = 1/c_N$ for every self-avoiding walk ω in S_N. We begin with a walk $\omega^{[0]}$ in S_N and apply some (randomized) procedure that changes $\omega^{[0]}$ to get another self-avoiding walk $\omega^{[1]}$; then we apply the same procedure to $\omega^{[1]}$ to get another walk $\omega^{[2]}$, and so on. In this way we generate a sequence of walks $\{\omega^{[n]} : n \geq 0\}$ such that (for sufficiently large n) the distribution of $\omega^{[n]}$ is arbitrarily close to π. This sequence of walks will be correlated, of course, but one hopes that the relevant correlations will decay quickly.

To make the preceding discussion more precise, we make the following definitions, which are fairly standard in probability textbooks:

Definition 9.1.1 *Let $\{X^{[t]} : t = 0, 1, \ldots\}$ be a Markov chain on a finite or countably infinite state space S. Let*

$$P(i,j) = \Pr\{X^{[t+1]} = j | X^{[t]} = i\} \qquad (t \geq 0, i, j \in S)$$

be the one-step transition probabilities of the chain, and for every nonnegative integer n let

$$P^n(i,j) = \Pr\{X^{[t+n]} = j | X^{[t]} = i\} \qquad (t \geq 0, i, j \in S)$$

be the n-step transition probabilities. (We only consider chains that are time-homogeneous, i.e. whose transition probabilities are independent of t.) The chain is said to be irreducible *if for every i and j in S there exists an $n > 0$ such that $P^n(i,j) > 0$ (i.e. every state can be reached from every state). An irreducible chain is said to have period p if p is the greatest common divisor of $\{n : P^n(i,i) > 0\}$ for every state i (or equivalently for at least one i). A chain which has period 1 is said to be* aperiodic.

We remark that P^n is simply the n-th matrix power of P. Notice that if an irreducible chain has $P(i,i) > 0$ for some i, then it is aperiodic.

The standard theory of Markov chains (see references in Notes) tells us the following about the long-term behaviour of an aperiodic irreducible chain $X^{[t]}$. First, the limit

$$\lim_{n \to \infty} P^n(i,j)$$

exists for every i and j in S, and this limit is independent of i; call it $\pi(j)$. Next, if S is finite, then

$$\sum_{j \in S} \pi(j) = 1 \qquad (9.1.3)$$

and

$$\sum_{i \in S} \pi(i)P(i, j) = \pi(j) \qquad \text{for every } j \text{ in } S; \qquad (9.1.4)$$

and moreover π is the *only* nonnegative solution of (9.1.3) and (9.1.4). Finally, if S is countably infinite, then there are two possibilities: either $\pi(j) = 0$ for every j, in which case (9.1.3) and (9.1.4) have no nonnegative solution; or else π is the unique solution of (9.1.3) and (9.1.4).

An important special case is the following: we say that a chain is *reversible* with respect to π if

$$\pi(i)P(i, j) = \pi(j)P(j, i) \qquad \text{for every } i \text{ and } j \text{ in } S. \qquad (9.1.5)$$

(In alternate terminology, (9.1.5) is called the *detailed balance condition*.) Note that if π is the uniform distribution, then reversibility is equivalent to symmetry of P. If a chain is reversible with respect to π, then (9.1.4) holds (to see this, sum (9.1.5) over i). In practice, almost all dynamic Monte Carlo procedures use reversible chains (or are a combination of several reversible chains, as in Section 9.5.2).

If (9.1.3) and (9.1.4) hold for an irreducible chain and some π, then the chain is said to be *positive recurrent*, and π is called its *equilibrium*, or *stationary*, *distribution*. In general, $\pi(j)$ is the fraction of time that the chain spends in state j, in the long run (irrespective of the initial state). Thus, if our chain $X^{[t]}$ is positive recurrent, and if we observe it for a sufficiently long time, then the data should be pretty representative of the distribution π. For example, this tells us that if the state space is S_N and we observe end-to-end distance of the walks $X^{[t]}$ for a sufficiently long time, then we will obtain a good estimate of the mean square displacement $\langle |\omega(N)|^2 \rangle$ computed according to π. This is essentially a consequence of the ergodic theorem, which tells us that for a real-valued function f on the state space of a positive recurrent chain,

$$\lim_{m \to \infty} \frac{1}{m} \sum_{t=1}^{m} f(X^{[t]}) = \sum_{i \in S} f(i)\pi(i)$$

with probability one (assuming that the right hand side, which is just the expectation of $f(\cdot)$ with respect to π, is absolutely convergent).

Let us now look at a particular example: an algorithm due to Verdier and Stockmayer (1962) which turns one self-avoiding walk into another by

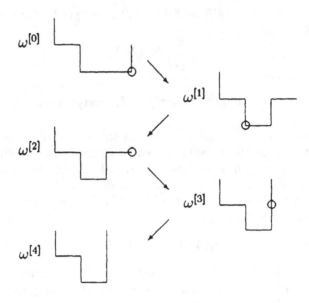

Figure 9.1: An example of the Verdier-Stockmayer algorithm in action. The circled site of $\omega^{[t]}$ corresponds to the randomly chosen I of Step 2. Observe that $\omega^{[2]} = \omega^{[1]}$ because the $\tilde{\omega}$ resulting from $\omega^{[1]}$ is not self-avoiding. Also observe that the $\tilde{\omega}$ resulting from $\omega^{[3]}$ in fact equals $\omega^{[3]}$.

moving one or two bonds of the walk. Briefly, it picks a site at random and tries to "flip" the two incident bonds if they form a right angle (or tries to wiggle the end bond if the chosen site is an endpoint of the walk). Here is a precise statement of the algorithm; a verbal description follows, and Figure 9.1 gives an illustration.

> *Verdier-Stockmayer (V-S) Algorithm.* This algorithm generates a Markov chain $\{\omega^{[t]} : t = 0, 1, \ldots\}$ on the state space $\mathcal{S}_N$ which is reversible with respect to the uniform distribution on $\mathcal{S}_N$.
>
> 1. Let $\omega^{[0]}$ be any self-avoiding walk in $\mathcal{S}_N$. Set $t = 0$.
> 2. Choose an integer I uniformly at random from $\{0, 1, \ldots, N\}$.
> 3. Define a new walk $\tilde{\omega} = (\tilde{\omega}(0), \ldots, \tilde{\omega}(N))$, which is not necessarily self-avoiding, as follows. First set $\tilde{\omega}(l) = \omega^{[t]}(l)$ for all $l \neq I$. Then:
>
>   (a) if $0 < I < N$, then set $\tilde{\omega}(I) = \omega^{[t]}(I - 1) +$

$(\omega^{[t]}(I+1) - \omega^{[t]}(I))$;

(b) if $I = N$, then set $\tilde{\omega}(N)$ equal to any neighbour of $\omega^{[t]}(N-1)$ except for $\omega^{[t]}(N-2)$ and $\omega^{[t]}(N)$, chosen at random;

(c) if $I = 0$, then set $\tilde{\omega}(0)$ equal to any neighbour of $\omega^{[t]}(1)$ except for $\omega^{[t]}(0)$ and $\omega^{[t]}(2)$, chosen at random. Then translate $\tilde{\omega}$ so that it begins at the origin.

4. If $\tilde{\omega}$ is self-avoiding, then set $\omega^{[t+1]} = \tilde{\omega}$; otherwise, set $\omega^{[t+1]} = \omega^{[t]}$.

5. Increase t by one and go to Step 2.

To visualize this algorithm, think of the N bonds of a walk ω as a sequence of N unit vectors $\Delta\omega(i) \equiv \omega(i) - \omega(i-1)$ $(i = 1, \ldots, N)$. Step 2 chooses a site $\omega(I)$ at random. Then Step 3 either interchanges the I-th bond with the $(I+1)$-th bond (if $0 < I < N$) or else randomly changes the first or last bond (if I is 0 or N). [Observe that in Step 3(a) we obtain $\Delta\tilde{\omega}(I) = \Delta\omega(I+1)$ and $\Delta\tilde{\omega}(I+1) = \Delta\omega(I)$.] Step 4 rejects the proposed walk ω if it is not self-avoiding.

To show that a certain probability distribution π is the equilibrium distribution of a Markov chain, we check both reversibility and irreducibility. First we shall show that the V-S algorithm is reversible (with respect to the uniform measure on $\mathcal{S}_N$). To do this, it suffices to check that P is symmetric, i.e.

$$P(\omega, \omega') = P(\omega', \omega) \qquad \text{whenever } \omega \neq \omega'. \qquad (9.1.6)$$

So suppose that ω and ω' are distinct walks in $\mathcal{S}_N$. If $P(\omega, \omega') = 0$ and $P(\omega', \omega) = 0$, then (9.1.6) holds, so assume without loss of generality that $P(\omega, \omega') > 0$. That is, if we start with ω, then there is a choice of I such that the walk $\tilde{\omega}$ obtained in Step 3 equals ω'. In this case, ω and ω' differ by either one or two bonds, and so there is a *unique* choice of I that transforms ω into ω'; denote this unique number by $i[\omega, \omega']$. Thus, since $\Pr\{I = i[\omega, \omega']\} = 1/(N+1)$, we have

$$P(\omega, \omega') = \begin{cases} \frac{1}{N+1} & \text{if } 0 < i[\omega, \omega'] < N \\ \frac{1}{(N+1)(2d-2)} & \text{if } i[\omega, \omega'] \text{ is 0 or } N \end{cases}$$

(the second line follows since there are $2d - 2$ ways to choose the new first or last bond). Now, if ω can be transformed into ω', then ω' can be transformed into ω; in particular, we have $i[\omega', \omega] = i[\omega, \omega']$. Thus $P(\omega', \omega)$ is given by the right hand side of the above equation, and so (9.1.6) holds.

There is a subtlety in the algorithm that makes reversibility so easy to prove. Consider a variation of the V-S algorithm in which we wait until a successful move occurs before recording the next observation: specifically, suppose that Steps 4 and 5 are replaced by

4'. If $\tilde{\omega}$ is not self-avoiding, then go to Step 2. If $\tilde{\omega}$ is self-avoiding, then set $\omega^{[t+1]} = \tilde{\omega}$, increase t by one, and go to Step 2.

Now there is no guarantee that (9.1.6) holds; the proof fails because in the new chain the one-step transition probability from ω to ω' is $P(\omega,\omega')/(1 - P(\omega,\omega))$ (here P refers to the probabilities in the original chain; observe that $1 - P(\omega,\omega)$ is the probability that a single attempt turns ω into something different). So we cannot have symmetry in the new chain unless $P(\omega,\omega)$ is the same for every ω in the original chain. This does not happen for the V-S algorithm, nor for any other interesting algorithm that we know of. Thus we see that in order to guarantee that we get the correct equilibrium distribution, it is vital to record the current walk *after every attempt*, whether the attempt results in a walk that is self-avoiding (a "success") or not (a "rejection").

We have seen that the original V-S algorithm is reversible, but unfortunately it is *not* irreducible. For example, there exist self-avoiding walks which are "frozen", i.e. they can never be changed by the V-S algorithm (see Figure 9.2). But the irreducibility difficulties are worse than just

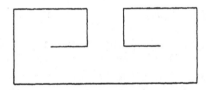

Figure 9.2: A 17-step self-avoiding walk in $\mathbf{Z}^2$ which is "frozen" with respect to the Verdier-Stockmayer algorithm.

having a few frozen walks. An *ergodicity class* of a Markov chain is defined to be a maximal subset A of the state space such that for every i and j in A, there exists a $n > 0$ such that $P^n(i,j) > 0$. Thus $\mathcal{S}_N$ is partitioned into many ergodicity classes, some of which contain a single walk. If $\omega^{[0]}$ is in a given ergodicity class, then we can view the V-S algorithm as producing a Markov chain whose equilibrium distribution is uniform *on that ergodicity class*, not on all of $\mathcal{S}_N$. As we shall see, this is a serious concern in principle, because the largest of the ergodicity classes is an exponentially small part of $\mathcal{S}_N$ as $N \to \infty$ (Theorem 9.4.2).

We conclude this introductory section with some remarks on the following problem, which is relevant to any computer program that works with self-avoiding walks: how fast can we check that a given walk is self-avoiding? To be precise, suppose that you are given a finite sequence of lattice sites $\omega(0), \omega(1), \ldots, \omega(N)$ such that $|\omega(i) - \omega(i-1)| = 1$ for every $i = 1, \ldots, N$. What is the most efficient way to check whether these $N+1$ sites are all distinct?

The most obvious algorithm is to look at every pair i and j such that $0 \le i < j \le N$ and check whether $\omega(i)$ equals $\omega(j)$. There are $N(N+1)/2$ such pairs, so the running time of this algorithm is $O(N^2)$. A different algorithm achieves a running time of $O(N)$ by using a "bit map". The idea behind this method is to simply draw a picture of the walk. For example, suppose we are working with N-step walks starting at the origin in $\mathbf{Z}^2$. The simplest bit map is a $(2N+1) \times (2N+1)$ array, indexed by (i, j), $-N \le i, j \le +N$, with all entries initially 0. Then every site of $\mathbf{Z}^2$ that can be reached by an N-step walk corresponds to an entry. For each $i = 0, 1, \ldots, N$ in turn, check the entry corresponding to the site $\omega(i)$: if the entry is 0 then change it to 1, but if the entry is already 1 then the walk is not self-avoiding. Afterward, go through the list of sites again to reset the entries to 0. The running time of this algorithm is clearly $O(N)$.

The disadvantage of a bit map is that it requires a lot of space: in $\mathbf{Z}^d$, it requires $O(N^d)$ words of computer memory. An alternative approach uses a data structure known as a "hash table". A set of N sites can be stored in a hash table of size $O(N)$ in such a way that we can check whether a given site is in the set in *average* time $O(1)$ — i.e. independent of N. Thus a hash table allows one to check self-avoidance in average time $O(N)$ using only $O(N)$ words of memory. Thus we have the satisfactory property that the amount of time and space needed to check self-avoidance are both proportional to what is required just to write down the walk. References about hash tables and their implementation for self-avoiding walk problems can be found in the Notes for this chapter.

9.2 Statistical considerations

In this section we shall survey some of the statistical problems associated with Monte Carlo methods. In particular, this will lead us to the important concept of *autocorrelation times* for dynamic methods.

9.2.1   Curve-fitting and linear regression

First we shall recall some elementary statistics. If a random variable Y is normally distributed with (unknown) mean m and (known) variance σ^2, then the probability that m lies in the (random) interval $[Y - 1.96\sigma, Y + 1.96\sigma]$ is about 0.95. Thus we say $Y \pm 1.96\sigma$ is a *95% confidence interval* for m. Often the variance is also unknown, and we have to compute an estimate $\hat{\sigma}^2$ of σ^2. In this case, $Y \pm 1.96\hat{\sigma}$ is only an approximate 95% confidence interval; for the usual estimates of the variance, the 1.96 should be replaced by a suitable number from a table of the Student's t distribution (for more details, consult the statistics references in the Notes).

Now let us consider the scenario described at the beginning of this chapter, in which we attempt to estimate ν from several data points $(N_i, \hat{Y}_i)$ $(i = 1, \ldots, m)$, where N_i is chosen in advance by the experimenter and $\hat{Y}_i$ is an estimate of $\langle |\omega(N_i)|^2 \rangle$ obtained by generating a large number of random N_i-step self-avoiding walks. Let σ_i^2 be the variance of $\hat{Y}_i$; since the variance is generally not known, we will in practice need to compute an *estimate* $\hat{\sigma}_i^2$ of σ_i^2 (we shall discuss how to do this below).

To estimate ν, we begin with the scaling relation

$$\langle |\omega(N)|^2 \rangle \sim AN^{2\nu}. \tag{9.2.1}$$

We can write this asymptotic relation as an equality with (infinitely many) "correction-to-scaling" terms:

$$\langle |\omega(N)|^2 \rangle = AN^{2\nu}(1 + BN^{-\Delta} + \cdots). \tag{9.2.2}$$

The exponents of the correction terms are strictly positive, and Δ is the smallest of them, i.e. $BN^{-\Delta}$ is the dominant correction term. (Like ν, these exponents are believed to depend only on the dimension. Other forms of corrections, such as logarithms, are also possible.) Our job is to fit a curve $Y = f(N)$ to the data; to do this in a meaningful way, we must only allow a small number of parameters in the family of curves. The obvious choices are either to eliminate all of the correction-to-scaling terms, giving the two-parameter family of curves

$$Y = AN^{2\nu}, \tag{9.2.3}$$

or else to eliminate all but the dominant correction term, giving the four-parameter family

$$Y = AN^{2\nu}(1 + BN^{-\Delta}). \tag{9.2.4}$$

The form (9.2.3) is appropriate if the N_i's are all large enough so that the actual corrections to scaling are smaller than the statistical errors in the data (i.e. smaller than σ_i). In general, however, we cannot expect this *a priori*. If we choose to work with (9.2.4), there is no guarantee that the best curve of this form will reflect the true value of Δ, since we do not know the size of the omitted correction terms (when N is small, these terms can be large, making it hard to see Δ from data corresponding to small N_i; but when N is large and the omitted terms are small, then the included term $BN^{-\Delta}$ is also small). The combination of all of the correction terms may very well show up in the data as a single "effective exponent" Δ_{eff}, which has no real relation to (9.2.2). Thus it is a very delicate business to try to estimate the true value of Δ. Rather, we may view the role of the parameter Δ in (9.2.4) as an aid to the extrapolation of a finite amount of data into the $N \to \infty$ asymptotic regime. (This represents a relatively cautious viewpoint which is definitely not universally accepted within the physics community.)

The standard statistical tool for fitting curves of the above forms to data is the method of least squares. Functions of the form (9.2.3) and (9.2.4) are examples of *regression functions*. Linear regression functions are the easiest to work with, so we begin by taking logarithms of the above two equations, obtaining

$$\log Y = \log A + 2\nu \log N \qquad (9.2.5)$$

and

$$\log Y = \log A + 2\nu \log N + BN^{-\Delta}, \qquad (9.2.6)$$

where the last term of (9.2.6) was obtained by the approximation $\log(1 + x) \approx x$ for x near 0. (If the reader accepts the viewpoint of the preceding paragraph that the parameter Δ should be regarded merely as an aid to extrapolation, then this approximation should cause no worries.)

Let us first focus on (9.2.5). Ordinary least squares estimation would tell us to estimate A and ν by the values that minimize the sum of squares

$$\sum_{i=1}^{m} (\log \hat{Y}_i - \log A - 2\nu \log N_i)^2.$$

This is not appropriate for us because an underlying assumption of this method is that the variance of $\log \hat{Y}_i$ is the same for every i. Instead, we should use *weighted least squares* estimation, weighting each term according to the inverse of its (estimated) variance, so that the Y_i's in which we have more confidence will have more say in determining the best fit. The general procedure is the following. Suppose that we observe independent random variables $U_1, \ldots, U_m$ where each U_i is normally distributed with mean $a + bM_i$ (where we know M_i and we want to estimate a and b) and variance v_i^2. (Our case corresponds to $b = \nu$, $a = \log A$, $U_i = \log \hat{Y}_i$ and

$M_i = 2 \log N_i$.) Then the weighted least squares estimates $\hat{a}$ and $\hat{b}$ are the values of a and b that minimize the weighted sum of squares

$$SS(a, b) \equiv \sum_{i=1}^{m} w_i (U_i - a - b M_i)^2, \tag{9.2.7}$$

where w_i is a positive "weight" (typically $1/v_i^2$, but not necessarily). The minimizing values are

$$\hat{b} = \frac{\sum w_i \sum w_i M_i U_i - \sum w_i M_i \sum w_i U_i}{\sum w_i \sum w_i M_i^2 - (\sum w_i M_i)^2} \tag{9.2.8}$$

and

$$\hat{a} = \frac{\sum w_i U_i - \hat{b} \sum w_i M_i}{\sum w_i}. \tag{9.2.9}$$

These are *unbiased estimators* of b and a (i.e. $E(\hat{b}) = b$ and $E(\hat{a}) = a$). Also, $\hat{b}$ and $\hat{a}$ are normally distributed with variances

$$\mathrm{Var}(\hat{b}) = \frac{\sum w_i}{\sum w_i \sum w_i M_i^2 - (\sum w_i M_i)^2}$$

and

$$\mathrm{Var}(\hat{a}) = \frac{\sum w_i M_i^2}{\sum w_i \sum w_i M_i^2 - (\sum w_i M_i)^2}.$$

The variances can be used to give statistical confidence intervals for a and b in the usual way. We can also formulate a test for the "goodness of fit" of our model: If the model is correct, and if the weights are given by $w_i = 1/v_i^2$, then the "residual sum of squares" $SS(\hat{a}, \hat{b})$ has a χ^2 distribution with $m - 2$ degrees of freedom.

When applying this theory in our Monte Carlo setting, we must first decide whether the estimates $\hat{Y}_i$ are normally distributed. Typically, $\hat{Y}_i$ is the average of a large number of observations

$$\hat{Y}_i = \frac{X_1 + \cdots + X_T}{T}.$$

In the case of a static method such as NRSS, the X_t's all come from different N_i-step walks, so they are i.i.d. (independent and identically distributed). The central limit theorem tells us that their average is normally distributed if T is large enough. (For an objective statistical test of normality, one can use the test of Shapiro and Wilk (1965); see Appendix A of Bratley, Fox, and Schrage (1987).) Moreover, in the i.i.d. case, the variance of $\hat{Y}_i$ is $\text{Var}(X_1)/T$, so we can estimate σ_i^2, the variance of $\hat{Y}_i$, by

$$\hat{\sigma}_i^2 = \frac{1}{T^2} \sum_{t=1}^{T} (X_t - \hat{Y}_i)^2.$$

The case of dynamic methods will be discussed in Section 9.2.2.

Suppose now that we believe that $\hat{Y}_i$ is approximately normally distributed, say with mean Y_i and variance σ_i^2. What can we say about $U_i = \log \hat{Y}_i$? Assume that σ_i is much smaller than Y_i, i.e. the uncertainty is relatively small compared to the magnitude of the quantity being estimated, as should be true in any good Monte Carlo experiment which tries to estimate something that can only be positive. Then U_i is approximately normally distributed with mean $\log Y_i$ and variance σ_i^2/Y_i^2. To see this, we write $\hat{Y}_i = Y_i + Z\sigma_i$, where Z is approximately normally distributed with mean 0 and variance 1; then

$$U_i = \log \left[Y_i \left(1 + \frac{Z\sigma_i}{Y_i} \right) \right] = \log Y_i + \log \left(1 + \frac{Z\sigma_i}{Y_i} \right) \approx \log Y_i + \frac{Z\sigma_i}{Y_i},$$

and the assertion follows. Thus $\hat{\sigma}_i^2/\hat{Y}_i^2$ is an estimate of the variance of U_i. Therefore, in the weighted least squares procedure described above, the appropriate choices of weights are $w_i = \hat{Y}_i^2/\hat{\sigma}_i^2$.

For completeness, we shall briefly describe weighted least squares estimation for more than two parameters. The framework of general linear regression is best expressed in matrix notation. Put the observed random variables ($\hat{Y}_1, \ldots, \hat{Y}_m$ in our case) into an $m \times 1$ column matrix $\mathbf{Y}$. Let β be a $p \times 1$ column vector containing the unknown parameters, and let $\mathbf{X}$ be a known $m \times p$ matrix; the model assumes that $E(\mathbf{Y}) = \mathbf{X}\beta$. (For example, in the model (9.2.6)[2]: p is 3; the entries of β are $\log A$, 2ν, and B; and the i-th row of $\mathbf{X}$ consists of the entries 1, $\log N_i$, and $N_i^{-\Delta}$.) Also let $\mathbf{V}$ be a known $m \times m$ positive definite matrix, which we assume to be the covariance matrix of $\mathbf{Y}$ (i.e. $\mathbf{V} = E[(\mathbf{Y} - \mathbf{X}\beta)(\mathbf{Y} - \mathbf{X}\beta)^T]$, where the T denotes transpose). The weighted least squares estimator is the vector $\hat{\beta}$ which minimizes

$$SS(\beta) = (\mathbf{Y} - \mathbf{X}\beta)^T \mathbf{V}^{-1} (\mathbf{Y} - \mathbf{X}\beta);$$

it is given by

$$\hat{\beta} = (\mathbf{X}^T \mathbf{V}^{-1} \mathbf{X})^{-1} \mathbf{X}^T \mathbf{V}^{-1} \mathbf{Y}.$$

---

[2] Observe that in the context of *linear* regression, we must assume a fixed value for Δ in this model. The most common choice is $\Delta = 1$; sometimes renormalization group calculations suggest other values, such as $\Delta = 1/2$.

Then $\hat{\beta}$ has a multidimensional normal distribution whose mean vector is the true β and whose covariance matrix is $(\mathbf{X}^T \mathbf{V}^{-1} \mathbf{X})^{-1}$. If the model is correct, then $SS(\hat{\beta})$ has a χ^2 distribution with $m - p$ degrees of freedom.

9.2.2 Autocorrelation times: statistical theory

When a dynamic Monte Carlo experiment is performed, the observations do not form an independent sequence, and so elementary statistical methods are often not applicable. In this section we shall address the problem of how to estimate the variance of the average of a large number of observations from a dynamic Monte Carlo experiment. Once we know how to perform such estimates, we can apply the regression theory outlined in Section 9.2.1.

To be specific, suppose that $\{\omega^{[t]} : t = 1, 2, \ldots\}$ is a stationary Markov chain. (A stochastic process is said to be stationary if, for every $k \geq 0$, the joint distribution of $(\omega^{[t]}, \ldots, \omega^{[t+k]})$ is the same for every t.) For a stationary Markov chain, the distribution of $\omega^{[t]}$ for any fixed time t must be the equilibrium distribution. Every positive recurrent Markov chain is asymptotically stationary; in practice, we can assume that a Markov chain is stationary if we discard enough initial observations so that the chain has had enough time to forget any influence of its initial state and has "reached equilibrium".

Let g be a real-valued function on the state space (e.g. $g(\omega) = |\omega(N)|^2$ if the state space is $\mathcal{S}_N$). Such a function is often called an "observable" in the physics literature. Let $\langle \cdot \rangle$ denote expected value with respect to the equilibrium distribution of the chain. Then we would like to estimate $\langle g \rangle$ by the estimator

$$\hat{Y} \equiv \hat{Y}[T] \equiv \frac{1}{T}[g(\omega^{[1]}) + \cdots + g(\omega^{[T]})].$$

Since the distribution of $\omega^{[t]}$ is the stationary distribution, it is easy to see that $\hat{Y}$ is an unbiased estimator of Y. But what is its variance? This question is addressed in the following lemma.

Lemma 9.2.1 *Suppose $\{X^{[t]}\}$ is a real-valued stationary process with finite second moment. Let*

$$\hat{Y}[T] = \frac{1}{T}(X^{[1]} + \cdots + X^{[T]}).$$

For each integer k, let $C_X(k)$ denote the covariance of $X^{[t]}$ and $X^{[t+k]}$ (observe that this is independent of t by stationarity; we are implicitly re-

stricting consideration to $t \geq 1$ *and* $t + k \geq 1$*). Let*

$$v = \sum_{k=-\infty}^{\infty} C_X(k),$$

and assume that this sum converges absolutely. Then

$$\lim_{T \to \infty} T \operatorname{Var}(\hat{Y}[T]) = v.$$

Proof. We have

$$\operatorname{Var}(\hat{Y}[T]) = \frac{1}{T^2} \sum_{s,t=1}^{T} \operatorname{Cov}(X^{[s]}, X^{[t]}) = \frac{1}{T^2} \sum_{k=-(T-1)}^{T-1} (T - |k|) C_X(k).$$

The result now follows from the dominated convergence theorem. □

In the notation of the above lemma, $C_X(0)$ is the variance of $X^{[t]}$, and $C_X(k)/C_X(0)$ is called the *autocorrelation function*. The ratio $v/[2C_X(0)]$ is called the *integrated autocorrelation time*, and is denoted $\tau_{int,X}$. When the $X^{[t]}$ are independent, $\tau_{int,X} = 1/2$.

Returning to our dynamic Monte Carlo algorithm, we shall take $X^{[t]} = g(\omega^{[t]})$ in the above lemma. We now write $C_g(k)$ for the covariance of $g(\omega^{[t]})$ and $g(\omega^{[t+k]})$, and the integrated autocorrelation time is

$$\tau_{int,g} = \sum_{k=-\infty}^{\infty} \frac{C_g(k)}{2C_g(0)} = \frac{1}{2} + \sum_{k=1}^{\infty} \frac{C_g(k)}{C_g(0)}. \tag{9.2.10}$$

The lemma tells us that

$$\operatorname{Var}(\hat{Y}[T]) \sim \frac{2}{T} \tau_{int,g} \operatorname{Var}(g(\omega^{[1]})) \quad \text{as } T \to \infty. \tag{9.2.11}$$

This asymptotic relation has a very useful intuitive interpretation. If the $\omega^{[t]}$'s (and hence the $X^{[t]}$'s) were *independent*, then the variance of the average $\hat{Y}[T]$ would be given by (9.2.11) with $2\tau_{int,g}$ replaced by 1. This means that if we are using a dynamic Monte Carlo method and we want to get an estimator with the same variance as one that samples T *independent* observations, then we need $2\tau_{int,g}T$ consecutive observations from the Markov chain. In other words, $2\tau_{int,g}$ is the number of observations from the chain that we need to get one "effectively independent" data point.

So far we have neglected the question of whether or not the series defining v in Lemma 9.2.1 converges absolutely. Fortunately, the answer is that it usually does; in fact, the terms $C_X(k)$ frequently decay exponentially.

The inverse of this decay rate is known as the *exponential autocorrelation time*. Specifically, given a real-valued function g on the state space of our stationary Markov chain $\{\omega^{[t]}\}$, we define its exponential autocorrelation time to be

$$\tau_{exp,g} \equiv \limsup_{k \to \infty} \frac{k}{-\log |C_g(k)|}; \qquad (9.2.12)$$

thus, the covariances $C_g(k)$ decay roughly like $\exp(-k/\tau_{exp,g})$. We also define the exponential autocorrelation time of the Markov chain to be

$$\tau_{exp} = \sup_g \tau_{exp,g}, \qquad (9.2.13)$$

where the sup is over all g such that $E(g(\omega^{[t]})^2)$ is finite. This means that τ_{exp} is the relaxation time of the slowest mode in the system. As we shall see in Section 9.2.3, τ_{exp} plays an important role in measuring the rate of convergence to equilibrium from an arbitrary initial distribution. The exponential autocorrelation time could be infinite (as in the BFACF algorithm of Section 9.6.1), but this is typically not the case. In particular, as we shall see in Section 9.2.3, τ_{exp} is finite whenever the state space is finite.

Given this theoretical description of the situation, we still need to find good statistical techniques for estimating the variance of $\hat{Y}[T]$, or, equivalently, for estimating $C_g(0)\tau_{int,g}$. This kind of problem has been the focus of much research in the field of time series, and we shall limit ourselves here to a very brief discussion; the Notes at the end of the chapter give some references for additional information.

One of the simplest procedures is the method of *batched means*. Given a long sequence of observations $X^{[1]}, \ldots, X^{[T]}$ of a stationary process, divide them into some relatively small number n of equal length subsequences, or "batches". Let $b = T/n$ be the number of observations in each batch, and let Y_i be the average of the i-th batch:

$$Y_i = \sum_{j=(i-1)b+1}^{ib} X^{[j]}/b.$$

If we assume that b is much larger than τ_{exp}, then the Y_i's are approximately independent and approximately normal, each with mean $E(X^{[1]})$ and variance $\approx v/b$ where v is defined as in Lemma 9.2.1 [see for example Theorem 20.1 of Billingsley (1968) or Corollary 1.5 of Kipnis and Varadhan (1986)]. Thus the overall average $\hat{Y}[T]$ is the average of the Y_i's, and we can estimate its variance using the sample variance of the Y_i's. For a "quick and dirty" method, this one is not bad. One serious drawback of course is the assumption that $b \gg \tau_{exp}$: in particular, the results of the procedure cannot be used as a check on the assumption after the fact.

A more developed approach is the spectral analysis of time series. Briefly, this tries to estimate the infinite sum v $(= 2C_g(0)\tau_{int,X})$ by estimating each term in the infinite series. By analogy with the usual estimator for covariance, we define the following estimator of $C_X(k)$:

$$\hat{C}_X(k) = \frac{1}{T-k} \sum_{j=1}^{T-k} (X^{[j]} - \hat{Y}[T])(X^{[j+k]} - \hat{Y}[T])$$

for $k = 0, 1, \ldots, T-1$. This is a biased estimator of $C_X(k)$, but it converges to $C_X(k)$ with probability one as $T \to \infty$ by the ergodic theorem. We next define the estimators of v

$$V_{T,m} = \hat{C}_X(0) + 2 \sum_{k=1}^{m} \hat{C}_X(k)$$

(the number m is chosen by the user). We don't insist on taking $m = T - 1$ because we believe that $C_X(k)$ is close to 0 when k is large, and so $\hat{C}_X(k)$ is mostly noise when k is large. (Heuristically: $\mathrm{Var}(\hat{C}_X(k)) = O(1/T)$, so $\mathrm{Var}(V_{T,T-1}) = O(1)$ — i.e. the uncertainty does not disappear as $T \to \infty$!) How should m be chosen? One reasonable way is the following "automatic windowing" procedure: Let m be the smallest integer such that $m \geq 10 V_{T,m}$. The factor 10 here is somewhat arbitrary, but the idea is that we want to make sure that we include contributions from terms that are up to several τ_{int}'s apart.

There is one more statistical issue that we must mention here, and that is the problem of *initialization bias*. In this section we have assumed that our observations come from a stationary process. Although Markov chains are *asymptotically* stationary, a simulation typically starts from a state which is *not* chosen according the equilibrium distribution. For example, in the case of self-avoiding walks, one might wish to start with a walk that is a straight line segment (for programming convenience). Thus, a simulation typically begins with an initial period which is "far from equilibrium", and it eventually "approaches equilibrium". The initial period must be removed from the data lest it introduce a bias to our estimates. Thus the experimenter must decide when the process has "reached equilibrium". The simplest procedure is to watch some observables over time until they all appear to have stabilized. There are also various statistical procedures that have been developed for removing initialization bias; see Bratley, Fox, and Schrage (1987) for a survey and references.

For concreteness, let us briefly consider the specific problem of choosing an initial state for a simulation of a Markov chain on the state space $\mathcal{S}_N$

of N-step self-avoiding walks. We could generate an initial walk using a static algorithm such as NRSS; although this would be slow, it has the theoretical advantage that we would then be starting the chain in equilibrium (exactly!), and so we would not have to worry about initialization bias at all. However, even for N around 200, it would be much faster to start with a straight walk and run until equilibrium is reached than it would be to generate a single walk by NRSS. But there are better static methods than NRSS; in particular, it is feasible to use dimerization (Section 9.3.2) to generate a single self-avoiding walk of two or three thousand steps in two dimensions (and even longer in three dimensions) to use as an initial state. (We remark that self-avoiding walks are one of the few interesting systems where there exists a feasible procedure for generating an initial state from the *exact* equilibrium distribution; nothing comparable is known for Ising-type models.)

9.2.3   Autocorrelation times: spectral theory and rigorous bounds

Consider an irreducible Markov chain with state space S, transition probabilities P, and equilibrium distribution π. Define the inner product of two complex-valued functions f and g on S to be

$$(f, g) \equiv \sum_{i \in S} \overline{f(i)} g(i) \pi(i); \tag{9.2.14}$$

the associated norm is

$$\|f\|_2 \equiv (f, f)^{1/2} = \left(\sum_{i \in S} |f(i)|^2 \pi(i) \right)^{1/2} \tag{9.2.15}$$

Let $l^2(\pi)$ denote the Hilbert space of the complex-valued functions f with $\|f\|_2$ finite. As usual, the norm of an operator T on $l^2(\pi)$ is given by

$$\|T\| \equiv \sup\{\|Tf\|_2 : f \in l^2(\pi), \|f\|_2 \le 1\}.$$

We view P as an operator on $l^2(\pi)$ by defining

$$(Pf)(i) = \sum_{j \in S} P(i, j) f(j).$$

The operator P is a contraction on $l^2(\pi)$, i.e. $\|P\| \le 1$. To prove this, we observe that $[(Pf)(i)]^2 \le (P(f^2))(i)$ for every i by the Schwarz inequality, and therefore

$$\|Pf\|_2^2 \le \sum_{i \in S} (P(f^2)(i)) \pi(i) = \|f\|_2^2 \tag{9.2.16}$$

[using (9.1.4) to get the equality].

Since $\|P\| \leq 1$, all of the eigenvalues of P lie on or inside the unit circle. Moreover, using Perron-Frobenius theory one can show the following [Šidák (1964)]: since the chain is irreducible, 1 is a simple eigenvalue of P, with the constant function 1 as an eigenfunction; and 1 is the only eigenvalue of P on the unit circle if and only if the chain is aperiodic.

Define the operator Π which maps $l^2(\pi)$ to the constant functions as follows:

$$(\Pi f)(i) = \sum_{j \in S} \pi(j) f(j) \qquad \text{for every } i;$$

thus $(\Pi f)(i)$ equals the expectation of f with respect to π. The basic convergence theory of Markov chains tells us that P^k converges to Π in a sense that will be made precise below. Observe that $\Pi^2 = \Pi$ and Π is self-adjoint [i.e. $(f, \Pi g) = (\Pi f, g)$], so Π is the orthogonal projection onto the space of constant functions. Also, $\Pi P = \Pi = P\Pi$ [by (9.1.4)], and so

$$(I - \Pi)P = P - \Pi = P(I - \Pi). \qquad (9.2.17)$$

We shall focus on the operator $P - \Pi$, which is 0 on the subspace of constant functions and equals P on the orthogonal complement of that subspace.

For the rest of this section, we shall also assume that the Markov chain is *reversible* with respect to π, i.e. that (9.1.5) holds. This implies that P is *self-adjoint* on $l^2(\pi)$:

$$
\begin{aligned}
(f, Pg) &= \sum_i \overline{f(i)} \sum_j P(i,j) g(j) \pi(i) \\
&= \sum_j \sum_i \overline{f(i)} g(j) P(j,i) \pi(j) = (Pf, g).
\end{aligned}
$$

Since a self-adjoint operator must have real spectrum, it follows from the fact that $\|P\| \leq 1$ that the spectrum of P is a subset of the interval $[-1, 1]$.

We shall now state a few facts from functional analysis. Let T be a bounded operator on a Hilbert space, and let $\sigma(T)$ be the spectrum of T. The *spectral radius* of T, denoted $r(T)$, is defined to be

$$r(T) \equiv \sup\{|\lambda| : \lambda \in \sigma(T)\};$$

it satisfies the well known "spectral radius formula"

$$r(T) = \lim_{n \to \infty} \|T^n\|^{1/n} = \inf_{n \geq 1} \|T^n\|^{1/n}.$$

Suppose now that T is also *self-adjoint*, so that $\sigma(T)$ is real. Then we in fact have

$$r(T) = \|T\| = \|T^n\|^{1/n} \qquad \text{for every } n \geq 1. \qquad (9.2.18)$$

The first equality is well-known [e.g. Theorem VI.6 of Reed and Simon (1972)]; the second equality follows from $r(T) \leq \|T^n\|^{1/n}$ (from the spectral radius formula), the inequality $\|T^n\| \leq \|T\|^n$, and the first equality. We also know

$$\inf \sigma(T) = \inf\{(f, Tf) : \|f\|_2 \leq 1\} \quad \text{and}$$
$$\sup \sigma(T) = \sup\{(f, Tf) : \|f\|_2 \leq 1\} \qquad (9.2.19)$$

(Yosida (1980), p.320); in particular, this implies the "Rayleigh-Ritz principle"

$$r(T) = \sup\{|(f, Tf)| : \|f\|_2 \leq 1\}. \qquad (9.2.20)$$

Finally, we have the relation

$$r(T) = \sup_{f} \limsup_{n \to \infty} |(f, T^n f)|^{1/n} \qquad (9.2.21)$$

which we shall prove in Section 9.7.1.

Now let us return to our Markov chain. Using the notation of Section 9.2.2, we find for the stationary Markov chain $\{\omega^{[t]}\}$ that

$$
\begin{aligned}
C_g(k) &= E[(g(\omega^{[t]}) - \langle g \rangle)(g(\omega^{[t+k]}) - \langle g \rangle)] \\
&= \sum_i \pi(i) \left[(g - \Pi g)(i) \right] \left[\sum_j P^k(i,j)(g - \Pi g)(j) \right] \\
&= ((I - \Pi)g, P^k(I - \Pi)g) \\
&= \begin{cases} (g, (P - \Pi)^k g) & \text{for } k \geq 1 \\ (g, (I - \Pi)g) & \text{for } k = 0, \end{cases}
\end{aligned}
\qquad (9.2.22)
$$

where we have used $I - \Pi = (I - \Pi)^2$ and (9.2.17) in the last step. By definition, $\limsup_{k \to \infty} |C_g(k)|^{1/k} = \exp(-1/\tau_{exp,g})$ and $\tau_{exp} = \sup_g \tau_{exp,g}$, so (9.2.22) and (9.2.21) imply that $\exp(-1/\tau_{exp}) = r(P - \Pi)$; equivalently,

$$\tau_{exp} = \frac{1}{-\log r(P - \Pi)}. \qquad (9.2.23)$$

Since $P - \Pi$ is self-adjoint,

$$r(P - \Pi) = \|P - \Pi\| = \|(P - \Pi)^k\|^{1/k} = \|P^k - \Pi\|^{1/k}. \qquad (9.2.24)$$

This implies that τ_{exp} also measures the exponential rate of convergence to equilibrium when the Markov chain is not started in equilibrium. In detail, consider the metric for probability measures on S defined by

$$\rho(\phi, \psi) = \sup\{|\sum_j f(j)\phi(j) - \sum_j f(j)\psi(j)| : \|f\|_2 \leq 1\}$$

[recall that $\| \cdot \|_2$ is the $l^2(\pi)$ norm of (9.2.15)]. If a Markov chain begins with the initial probability distribution ϕ at time 0, then at time k its distribution is given by the measure $(\phi P^k)(j) \equiv \sum_i \phi(i) P^k(i,j)$. For any f in $l^2(\pi)$, we have

$$
\begin{aligned}
&\left| \sum_j f(j)(\phi P^k)(j) - \sum_j f(j)\pi(j) \right| \\
={}& \left| \sum_{i,j} [\phi(i) - \pi(i)][P^k(i,j) - \pi(j)] f(j) \right| \\
\leq{}& \sup\Big\{ \sum_i [\phi(i) - \pi(i)] h(i) : \|h\|_2 \leq \|P^k - \Pi\| \, \|f\|_2 \Big\},
\end{aligned}
$$

and hence

$$
\rho(\phi P^k, \pi) \leq \|P^k - \Pi\| \rho(\phi, \pi) = \exp(-k/\tau_{exp})\rho(\phi, \pi). \tag{9.2.25}
$$

Equation (9.2.25) has the following practical interpretation: it tells us that if we begin from an initial distribution which is different from π and run the Markov chain for $10\tau_{exp}$ iterations, say, then the deviation from equilibrium (with respect to the metric ρ) is at most e^{-10} (about 0.00004) times the initial deviation. On the one hand, it is usually very difficult to get information about the size of τ_{exp} (either rigorously or numerically), so this is rarely a practical criterion for ensuring that the simulation has "reached equilibrium". On the other hand, the convergence to equilibrium could in fact be much faster than the upper bound of (9.2.25) indicates, so not knowing τ_{exp} may not be a real disadvantage. Ultimately, one has to analyze the data to determine empirically when the process is sufficiently close to equilibrium (see the discussion at the end of Section 9.2.2).

We remark that when the state space S is finite, then the spectrum of $P - \Pi$ is a finite subset of $(-1, 1)$ (assuming aperiodicity), and in particular τ_{exp} must be finite.

Up to now, we have been talking about the spectral radius of $P - \Pi$, but in Monte Carlo work one is usually just interested in the spectrum near $+1$ rather than near -1. An eigenvalue at -1 causes τ_{exp} to be infinite, but for a trivial reason: it happens if and only if the Markov chain is periodic with an even period, and so $\rho(\phi P^k, \pi)$ typically does not even converge to 0 because the chain always remembers which part of the state space it started in. But this does not prevent the *averages* $\hat{Y}[T]$ from converging rapidly to the correct values. So let us define the *modified autocorrelation time*

$$
\tau'_{exp} = \frac{1}{-\log[\sup \sigma(P - \Pi)]}. \tag{9.2.26}
$$

Then it will be shown in Section 9.7.1 that for every g in $l^2(\pi)$

$$\tau_{int,g} \leq \frac{1}{2}\left(\frac{1 + \exp[-1/\tau'_{exp}]}{1 - \exp[-1/\tau'_{exp}]}\right) = \tau'_{exp}\left[1 + O(1/\tau'_{exp})\right] \qquad (9.2.27)$$

for τ'_{exp} bounded away from 0.

The following result will be proven in Section 9.7.1. Its corollary below will be used a number of times in this chapter (see Sections 9.4.1, 9.5.1, and 9.6.1). We remind the reader that the covariances $C_g(k)$ and the various autocorrelation times are always defined in terms of the *stationary* Markov chain corresponding to P and π.

Proposition 9.2.2 *Suppose that P is reversible with respect to π. Then for any nonconstant g in $l^2(\pi)$,*

$$\tau_{int,g} \geq \frac{1}{2}\left(\frac{1 + \rho_g(1)}{1 - \rho_g(1)}\right) = \frac{1}{1 - \rho_g(1)} - \frac{1}{2},$$

where

$$\rho_g(1) = \frac{C_g(1)}{C_g(0)}.$$

Corollary 9.2.3 *Suppose that P is reversible with respect to π, and let g be a function in $l^2(\pi)$. Assume that there is a finite constant A such that $|g(i) - g(j)| < A$ whenever $P(i,j) > 0$ (i.e. the value of g can never change by more than A during a single step of the Markov chain). Then*

$$\tau_{int,g} \geq \frac{2C_g(0)}{A^2} - \frac{1}{2}.$$

Proof. First we list the following identities, which may be verified by direct calculation:

$$
\begin{aligned}
C_g(0)(1 - \rho_g(1)) &= C_g(0) - C_g(1) \\
&= (g, (I - P)g) \\
&= \frac{1}{2}\sum_{i,j}\pi(i)P(i,j)|g(i) - g(j)|^2. \qquad (9.2.28)
\end{aligned}
$$

From (9.2.28), we see that $C_g(0)(1 - \rho_g(1)) \leq A^2/2$. The result now follows immediately from Proposition 9.2.2.   □

As a further application of the identities (9.2.28), we have the following result:

Proposition 9.2.4 *Suppose that P_1 and P_2 are transition probabilities of two Markov chains which are reversible with respect to the same π, and assume that $P_1(i,j) \geq P_2(i,j)$ whenever $i \neq j$. Then their respective modified autocorrelation times satisfy $\tau'_{exp}(P_1) \leq \tau'_{exp}(P_2)$.*

Proof. For $k = 1, 2$, we see from (9.2.19) that

$$\sup \sigma(P_k - \Pi) = \sup\{(f, (I - \Pi)f) - (f, (I - P_k)f) : \|f\|_2 \leq 1\}.$$

In view of (9.2.28), this implies that $\sup \sigma(P_1 - \Pi) \leq \sup \sigma(P_2 - \Pi)$. The proposition then follows from (9.2.26).   □

Remark. Caracciolo, Pelissetto, and Sokal (1990) prove several generalizations of Proposition 9.2.4, including the result due to Peskun (1973) that the same hypotheses imply that $\tau_{int,f}(P_1) \leq \tau_{int,f}(P_2)$ for every f. The intuition behind these results is clear: since P_1 makes more transitions than P_2, it approaches equilibrium faster.

9.3   Static methods

In this section we shall discuss a number of static Monte Carlo algorithms. These algorithms generate either a sequence of independent self-avoiding walks or a sequence of independent batches of self-avoiding walks (the walks within each batch possibly being highly correlated).

9.3.1   Early methods: strides and biased sampling

Two methods of generating independent sequences were discussed in Section 9.1, namely Elementary Simple Sampling and Non-Reversed Simple Sampling; both were seen to require an exponentially large amount of computer time for each self-avoiding walk generated. A natural generalization of these methods uses "strides" to build walks instead of single steps. An m-step stride is a self-avoiding walk of length m. For the following algorithm, let m be a fixed nonnegative integer.

> *m-Step Stride Method (SM(m)).* This algorithm generates a self-avoiding walk of length km (k an integer). It requires a list $\psi[1], \ldots, \psi[c_m]$ of all m-step self-avoiding walks.
>
> 1. Set W to be the 0-step walk consisting of the single site at the origin. Set $i = 0$.
> 2. Increase i by one. Choose an integer J uniformly at random from $\{1, \ldots, c_m\}$. Redefine W to be $W \circ \psi[J]$, the concatenation of $\psi[J]$ to the current W.

3. If W is not self-avoiding, then go back to Step 1. Otherwise, go to Step 2 if $i < k$, and stop if $i = k$.

The average amount of computer time required to generate one N-step self-avoiding walk using this algorithm is

$$T_{SM(m)} = \frac{(c_m)^{N/m}}{c_N} = \left(\frac{c_m^{1/m}}{\mu} \right)^{N + o(N)}$$

This still grows exponentially in N, but at a slow rate if m is large. Of course, the larger m is, the more overhead must be invested in preparing and storing the list of all m-step walks.

One easy way to improve the Stride Method (for a given m) is in Step 2 to choose $\psi[J]$ from among only those walks whose first bond is not in the direction opposite to the last bond of the current W. A more sophisticated approach, requiring additional work in advance, is the following. For each $i = 1, \ldots, c_m$, make a list L_i containing all values of j such that $\psi[i] \circ \psi[j]$, the concatenation of $\psi[j]$ to $\psi[i]$, is self-avoiding. Then in Step 2 only choose the next J from the list L_J corresponding to the current J. Unfortunately, since the lists do not all have the same length, this will not generate walks with uniform distribution on $\mathcal{S}_{km}$ unless we exercise some caution. Specifically, we could let $L = \max_i |L_i|$ (where $|L_i|$ denotes the length of the list L_i), and replace Steps 1 and 2 above as follows.

1′. Choose $J(1)$ uniformly at random from $\{1, \ldots, c_m\}$. Set $W = \psi[J(1)]$ and set $i = 1$.

2′. Increase i by one. Choose $J(i)$ uniformly at random from $\{1, \ldots, L\}$. If $J(i) > |L_{J(i-1)}|$, then go back to Step 1′ and start over; otherwise, redefine W to be the concatenation of the $J(i)$-th walk on the list $L_{J(i-1)}$ to the current W.

Again we see the usefulness of a "rejection" step (occurring here when $J(i) > |L_{J(i-1)}|$) in producing the desired distribution. Without this, our method would suffer from the same flaw as the MSAW algorithm of Section 9.1. Having sounded these warnings, let us now say that all is not necessarily lost if we generate self-avoiding walks with a nonuniform distribution, for we can still estimate interesting quantities by reweighting our observations, as we shall now explain.

Suppose that $\omega^{[1]}, \ldots, \omega^{[m]}$ is an i.i.d. sample from $\mathcal{S}_N$ with a common known probability distribution

$$q(v) \equiv \Pr\{\omega^{[1]} = v\} \qquad (v \in \mathcal{S}_N)$$

which is not uniform but is strictly positive for every v (for example they could be generated by the MSAW algorithm). Suppose that we wish to estimate some quantity $\langle f(\omega)\rangle_N$, where f is a real-valued function on $\mathcal{S}_N$ and $\langle \cdot \rangle_N$ denotes the expectation with respect to the uniform distribution of ω on $\mathcal{S}_N$. If we define the reweighted average

$$Y_m^f \equiv \frac{1}{m} \sum_{i=1}^{m} \frac{f(\omega^{[i]})}{q(\omega^{[i]})},$$

then the expectation of Y_m^f is $c_N \langle f(\omega)\rangle_N$ (that is, Y_m^f/c_N is an *unbiased estimator* of $\langle f(\omega)\rangle_N$). This is because

$$E\left(\frac{f(\omega^{[i]})}{q(\omega^{[i]})}\right) = \sum_{v \in \mathcal{S}_N} \left(\frac{f(v)}{q(v)}\right) q(v) = c_N \left(\frac{1}{c_N} \sum_{v \in \mathcal{S}_N} f(v)\right). \qquad (9.3.1)$$

In particular, if we take f identically 1, then Y_m^1 is an unbiased estimator of c_N. Since the $\omega^{[i]}$'s are i.i.d., the strong law of large numbers guarantees that Y_m^f converges to $c_N \langle f(\omega)\rangle_N$ as $m \to \infty$, with probability one. Therefore, if we define the ratio

$$R_m^f \equiv Y_m^f/Y_m^1,$$

then R_m^f converges to $\langle f(\omega)\rangle_N$ as $m \to \infty$, with probability 1.

This theory can be applied to the case of walks generated by the MSAW algorithm, once we compute the function q. This was done for two examples of four-step walks in the paragraph following the statement of the algorithm in Section 9.1. In general, suppose that $v = (v(0), \ldots, v(N))$ is a self-avoiding walk. For each $i = 0, \ldots, N-1$, let t_i be the number of neighbours of $v(i)$ that are not in the set $\{v(0), \ldots, v(i-1)\}$. Then $q(v)$ is the product of the reciprocals of $t_0, \ldots, t_{N-1}$.

This method is often referred to as "inversely restricted sampling" or "biased sampling"; it is closely related to "importance sampling" [see for example Hammersley and Handscomb (1964) or Bratley, Fox and Schrage (1987)]. It was originally used by Rosenbluth and Rosenbluth (1955) for the function $f(\omega) = |\omega(N)|^2$. Earlier, Hammersley and Morton (1954) had used a slight variant of Y_m^1 to estimate c_N.

Biased sampling has some apparent drawbacks:

- Long walks will eventually become "trapped"; this could lead to many attempts being necessary to generate a single walk, unless we had a mechanism of avoiding steps that would lead into a trap. (We remark that a Monte Carlo study by Hemmer and Hemmer (1984) concluded that walks in $\mathbf{Z}^2$ survive for 71 steps before being trapped, on average.)

- The estimator R_m^J is not unbiased in general. However, McCrackin (1972) showed that $E(R_m^J) - \langle f(\omega) \rangle_N$ is of order m^{-1}, and hence for large m the difference is negligible compared with the ubiquitous $m^{-1/2}$ statistical error inherent in i.i.d. sampling schemes.

- The weights $1/q$ vary considerably, and a typical experiment is likely to end up with most of the overall weight coming from a very small fraction of the observations [Hammersley and Handscomb (1964), Batoulis and Kremer (1988)]. That is, the variance of the estimator R_m^J is likely to be uncomfortably large for any practical value of m. One might try to improve this situation by a variant of importance sampling, in which the possibilities in Step 2 of the MSAW algorithm are weighted so that the walk is encouraged to spread out faster (the original MSAW produces walks that tend to be more compact than typical self-avoiding walks). However, any such reweighting method where the distribution being sampled is substantially different from the desired (uniform) distribution could quite easily encounter the same problems, and the situation is rather delicate. Some work in this direction is surveyed in Kremer and Binder (1988, Sec. 2.1.2).

9.3.2   Dimerization

A different method of generating self-avoiding walks uniformly on $\mathcal{S}_N$ is *dimerization*, which is essentially a recursive procedure. The idea is that if we wish to generate an N-step self-avoiding walk, then we generate two independent $(N/2)$-step self-avoiding walks ("dimers") and try to concatenate them. If the result is self-avoiding, we are done; otherwise, we discard both dimers and start again. To generate each of the $(N/2)$-step walks, we generate two $(N/4)$-step walks and try to concatenate them, and so on. The recursion can stop at the k-th level if there is a fast way to generate self-avoiding walks of length $N/2^k$. For example, 10-step walks are easy to generate by Non-Reversed Simple Sampling, so only three levels are needed to create an 80-step walk by dimerization. We can express this as the following recursive procedure.

> $DIM(N)$. This procedure generates one N-step self-avoiding walk ω uniformly from $\mathcal{S}_N$. Here N_0 is a fixed small integer (e.g. $N_0 = 10$).
>
> 1. If $N \leq N_0$, then generate an N-step walk ω by NRSS and then stop.
> 2. ($N > N_0$) Set $N_1 = \lfloor N/2 \rfloor$ and $N_2 = N - N_1$.
> 3. Recursively perform $DIM(N_1)$ and $DIM(N_2)$, yielding the self-avoiding walks ω^1 and ω^2 respectively.

4. Set $\omega = \omega^1 \circ \omega^2$, the concatenation of ω^2 to ω^1. If ω is self-avoiding, then stop; otherwise, return to Step 2 and start over.

We remark that NRSS in Step 1 could be replaced by any other method that generates self-avoiding walks uniformly.

We shall use the following lemma to see that the end product ω is in fact uniformly distributed, as well as to investigate the efficiency of dimerization.

Lemma 9.3.1 *Let M and N be positive integers. Let $v^1, v^2, \ldots$ be independent self-avoiding walks uniformly distributed on S_M, and let $\varphi^1, \varphi^2, \ldots$ be independent self-avoiding walks uniformly distributed on S_N. For each $i \geq 1$, let ψ^i denote the concatenation of φ^i to v^i. Let τ be the smallest i such that ψ^i is self-avoiding. Then ψ^τ is uniformly distributed on S_{M+N}, and*

$$E(\tau) = \frac{c_M c_N}{c_{N+M}}. \tag{9.3.2}$$

Proof. For any fixed i we have

$$\Pr\{\psi^i \text{ is self-avoiding}\} = \frac{c_{M+N}}{c_M c_N};$$

call this quantity p. Then τ has a geometric distribution, i.e.

$$\Pr\{\tau = i\} = (1-p)^{i-1}p \qquad (i \geq 1)$$

so $E(\tau) = 1/p$, which proves (9.3.2). Now let ω be any fixed $(M+N)$-step self-avoiding walk, and let ω' and ω'' be the unique M-step and N-step walks whose concatenation $\omega' \circ \omega''$ is ω. Then

$$
\begin{aligned}
\Pr\{\psi^\tau = \omega\} &= \sum_{i=1}^{\infty} \Pr\{\tau = i \text{ and } \psi^i = \omega\} \\
&= \sum_{i=1}^{\infty} (1-p)^{i-1} \Pr\{v^i = \omega' \text{ and } \varphi^i = \omega''\} \\
&= \sum_{i=1}^{\infty} (1-p)^{i-1} \frac{1}{c_M} \frac{1}{c_N} \\
&= \frac{1}{c_{N+M}},
\end{aligned}
$$

which proves the lemma.   □

This lemma shows that in the procedure $\text{DIM}(N)$, the final walk ω is uniformly distributed provided that the walks ω^1 and ω^2 are uniformly distributed. We know that this will be true if N_1 and N_2 are small enough (since Step 1 is completely reliable), and so the uniformity of ω follows by induction on the number of levels in the recursion.

We now shall discuss the efficiency of dimerization, under the scaling assumption (1.1.4), i.e.

$$c_N \sim A\mu^N N^{\gamma-1}.$$

For simplicity, we assume $N = 2^k N_0$, where k is the number of levels of recursion. Let $T_{DIM(N)}$ denote the expected amount of time for the procedure $DIM(N)$ to produce a walk. By (9.3.2), the average number of pairs of $(N/2)$-step walks that must be generated before we get a pair whose concatenation is self-avoiding is $(c_{N/2})^2/c_N$, which is asymptotic to $A(N/4)^{\gamma-1}$ by the above scaling assumption. This gives us the recursive relation

$$T_{DIM(N)} \sim BN^{\gamma-1}(2T_{DIM(N/2)})$$

(where $B = A/4^{\gamma-1}$). (We have omitted the amount of time required to check whether the two dimers intersect each other, but since this time is $O(N)$, it will be seen to be negligible compared with $2T_{DIM(N/2)}$, the time required to generate the two dimers.) Iterating this relation k times (and assuming the approximate validity of our scaling assumption all the way down to N_0) yields

$$T_{DIM(N)} \approx \frac{(2BN^{\gamma-1})^k}{2^{(\gamma-1)k(k-1)/2}} T_{DIM(N_0)} = d_0 N^{d_1 \log_2 N + d_2}, \qquad (9.3.3)$$

where the d_i are independent of N:

$$d_1 = \frac{\gamma-1}{2}, \qquad d_2 = \frac{\gamma-1}{2} + \log_2(2B) = \frac{5-3\gamma}{2} + \log_2 A,$$

and d_0 depends on N_0. We thus conclude that the growth of $T_{DIM(N)}$ is slower than exponential in N. We also notice that the anticipated values for d_1 are small: according to (1.1.11), we expect d_1 to be 11/64 in two dimensions, 0.081... in three, and 0 in four or more dimensions. In particular, since it is known rigorously that $\gamma = 1$ in five or more dimensions (see Theorem 6.1.1), the above argument can be made into a rigorous proof that $T_{DIM(N)}$ grows *polynomially* in five or more dimensions. We also note that d_2 is small in high dimensions: in $d = 5$ we have the rigorous bound $d_2 = 1 + \log_2 A \le 1 + \log_2 1.493 \le 1.58$, and it is even smaller for $d \ge 6$ (see Remark following Theorem 6.1.1).

It is tempting to try to squeeze more data out of dimerization than just the information contained in the final N-step walk. For example, to estimate ν as described at the beginning of Section 9.1, one might try to use all the generated subwalks to get estimates of $\langle |\omega(N/2^i)|^2| \rangle_{N/2^i}$ for $i = 0, \ldots, k$.

This will give an unbiased estimate for each i, but the $k+1$ estimates will be mutually correlated; this makes it difficult to find a confidence interval for ν using classical linear regression theory (Section 9.2.1). Things look better if we are trying to estimate γ. For $n = N, N/2, \ldots, N/2^{k-1}$, let $\tau_n[j]$ denote the number of attempts needed to produce the j-th n-step self-avoiding walk (i.e. the number of pairs of $(n/2)$-step walks that are concatenated after the $(j-1)$-th success until the j-th success). As discussed above,

$$E(\tau_n[j]) \sim \frac{A}{4^{\gamma-1}} n^{\gamma-1},$$

so one could try using linear regression here. If we think of repeating DIM(N) indefinitely to produce an infinite sequence of N-step self-avoiding walks, then one can easily see that all of the random variables $\tau_n[j]$ ($n = N, N/2, \ldots, N/2^{k-1}$, $j \geq 1$) are *independent*. (This is essentially because the number of attempts needed to generate an n-step walk is independent of the walk itself.)

Suppose that we wish to generate m N-step walks by dimerization and use the $\tau_n[j]$ data to estimate γ. At the top level, we get m independent observations of $\tau_N[1]$. At the next level, we get a random number of independent copies of $\tau_{N/2}[\cdot]$: in fact, this random number is exactly $2(\tau_N[1] + \cdots + \tau_N[m])$. Thus there is some dependence between the data at different levels, but one can argue that it is negligible when m is large. A more serious difficulty with this scheme is its efficiency. It produces much more data for small n than for large n (in fact, more than twice as much data for $N/2^{i+1}$ than for $N/2^i$), but this is where we have the least confidence in our scaling assumption. So it is not clear how useful this method can be for estimating γ.

9.3.3 Enrichment

The enrichment method attempts to overcome the high attrition rate of simple sampling by reusing intermediate-length walks many times. This method was originally used by Wall and Erpenbeck (1959). The basic procedure requires two integer parameters, s and t. We first attempt to generate s-step self-avoiding walks by NRSS (or a similar method). Each time that we get an s-step walk, we make t (identical) copies of it and we attempt to extend each copy independently by NRSS to length $2s$. Similarly, each time that we get a self-avoiding walk of length $2s, 3s, \ldots$, we make t copies of that walk, each of which then evolves independently. The result will be a collection of self-avoiding walks of various lengths (all multiples of s). There will be a great deal of correlation between some of these walks, because they will have exactly the same first s (or $2s$, or $3s, \ldots$) steps;

but any two walks which are not extensions of copies of the same initial s-step walk will be statistically independent. Thus the enrichment method produces several independent groups of self-avoiding walks, but the walks within each group are highly correlated. Finding the correct statistical approach to handling these correlations remains an open problem.

Let M_{ks} denote the number of ks-step walks that are produced while performing this method. Then M_s is the number of independent groups; this number in practice is likely to be fixed in advance by the experimenter (of course, M_{ks} is random for $k \geq 2$). In the subsequent analysis, we shall assume for convenience that $M_s = 1$. The probability that a single attempt to extend a ks-step walk to a $(k + 1)s$-step walk succeeds is

$$\frac{c_{(k+1)s}}{(2d - 1)^s c_{ks}} \sim \left(\frac{\mu}{2d - 1}\right)^s.$$

We can think of M_s, M_{2s}, ... as a *branching process* in which M_{ks} represents the number of "individuals" alive in the k-th generation, and each individual reproduces independently, the number of offspring of an individual being a binomial random variable with parameters t and $p \approx (\mu/(2d - 1))^s$. No individual survives more than one generation. We can also think of every individual having t offspring, but each offspring only having probability p of reaching maturity. For more about branching processes, see for example Feller (1968) or Karlin and Taylor (1975). Strictly speaking, p is different for each generation, so we really have a time-inhomogeneous branching process. However we are not going to prove anything rigorously here, and it will be convenient to ignore this fact.

Given the number of js-step walks, the expected number of $(j+1)s$-step walks produced is

$$E(M_{(j+1)s}|M_{js}) = tpM_{js},$$

and by induction we conclude that $E(M_{ks}) = (tp)^k$. If $tp < 1$, then $E(M_{ks})$ decays exponentially: i.e. the branching process dies out exponentially fast. In this case, we do not expect to observe many long walks, and this method should not be much of an improvement over ordinary NRSS. If $tp > 1$, then there is a positive probability of a population explosion: that is, of M_{ks} increasing exponentially forever. This will lead to an enormous group of highly correlated walks. If $tp = 1$, then the branching process is "critical": it will die out eventually, but the expected time until this happens is infinite. This should produce some large walks, but there can be no population explosion of a single group. The preceding intuitive arguments are supported by the theory of branching processes. (This three-way classification is a hallmark of critical phenomena; in fact, the above branching process is essentially the same as percolation on an infinite tree in which every site has $t + 1$ neighbours.) From this discussion, we conclude that the best choice of parameters is to take t equal to $1/p$, i.e.

$$t \approx \left(\frac{2d-1}{\mu} \right)^s.$$

One can improve the enrichment method by combining it with the dimerization approach, as follows. Suppose that a self-avoiding walk ω of length ks has just been generated. Make t copies of this walk. For each copy, generate an s-step self-avoiding walk (by NRSS or some other method) completely independently of ω, and then try to concatenate it with ω. If the result has no intersections, then we have successfully produced a $(k + 1)s$-step self-avoiding walk, which we can now copy t times, and so on; otherwise, the attempt fails, and this copy of ω is no longer used. The probability of a success for such an attempt is

$$\frac{c_{(k+1)s}}{c_{ks} c_s} \sim \frac{(1 + \frac{1}{k})^{\gamma - 1}}{A s^{\gamma - 1}}$$

[by the usual scaling assumption (1.1.4)]. The above discussion then suggests taking t to be the inverse of this probability. (Note that allowing t to vary with k does not bias our results, whereas allowing t to depend upon the generated walks could easily introduce significant biases.) This method appears to be significantly more efficient than ordinary enrichment, but of course it still has the problem that walks within groups are highly correlated. Variants of this method have been used by Grishman (1973) and Rapaport (1985).

A closely related method has been proposed by Redner and Reynolds (1981). Its philosophy is a bit different, in that it estimates the susceptibility and other generating functions directly. A simple version of their method may be stated as follows.

Redner-Reynolds Algorithm. This algorithm generates random sets of self-avoiding walks $A_i \subset S_i$ $(i \geq 0)$. It requires a parameter z between 0 and 1. We denote the $2d$ (positive and negative) unit vectors of $\mathbf{Z}^d$ by $e_1, \ldots, e_{2d}$.

1. Let A_0 be the set consisting of the 0-step walk at the origin. Set $i = 0$. (Initially, A_k is empty for every $k \geq 1$.)
2. Independently, for each walk ω in A_i, and for each $j = 1, \ldots, 2d$: With probability $1 - z$, do nothing; otherwise (i.e. with probability z) try to add a step e_j to ω, and if the result is self-avoiding, then put it in A_{i+1}.
3. Increase i by one and go back to Step 2.

The algorithm stops when some A_i is empty. This algorithm is essentially a direct exact enumeration procedure in which each possibility is only pursued with probability z. Any given N-step self-avoiding walk is generated with probability z^N and so the expected cardinality of A_N is $c_N z^N$. Thus the total number of generated walks is an unbiased estimator for the susceptibility:

$$E\left(\sum_{N=0}^{\infty} |A_N|\right) = \chi(z).$$

In particular, for the interesting case $z < z_c = \mu^{-1}$, the Redner-Reynolds algorithm terminates in finite time with probability one. One can just as easily get estimates of other quantities: for example, the sum of the squares of the end-to-end distances of all of the generated walks is an unbiased estimator of $\chi(z)\xi_2(z)^2$, where $\xi_2(z)$ is the correlation length of order two defined in (1.3.18).

9.4 Length-conserving dynamic methods

In this section we shall look at dynamic Monte Carlo methods that generate walks having a fixed number of steps N. Each method of this type corresponds to a Markov chain that takes a self-avoiding walk and tries to change it in a random way to get another self-avoiding walk of the same length. The Verdier-Stockmayer algorithm, described in Section 9.1, is an example of such a method.

The algorithms that we shall consider in this section are of the following form.

> Generic Fixed-Length Dynamic Algorithm. Generates a Markov chain $\{\omega^{[t]} : t = 0, 1, \ldots\}$ on the state space $\mathcal{S}_N$ which is reversible with respect to the uniform distribution on $\mathcal{S}_N$.
>
> 1. Let $\omega^{[0]}$ be any self-avoiding walk in $\mathcal{S}_N$. Set $t = 0$.
> 2. Use a certain randomized procedure to define a new walk $\tilde{\omega} = (\tilde{\omega}(0), \ldots, \tilde{\omega}(N))$, which is not necessarily self-avoiding.
> 3. If $\tilde{\omega}$ is self-avoiding, then set $\omega^{[t+1]} = \tilde{\omega}$; otherwise, set $\omega^{[t+1]} = \omega^{[t]}$.
> 4. Increase t by one and go to Step 2.

Usually, it will be fairly routine to check reversibility, but questions about irreducibility (ergodicity) may require some work.

Before going on, we first make some remarks about conventions for this section. We shall always use N to denote the length of the walks being generated; N is an arbitrary integer which has been fixed (by the person running the experiment). The state space of the corresponding Markov chain is S_N. If the algorithm changes the first part of the current walk, then its initial point may no longer be the origin (as in Step 3(c) of the V-S algorithm); in such a case, we will always implicitly assume that the resulting walk is translated so that its initial step is the origin, thereby staying in the set S_N. (Alternatively, we can think of S_N as the set of equivalence classes of all N-step self-avoiding walks modulo translation; then the starting point of a generated walk is irrelevant, so there is no need to worry about translating back to the origin.) The transition probabilities will always be written $P(\cdot, \cdot)$.

9.4.1 Local algorithms

A *local* algorithm operates on walks by attempting to change only a few contiguous sites (and bonds) of the current walk at a time. The Verdier-Stockmayer algorithm is the prototype of this class of methods. Typically, a local algorithm chooses a small subwalk of the current walk at random, and attempts to replace it with a different (self-avoiding) subwalk having the same length and the same endpoints (unless the chosen subwalk includes an endpoint of the entire walk, in which case that endpoint may move). We keep the new walk if it is self-avoiding and reject it otherwise. The subwalk that we delete may uniquely determine the subwalk that replaces it (as in Step 3(a) of the V-S algorithm); alternatively, each possible subwalk may have a corresponding list of possible replacements, from which one must be chosen at random (as in Steps 3(b) and 3(c) of the V-S algorithm). Some examples are given in Figure 9.3.

The main theoretical result about these algorithms is that *none* of them is irreducible: in fact, for any given initial self-avoiding walk, the number of different walks that can be obtained from this walk by such an algorithm is exponentially smaller than c_N (for large N). Before we prove this, we shall first make our terms more precise.

Let $k \geq 1$ be a fixed integer, and let ω and ω' be N-step walks. Then we say that ω *can be transformed into* ω' *by a* k-*site move* if there exists an i ($0 \leq i \leq N - k + 1$) such that $\omega(j) = \omega'(j)$ for every $j = 0, 1, \ldots, i-1, i+k, \ldots, N$—that is, if ω and ω' are the same except for at most k contiguous sites. (Observe that the initial points of ω and ω' may be different if $i = 0$; similarly for their last points if $i = N - k + 1$.) We say that an algorithm is a k-*site algorithm* if the following holds: $P(\omega, \omega') > 0$ only if ω can be transformed into ω' by a k-site move. Thus the V-S algorithm is a 1-site algorithm. Finally, a length-conserving algorithm is said to be *local* if it is a k-site algorithm for some finite k. (Here, k must be independent of N; the term "algorithm" technically refers to a collection of algorithms, one

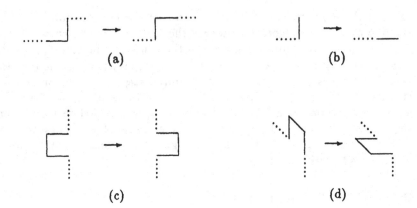

Figure 9.3: Some local length-conserving transformations. The transformation of (b) depicts the movement of an endpoint. The Verdier-Stockmayer algorithm uses (a) and (b), which are "one-site moves". Transformations (c) and (d) are "two-site moves" [(d) is shown in three dimensions].

for each N, but they are really all defined by exactly the same rules, so we use the word algorithm in the singular.)

We can define the "most general" k-site algorithm according to the recipe for the Generic Fixed-Length Dynamic Algorithm, where Step 2 is designed to allow transitions to any $\tilde{\omega}$ into which $\omega^{[t]}$ can be transformed by a k-site move. It is not hard to guarantee reversibility [Equation (9.1.6)]; for example, we can use the following rule.

2. Choose I uniformly at random from $\{0, 1, \ldots, N - k + 1\}$. Set $\tilde{\omega}(l) = \omega^{[t]}(l)$ for every $l < I$ and every $l \geq I + k$. If $0 < I < N - k + 1$, then randomly choose a $(k+1)$-step self-avoiding walk ω^* from among those walks in S_{k+1} satisfying $\omega^*(k+1) - \omega^*(0) = \omega^{[t]}(I + k) - \omega^{[t]}(I - 1)$. If I is 0 or $N - k + 1$, then randomly choose a k-step walk ω^* from S_k. Then $\tilde{\omega}(I), \ldots, \tilde{\omega}(I + k - 1)$ are obtained by translating ω^* so that it begins at $\omega^{[t]}(I - 1)$ (or, if $I = 0$, so that it ends at $\omega^{[t]}(k)$).

Then for any two distinct N-step self-avoiding walks ω and ω',

$$P(\omega, \omega') = \frac{1}{N - k + 2} \left[\sum_{i=1}^{N-k} F_i(\omega, \omega') \frac{1}{c_{k+1}(\omega(i - 1), \omega(i + k))} \right.$$

$$+(F_0(\omega,\omega') + F_{N-k+1}(\omega,\omega'))\frac{1}{c_k}\Big]$$
$$= P(\omega',\omega),$$

where $F_i(\omega,\omega')$ is 1 if $\omega(l) = \omega'(l)$ for every $l < i$ and every $l \geq i + k$, and it is 0 otherwise. Thus we see that P is symmetric, and moreover that $P(\omega,\omega') > 0$ if and only if ω can be transformed into ω' by a k-site move. We shall call this algorithm the *Maximal k-Site Algorithm* (MAX(k)).

Observe that two N-step walks ω and v are in the same ergodicity class of MAX(k) if and only if there exists a finite sequence of N-step walks $\omega \equiv \omega^{(0)}, \omega^{(1)}, \ldots, \omega^{(m)} \equiv v$ such that $\omega^{(i)}$ can be transformed into $\omega^{(i+1)}$ by a k-site move for every $i = 0, \ldots, m - 1$. In particular, any ergodicity class of any other k-site algorithm is contained in an ergodicity class of MAX(k).

It is not hard to see that the Verdier-Stockmayer algorithm is not irreducible in general. In $\mathbf{Z}^2$, the 17-step walk $ENW^2S^2E^5N^2W^2SE$ (Figure 9.2 in Section 9.1) cannot be transformed into any other self-avoiding walk by a 1-site move. We say that this walk is *frozen* (with respect to 1-site algorithms). In $\mathbf{Z}^3$, the V-S algorithm is not irreducible because of knot-like configurations: Figure 9.4 shows a 20-step walk which is in a different ergodicity class from, say, the straight walk for any 1-site or 2-site algorithm.

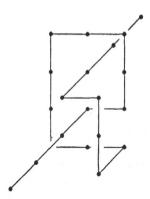

Figure 9.4: A knot-like walk in $\mathbf{Z}^3$ which cannot be transformed into a straight walk using 1-site or 2-site moves.

The observation that 1-site algorithms have frozen configurations in $\mathbf{Z}^2$ is generalized in the next theorem.

Theorem 9.4.1 *Let $d = 2$. For any integer $k \geq 1$ and any $r \geq k$, there exists a $(6r + 17)$-step self-avoiding walk which cannot be transformed into any other $(6r + 17)$-step walk by k-site moves.*

The idea of the proof is the following construction. Let $\psi^{(r)}$ be the $(6r+17)$-step walk

$$N^r ES^{r+1} W^2 N^{r+2} E^5 S^{r+2} W^2 N^{r+1} ES^r$$

(see Figure 9.5). If $r \geq k$, then $\psi^{(r)}$ is frozen under k-step moves. The details of the proof are given in Section 9.7.2. We remark that the conclu-

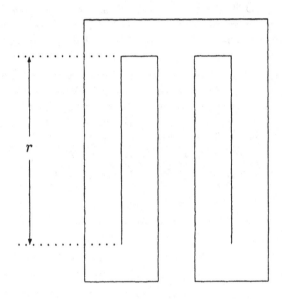

Figure 9.5: The walk $\psi^{(r)}$ from the proof of Theorem 9.4.1.

sions of this theorem are not restricted to lengths of the form $N = 6r + 17$. In fact, for every k it is true that for all sufficiently large N there exists an N-step self-avoiding walk which is frozen with respect to k-site algorithms [Madras and Sokal (1987)].

The next theorem discusses the cardinality of the largest ergodicity class (or CLEC for short) of local algorithms. It proves that for $d = 2$ or 3 the CLEC is exponentially smaller than the cardinality of the entire state space. Thus, even if we ran a Monte Carlo experiment for an infinitely long time using a local algorithm, we would only observe a small fraction of all N-step self-avoiding walks.

Theorem 9.4.2 *Let* $d = 2$ *or 3, and let* k *be a positive integer. Let* $CLEC_{k,N}$ *be the cardinality of the largest ergodicity class of MAX(k) (for* N*-step walks). Then*

$$\limsup_{N \to \infty} (CLEC_{k,N})^{1/N} < \mu.$$

The proof of this theorem relies on Kesten's Pattern Theorem. The idea is that there are certain patterns that cannot be changed by k-site moves, and these patterns can occur many times on a self-avoiding walk. (Of course, the pattern depends on k.) A walk on which many such patterns occur must be in a small ergodicity class, since only some parts of the walk are able to change. But such patterns must occur many times on all but exponentially few walks, so those walks which are most able to change are necessarily in a small ergodicity class. The full proof is given in Section 9.7.2 for two dimensions. The proof will work in any dimension, as long as the existence of these special patterns is proven. This has been done in three dimensions by Madras and Sokal (1987), but it has not been done in four or more dimensions.

The practical implications of the nonergodicity (i.e. lack of irreducibility) of local algorithms are somewhat controversial. On the one hand, if your sole wish is to study "static" properties of a single self-avoiding walk (or a linear polymer), then the nonergodicity of local algorithms together with their long autocorrelation times (see below) should convince you to look at other algorithms. On the other hand, if you are interested in the *dynamic* properties of real polymers, then local moves are a better model for how real polymers move than are, say, the pivots of Section 9.4.3. Also, in more complicated systems (e.g. many polymers, or strong attractive interactions between monomers) other methods may be infeasible, and so one has little choice but to use local moves and hope that the systematic bias due to nonergodicity is negligible.

To conclude our discussion of local algorithms, we shall briefly discuss their autocorrelation times. Technically, they should be infinite, since nonergodicity prevents us from ever reaching the desired equilibrium distribution; so instead our discussion will apply either to the Markov chain whose state space is the ergodicity class of the straight walk, or to a Markov chain which allows self-intersecting walks (perhaps with reduced probability).

For each N-step walk ω, let $g(\omega)$ denote the mean distance between pairs of sites on ω:

$$g(\omega) = \frac{1}{N(N+1)} \sum_{i \neq j} |\omega(i) - \omega(j)|.$$

Then under the usual scaling assumption that the distribution of $g(\omega)$ scales like N^{ν},

$$\tau_{int,g} \geq \text{const.} N^{2+2\nu}. \tag{9.4.1}$$

This follows from Corollary 9.2.3, since the variance $C_g(0)$ of $g(\omega)$ scales like $N^{2\nu}$, and since $|g(\omega) - g(\omega')| = O(1/N)$ whenever ω can be transformed into ω' by a local move. The same lower bound also holds for τ'_{exp}, by (9.2.27). It is generally believed that τ'_{exp} and τ_{exp} are in fact proportional to $N^{2+2\nu}$ for local algorithms that allow a wide enough class of moves [see Kremer and Binder (1988) for a discussion; note that their definition of τ differs from ours by a factor of N]. In the "mean-field" case of the VS algorithm applied to ordinary random walks (with $\nu = 1/2$), one can show that τ'_{exp} scales like $N^3 = N^{2+2\nu}$ [see Appendix 4.I of Doi and Edwards (1986)].

9.4.2   The "slithering snake" algorithm

A different kind of length-conserving dynamic algorithm was devised by Kron (1965) and by Wall and Mandel (1975) [see also Kron *et al.* (1967) and Mandel (1979)]. The basic move of the algorithm is to remove a bond from one end of the current walk while simultaneously trying to add a bond to the other end (rejecting the result if it is not self-avoiding). For an explicit description, use the following procedure as Step 2 in the Generic Fixed-Length Dynamic Algorithm.

> 2. Generate a random variable X which equals 0 with probability 1/2 and equals N with probability 1/2. If $X = 0$, then let Y be one of the $2d$ nearest neighbours of $\omega^{[t]}(0)$ (chosen uniformly at random), and set $\tilde{\omega} = (Y, \omega^{[t]}(0), \ldots, \omega^{[t]}(N-1))$. If $X = N$, then let Y be one of the $2d$ nearest neighbours of $\omega^{[t]}(N)$, and set $\tilde{\omega} = (\omega^{[t]}(1), \ldots, \omega^{[t]}(N), Y)$.

The nature of these moves has earned this algorithm and its variants the names "slithering snake" and "reptation" (the latter term is also used in polymer dynamics to describe similar motions of real polymers). This algorithm is reversible, but it is not irreducible: for example the walk of Figure 9.2 in Section 9.1 is frozen with respect to the slithering-snake algorithm in $\mathbb{Z}^2$. In fact, for sufficiently large N, it turns out that a positive fraction of all N-step walks are frozen, because there is a positive probability that both ends of the walk are "trapped" and cannot be extended by a single step in any direction. To be more precise, let Φ_N denote the set of all walks in $\mathcal{S}_N$ which are frozen with respect to the slithering-snake algorithm (that is, ω is in Φ_N if and only if the ergodicity class containing ω has cardinality one). Using the terminology of Definitions 7.1.2 and 7.4.1, let P be a proper front pattern with the property that the $2d$ nearest neighbours of the first site of P are all sites of P. Let R be the walk whose sites are the sites of P in reverse order (see Figure 9.6; note that R is a proper tail pattern). Then any self-avoiding walk that begins with the pattern P and ends with the pattern R must be frozen; i.e. $S_N(P, R) \subset \Phi_N$. Therefore (7.4.7) implies that

$$\liminf_{N\to\infty} \frac{|\Phi_N|}{c_N} > 0. \tag{9.4.2}$$

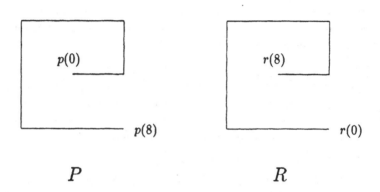

Figure 9.6: The proper front pattern $P = (p(0), \ldots, p(8))$ and the proper tail pattern $R = (r(0), \ldots, r(8))$. Any two-dimensional self-avoiding walk beginning with P and ending with R is frozen with respect to the slithering-snake algorithm.

Observe that although the intuitive description of the slithering-snake algorithm only involves moving one bond at a time, it is *not* a local algorithm by the definition of Section 9.4.1, because every site changes its position on the walk at every successful attempt [that is, $\omega^{[t]}(i)$ corresponds to $\omega^{[t+1]}(i \pm 1)$]. To emphasize the difference, we note that the analogue of (9.4.2) is false for local algorithms (since Kesten's Pattern Theorem 7.2.3 implies that most long walks contain many places where at least a single 1-site move can be made), and also that the analogue of Theorem 9.4.2 is false for the slithering-snake algorithm (since for example all N-step bridges are in the same ergodicity class as the straight self-avoiding walk, and $\lim_N (b_N/c_N)^{1/N} = 1$.) A better lower bound for the size of the largest ergodicity class is the following:

Proposition 9.4.3 *In the slithering-snake algorithm, denote by $\mathcal{E}_N$ the ergodicity class containing the N-step walk from the origin to $(N, 0, \ldots, 0)$. Then $|\mathcal{E}_N| \geq c_{2N}^{1/2}$.*

Remark. This bound is indeed better than $|\mathcal{E}_N| \geq b_N$ because $c_{2N}^{1/2} \geq \mu^N \geq b_N$ [by (1.2.10) and (1.2.17)].

Proof. Let $\mathcal{E}_N^*$ denote the set of all N-step walks in $\mathcal{S}_N$ which can be extended (possibly from both ends) to a $2N$-step self-avoiding walk; that is, ω is in $\mathcal{E}_N^*$ if and only if there is a walk $\varrho \in \mathcal{S}_{2N}$ such that ω occurs at some step of ϱ. Since every $2N$-step self-avoiding walk is the concatenation of two walks in $\mathcal{E}_N^*$, we see that $c_{2N} \leq |\mathcal{E}_N^*|^2$. Thus the proposition will be proved if we can show that $\mathcal{E}_N^*$ is contained in $\mathcal{E}_N$.

To complete the proof, let $\omega \in \mathcal{E}_N^*$, and let $\varrho \in \mathcal{S}_{2N}$ such that ω occurs at some step of ϱ. Let $\varrho(j)$ be the lexicographically largest site of ϱ [so ϱ lies in the half-space $x_1 \leq \varrho_1(j)$]. Now let v be the self-avoiding walk $(\varrho(j), \ldots, \varrho(j+N))$ if $j \leq N$, or $(\varrho(j-N), \ldots, \varrho(j))$ if $j > N$ (so v is an N-step subwalk of ϱ and has $\varrho(j)$ as an endpoint). Observe that ω and v are in the same ergodicity class, since ω can be transformed into v by "slithering" along the path ϱ. Since v lies in the half-space $x_1 \leq \varrho_1(j)$ and has one endpoint at $\omega(j)$ on the boundary of this half-space, it can be transformed into the straight walk whose endpoints are $\omega(j)$ and $\omega(j) + (N, 0, \ldots, 0)$. Therefore v is in $\mathcal{E}_N$, and hence so is ω. This completes the proof. $\square$

Proposition 9.4.3 and (9.4.2) imply that for sufficiently large N, the cardinality of the largest ergodicity class of the slithering-snake algorithm on $\mathcal{S}_N$, $CLEC_{SS,N}$, satisfies

$$aN^{-(\gamma-1)/2} \leq \frac{CLEC_{SS,N}}{c_N} \leq 1 - \epsilon$$

for some positive constants a and ϵ. Of course, the lower bound is only rigorous if we can prove the expected scaling behaviour $c_N \sim A\mu^N N^{\gamma-1}$. We do know that γ exists and equals 1 in five or more dimensions (see Section 6.1), so there $CLEC_{SS,N}/c_N$ stays bounded away from both 0 and 1; it is not known whether this ratio goes to 0 in 2, 3, or 4 dimensions.

9.4.3 The pivot algorithm

The preceding dynamic algorithms only attempt to move a few bonds at a time. In contrast, the *pivot algorithm* attempts to move large pieces of the walk at every iteration. These big moves are more likely to be rejected than are local moves, but a success is typically rewarded by a large change in global observables such as end-to-end distance.

The pivot algorithm picks a "pivot site" at random on the current walk, breaks the walk into two pieces at that site, and then applies a randomly chosen symmetry operation of $\mathbf{Z}^d$ to one piece, using the pivot site as the origin. As usual, the result is accepted if and only if it is self-avoiding. This algorithm was originally used by Lal (1969), and has subsequently been rediscovered by several authors (see the Notes at the end of this chapter). As we shall see, the pivot algorithm is remarkably efficient for the investigation of global observables: it requires about $O(N \log N)$ computer time to generate an "effectively independent" observation. (This is about as good as one has the right to expect, since it takes time $O(N)$ just to write down an N-step walk!)

To give a formal description of the pivot algorithm, let us first consider the symmetry group of $\mathbf{Z}^d$. To be precise, let $\mathcal{G}_d$ be the set of orthogonal linear transformations of $\mathbf{R}^d$ which leave the lattice $\mathbf{Z}^d$ invariant. In two dimensions, $\mathcal{G}_2$ has eight members: two axis reflections, two diagonal reflections, rotations by $\pm \pi/2$ and π, and the identity. For general d, a transformation g in $\mathcal{G}_d$ is completely determined by its action on the d positive unit vectors $e_1, \ldots, e_d$ of $\mathbf{Z}^d$. Since each $g(e_i)$ must be a unit vector of $\mathbf{Z}^d$, g can be uniquely specified by a permutation π of $\{1, \ldots, d\}$ and numbers $\epsilon_1, \ldots, \epsilon_d = \pm 1$ via the relations

$$g(e_i) = \epsilon_i e_{\pi(i)}. \tag{9.4.3}$$

Thus $\mathcal{G}_d$ has $2^d d!$ members. Next, observe that each g in $\mathcal{G}_d$ leaves the origin fixed (since g is a linear transformation). For every g in $\mathcal{G}_d$ and x in $\mathbf{Z}^d$, define g_x to be the corresponding affine transformation that leaves x fixed, i.e.

$$g_x(y) = g(y - x) + x \quad \text{for every } y \in \mathbf{Z}^d.$$

We can now describe the basic version of the pivot algorithm by using the following Step 2 in the Generic Fixed-Length Dynamic Algorithm.

2. Choose an integer I uniformly at random from $\{0, 1, \ldots, N - 1\}$. Set $x = \omega^{[t]}(I)$ (the "pivot site"). Choose a G uniformly at random from $\mathcal{G}_d$. Set $\tilde{\omega}(l) = \omega^{[t]}(l)$ for every $l \leq I$ and $\tilde{\omega}(l) = G_x(\omega^{[t]}(l))$ for every $l > I$.

As we shall see, this procedure is reversible and irreducible. We can get variants of this algorithm if we choose I or G from some nonuniform distribution. We shall also discuss irreducibility and reversibility of these variants below. As a different kind of variant, we could always pivot the *shorter* part of the walk, leaving the longer part fixed. This should improve the efficiency of the algorithm without changing the Markov chain in any important way.

It is not hard to check reversibility with respect to the uniform distribution on S_N. Suppose that ω and ω' are distinct self-avoiding walks such that $P(\omega, \omega') > 0$. There could be several ways to get from ω to ω': specifically, suppose that there are m possible pairs $(i^{(j)}, g^{(j)})$ $(1 \leq j \leq m)$ such that applying the operation $g^{(j)}$ to ω with pivot site $\omega(i^{(j)})$ will produce ω'. Then

$$P(\omega, \omega') = \sum_{j=1}^{m} \Pr\{I = i^{(j)}\} \Pr\{G = g^{(j)}\}.$$

Observe that applying the operation $(g^{(j)})^{-1}$ to ω' with pivot site $\omega'(i^{(j)})$ will produce ω. Therefore, we see from the above equation that $P(\omega, \omega') = P(\omega', \omega)$ in the original algorithm, as well as in any variant that satisfies

$$\Pr\{G = g\} = \Pr\{G = g^{-1}\} \qquad \text{for every } g \text{ in } \mathcal{G}_d.$$

We shall now consider the irreducibility of the pivot algorithm and also of variants which choose I and G from possibly nonuniform distributions. First of all, since the angle between the i-th and $(i+1)$-th step of the walk can only change when $I = i$, such a variant cannot be irreducible unless we require $\Pr\{I = i\} > 0$ for every $i = 1, \ldots, N-1$. Also, if $\Pr\{I = 0\} = 0$, then irreducibility fails because the direction of the first step never changes. (Of course, if the observables being measured are invariant with respect to the symmetries of the lattice, then it cannot hurt to take $\Pr\{I = 0\} = 0$.) Thus the interesting questions about irreducibility of the variants arise when some symmetries are allowed to have zero probability. The following result holds in every dimension $d \geq 2$.

Theorem 9.4.4 *The pivot algorithm is irreducible, as is any variant which gives nonzero probability to all d reflections through coordinate hyperplanes* $x_i = 0$ *and to all rotations by* $\pm\pi/2$ *(which leave* $d - 2$ *axes fixed). In fact, any walk in* S_N *can be transformed into a straight walk by some sequence of at most* $2N - 1$ *such pivots.*

The proof will be given in Section 9.7.3. The basic idea is that if we consider a snug box around a walk, then we can try to "unfold" the walk by performing a reflection through one of the faces of the box.

The above theorem remains true if we replace $\pm\pi/2$ rotations by any set of symmetries that contains, for every distinct i and j in $\{1, \ldots, d\}$, a symmetry that sends e_i to e_j and another that sends e_i to $-e_j$ (for example, the set of all reflections through hyperplanes $x_i = x_j$ or $x_i = -x_j$). The proof is the same. It is clear that some such set of symmetries must be used; notice that if we *only* allowed reflections through coordinate hyperplanes, then we could never change the angle between consecutive steps, and so the total number of right-angle turns in the walk could never change (in particular, straight walks would be frozen).

Some additional results about irreducibility of variants in two dimensions are known. If a variant gives nonzero probability to the three rotations $\pm\pi/2$ and π, then it is irreducible [see Section 3.5 of Madras and Sokal (1988)]. A variant is *not* irreducible if we only allow rotations by π (since the number of right-angle turns cannot change) or if we only allow rotations by $\pm\pi/2$ [a counterexample for $N = 223$ is shown on p. 139 of Madras and Sokal (1988)]. Finally, if we only allow the two diagonal reflections, then we *do* have irreducibility—in fact, any walk in $\mathcal{S}_N$ having exactly k right-angle turns can be transformed into a straight walk by some sequence of k diagonal reflections [Madras, Orlitsky, and Shepp (1990)]. As a consequence of this last result, we have

Corollary 9.4.5 *Let $d = 2$. For the transition probability P of the original pivot algorithm, $P^{2N-1}(\omega,\omega') > 0$ for every ω and ω' in $\mathcal{S}_N$.*

This means that the "diameter" of the state space of the two-dimensional pivot algorithm is at most $2N - 1$ ($N - 1$ pivots to straighten out ω, 1 pivot at the origin, and then $N - 1$ to make ω').

Now that we have seen that the pivot algorithm is a *valid* method (since it is reversible and irreducible), it is is time to discuss why it is a *good* algorithm. Only a limited part of this discussion will be based on rigorous proofs; the rest will consist of nonrigorous arguments (scaling theory, etc.) supported by numerical evidence from computer experiments.

The intuitive picture, which we shall elaborate upon below, is the following. Firstly, since a pivot makes a large-scale change in a walk, it is reasonable to expect that we will obtain an "effectively independent" configuration (at least with respect to global observables) after relatively few successful pivots. It will turn out that "relatively few" means about $\log N$. Secondly, the probability of a particular pivot being accepted will tend to 0 as $N \to \infty$, but as some power law N^{-p}. Since there are no frozen configurations, this probability cannot decay faster than N^{-1}, and so $0 \le p \le 1$. (Numerically, p is estimated to be about 0.19 in two dimensions and 0.11 in three.) Thus one expects a successful pivot in every N^p attempts. Recalling the discussion following (9.2.11), we infer from these first two points that the integrated autocorrelation time for a global observable should be about $N^p \log N$. Finally, we also have to include the average amount of computer time required per attempted pivot. The amount of work—checking for intersections, updating arrays, etc.—is at worst proportional to N; so suppose that the amount of computer time per attempt is on the order of N^q. Therefore the amount of computer time required per *successful* attempt is N^{p+q}, and the amount of computer time required per "effectively independent" observation of a global observable is $N^{p+q} \log N$. We shall argue below that $p + q = 1$.

In the remainder of the section we shall elaborate on the intuitive argument described above. As a guide for the first part, which says that relatively few successful pivots are needed to get an "effectively independent" observation of a global observable, we can consider a simpler model: the pivot algorithm applied to ordinary random walk. That is, the state space is now $\mathcal{S}_N^o$ (the set of all $(2d)^N$ ordinary walks), and the pivot algorithm now does not care about self-avoidance (so in Step 3 of the Generic Algorithm, we always set $\omega^{[t+1]} = \tilde{\omega}$). For this model, we can do exact calculations to prove rigorously that the integrated autocorrelation time $\tau_{int,g}^o$ for the global observable $g(\omega) = |\omega(N)|^2$ is asymptotic to $2 \log N$ as $N \to \infty$ (see Proposition 9.7.1 in Section 9.7.3). The same conclusion holds (except for a constant factor) for the global observables $\omega_i(N)$ and the squared radius of gyration [Madras and Sokal (1988)].

It is important to observe that the situation is quite different for the *exponential* autocorrelation times of the ordinary random walk: in particular, $\tau_{exp,g}^o$ is asymptotically equal to N as $N \to \infty$ for the global observable $g(\omega) = |\omega(N)|^2$ (Proposition 9.7.1). In fact, the exponential autocorrelation time for the entire chain, τ_{exp}^o, is also asymptotically proportional to N [Madras and Sokal (1988)]. It is easy to understand the situation for *local* observables: consider for example the angle between the 15-th and 16-th steps. The probability that this changes in a particular pivot is $1/N$ times a constant, since the angle can only change when the pivot site is $\omega(15)$, which happens with probability $1/N$. So both the integrated and exponential autocorrelation times for this observable should behave like N.

To summarize: global characteristics of walks tend to correspond to short modes of this system, while the long modes tend to be orthogonal to the quantities of interest. This emphasizes how the pivot algorithm is specially designed for looking at global quantities. It is reasonable to expect this to carry over to the self-avoiding case as well, and results of simulations seem to indicate that this is indeed what happens. However, proving such claims rigorously remains an open and apparently difficult problem.

We now turn to the amount of computer time required per attempted pivot, and its behaviour as N increases. The main issue is how long it takes to discover whether or not the proposed walk $\tilde{\omega}$ is self-avoiding. If we compute all of $\tilde{\omega}$ and then check for intersections, then each attempted pivot requires time proportional to N. But we can do better by looking for self-intersections as we compute $\tilde{\omega}$, so that we can stop early if one is found. We expect that $\tilde{\omega}$ is most likely to intersect itself in the vicinity of the pivot site, so we first compute $\tilde{\omega}$ at the pivot site, and then move outwards towards both ends of the walk simultaneously, computing $\tilde{\omega}$ and checking for self-intersections as we go. We shall now make the description of this procedure more precise. In doing so, it will be convenient to use the following notation for integers $a \leq b$ satisfying $a \leq N$ and $b \geq 0$:

$$\omega[a, b] \equiv (\omega(\max\{a, 0\}), \omega(\max\{a, 0\} + 1), \ldots, \omega(\min\{b, N\})).$$

Consider the following procedure for a single attempt of the pivot algorithm (where $\omega^{[t]}$ is the current walk).

(a) Choose the pivot site I and the symmetry G at random. Set $x = \omega^{[t]}(I)$, $j = 1$, and $\tilde{\omega}(I) = \omega^{[t]}(I)$.

(b) Set $\tilde{\omega}(I + j) = G_x(\omega^{[t]}(I + j))$ (if $I + j \leq N$) and set $\tilde{\omega}(I - j) = \omega^{[t]}(I - j)$ (if $I - j \geq 0$).

(c) If $I + j \leq N$, then check to see if $\tilde{\omega}(I + j)$ is in the set of sites $\tilde{\omega}[I - j + 1, I + j - 1]$. If it is, then the current attempt fails, so stop; otherwise, continue.

(d) If $I - j \geq 0$, then check to see if $\tilde{\omega}(I - j)$ is in the set of sites $\tilde{\omega}[I - j + 1, I + j]$. If it is, then the current attempt fails, so stop; otherwise, continue.

(e) If $j < \max\{N - I, I\}$, then increase j by one and go to Step (b). Otherwise, the current attempt has succeeded, so set $\omega^{[t+1]} = \tilde{\omega}$ and stop.

Steps (a) and (b) can be performed in time $O(1)$ (i.e. independent of N). In addition, Steps (c) and (d) can also be performed in average time $O(1)$ with the use of a bit map or a hash table (see the discussion at the end of Section 9.1), as follows. We begin with an empty bit map (or hash table); at each step, it will contain the sites of $\tilde{\omega}$ that have already been computed. As each new site of $\tilde{\omega}$ is computed, we check to see whether its location is still vacant in the bit map; if so, then we add this site to the bit map, but otherwise we stop because we have found a self-intersection. In the case of a success, Step (e) requires time $O(N)$ for recording $\omega^{[t+1]}$ and reinitializing the bit map. In summary, we see that the total amount of work is proportional to the number of times that Step (b) is performed (i.e. the number of times through the "loop"). Define the random variable $H(\omega)$ to be the smallest value of j such that $\tilde{\omega}[I - j, I + j]$ is not self-avoiding (and set $H(\omega) = N$ if $\tilde{\omega}[0, N]$ is self-avoiding). Thus the amount of work per attempt is of order $E(H(\omega))$. Evidently this is at most $O(N)$, but we can improve this bound by the following heuristic argument. First we have

$$\Pr\{H(\omega) > k\} \quad = \quad \Pr\{\tilde{\omega}[I - k, I + k] \text{ is self-avoiding}\}$$

$\approx$ Pr$\{$ a $2k$-step self-avoiding walk pivoted at its

midpoint is again self-avoiding$\}$

$\sim$ const.k^{-p}

where p is the exponent discussed above. We can now estimate the expectation of $H(\omega)$:

$$E(H(\omega)) = \sum_{k=0}^{N} \Pr\{H(\omega) > k\} \approx N^{1-p}.$$

Therefore the average amount of work per attempt is of order N^{1-p}, and so $p+q = 1$ as anticipated. This heuristic argument does in fact agree with computational experience.

This completes our discussion of why the integrated autocorrelation times for global observables are believed to be $O(N \log N)$ for the pivot algorithm.

9.5   Variable-length dynamic methods

In this section we shall discuss two dynamic methods whose state spaces include self-avoiding walks of various lengths. The Berretti-Sokal algorithm is the conceptually simplest such method: its state space is the set of all self-avoiding walks. The "join-and-cut" algorithm has as its state space the set of all pairs of self-avoiding walks whose lengths sum to some fixed number N. A third method, the BFACF algorithm, will be discussed in Section 9.6.1: its state space is the set of all self-avoiding walks with specified endpoints 0 and x for some fixed point x in $\mathbf{Z}^d$.

When using variable-length methods, the statistical analysis of the data can be more complicated than our discussion in Section 9.2 indicated. In that section, we assumed that the estimates from different values of N were independent. While this is true for fixed-length methods, where different values of N correspond to different simulations, it will be false for the algorithms of the present section. Berretti and Sokal (1985) show how to use maximum-likelihood estimation for their variable-length algorithm; the techniques developed there can be adapted to other algorithms.

9.5.1   The Berretti-Sokal algorithm

The Berretti-Sokal algorithm is designed to sample from the set of all self-avoiding walks of all possible lengths. It will be defined precisely below, but the basic idea is that at each step you either delete the last bond of the walk or else you attempt to increase the length of the walk by adding a bond to the end (rejecting the attempt if the result is not self-avoiding). The state space is

$$\mathcal{S} \equiv \bigcup_{N=0}^{\infty} \mathcal{S}_N,$$

which is infinite, so we cannot ask for uniform probabilities on all walks. It is natural, however, to ask for uniform probabilities within each $\mathcal{S}_N$. The Berretti-Sokal algorithm simulates walks in the "canonical ensemble" (in contrast to the fixed-length "microcanonical ensemble"). This requires a parameter $z > 0$ (as in the Redner-Reynolds algorithm of Section 9.3.3). Each N-step self-avoiding walk is given a weight (i.e. a relative probability) of z^N. The sum of all the weights of walks in $\mathcal{S}$ is just the susceptibility $\chi(z)$. Using this weight to normalize the probabilities, we obtain the probability distribution

$$\pi(\omega) \equiv \pi_z(\omega) = \frac{z^{|\omega|}}{\chi(z)}. \tag{9.5.1}$$

Of course, this only makes sense if $\chi(z)$ is finite, so we shall henceforth assume that

$$0 < z < z_c = \mu^{-1}.$$

(In physical terminology, z is the "fugacity per bond", and $\chi(z)$ plays the role of a "partition function"; also, π is a "Gibbs distribution".) Observe that π is a genuine probability distribution on $\mathcal{S}$. The mean square displacement of a walk chosen at random from this distribution is

$$\sum_{\omega \in \mathcal{S}} |\omega(|\omega|)|^2 \pi(\omega) = \sum_{\omega \in \mathcal{S}} \frac{|\omega(|\omega|)|^2 z^{|\omega|}}{\chi(z)} = \xi_2(z)^2,$$

which is the square of the correlation length of order 2. Thus we can obtain information about the critical exponent ν_2, which is believed to equal ν. Moreover, the canonical ensemble is a natural setting for studying μ and γ, since the fraction of time that the Markov chain spends in $\mathcal{S}_N$ (i.e. the fraction of time that an N-step self-avoiding walk is observed) is

$$\sum_{\omega: |\omega|=N} \pi(\omega) = \frac{c_N z^N}{\chi(z)} \sim \frac{A(\mu z)^N N^{\gamma-1}}{\chi(z)}.$$

We shall use $\langle \cdot \rangle_z$ to denote expectation with respect to π_z. For future reference, we note that the mean length of a walk is

$$\langle N \rangle_z = \frac{\sum_{N=0}^{\infty} N c_N z^N}{\sum_{N=0}^{\infty} c_N z^N} \simeq (z_c - z)^{-1} \tag{9.5.2}$$

[under the usual scaling assumptions, arguing as we did for (1.3.11)]. In particular, the mean length diverges as z increases to z_c.

We now state the algorithm of Berretti and Sokal (1985).

> *Berretti-Sokal (B-S) Algorithm.* This algorithm generates a Markov chain $\{\omega^{[t]}\}$ on the state space S which is reversible with respect to π_z.
>
> 1. Let $\omega^{[0]}$ be any self-avoiding walk in S. Set $t = 0$.
> 2. Let $N = |\omega^{[t]}|$. Generate a random variable X which is $+1$ with probability $2dz/(1 + 2dz)$ and -1 with probability $1/(1 + 2dz)$. If $X = +1$, then go to Step 3; if $X = -1$, then go to Step 4.
> 3. Try to add a step to $\omega^{[t]}$: Choose one of the $2d$ nearest neighbours of $\omega^{[t]}(N)$ uniformly at random; call this point Y. If Y is not already a site of $\omega^{[t]}$, then set $\omega^{[t+1]} = (\omega^{[t]}(0), \ldots, \omega^{[t]}(N), Y)$; if Y is a site of $\omega^{[t]}$, then set $\omega^{[t+1]} = \omega^{[t]}$. Increase t by one and go to Step 2.
> 4. Delete the last step of $\omega^{[t]}$: If $N > 0$, then set $\omega^{[t+1]} = (\omega^{[t]}(0), \ldots, \omega^{[t]}(N - 1))$; if $N = 0$, then set $\omega^{[t+1]} = \omega^{[t]}$ (the 0-step walk). Increase t by one and go to Step 2.

It is easy to see that the Markov chain corresponding to the B-S algorithm is irreducible: any N-step walk can be transformed into the 0-step walk in N iterations, and vice versa. Now let us check reversibility. Let ω be an N-step self-avoiding walk, and let ω' be an $(N+1)$-step self-avoiding walk which can be obtained by adding a single step to ω. Then

$$\pi(\omega)P(\omega, \omega') = \frac{z^N}{\chi(z)} \left(\frac{2dz}{1 + 2dz} \cdot \frac{1}{2d} \right)$$

and

$$\pi(\omega')P(\omega', \omega) = \frac{z^{N+1}}{\chi(z)} \frac{1}{1 + 2dz},$$

which implies that

$$\pi(\omega)P(\omega, \omega') = \pi(\omega')P(\omega', \omega).$$

For all other choices of distinct ω and ω', both sides of the above equation are 0. And of course the equation is trivial when $\omega = \omega'$. This proves reversibility with respect to π_z.

We now turn our attention to the autocorrelation times of the Berretti-Sokal algorithm. Before summarizing what is rigorously known, we shall give a heuristic argument which provides a pretty good intuition for what is happening. The first claim is that the autocorrelation times should be of the same order as the average time required to reach the 0-step walk from a typical initial walk in the state space. This is because before the 0-step walk is reached, the Markov chain still remembers the first steps of the initial walk, but the chain forgets everything once the 0-step walk is reached. Next, consider the process $N(t) \equiv |\omega^{[t]}|$, i.e. the length of the walk at time t. One expects this process to behave more or less like a random walk on the nonnegative integers having transition probabilities

$$
P(i, i+1) = \frac{2dz}{2dz+1} \frac{1}{2d} \mu = \frac{\mu z}{2dz+1}
$$

$$
P(i, i) = \frac{2dz - \mu z}{2dz+1}
$$

$$
P(i, i-1) = \frac{1}{2dz+1}
$$

for moderately large i (the factor μ in the first line is an approximation of c_{i+1}/c_i, the number of ways in which an average i-step self-avoiding walk can be extended by a single step). This random walk has a drift of $(\mu z - 1)/(2dz+1)$, which is *negative*. Thus the expected time for the process to go from a state N_0 to the state 0 is about N_0 divided by the magnitude of the drift. Finally, suppose that the initial walk $\omega^{[0]}$ is drawn at random from the equilibrium distribution π; then the expected time to reach 0 is about

$$
\langle N(0) \rangle_z \frac{2dz+1}{1 - \mu z} ;
$$

by (9.5.2), this is asymptotically proportional to $\langle N \rangle_z^2$ as $z \to z_c = \mu^{-1}$. Thus we conclude from our heuristic argument that τ_{exp} should scale like $\langle N \rangle_z^2$ [i.e. like $(z_c - z)^{-2}$].

This argument does quite well in several respects. First, one can do exact calculations when the B-S algorithm is applied to ordinary random walks [for which the state space is $\cup_N S_N^\circ$, and we take $0 < z < z_c = (2d)^{-1}$]. In this case, $N(t)$ is *exactly* a random walk with drift, and the integrated autocorrelation time of this observable can be shown to scale like $\langle N \rangle_z^2$ [see Appendix A of Berretti and Sokal (1985)]. Secondly, the random-walk-with-drift approximation is in fact a lower bound for the actual chain: an application of Corollary 9.2.3 (with $g = N$, $A = 1$, and the assumption that the probability distribution of N, in particular its standard deviation, scales like $\langle N \rangle_z$) shows that

$$
\tau_{int,N} \geq \text{const.} \langle N \rangle_z^2; \tag{9.5.3}
$$

by (9.2.27), this is also a lower bound for τ'_{exp}. Thirdly, Sokal and Thomas (1989) proved a rigorous upper bound, subject to the assumption that c_N

scales like $\mu^N N^{\gamma-1}$, that

$$\tau'_{exp} \le \text{const.} \langle N \rangle_z^{1+\gamma}. \, . \tag{9.5.4}$$

The exponent $1 + \gamma$ is near 2 in all dimensions (and in fact equals 2 when $d \ge 5$; see Section 6.1), so the above two bounds place pretty narrow limits on the scaling behaviour of τ'_{exp}. The exact behaviour remains an open question. (We remark that the proof of Sokal and Thomas also works for the B-S algorithm applied to ordinary random walks, where $\gamma = 1$.)

Lastly, we mention a slightly weaker bound derived by Lawler and Sokal (1988), using very different methods:

$$\tau'_{exp} \le \text{const.} \langle N \rangle_z^{2\gamma}. \tag{9.5.5}$$

Their main tool is a general version of Cheeger's inequality, which in its original form was a lower bound on the second smallest eigenvalue of the Laplacian on a compact Riemannian manifold [Cheeger (1970)]. Cheeger's inequality has recently found a wide range of applications in problems involving rates of convergence to equilibrium in Markov chains [see Diaconis and Stroock (1991) and references therein, as well as in Lawler and Sokal (1988)].

Finally, we note that one could implement a variant of the B-S algorithm in which one is allowed to add or delete steps from *either* end of the walk. We can regard this as a combination of the B-S and the "slithering snake" algorithm. The resulting algorithm should behave very much like the B-S algorithm. A form of this variant was used by Kron *et al.* (1967).

9.5.2   The join-and-cut algorithm

The join-and-cut algorithm was invented by Caracciolo, Pelissetto, and Sokal (1992) as an efficient method for estimating the exponent γ. This algorithm works on a rather different state space: the set of all *pairs* of self-avoiding walks whose combined length is fixed. To formalize the definition, let M be a fixed positive integer. We define T_M to be the set of all pairs (ψ, φ) of self-avoiding walks such that $|\psi| + |\varphi| = M$:

$$T_M \equiv \bigcup_{m=0}^{M} S_m \times S_{M-m}.$$

We shall see that the equilibrium distribution of the algorithm is uniform on T_M, and hence the distribution of the length of the first walk in the pair is

$$\Pr\{|\psi| = k\} = \frac{c_k c_{M-k}}{|T_M|} \approx k^{\gamma-1}(M-k)^{\gamma-1},$$

from which one can try to estimate γ.
The algorithm is as follows.

The Join-and-Cut Algorithm. This algorithm generates a Markov chain $\{X^{[t]}\} = \{(\psi^{[t]}, \varphi^{[t]})\}$ on the state space $\mathcal{T}_M$ which is reversible with respect to the uniform distribution on $\mathcal{T}_M$.

1. Let $X^{[0]} = (\psi^{[0]}, \varphi^{[0]})$ be any pair of self-avoiding walks in $\mathcal{T}_M$. Set $t = 0$.

2. Apply one iteration of the pivot algorithm (see Section 9.4.3) to $\psi^{[t]}$, obtaining $\hat{\psi}$. Then apply one iteration of the pivot algorithm to $\varphi^{[t]}$, obtaining $\hat{\varphi}$. (Alternatively, with the hope of reducing autocorrelation times, we could replace "one iteration" by "some fixed number n_{piv} of iterations", and "pivot algorithm" by "some length-conserving ergodic algorithm whose equilibrium distribution is uniform".)

3. *(Join)* Let $\zeta = \hat{\psi} \circ \hat{\varphi}$ be the concatenation of $\hat{\varphi}$ to $\hat{\psi}$.

4. *(Cut)* Choose J uniformly at random from $\{0, \ldots, M\}$. Set $\psi' = (\zeta(0), \ldots, \zeta(J))$ and $\varphi' = (\zeta(J), \ldots, \zeta(M))$. If both ψ' and φ' are self-avoiding, then set $\psi^{[t+1]} = \psi'$ and $\varphi^{[t+1]} = \varphi'$; otherwise, set $\psi^{[t+1]} = \hat{\psi}$ and $\varphi^{[t+1]} = \hat{\varphi}$.

5. Increase t by one and go to Step 2.

We emphasize that in Step 3 one does not need to check whether the walk ζ is self-avoiding. For purposes of comparison, however, let us consider also a variant of the join-and-cut algorithm in which we do perform this check. Specifically, this variant is obtained by replacing Steps 3 and 4 by the following:

3'. *(Join)* Let $\zeta = \hat{\psi} \circ \hat{\varphi}$ be the concatenation of $\hat{\varphi}$ to $\hat{\psi}$. If ζ is self-avoiding, then go to Step 4; otherwise, set $\psi^{[t+1]} = \hat{\psi}$ and $\varphi^{[t+1]} = \hat{\varphi}$ and go to Step 5.

4'. *(Cut)* Choose J uniformly at random from $\{0, \ldots, M\}$. Set $\psi^{[t+1]} = (\zeta(0), \ldots, \zeta(J))$ and $\varphi^{[t+1]} = (\zeta(J), \ldots, \zeta(M))$.

(Observe that whenever Step 4' is performed, the resulting $\psi^{[t+1]}$ and $\varphi^{[t+1]}$ are necessarily self-avoiding.)

The transition probability matrix P of the join-and-cut algorithm can be expressed as the product of two transition matrices P_a and P_b, which correspond respectively to Step 2 and to Steps 3 and 4 of the algorithm. To describe P_a and P_b more precisely, let Q be the transition matrix of the ergodic length-conserving algorithm used in Step 2 for single walks, defined with respect to the state space $\mathcal{S}$ of all self-avoiding walks (thus the ergodic classes of Q are precisely the sets S_N, and $Q(\omega, \omega') = 0$ whenever $|\omega| \neq |\omega'|$). Then

$$P_a((\psi_1, \varphi_1), (\psi_2, \varphi_2)) = Q(\psi_1, \psi_2)Q(\varphi_1, \varphi_2)$$

(for (ψ_i, φ_i) in T_M, $i = 1, 2$). If the length-conserving algorithm is reversible (i.e. if Q is symmetric), then so is P_a; more generally, if the restriction of Q to each S_N has the uniform distribution as a stationary distribution, then the same is true for the restriction of P_a to each $S_{N_1} \times S_{N_2}$. To describe P_b, suppose that (ψ_1, φ_1) and (ψ_2, φ_2) are distinct members of T_M whose concatenations $\psi_1 \circ \varphi_1$ and $\psi_2 \circ \varphi_2$ are the same; then

$$P_b((\psi_1, \varphi_1), (\psi_2, \varphi_2)) = \frac{1}{M+1}.$$

All other entries of P_a are 0, except for those on the main diagonal (which represent either a rejection in Step 4 or else the choice $J = |\psi^{[t]}|$). Clearly P_b is symmetric.

Unfortunately, the product of two symmetric matrices is not in general symmetric; therefore, even if P_a is symmetric (as it is when we use the pivot algorithm in Step 2), the product $P_a P_b$ cannot be expected to be symmetric. Thus the join-and-cut algorithm is *not* reversible in general. However its equilibrium distribution is nevertheless uniform on T_M, because both P_a and P_b have the constant vector as a left eigenvector, and hence so does their product. The failure of reversibility is due to the fact that a single iteration of the algorithm consists of two stages whose order matters: doing the pivoting followed by the join-and-cut steps. We remark that the variant of the join-and-cut algorithm corresponding to the transition matrix $P = \frac{1}{2}P_a + \frac{1}{2}P_b$ *would* be reversible (in this variant, at each iteration one randomly decides either to do Step 2 or else to do Steps 3 and 4).

It is easy enough to prove that the join-and-cut algorithm is irreducible, as follows. For any length N, let ρ_N be the N-step walk with $\rho_N(i) = (i, 0, \ldots, 0)$ for every i. Given any ψ in S_m and any φ in S_{M-m}, there exists a T such that $Q^T(\psi, \rho_m) > 0$ and $Q^T(\varphi, \rho_{M-m}) > 0$ (assuming that the restriction of Q to S_m is aperiodic, as it is in the case of the pivot algorithm). Since it is possible to pick $J = m$ on T consecutive iterations, we see that $P^T((\psi, \varphi), (\rho_m, \rho_{M-m}))$ and $P^T((\rho_m, \rho_{M-m}), (\psi, \varphi))$ are both nonzero. The concatenation of ρ_m and ρ_{M-m} is ρ_M, which may be cut successfully at any point, so the irreducibility of the algorithm follows.

It is possible to get some insight into the efficiency of the join-and-cut algorithm by a combination of rigorous analysis, scaling arguments and numerical work. We shall only give a brief description of some of these results. The reader is referred to Caracciolo *et al.* (1992) for more details.

First, let us estimate W, the amount of computer work that is required for a typical attempt to join and cut (that is, for Steps 3 and 4). For a given $\hat{\psi}$, $\hat{\varphi}$, and J (as produced by Step 2 and subsequently by Step 4), let $n = |\hat{\psi}|$ and let $L = |J - n|$. Then the attempt to join and cut may be described as an attempt to transfer the last L steps of $\hat{\psi}$ to the front of $\hat{\varphi}$ if $J < n$ (or vice versa if $J > n$). Roughly speaking this is like an attempt to concatenate two independent self-avoiding walks of lengths L and $M - n$ (or L and n). Thus if we start looking for self-intersections at the joining point and work our way outwards, then the probability that we will not have found one before k steps of both walks have been checked is approximately the same as the probability that two independent k-step self-avoiding walks can be concatenated successfully:

$$\Pr\{W > k\} \approx \frac{c_{2k}}{c_k^2} \sim \text{const.} k^{-(\gamma-1)}. \qquad (9.5.6)$$

We have not included any n-dependence in (9.5.6) because we expect it to disappear when we average over n (since n, $M - n$, and L all typically have order of magnitude M). Therefore the average amount of work required for Steps 3 and 4 should be

$$E(W) = \sum_{k=0}^{M} \Pr\{W > k\} \approx M^{2-\gamma}.$$

Recall from Section 9.4.3 that when applying the pivot algorithm to $\mathcal{S}_N$ the average work per pivot should scale like N^{1-p}. This implies that the expected amount of work for one complete iteration of the join-and-cut algorithm, in which Step 2 consists of doing some fixed number n_{piv} of pivots on each of $\psi^{[t]}$ and $\varphi^{[t]}$, is

$$n_{piv} M^{1-p} + M^{2-\gamma}.$$

By all evidence, $p < \gamma - 1$ in two and three dimensions, and so $1 - p > 2 - \gamma$; this implies that the most of the computer work in the join-and-cut algorithm is used in the pivoting step, even when $n_{piv} = 1$.

Suppose for the moment that n_{piv} is very large. Then the join-and-cut algorithm can be thought of as an "idealized algorithm", in which Step 2 actually produces walks $\hat{\psi}$ and $\hat{\varphi}$ that are independent of $\psi^{[t]}$ and $\varphi^{[t]}$. This idealized algorithm is more amenable to rigorous analysis: Caracciolo et al. (1992) prove that the exponential autocorrelation time is at most $M^{\gamma-1}$ (under the usual scaling assumption $c_N \sim A\mu^N N^{\gamma-1}$). This is done by showing that $\tau_{exp} \simeq M^{\gamma-1}$ for the variant that uses the idealized Step

2 in conjunction with Steps 3′ and 4′ described earlier, and then appealing to Proposition 9.2.4 (although some extra work is needed, since these algorithms are not reversible).

Now consider the original join-and-cut algorithm with $n_{piv} = 1$. Since the idealized algorithm should be more efficient than the actual algorithm with respect to the observable $n = |\psi^{[t]}|$, and more generally observables $g(n)$ that depend only on n, we shall define the exponent h by

$$\tau_{int,g(n)} \simeq M^{\gamma-1+h},$$

and we can expect that h is positive and hope that it is small. Combining this with the discussion above, we conclude that the amount of computer time per effectively independent observation scales like $M^{\gamma-1+h}M^{1-p}$, i.e. like M^κ where $\kappa \equiv \gamma - p + h$. Using the conjectured values of γ and p from (1.1.11) and Section 9.4.3 respectively, the bound $\kappa \geq \gamma - p$ becomes (approximately) $\kappa \geq 1.15$ in $\mathbf{Z}^2$, $\kappa \geq 1.05$ in $\mathbf{Z}^3$, and $\kappa \geq 1$ in four or more dimensions. Caracciolo *et al.* (1992) argue that in fact κ should equal 1 in four or more dimensions, which would virtually make this an optimal algorithm there. They also report the results of Monte Carlo runs which lead them to estimate that κ is about 1.5 in two dimensions, which is significantly better than the Berretti-Sokal algorithm [compare (9.5.3)].

9.6   Fixed-endpoint methods

This section will discuss some dynamic Monte Carlo methods that generate self-avoiding walks with endpoints that have been specified in advance.

First we shall describe the relevant state spaces. For each x in $\mathbf{Z}^d$ ($x \neq 0$), we denote by $\mathcal{S}_N(x)$ the set of all N-step self-avoiding walks ω having $\omega(0) = 0$ and $\omega(N) = x$. In this section, we shall always assume that N and $\|x\|_1$ have the same parity, since $\mathcal{S}_N(x)$ is empty otherwise. Also, we denote by $\mathcal{S}(x) = \cup_N \mathcal{S}_N(x)$ the set of all self-avoiding walks having endpoints 0 and x. When generating walks with fixed length and fixed endpoints, then we want to sample from the uniform distribution on $\mathcal{S}_N(x)$:

$$\pi(\omega) \equiv \pi_N^x(\omega) = \frac{1}{c_N(0,x)} \qquad \text{for every } \omega \text{ in } \mathcal{S}_N(x). \tag{9.6.1}$$

When generating walks from the variable-length fixed endpoint ensemble, the situation is similar to that of Section 9.5.1. In addition to specifying the endpoint x, we also specify a parameter z (the "fugacity per bond") between 0 and $z_c = \mu^{-1}$. We sample from the Gibbs distribution

$$\pi(\omega) \equiv \pi_z^x(\omega) = \frac{1}{\Xi(z, x)} |\omega| z^{|\omega|} \quad \text{for every } \omega \text{ in } \mathcal{S}(x), \qquad (9.6.2)$$

where $\Xi(z, x)$ is the normalizing constant

$$\Xi(z, x) = \sum_{N=0}^{\infty} N z^N c_N(0, x) = z \frac{\partial}{\partial z} G_z(0, x). \qquad (9.6.3)$$

The variable-length ensemble is the natural choice for studying the critical exponent α_{sing}, defined in (1.4.13) by

$$c_N(0, x) \sim B \mu^N N^{\alpha_{sing}-2}$$

This is because the fraction of time that the observed walk has length n is

$$\sum_{\omega \in \mathcal{S}_n(x)} \pi(\omega) = \frac{n c_n(0, x) z^n}{\Xi(z, x)} \sim \frac{B}{\Xi(z, x)} (\mu z)^n n^{\alpha_{sing}-1}, \qquad (9.6.4)$$

and so we can estimate α_{sing} by fitting a distribution of this form to the observed data (for fixed z near μ^{-1} and fixed x).

We remark that the multiplicative factor $|\omega|$ in (9.6.2) is not there for any deep reason, but only because this is what the algorithm of Section 9.6.1 naturally gives. By modifying the algorithm, one could get a different π, but there does not appear to be a good reason to do so.

Recall from Definition 3.2.1 that when $\|x\|_1 = 1$, we can associate each walk in $\mathcal{S}_N(x)$ with an $(N + 1)$-step self-avoiding polygon. Thus any of the methods discussed in this section can be used to study self-avoiding polygons simply by fixing x to be a nearest neighbour of the origin. In this case, we say that we are working with the ensemble of "rooted" polygons: there is a particular bond (the one joining x to the origin) which must occur in every polygon of the state space. It is also possible to work with the ensemble of "unrooted" polygons, where each bond of the current polygon is allowed to change during the iteration of the algorithm. Then the state space is the set of all polygons on the lattice (or their equivalence classes up to translation). There is little difference between the two ensembles in practice, aside from a factor of $N+1$ in their cardinalities [recall (3.2.1)] and the orientation of the rooted bond (which is irrelevant for most simulations). However, the Markov chains that are defined on the two ensembles are different in a non-trivial way; for example, a proof of irreducibility for the unrooted ensemble may not work for the rooted ensemble.

9.6.1   The BFACF algorithm

We shall first discuss an algorithm due to Berg, Foerster, Aragão de Carvalho, Caracciolo, and Fröhlich (for references, see the Notes at the end of the chapter). This algorithm uses transitions of a local nature to generate walks in the variable-length fixed endpoint ensemble according to the distribution given by (9.6.2) and (9.6.3).

The elementary transformations for this algorithm are depicted in Figure 9.7.   Each transformation is determined by choosing a bond of the

Figure 9.7: The elementary transformations of the BFACF algorithm.

current walk [say the bond from $\omega(i)$ to $\omega(i+1)$] and one of the $2d-2$ lattice directions perpendicular to the bond (let e be a unit vector in the chosen direction). Let x and y denote the lattice points $\omega(i) + e$ and $\omega(i+1) + e$ respectively. The transformation then moves the chosen bond by unit distance in the direction e, so that its new endpoints are x and y. Then there are now three possibilities, as illustrated in Figure 9.8:

(a) if the original ω had $\omega(i-1) \neq x$ and $\omega(i+2) \neq y$, then we add two bonds: the new walk is $\tilde\omega = (\omega(0), \ldots, \omega(i), x, y, \omega(i+1), \ldots, \omega(|\omega|))$, and $|\tilde\omega| = |\omega| + 2$;

(b) if the original ω had $\omega(i-1) = x$ and $\omega(i+2) = y$, then we remove two bonds: the new walk is $\tilde\omega = (\omega(0), \ldots, \omega(i-1), \omega(i+2), \ldots, \omega(|\omega|))$, and $|\tilde\omega| = |\omega| - 2$;

(c) if the original ω had $\omega(i-1) \neq x$ and $\omega(i+2) = y$ [or, respectively, $\omega(i-1) = x$ and $\omega(i+2) \neq y$], then the new walk is $\tilde\omega = (\omega(0), \ldots, \omega(i), x, \omega(i+2), \ldots, \omega(|\omega|))$ [respectively, $\tilde\omega = (\omega(0), \ldots, \omega(i-1), y, \omega(i+1), \ldots, \omega(|\omega|))$]. Here, $|\tilde\omega| = |\omega|$.

[If the chosen bond is the first bond of the walk, then $i = 0$ and we always have $\omega(i-1) \neq x$. Similarly, $\omega(|\omega| + 1) \neq y$ always.] We shall write ΔN to denote $|\tilde\omega| - |\omega|$, the change in the number of bonds of the walk for each possibility; we shall say that (a), (b), and (c) are $\Delta N = +2$, -2, and 0 transformations respectively.

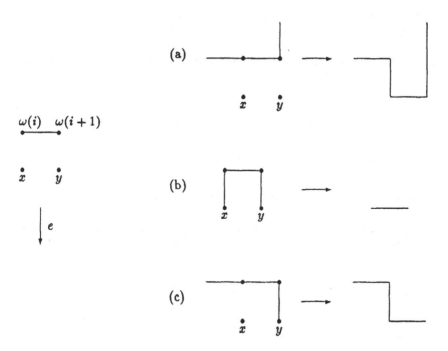

Figure 9.8: The three possibilities for a BFACF move, in detail.

To complete the definition of the BFACF algorithm, we need three numbers $p(+2)$, $p(-2)$, and $p(0)$ between 0 and 1 (they also must satisfy certain other conditions; see below).

BFACF Algorithm. This algorithm generates a Markov chain $\{\omega^{[t]}\}$ on the state space $\mathcal{S}(x)$.

1. Let $\omega^{[0]}$ be any walk in $\mathcal{S}(x)$. Set $t = 0$.
2. Choose an integer I uniformly at random from $\{0, 1, \ldots, |\omega^{[t]}| - 1\}$.
3. Consider the $2d - 2$ walks $\tilde{\omega}$ that would be obtained by moving the I-th bond of $\omega^{[t]}$ in one of the directions perpendicular to the vector $\omega^{[t]}(I + 1) - \omega^{[t]}(I)$. Choose one of these walks at random, with probabilities $p(|\tilde{\omega}| - |\omega^{[t]}|)$. (If these $2d - 2$ probabilities add up to $q < 1$, then also choose $\tilde{\omega} = \omega^{[t]}$ with probability $1 - q$.)
4. If $\tilde{\omega}$ is self-avoiding, then set $\omega^{[t+1]} = \tilde{\omega}$; otherwise, set $\omega^{[t+1]} = \omega^{[t]}$.
5. Increase t by one and go to Step 2.

The necessary constraints on $p(+2)$, $p(-2)$, and $p(0)$ are given by the following lemma.

Lemma 9.6.1 *The BFACF algorithm is well-defined and is reversible with respect to π_z^x if and only if the following constraints are satisfied:*

$$p(+2) = z^2 p(-2), \tag{9.6.5}$$

$$p(+2) \leq \frac{z^2}{1 + (2d - 3)z^2}, \tag{9.6.6}$$

and

$$2p(0) + (2d - 4)p(+2) \leq 1. \tag{9.6.7}$$

Proof. First we consider reversibility. Suppose that ω and $\tilde{\omega}$ are distinct walks in $\mathcal{S}(x)$ such that $P(\omega, \tilde{\omega}) > 0$. On the one hand, if $|\omega|$ and $|\tilde{\omega}|$ differ by 2, then

$$P(\omega, \tilde{\omega}) = \frac{1}{|\omega|} p(|\tilde{\omega}| - |\omega|);$$

the condition (9.1.5) for reversibility in this case reduces to (9.6.5). On the other hand, if $|\omega| = |\tilde{\omega}|$, then

$$P(\omega, \tilde{\omega}) = \frac{2}{|\omega|} p(0),$$

since there are two possible choices of bond of ω that can produce $\tilde{\omega}$ [for example, in (c) of Figure 9.8, we get the same result by choosing the bond joining $\omega(i + 1)$ to $\omega(i + 2) = y$ and moving it in the direction $x - y$]; the reversibility condition imposes no additional constraint in this case.

Next, we note that the algorithm is well-defined if and only if the sum of the $2d - 2$ probabilities in Step 3 does not exceed 1. There are several possibilities to consider, depending upon the relative orientations of the I-th, $(I - 1)$-th and $(I + 1)$-th bonds of $\omega^{[t]}$ (see Figure 9.9):

(i) All $2d - 2$ directions yield $\Delta N = +2$: This requires $(2d - 2)p(+2) \leq 1$.

(ii) One direction yields $\Delta N = 0$, while the others yield $\Delta N = +2$: This requires $p(0) + (2d - 3)p(+2) \leq 1$.

(iii) Two directions yield $\Delta N = 0$, while the others yield $\Delta N = +2$: This requires $2p(0) + (2d - 4)p(+2) \leq 1$.

(iv) One direction yields $\Delta N = -2$, while the others yield $\Delta N = +2$: This requires $p(-2) + (2d - 3)p(+2) \leq 1$.

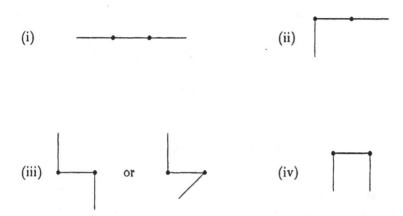

Figure 9.9: Proof of Lemma 9.6.1: relative orientations of three consecutive bonds.

The inequality of (ii) is redundant, since it follows from those of (i) and (iii). Next, substituting (9.6.5) into the inequality of (iv) gives

$$(z^{-2} + (2d - 3))p(+2) \leq 1, \qquad (9.6.8)$$

which is stronger than the inequality of (i) since $z \leq z_c < 1$. The inequality (9.6.8) is the same as (9.6.6), and the inequality of (iii) is the same as (9.6.7), so the lemma is proven. □

Now that we have a continuum of possible parameter values for a valid BFACF algorithm, we want to find the "best" choices of $p(+2)$, $p(-2)$, and $p(0)$ (for a given fixed z). Intuitively, we should prefer large values of these probabilities, so as to reduce the probability of "null transitions" (i.e. the quantity $1 - q$ described in Step 3 of the algorithm). Indeed, as we saw in Proposition 9.2.4 and the remark that follows it, increasing the off-diagonal elements of the transition matrix can only decrease the auto-correlation times (or at worst leave them unchanged). In two dimensions, the situation is easy: the constraint (9.6.7) simplifies to $p(0) \leq 1/2$, so the three probabilities can be maximized simultaneously:

$$p(+2) = z^2/(1 + z^2), \quad p(-2) = 1/(1 + z^2), \quad p(0) = 1/2. \qquad (9.6.9)$$

In three or more dimensions, the constraint (9.6.7) forces a tradeoff between $p(0)$ and $p(+2)$. The standard choice is the point determined by the intersection of the equalities corresponding to (9.6.6) and (9.6.7), which is

$$p(+2) = \frac{z^2}{1 + (2d - 3)z^2}, \qquad p(-2) = \frac{1}{1 + (2d - 3)z^2},$$

$$p(0) = \frac{1 + z^2}{2[1 + (2d - 3)z^2]}. \tag{9.6.10}$$

Observe that setting $d = 2$ in (9.6.10) gives the values of (9.6.9). Caracciolo, Pelissetto, and Sokal (1990) have proven rigorously that (9.6.10) is close to optimal in every dimension.

Let us now turn to the problem of irreducibility. In two dimensions, the algorithm is irreducible for every $x \neq 0$ (see Theorem 9.7.2). In three dimensions, the algorithm is *not* irreducible if $||x||_\infty = 1$ (in particular, for the case of self-avoiding polygons). This is essentially because of knots: Consider the closed curve defined by the steps of the current walk of $\mathcal{S}(x)$ and by the line segment joining x to the origin. Each possible BFACF transformation may be viewed as the result of a continuous deformation of this closed curve during which it never crosses itself. In the terminology of topology, we say that the result of a BFACF transformation is *ambient isotopic* to the initial curve. Thus, for $||x||_\infty = 1$, the walks in any given ergodicity class of the BFACF algorithm must all correspond to the same knot type. The converse assertion, that the ergodicity classes correspond precisely to knot classes, has been proven by Janse van Rensburg and Whittington (1991) for the special case of unrooted polygons by showing that "Reidemeister moves" on knots can be achieved using BFACF moves. When $||x||_\infty \geq 2$ in three dimensions, then the BFACF algorithm is irreducible [Janse van Rensburg (1992a)].

We conclude our discussion of the BFACF algorithm with a look at its autocorrelation times. These tend to be large, and it is not hard to identify one of the reasons: the "area" determined by a walk is a very slow mode. (The meaning of "area" is obvious in the case $d = 2$, $||x||_1 = 1$; in general, consider a fixed walk ζ from 0 to x and let $a(\omega)$ be the minimum area of a lattice surface whose boundary is the union of ω and ζ.) The problem is that an N-step walk ω can have $a(\omega)$ of order N^2; since a single BFACF move can only change a by one unit, such a configuration can survive a very long time before being changed into something substantially different. In particular, we can apply Corollary 9.2.3 with $A = 1$ to obtain

$$\tau_{int,a} \geq \text{const.} \langle N \rangle^{4\nu} \tag{9.6.11}$$

(under the usual assumption that the probability distribution of a, and in particular its standard deviation, scales like $N^{2\nu}$). The slowness of a

to change for certain configurations was exploited further by Sokal and Thomas (1988), who proved the unsettling result that the exponential autocorrelation time of the BFACF algorithm is *infinite* (see Theorem 9.7.4).

9.6.2 Nonlocal methods

In Section 9.6.1, we saw that the BFACF algorithm has rather long autocorrelation times. Recalling that the pivot algorithm is more efficient than local algorithms in the free-endpoint ensemble (recall Section 9.4), it is clearly desirable to try to find large-scale transformations of self-avoiding walks that work in the fixed-endpoint ensemble. The transformations of the pivot algorithm of Section 9.4.3 do not leave both endpoints fixed, in general; however, other fixed-length transformations that use one or two "pivot sites" have been used with some success.

Fixed-length transformations have been used in the ensemble $\mathcal{S}_N(x)$ to study properties such as the radius of gyration or knottedness, particularly in the case of self-avoiding polygons ($\|x\|_1 = 1$). They have also been used in the variable-length ensemble $\mathcal{S}(x)$ together with BFACF moves in the hope of obtaining a more efficient algorithm for this ensemble.

We now describe fixed-length transformations which leave both endpoints fixed (see Figure 9.10). In these descriptions, ω is always an N-step self-avoiding walk.

1. *Inversion:* For integers k and l ($0 \le k < l \le N$), define the new walk $\tilde{\omega}$ by

$$\tilde{\omega}(i) = \begin{cases} \omega(k) + \omega(l) - \omega(k+l-i) & \text{if } k \le i \le l \\ \omega(i) & \text{otherwise.} \end{cases}$$

Thus the subwalk $(\tilde{\omega}(k), \ldots, \tilde{\omega}(l))$ is the inversion through the point $(\omega(k) + \omega(l))/2$ of the points $(\omega(l), \ldots, \omega(k))$. Another way to view inversion is by the sequence of bonds $\Delta\omega(i) \equiv \omega(i) - \omega(i-1)$. Then the bonds of $\tilde{\omega}$ are $\Delta\omega(1), \Delta\omega(2), \ldots, \Delta\omega(k), \Delta\omega(l), \Delta\omega(l-1), \ldots, \Delta\omega(k+2), \Delta\omega(k+1), \Delta\omega(l+1), \ldots, \Delta\omega(N)$.

2. *Cyclic permutation:* For an integer i ($0 < i < N$), define the new walk $\tilde{\omega}$ by breaking ω into two pieces at $\omega(i)$ and then concatening the two pieces in the other order. Thus the bonds of $\tilde{\omega}$ are $\Delta\omega(i+1), \Delta\omega(i+2), \ldots, \Delta\omega(N), \Delta\omega(1), \Delta\omega(2), \ldots, \Delta\omega(i)$.

3. *Lattice symmetries:* Using the notation of Section 9.4.3, let $g \in \mathcal{G}_d$ be a lattice symmetry. Let k and l be integers ($0 \le k < l \le N$) and let $x = \omega(k)$. If $g_x(\omega(l)) = \omega(l)$, then we get a new walk $\tilde{\omega}$ by applying this symmetry to

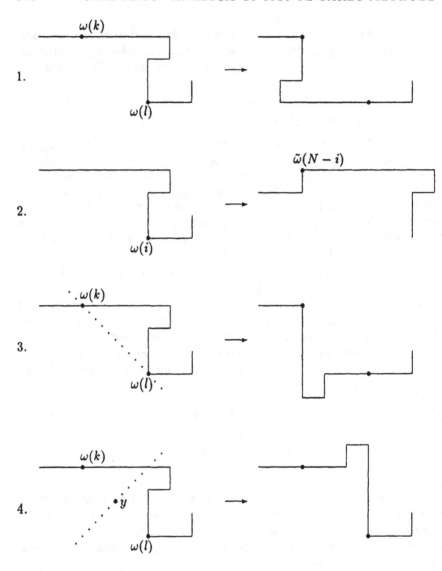

Figure 9.10: Length-preserving fixed-endpoint transformations: 1. inversion; 2. cyclic permutation; 3. reflection through line of slope −1; 4. reversing reflection through line of slope +1, where y is the midpoint between $\omega(k)$ and $\omega(l)$.

the part of ω between k and l:

$$\tilde{\omega}(i) = \begin{cases} g_x(\omega(i)) & \text{if } k \leq i \leq l \\ \omega(i) & \text{otherwise.} \end{cases}$$

Observe that the two "pivot sites" $\omega(k)$ and $\omega(l)$ are both fixed by g_x.

4. *Reversing lattice symmetries:* Again, let g be a lattice symmetry and let k and l be integers ($0 \leq k < l \leq N$). Now suppose that there exists a y such that $g_y(\omega(k)) = \omega(l)$ and $g_y(\omega(l)) = \omega(k)$. Then we can get a new walk $\tilde{\omega}$ by applying this symmetry to the part of ω between k and l, and reversing the order in which the sites appear in this part:

$$\tilde{\omega}(i) = \begin{cases} g_y(\omega(k + l - i)) & \text{if } k \leq i \leq l \\ \omega(i) & \text{otherwise.} \end{cases}$$

Dubins, Orlitsky, Reeds, and Shepp (1988) proposed (and proved the irreducibility of) an algorithm for unrooted polygons of fixed length in two dimensions. The DORS algorithm, as we shall call it, uses only inversion (1 above) and reversing diagonal reflection (4 above). The latter move may be described in words as follows. Choose two sites on the polygon such that the line segment L joining them makes an angle of $\pm\pi/4$ with the coordinate directions. Break the polygon into two pieces by cutting it at the two chosen sites, and reflect one of the pieces through the line which is the perpendicular bisector of L.

To prove that the DORS algorithm is irreducible, one shows first that inversions suffice to transform any polygon into a rectangle, and then that any rectangle may be transformed into any other rectangle by an inversion and a reversing reflection. The details of the proof are given in Section 9.7.4 (Theorem 9.7.3). Notice that an inversion does not change the number of bonds parallel to the x_1-axis, and so inversions alone do not suffice for ergodicity.

For the general fixed-length fixed-endpoint ensemble $\mathcal{S}_N(x)$ in two dimensions, the transformations of the DORS algorithm also provide an irreducible algorithm, but the proof is more involved [Madras, Orlitsky, and Shepp (1990)]. In higher dimensions, these transformations are not enough because if the initial walk is contained in the hyperplane $x_1 = 0$, say, then all of the resulting walks will lie in the same hyperplane.

To ensure irreducibility in $\mathcal{S}_N(x)$ in three or more dimensions, it suffices to use inversions (1 above), diagonal reflections (3 above), and reversing diagonal reflections (4 above). Here, a "diagonal reflection" is a reflection through a hyperplane which makes angles of $\pm\pi/4$ with two coordinate directions and angles of 0 with the remaining $d - 2$ directions. Irreducibility is proven by a lengthy argument that uses induction on the number of dimensions. The proof in d dimensions, even for $\|x\|_1 = 1$, requires knowledge of irreducibility of all fixed-length fixed-endpoint ensembles in $d - 1$ dimensions. For details, see Madras *et al.* (1990).

Caracciolo, Pelissetto, and Sokal (1990) introduced an algorithm for the variable-length fixed-endpoint ensemble $S(x)$, which uses inversion and cyclic permutation in addition to the usual BFACF transformations. This algorithm is irreducible in every dimension [Madras *et al.* (1990)].

9.7   Proofs

This section contains the longer proofs and calculations that have been deferred from the preceding sections of this chapter. The subsections may be read in any order.

9.7.1   Autocorrelation times

In this section we shall provide several arguments which were postponed from our discussion of the spectral theory of autocorrelation times that was begun in Section 9.2.3.

Let T be a self-adjoint contraction operator on $l^2(\pi)$. Then the spectrum $\sigma(T)$ is a subset of the interval $[-1, 1]$. The Spectral Theorem [see for example Reed and Simon (1972)] tells us that there is a *spectral measure E* such that

$$T = \int_{[-1,1]} \lambda \, dE(\lambda);$$

in fact, for every positive integer k we have

$$T^k = \int_{[-1,1]} \lambda^k \, dE(\lambda).$$

Recall that for every Borel subset A of $[-1, 1]$, $E(A)$ is a projection operator; in particular, $E(\emptyset) = 0$ and $E([-1, 1]) = I$. Also, for every g in $l^2(\pi)$ we define E_g by

$$E_g(A) \equiv (g, E(A)g) = \|E(A)g\|_2^2 \quad \text{for Borel sets } A \subset [-1, 1].$$

Then E_g is a positive measure and

$$(g, T^k g) = \int_{[-1,1]} \lambda^k \, dE_g(\lambda) \tag{9.7.1}$$

for every positive integer k.

We can use this representation to prove

$$r(T) = \sup_f \limsup_{n \to \infty} |(f, T^n f)|^{1/n},$$

which is Equation (9.2.21). Let $q(T)$ denote the right hand side of the above equation. Since $r(T) = \|T\|$, we clearly have $r(T) \geq q(T)$. Thus it suffices to prove the reverse inequality. Choose t so that $0 < t < r(T)$ and let $A[t] = \{\lambda : t < |\lambda| \leq 1\}$. We claim that there is a g in $l^2(\pi)$ such that $E_g(A[t]) > 0$. If not, then for every g

$$|(g, Tg)| \leq \left| \int_{-t}^{t} \lambda \, dE_g(\lambda) \right| \leq t E_g([-1, 1]) = t\|g\|_2^2,$$

which contradicts $t < r(T) = \|T\|$, so the claim is true. For g as in the claim, we have for any even n that

$$(g, T^n g) \geq \int_{A[t]} |\lambda|^n \, dE_g(\lambda) \geq t^n E_g(A[t]),$$

which implies that $q(T) \geq t$. Since this holds whenever $0 < t < r(T)$ we have $q(T) \geq r(T)$, so Equation (9.2.21) is proven.

We now return to our Markov chain, with $T = P - \Pi$ and E the corresponding spectral measure. Let g be a function in $l^2(\pi)$ and let $h = (I - \Pi)g$ be its projection onto the space of functions with mean 0. Since g and h differ by a constant, we know that $C_g(k) = C_h(k)$ for every $k \geq 0$. By (9.2.22) and (9.7.1),

$$C_g(k) = C_h(k) = \int_{[-1,1]} \lambda^k dE_h(\lambda) \quad \text{for every } k \geq 0$$

(where we interpret $0^0 = 1$). Using this in (9.2.10), along with the identity

$$\frac{1}{2} + \sum_{k=1}^{\infty} \lambda^k = \frac{1}{2}\left(\frac{1+\lambda}{1-\lambda}\right),$$

we obtain

$$\tau_{int,g} = \tau_{int,h} = \frac{\frac{1}{2}\int_{[-1,1]}\left(\frac{1+\lambda}{1-\lambda}\right) dE_h(\lambda)}{\int_{[-1,1]} dE_h(\lambda)}. \tag{9.7.2}$$

The support of E_h lies in $[-1, s]$, where $s = \sup \sigma(P - \Pi)$; together with (9.7.2) and the fact that $(1 + \lambda)/(1 - \lambda)$ is increasing for λ in $[-1, 1)$, this tells us that

$$\tau_{int,g} = \tau_{int,h} \leq \frac{1}{2}\left(\frac{1+s}{1-s}\right).$$

By (9.2.26), this proves (9.2.27).

We can now give a quick proof of Proposition 9.2.2 from Section 9.2.3, which says that $\tau_{int,g} \geq \frac{1}{2}(1+\rho_g(1))/(1-\rho_g(1))$, where $\rho_g(1) = C_g(1)/C_g(0)$.

Proof of Proposition 9.2.2.   Let g be a nonconstant function in $l^2(\pi)$, and let $h = (I - \Pi)g$. Then E_h is a finite measure that is not identically 0, and so $E_h/\int dE_h(\lambda)$ is a probability measure. The function $\lambda \mapsto (1 + \lambda)/(1 - \lambda)$ is convex, so Jensen's inequality implies that the right hand side of (9.7.2) is bounded below by $\frac{1}{2}(1 + \rho_h(1))/(1 - \rho_h(1))$, where

$$\rho_h(1) = \frac{\int \lambda \, dE_h(\lambda)}{\int dE_h(\lambda)} = \frac{C_h(1)}{C_h(0)} = \frac{C_g(1)}{C_g(0)}.$$

This proves the proposition. $\square$

9.7.2   Local algorithms

We shall begin by proving the two theorems about irreducibility of k-site algorithms from Section 9.4.1. Theorem 9.4.1 states that in two dimensions there are frozen $(6r + 17)$-step walks for every $r \geq k$. Theorem 9.4.2 states that for $d = 2$ or 3 and for sufficiently large N, the cardinality of the largest ergodicity class of any k-site algorithm is less than $e^{-aN} c_N$ for some $a > 0$. We give the proof only for $d = 2$, as discussed in Section 9.4.1.

Proof of Theorem 9.4.1.   Let $d = 2$. For each positive integer r, let $\psi^{(r)}$ be the $(6r + 17)$-step walk

$$N^r E S^{r+1} W^2 N^{r+2} E^5 S^{r+2} W^2 N^{r+1} E S^r$$

(see Figure 9.5 in Section 9.4.1). We shall show that if $r \geq k$ then $\psi^{(r)}$ is frozen under k-step moves.

Let $N = 6r+17$. Let $\mathcal{B}$ be the set of all sites of $\psi^{(r)}$, so that $\mathcal{B}$ consists of an $(r + 2) \times 6$ rectangle of sites of $\mathbf{Z}^2$. Consider removing any k contiguous sites $\psi^{(r)}(I), \ldots, \psi^{(r)}(I + k - 1)$ from $\psi^{(r)}$. We want to find k distinct sites $a_1, \ldots, a_k$ such that: $|a_j - a_{j+1}| = 1$ for $j = 1, \ldots, k - 1$; each a_j is in the set

$$\mathcal{D} \equiv (\mathbf{Z}^2 \setminus \mathcal{B}) \cup \{\psi^{(r)}(I), \ldots, \psi^{(r)}(I + k - 1)\};$$

$|a_1 - \psi^{(r)}(I - 1)| = 1$ if $I > 0$; and $|a_k - \psi^{(r)}(I + k)| = 1$ if $I < N - k + 1$. If the only choice for each a_j is $\psi^{(r)}(I + j - 1)$, then we can conclude that $\psi^{(r)}$ is indeed frozen.

If $I = 0$, then the removed sites all lie on a vertical line (since $r \geq k$). Moreover, the only nearest neighbour of $\psi^{(r)}(k)$ that is in $\mathcal{D}$ is $\psi^{(r)}(k - 1)$, so we must take a_k to be this site. Similarly, the only choice for a_j is

$\psi^{(r)}(j-1)$ for each j, so no changes are possible when $I = 0$. Similarly, no changes are possible when $I = N - k + 1$.

Suppose now that $0 < I < N - k + 1$. The proof now is essentially by inspection. Look at all possibilities for $\psi^{(r)}(I - 1)$ and $\psi^{(r)}(I + k)$ (in particular, whether or not they are on the boundary of the box $\mathcal{B}$), and look at how to connect these points by a $(k + 1)$-step self-avoiding walk which only passes through points of $\mathcal{D}$. On the one hand, if at least one of these two points are on the boundary of $\mathcal{B}$, then there is only one possible walk (in fact, $k+1$ is the length of the shortest such walk that joins these points). On the other hand, if neither point is on the boundary of $\mathcal{B}$, then the only points of $\mathcal{D}$ that can be reached are precisely $\psi^{(r)}(I), \ldots, \psi^{(r)}(I + k - 1)$, and there is no choice but to take them in their correct order (since they lie on either one or two vertical lines, and since all k of them must be used). □

Proof of Theorem 9.4.2. Let $d = 2$ and fix k. Let P be the $(10k + 39)$-step pattern

$$N^{k+2}W^3S^{k+1}EN^kES^{k+1}W^3N^{k+3}E^9S^{k+3}W^3N^{k+1}ES^kEN^{k+1}W^3S^{k+2}$$

(see Figure 9.11), and let $L = 10k + 39$. An argument similar to the proof of Theorem 9.4.1 shows that if P occurs at the m-th step of a given self-avoiding walk ω, then P must occur at the m-th step of every walk that is in the same ergodicity class as ω.

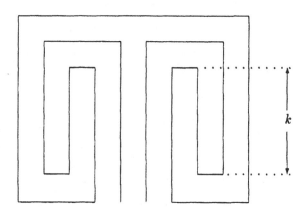

Figure 9.11: The pattern P from the proof of Theorem 9.4.2.

For every integer $t \geq 0$, and for every sequence $0 \leq m_1 < m_2 < \ldots < m_t < N$, let $\mathcal{E}_N(m_1, m_2, \ldots, m_t)$ denote the set of walks in $\mathcal{S}_N$ such that P

occurs at the m_j-th step of ω for every $j = 1, \ldots, t$ and nowhere else in ω. (For $t = 0$, this is the set of walks on which P does not occur.) Notice that successive occurrences of P in ω cannot overlap, and so we always have $m_j - m_{j-1} > L$ whenever $\mathcal{E}_N(m_1, m_2, \ldots, m_t)$ is nonempty. For each $t \geq 0$, let

$$M(t, k, N) = \max\{|\mathcal{E}_N(m_1, m_2, \ldots, m_t)| : 0 \leq m_1 < \ldots < m_t < N\}.$$
$$\text{(9.7.3)}$$

By the conclusion of the preceding paragraph, each ergodic class is contained in some $\mathcal{E}_N(m_1, m_2, \ldots, m_t)$, and so

$$CLEC_{k,N} \leq \max_{t \geq 0} M(t, k, N). \qquad (9.7.4)$$

Since P is a proper internal pattern, Kesten's Pattern Theorem 7.2.3 tells us that there exists an $a > 0$ such that P must occur at least aN times on "almost all" N-step walks, i.e.

$$\limsup_{N \to \infty} (c_N[aN, P])^{1/N} < \mu. \qquad (9.7.5)$$

Therefore

$$\limsup_{N \to \infty} \left[\max_{0 \leq t \leq aN} M(t, k, N) \right]^{1/N} < \mu. \qquad (9.7.6)$$

Next, we claim that for any $t \geq 0$ and any sequence $0 \leq m_1 < \ldots < m_t < N$,

$$\mathcal{E}_N(m_1, m_2, \ldots, m_t) \leq c_{N-t(L-1)}. \qquad (9.7.7)$$

To see this, define the function f from $\mathcal{E}_N(m_1, m_2, \ldots, m_t)$ to $\mathcal{S}_{N-t(L-1)}$ which removes each occurrence of P and replaces it by a single bond. Since f is one-to-one, the bound (9.7.7) follows. Therefore

$$\limsup_{N \to \infty} \left[\max_{t \geq aN} M(t, k, N) \right]^{1/N} \leq \limsup_{N \to \infty} \left[\max_{t \geq aN} c_{N-t(L-1)} \right]^{1/N}$$
$$= \mu^{1 - a(L-1)} < \mu. \qquad (9.7.8)$$

The theorem now follows from (9.7.8) and (9.7.6).   □

9.7.3   The pivot algorithm

We begin with a proof of the irreducibility of the pivot algorithm and some of its variants, which is asserted in Theorem 9.4.4. More precisely, this theorem says that any walk in $\mathcal{S}_N$ can be transformed into a straight walk by a sequence of at most $2N - 1$ pivots, each of which is either a reflection through a coordinate hyperplane or a rotation by $\pm \pi/2$.

Proof of Theorem 9.4.4.   We begin with some notation. For each N-step self-avoiding walk ω and for each $j = 1, \ldots, d$, let

$$m_j^1(\omega) = \min\{\omega_j(k) : k = 0, 1, \ldots, N\} \qquad (9.7.9)$$

and

$$m_j^2(\omega) = \max\{\omega_j(k) : k = 0, 1, \ldots, N\} \qquad (9.7.10)$$

denote the minimum and maximum values of the j-th coordinate of the sites of ω, and let

$$M_j(\omega) = m_j^2(\omega) - m_j^1(\omega) \qquad (9.7.11)$$

denote the extension of ω in the j-th coordinate direction. Let $\mathcal{B}(\omega)$ denote the smallest rectangular box containing ω, i.e.

$$\mathcal{B}(\omega) = \{x \in \mathbf{Z}^d : m_j^1(\omega) \leq x \leq m_j^2(\omega) \text{ for all } j = 1, \ldots, d\}, \qquad (9.7.12)$$

and let

$$D(\omega) = M_1(\omega) + \cdots + M_d(\omega) \qquad (9.7.13)$$

denote the l^1 diameter of $\mathcal{B}(\omega)$. A *face* of $\mathcal{B}(\omega)$ is any set of the form $\{x \in \mathcal{B}(\omega) : x_j = m_j^i(\omega)\}$ for some $i = 1, 2$ and some $j = 1, \ldots, d$. Finally, let

$$A(\omega) = |\{k : 0 < k < N \text{ and } \omega(k) = \frac{1}{2}[\omega(k-1) + \omega(k+1)]\}| \qquad (9.7.14)$$

denote the number of straight internal angles of ω.

The strategy of the proof is the following. Observe that for every N-step self-avoiding walk ω, we have $0 \leq D(\omega) \leq N$ and $0 \leq A(\omega) \leq N - 1$, and moreover $D(\omega) + A(\omega) = 2N - 1$ if and only if ω is a straight walk. It suffices to show that if ω is not straight, then there exists another self-avoiding walk $\tilde{\omega}$ such that $D(\tilde{\omega}) + A(\tilde{\omega}) > D(\omega) + A(\omega)$ and $\tilde{\omega}$ can be obtained from ω by either a single reflection through a coordinate hyperplane or a single rotation by $\pm \pi/2$. Specifically, we shall show that if there is a face of $\mathcal{B}(\omega)$ which contains neither of the endpoints $\omega(0)$ nor $\omega(N)$, then a reflection through that face will increase D but not change A; and if no such face exists, then there exists a rotation that increases A by one without decreasing D.

We now give the details. Consider an arbitrary N-step self-avoiding walk ω that is not straight. We shall consider two cases separately. Since ω is fixed, we shall write $\mathcal{B}$ for $\mathcal{B}(\omega)$ and m_j^i for $m_j^i(\omega)$.

Case I. Suppose that there exists an $i \in \{1, 2\}$ and a $j \in \{1, \ldots, d\}$ such that neither $\omega(0)$ nor $\omega(N)$ lies in the face $\{x \in \mathcal{B} : x_j = m_j^i\}$. Let t be the smallest index such that $\omega(t)$ lies in this face. Now let $\tilde{\omega}$ be the walk obtained by reflecting $\omega(t+1), \ldots, \omega(N)$ through the hyperplane $x_j = m_j^i$: that is, $\tilde{\omega}(k) = \omega(k)$ for each $k \leq t$, while the coordinates of $\tilde{\omega}(k)$ for $k > t$ are given by

$$\tilde{\omega}_l(k) = \begin{cases} 2m_j^i - \omega_j(k) & \text{if } l = j \\ \omega_l(k) & \text{if } l \neq j. \end{cases} \tag{9.7.15}$$

(See Figure 9.12.)   It is not hard to see that $\tilde{\omega}$ is self-avoiding, and that

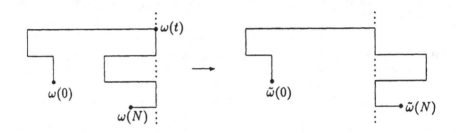

Figure 9.12: Case I of the proof of Theorem 9.4.4: reflection through the hyperplane denoted by the dotted line.

$A(\tilde{\omega}) = A(\omega)$ [notice that both $\tilde{\omega}$ and ω have right angles at $\omega(t)$]. Let us now show that $D(\tilde{\omega}) > D(\omega)$. Writing $M_j(\omega[r, s])$ to denote the extension of the subwalk $(\omega(r), \ldots, \omega(s))$ in the j-th coordinate direction, we see that

$$M_j(\omega) = \max\{M_j(\omega[0, t]), M_j(\omega[t, N])\} \tag{9.7.16}$$

and

$$M_j(\tilde{\omega}) = M_j(\omega[0, t]) + M_j(\omega[t, N]). \tag{9.7.17}$$

Since $\omega_j(0) \neq m_j^i$ and $\omega_j(N) \neq m_j^i$, both $M_j(\omega[0, t])$ and $M_j(\omega[t, N])$ are strictly positive, and so we conclude that $M_j(\tilde{\omega}) > M_j(\omega)$. Since $M_l(\tilde{\omega}) = M_l(\omega)$ whenever $l \neq k$, this proves that $D(\tilde{\omega}) > D(\omega)$, and hence that $D(\tilde{\omega}) + A(\tilde{\omega}) > D(\omega) + A(\omega)$. This completes the proof for Case I.

Case II. Suppose that ω is not covered by Case I; that is, suppose that every face of $\mathcal{B}$ contains at least one endpoint. This means that $\omega(0)$ and $\omega(N)$ are in diagonally opposite corners of the box $\mathcal{B}$. Since ω is not straight, let q be the largest index such that ω has a right angle at $\omega(q)$:

$$q = \max\{k : 0 < k < N \text{ and } \omega(k) \neq \frac{1}{2}[\omega(k-1) + \omega(k+1)]\}. \tag{9.7.18}$$

Since $\omega(N)$ is in a corner of $\mathcal{B}$, we will be able to perform a $\pm\pi/2$ rotation to straighten out the angle at $\omega(q)$. To be precise: the sites $\omega(q), \ldots, \omega(N)$ lie on a straight line perpendicular to the line segment joining $\omega(q-1)$ to $\omega(q)$. Let α be the coordinate such that $\omega_\alpha(q-1) \neq \omega_\alpha(q)$, and let β be the coordinate such that $\omega_\beta(q) \neq \omega_\beta(N)$. Observe that $\alpha \neq \beta$. Now we can define a new walk $\tilde{\omega}$ by choosing $\omega(q)$ as a pivot site and performing a rotation in the (x_α, x_β)-plane to get a straight angle at $\tilde{\omega}(q) = \omega(q)$. (See Figure 9.13.) The resulting walk has $\tilde{\omega}(q-1), \tilde{\omega}(q), \ldots, \tilde{\omega}(N)$ all

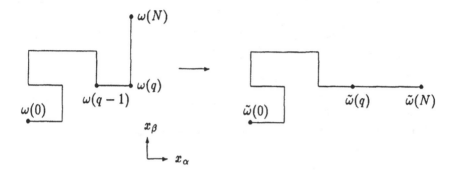

Figure 9.13: Case II of the proof of Theorem 9.4.4: rotation by $-\pi/2$. Also shown are the coordinate directions x_α and x_β.

on a straight line. Since $\tilde{\omega}(0), \ldots, \tilde{\omega}(q-1)$ are on the opposite side of the hyperplane $x_\alpha = \tilde{\omega}_\alpha(q)$ from $\tilde{\omega}(q+1), \ldots, \tilde{\omega}(N)$, we see that $\tilde{\omega}$ is self-avoiding. We also see that

$$M_\alpha(\tilde{\omega}) = M_\alpha(\omega) + N - q, \qquad (9.7.19)$$

that

$$M_\beta(\tilde{\omega}) = M_\beta(\omega[0,q]) \geq M_\beta(\omega) - (N - q), \qquad (9.7.20)$$

and that $M_j(\tilde{\omega}) = M_j(\omega)$ for all $j \neq \alpha, \beta$. Therefore $D(\tilde{\omega}) \geq D(\omega)$. Also, we clearly have $A(\tilde{\omega}) = A(\omega) + 1$, and hence that $D(\tilde{\omega}) + A(\tilde{\omega}) > D(\omega) + A(\omega)$. This completes the proof for Case II, and we are done. $\square$

Next we consider the pivot algorithm applied to the ordinary random walk, without ever checking for intersections. To be precise, the state space of the algorithm is the set $\mathcal{S}_N^o$ of all $(2d)^N$ ordinary N-step walks starting at the origin, and the Generic Fixed-Length Dynamic Algorithm (from the beginning of Section 9.4) is implemented using the usual Step 2 for the pivot algorithm as described in Section 9.4.3, but in Step 3 $\omega^{[t+1]}$ is *always* set equal to $\tilde{\omega}$. For a given real-valued function g on $\mathcal{S}_N^o$, let $\tau_{exp,g}^o$ and $\tau_{int,g}^o$

respectively denote the exponential and integrated autocorrelation times of g with respect to this algorithm [as defined in (9.2.12) and (9.2.10)].

Proposition 9.7.1 *For each N, define the function $r^2 \equiv r_N^2$ on the set $\mathcal{S}_N^o$ by $r_N^2(\omega) = |\omega(N)|^2$, the squared end-to-end distance of $\omega \in \mathcal{S}_N^o$. Then as $N \to \infty$*

$$\tau_{exp,r^2}^o \sim N \tag{9.7.21}$$

and

$$\tau_{int,r^2}^o \sim 2 \log N. \tag{9.7.22}$$

Proof. Fix N. Using the notation $\Delta\omega^{[t]}(i) = \omega^{[t]}(i) - \omega^{[t]}(i-1)$ to denote the i-th step of the walk $\omega^{[t]}$, define $A_{ij}^{[t]}$ to be the dot product of the i-th and j-th steps of $\omega^{[t]}$:

$$A_{ij}^{[t]} = \Delta\omega^{[t]}(i) \cdot \Delta\omega^{[t]}(j)$$

for $1 \le i, j \le N$. Then by expanding the square we have

$$r^2(\omega^{[t]}) = \left| \sum_{i=1}^N \Delta\omega^{[t]}(i) \right|^2 = N + 2 \sum_{1 \le i < j \le N} A_{ij}^{[t]}. \tag{9.7.23}$$

In equilibrium, $\omega^{[t]}$ is uniformly distributed on $\mathcal{S}_N^o$, and so the N steps of $\omega^{[t]}$ are independent and uniformly distributed on the set of the $2d$ (positive and negative) unit vectors of $\mathbf{Z}^d$. By symmetry we have

$$E(A_{ij}^{[t]}) = 0 \quad \text{whenever } i \ne j, \tag{9.7.24}$$

and also

$$E[(A_{ij}^{[0]})^2] = \Pr\{|A_{ij}^{[0]}| = 1\} = \frac{1}{d}. \tag{9.7.25}$$

We also have that if $t \ge 0$, $i < j$, and $k < l$, then

$$E(A_{ij}^{[0]} A_{kl}^{[t]}) = 0 \quad \text{unless } i = k \text{ and } j = l. \tag{9.7.26}$$

Consider the first iteration of the pivot algorithm with initial walk $\omega^{[0]}$. For a given k between 1 and N, a necessary condition for the direction of the k-th step to change is that the chosen pivot site I is less than k. In fact, if G is the chosen symmetry, then

$$\Delta\omega^{[1]}(k) = \begin{cases} \Delta\omega^{[0]}(k) & \text{if } I \ge k \\ G\left(\Delta\omega^{[0]}(k)\right) & \text{if } I < k. \end{cases} \tag{9.7.27}$$

Therefore if I is not in the interval $[i,j)$, then $A_{ij}^{[1]} = A_{ij}^{[0]}$. Also, since G is chosen uniformly at random from $\mathcal{G}_d$, the vector $G\left(\Delta\omega^{[0]}(k)\right)$ is uniformly distributed on the set of unit vectors of $\mathbf{Z}^d$; moreover, it is *independent* of the entire walk $\omega^{[0]}$. In particular, we see that if I is in the interval $[i,j)$, then $A_{ij}^{[1]}$ is independent of $A_{ij}^{[0]}$.

Let $Q \equiv Q_{i,j,t}$ be the event that at least one of the pivot sites of the first t iterations is in the interval $[i,j)$. Then as in the preceding paragraph we see that conditioned on the occurrence of Q the quantities $A_{ij}^{[t]}$ and $A_{ij}^{[0]}$ are independent; hence by (9.7.24),

$$E(A_{ij}^{[0]} A_{ij}^{[t]}|Q) = 0. \tag{9.7.28}$$

If Q does not occur, then $A_{ij}^{[t]} = A_{ij}^{[0]}$, and hence by (9.7.25)

$$E(A_{ij}^{[0]} A_{ij}^{[t]}|Q^c) = E((A_{ij}^{[0]})^2|Q^c) = E((A_{ij}^{[0]})^2) = \frac{1}{d}. \tag{9.7.29}$$

Since the probability of Q^c is $[1-(j-i)/N]^t$, we see from (9.7.28), (9.7.29) and (9.7.24) that

$$\mathrm{Cov}(A_{ij}^{[0]}, A_{ij}^{[t]}) = \frac{1}{d}\left(1 - \frac{j-i}{N}\right)^t \quad \text{for } i < j \text{ and } t \geq 0. \tag{9.7.30}$$

Using (9.7.23), (9.7.26) and (9.7.30), we see that for every $t \geq 0$

$$\begin{aligned}
C_{r^2}(t) \equiv \mathrm{Cov}(r^2(\omega^{[0]}), r^2(\omega^{[t]})) &= 4 \sum_{1 \leq i < j \leq N} \mathrm{Cov}(A_{ij}^{[0]}, A_{ij}^{[t]}) \\
&= \frac{4}{d}\sum_{m=1}^{N-1}(N-m)\left(1-\frac{m}{N}\right)^t \\
&= \frac{4N}{d}\sum_{m=1}^{N-1}\left(1-\frac{m}{N}\right)^{t+1}. \tag{9.7.31}
\end{aligned}$$

The $m = 1$ term in (9.7.31) is dominant; in fact,

$$\frac{4N}{d}\left(1-\frac{1}{N}\right)^{t+1} \leq C_{r^2}(t) \leq \frac{4N(N-1)}{d}\left(1-\frac{1}{N}\right)^{t+1},$$

and so the definition of exponential autocorrelation time in (9.2.12) implies that

$$\tau_{exp,r^2}^o = \frac{-1}{\log\left(1-\frac{1}{N}\right)} = N + O(1), \tag{9.7.32}$$

which proves (9.7.21). Next, (9.7.31) tells us that $C_{r^2}(0) = 2N(N-1)/d$, and putting this and (9.7.31) into the definition of integrated autocorrelation time [recall (9.2.10)] yields

$$
\begin{aligned}
\tau^o_{int,r^2} &= \frac{1}{2} + \frac{d}{2N(N-1)} \sum_{t=1}^{\infty} \frac{4N}{d} \sum_{m=1}^{N-1} \left(1 - \frac{m}{N}\right)^{t+1} \\
&= \frac{1}{2} + \frac{2}{N-1} \sum_{m=1}^{N-1} \frac{\left(1 - \frac{m}{N}\right)}{\frac{m}{N}} \\
&= \frac{1}{2} + \frac{2}{N-1} \left[N \sum_{m=1}^{N-1} \frac{1}{m} - (N-1) \right] \\
&= 2 \log N + O(1),
\end{aligned}
\tag{9.7.33}
$$

which proves (9.7.22).   □

9.7.4 Fixed-endpoint methods

In this section we shall prove three results. Theorems 9.7.2 and 9.7.3 prove the irreducibility of the BFACF and DORS algorithms, respectively, in two dimensions. Finally, Theorem 9.7.4 proves that the exponential autocorrelation time of the BFACF algorithm is infinite.

We first establish some terminology that will be needed for the proofs of the first two theorems. Every self-avoiding polygon $\mathcal{P}$ in $\mathbf{Z}^2$ forms a simple closed curve, and hence has an inside and an outside (in the sense of the Jordan curve theorem). If v is a site of $\mathcal{P}$, then we say that v is *convex*, *concave*, or *straight* according as to whether the inside angle of $\mathcal{P}$ at v is 90°, 270°, or 180°.

Theorem 9.7.2 *For every nonzero endpoint x in $\mathbf{Z}^2$, the BFACF algorithm is irreducible on $\mathcal{S}(x)$.*

Proof. We begin by looking at a special case in which the endpoint is on the x_1-axis. For every integer $L > 0$, let $\rho^{(L)}$ denote the straight L-step walk from $(0,0)$ to $(L,0)$. For $N > L$, let $\mathcal{S}^*_N((L,0))$ be the set of N-step self-avoiding walks beginning at $(0,0)$ and ending at $(L,0)$ such that none of the sites $(1,0),(2,0),\ldots,(L-1,0)$ is occupied by a site of ω. We shall henceforth assume implicitly that N and L have the same parity, since otherwise $\mathcal{S}^*_N((L,0))$ is empty. Observe that $\mathcal{S}^*_N((L,0)) \subset \mathcal{S}_N((L,0))$ whenever $N > L > 0$, with equality if $L = 1$. Every walk ω in $\mathcal{S}^*_N((L,0))$ has an associated $(N+L)$-step self-avoiding polygon $\mathcal{P} \equiv \mathcal{P}(\omega)$ whose bonds are the N bonds of ω together with the L bonds of $\rho^{(L)}$.

Our first goal is to prove the following:

Claim A: Suppose that $N > L > 0$ and ω is in $\mathcal{S}_N^*((L,0))$. Then it is possible to transform ω into the straight walk $\rho^{(L)}$ by BFACF moves in such a way that none of the intermediate walks obtained in this process has a site lying outside of (the original) $\mathcal{P}(\omega)$.

Claim A implies irreducibility in the case $\|x\|_1 = 1$ (upon taking $L = 1$), and the approach that we take to prove it will also be used in the proof for general x.

To prove Claim A, we need some additional terminology. If ω is in $\mathcal{S}_N^*((L,0))$, then we say that the subwalk $\omega[i,j] \equiv (\omega(i),\ldots,\omega(j))$ $(0 \le i < j \le N)$ is a *U-turn* of ω if $j - i \ge 3$, $\omega[i+1, j-1]$ lies on a straight line, and the sites $\omega(i + 1)$ and $\omega(j - 1)$ are both convex sites of $\mathcal{P}(\omega)$. We say that $\omega(k)$ is an *obstruction* of the U-turn $\omega[i,j]$ if $\omega(k)$ is on the line segment whose endpoints are $\omega(i)$ and $\omega(j)$, and $k \ne i,j$. We say that the U-turn $\omega[i,j]$ is *unobstructed* if it has no obstructions. Observe that if $\omega[i,j]$ is

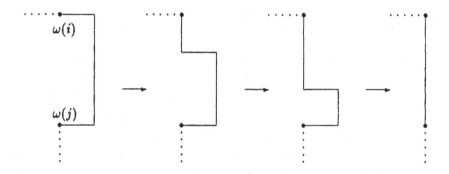

Figure 9.14: How the BFACF algorithm can use the presence of an unobstructed U-turn to shorten the length of a walk by 2.

an unobstructed U-turn of ω, then ω can be transformed into a walk ω' of length $N - 2$ using $j - i - 2$ BFACF moves as in Figure 9.14. Moreover, if $N - 2 = L$ then $\omega' = \rho^{(L)}$, while if $N - 2 > L$ then $\omega' \in \mathcal{S}_{N-2}^*((L,0))$. So Claim A will be proven if we can prove the following:

Claim B: For every $N > L > 0$, every walk in $\mathcal{S}_N^*((L,0))$ contains an unobstructed U-turn.

We now prove Claim B by induction on N.

Let $P(N)$ be the assertion that whenever L satisfies $N > L > 0$, every walk in $\mathcal{S}_N^*((L,0))$ contains an unobstructed U-turn. To start the induction, we note that $P(3)$ and $P(4)$ are clearly true. Let $N \ge 5$, and assume that

$P(n)$ is true for every $n < N$. Let ω be an arbitrary walk in $\mathcal{S}_N^*((L,0))$ for some L, with associated polygon $\mathcal{P}$. It is not hard to see that ω always contains a U-turn $\omega[i,j]$. (First observe that for every self-avoiding polygon, the number of convex sites exceeds the number of concave sites by exactly 4, because the sum of the signed inside angles must be exactly $+360°$. Therefore there must exist integers a and b with $0 < a < b < N$ such that $\omega(a)$ and $\omega(b)$ are both convex sites of $\mathcal{P}$ and such that if $b > a+1$ then the intervening sites $\omega(a+1), \ldots, \omega(b-1)$ are all straight sites of $\mathcal{P}$. Then $\omega[a-1,b+1]$ is a U-turn.) If it is unobstructed, then we are done, so assume that it has an obstruction $\omega(k)$. Then there exists an l satisfying $i+1 < l < j-1$ such that $||\omega(k)-\omega(l)||_1 = 1$. Suppose that $0 \le k < i$ (the same argument will work if $j < k \le N$). Let ζ denote the subwalk $\omega[k,l]$; since ζ has endpoints that are nearest neighbours, we can let $\mathcal{Q}$ denote its associated polygon. Observe that the bond $(\omega(k),\omega(l))$ lies inside $\mathcal{P}$, since we know that $\omega(i)$ and $\omega(j)$ are convex sites of $\mathcal{P}$. Therefore the inside of $\mathcal{Q}$ is a subset of the inside of $\mathcal{P}$, and hence all the sites of ω that are not part of ζ must lie outside of $\mathcal{Q}$. The inductive assumption tells us that ζ contains an unobstructed U-turn, and the observation of the preceding sentence guarantees that this must also be an unobstructed U-turn for ω. Therefore $P(N)$ is true.

We have now proven Claims B and A. To complete the proof of the theorem, consider the case of a general site x. Now the terms "inside" and "outside" are not meaningful, so we first need to find something to use in the place of U-turns. Let ω be an arbitrary walk in $\mathcal{S}(x)$, and let $N = |\omega|$. We say that the subwalk $\omega[i,j]$ $(0 \le i < j \le N)$ is a *C-turn* of ω if $j - i \ge 3$, $\omega[i+1,j-1]$ lies on a straight line that is perpendicular to the steps $\Delta\omega(i+1)$ and $\Delta\omega(j)$, and $\Delta\omega(i+1) = -\Delta\omega(j)$. (Observe that in the case $||x||_1 = 1$, every U-turn is a C-turn.) We define *obstruction* for C-turns exactly as we did for U-turns. The walk ω has no C-turns if and only if it has minimal length, i.e. $N = ||x||_1$, and it is easy to see that any minimal length walk can be transformed into any other by BFACF moves. So suppose $N > ||x||_1$; to prove the theorem, we need to show that it is always possible to reduce the length of ω using BFACF moves. Analogously to the case $||x||_1 = 1$, it suffices to prove that ω must have a C-turn with no obstructions.

Let $\omega[I,J]$ be a smallest C-turn of ω (i.e., satisfying $J - I \le j - i$ for every other C-turn $\omega[i,j]$). If $\omega[I,J]$ has no obstructions, then we are done, so assume $\omega[I,J]$ has one or more obstructions. It is not hard to see that one of these obstructions must be an endpoint of ω (because otherwise the obstructions would have to be part of a C-turn that is smaller than $\omega[I,J]$, which contradicts our choice of I and J). Without loss of generality, assume that $\omega(0)$ is an obstruction of $\omega[I,J]$ [the same argument will work for $\omega(N)$]. Let M be the (unique) integer such that $I + 1 <$ $M < J - 1$ and $\|\omega(0) - \omega(M)\|_1 = 1$. Let ζ denote the subwalk $\omega[0, M]$; since ζ has endpoints that are nearest neighbours, we can let Q denote its associated polygon. There are two cases that could occur: either the sites $\omega(M + 1), \ldots, \omega(N)$ all lie outside Q, or else they all lie inside Q (there are no other possibilities because the subwalks $\omega[0, M]$ and $\omega[M + 1, N]$ cannot intersect). We shall consider these two cases in turn.

Case I: The sites $\omega(M + 1), \ldots, \omega(N)$ all lie outside Q. By Claim B, we see that ζ has an unobstructed U-turn; since the sites of ω that are not part of ζ all lie outside Q, this must be an unobstructed C-turn of ω.

Case II: The sites $\omega(M + 1), \ldots, \omega(N)$ all lie inside Q. Let $u = \Delta\omega(N)$ be the direction of the last step of ω. Let

$$L = \min\{l > 0 : \omega(N) + lu \text{ is a site of } \omega\}$$

(note that $L < \infty$ because $\omega(N)$ lies inside Q). Let t be the integer such that $\omega(t) = \omega(N) + Lu$. Observe that the subwalk $\omega[t, N]$ is (the translation and rotation/reflection of) a walk in $\mathcal{S}_{N-t}^*((L, 0))$. Let $\mathcal{R}$ be the polygon consisting of the bonds of $\omega[t, N]$ and the straight line segment joining $\omega(N)$ to $\omega(t)$. Then the inside of $\mathcal{R}$ is a subset of the inside of Q. In particular, the sites $\omega(0), \ldots, \omega(t - 1)$ all lie outside $\mathcal{R}$. Now Claim B shows that the subwalk $\omega[t, N]$ has an unobstructed U-turn, and since the rest of ω lies outside $\mathcal{R}$, this must also be an unobstructed C-turn of the entire walk ω. This proves the theorem in Case II. $\square$

Recall that the state space of the DORS algorithm is the set of equivalence classes of N-step self-avoiding polygons in $\mathbf{Z}^2$ (Definition 3.2.2). Recall also that these polygons are "unrooted", as opposed to the set of polygons associated with $\mathcal{S}_{N-1}(e)$ (where $\|e\|_1 = 1$) which are "rooted" by the bond $(0, e)$ which can never be moved.

Theorem 9.7.3 *For every even N, the DORS algorithm is irreducible for unrooted N-step polygons in two dimensions. In fact, if Q_1 and Q_2 are N-step polygons in $\mathbf{Z}^2$, then there is a sequence of at most $2N - 2$ transformations that transforms Q_1 into Q_2.*

Proof. Let $c(\mathcal{P})$ denote the total number of convex and concave sites on the polygon $\mathcal{P}$. A *rectangle* is a polygon $\mathcal{P}$ that has $c(\mathcal{P}) = 4$. The theorem is an immediate consequence of the following two facts.

A. Any polygon $\mathcal{P}$ that is not a rectangle can be transformed into some other polygon Q having $c(Q) = c(\mathcal{P}) - 2$ using at most two transformations.

B. Any rectangle can be transformed into any other rectangle using one inversion and one reversing diagonal reflection.

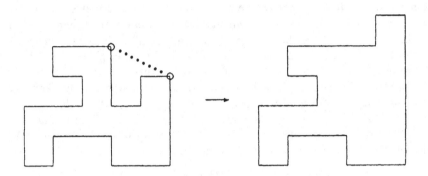

Figure 9.15: Proof of Theorem 9.7.3: a polygon with a supporting chord that is not parallel to a coordinate axis (left). Using the chord's endpoints as pivot sites for an inversion decreases the number of right angles by 2 (right).

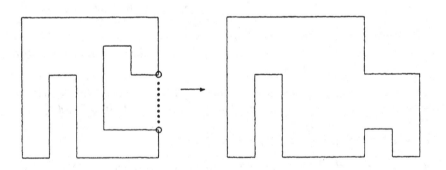

Figure 9.16: Proof of Theorem 9.7.3: a polygon whose two supporting chords are each parallel to a coordinate axis (left). Using one chord's endpoints as pivot sites for an inversion yields a polygon with a diagonal supporting chord (right).

We shall prove A first. A line segment is said to be a *supporting chord* of $\mathcal{P}$ if its endpoints are both on $\mathcal{P}$, its interior points are all on the outside of $\mathcal{P}$, and it is contained in the boundary of the convex hull of $\mathcal{P}$. (See Figure 9.15.) Observe that any polygon that is not a rectangle has a supporting chord.

Suppose that a polygon $\mathcal{P}$ has a supporting chord that is not parallel to either coordinate axis. It is not hard to see that performing an inversion on $\mathcal{P}$ with pivot sites chosen to be the endpoints of this supporting chord will yield a self-avoiding polygon $\mathcal{Q}$ with $c(\mathcal{Q}) = c(\mathcal{P}) - 2$ (see Figure 9.15). Next, suppose that $\mathcal{P}$ is not a rectangle but each of its supporting chords is parallel to a coordinate axis (see Figure 9.16). Performing an inversion on $\mathcal{P}$ with pivot sites chosen to be the endpoints of some supporting chord will yield a self-avoiding polygon $\mathcal{P}'$ with $c(\mathcal{P}') = c(\mathcal{P})$, and it is not hard to see that this $\mathcal{P}'$ will have a supporting chord that is not parallel to either coordinate axis. Thus we have proven A.

Finally we turn to fact B, whose proof is illustrated in Figure 9.17. Let

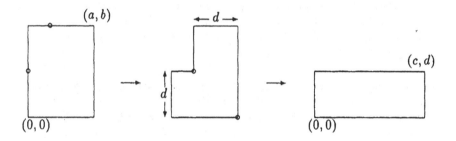

Figure 9.17: The DORS algorithm transforming one rectangle into another, using an inversion followed by a reversing diagonal reflection. The pivot sites are denoted by circles.

$\mathcal{R}_1$ and $\mathcal{R}_2$ be two N-step rectangles. Assume that the corners of $\mathcal{R}_1$ are at $(0,0)$, $(a,0)$, (a,b), and $(0,b)$, while the corners of $\mathcal{R}_2$ are at $(0,0)$, $(c,0)$, (c,d), and $(0,d)$, where a, b, c, and d are all positive and $a + b = c + d$. Without loss of generality, we can assume that $c > a \geq d$. Performing an inversion on $\mathcal{R}_1$ with pivot sites $(0,d)$ and $(a - d, b)$ gives a polygon $\mathcal{U}$, which in turn can be transformed into $\mathcal{R}_2$ by performing a reversing diagonal reflection with pivot sites $(a, 0)$ and $(a - d, d)$. $\square$

Theorem 9.7.4 *The exponential autocorrelation time τ_{exp} for the BFACF algorithm is infinite (for every x and z).*

Proof. Fix x and z. Let ϕ and ψ be two points (walks) in the state space, and let $T[\phi, \psi]$ be the smallest value of n such that $P^n(\phi, \psi) > 0$; i.e. $T[\phi, \psi]$ is the smallest time in which it is possible to get from ϕ to ψ. Let I_ϕ and I_ψ be the indicator functions of the singletons $\{\phi\}$ and $\{\psi\}$. Consider any $k < T[\phi, \psi]$. On the one hand, using the inner product defined in (9.2.14),

$$(I_\phi, (P^k - \Pi)I_\psi) = P^k(\phi, \psi)\pi(\phi) - \pi(\phi)\pi(\psi) = -\pi(\phi)\pi(\psi).$$

On the other hand, using (9.2.23) and (9.2.24), we have

$$|(I_\phi, (P^k - \Pi)I_\psi)| \le \|I_\phi\|_2 \|P^k - \Pi\| \, \|I_\psi\|_2 = (\pi(\phi)\pi(\psi))^{1/2} \exp[-k/\tau_{exp}].$$

Combining the above two observations, rearranging and taking $k = T[\phi, \psi] - 1$ gives

$$\tau_{exp} \ge \frac{2(T[\phi, \psi] - 1)}{-\log(\pi(\phi)\pi(\psi))}. \tag{9.7.34}$$

This inequality says that if there are two states that are far apart, but not too unlikely, then τ_{exp} must be large.

To apply (9.7.34), consider $N \gg \|x\|_1$. Let ϕ be a shortest walk from 0 to x, and let $\psi = \psi^{[N]}$ be a walk of length N which does not intersect ϕ and whose shape is approximately square; this means that the area of the smallest surface whose boundary is the union of ϕ and $\psi^{[N]}$ is approximately $N^2/16$. Since the BFACF algorithm only modifies a walk by adding and removing bonds around a single lattice square, this surface area cannot change by more than 1 in a single iteration. Therefore $T[\phi, \psi^{[N]}] \simeq N^2$. Also, $\pi(\phi) = \|x\|_1 z^{\|x\|_1}/\Xi(z, x)$ and $\pi(\psi^{[N]}) = N z^N/\Xi(z, x)$, so the right side of (9.7.34) behaves like a constant times N as $N \to \infty$. Therefore, since N can be arbitrarily large, we must have $\tau_{exp} = +\infty$. □

9.8   Notes

Section 9.1. One of the best general overviews of Monte Carlo methods is Hammersley and Handscomb (1964), whose age has done remarkably little to diminish its appeal. Bratley, Fox and Schrage (1987) is a more recent general reference to various theoretical and practical issues in simulation and Monte Carlo. Binder and Heermann (1988) is a useful step-by-step guide to the practical aspects of Monte Carlo experiments in statistical mechanics. Kremer and Binder (1988) is a detailed survey of Monte Carlo methods for polymers in general. Sokal (1991) is a review on the problem of critical slowing-down.

General references for the theory and applications of Markov chains include Feller (1968), Karlin and Taylor (1975), and Nummelin (1984).

These authors and others usually say that a Markov chain is *ergodic* if it is irreducible, positive recurrent, and aperiodic. Most chains arising in Monte Carlo are positive recurrent and aperiodic, and for these chains questions of ergodicity are equivalent to questions of irreducibility. Although the Monte Carlo literature tends to use the term "ergodicity" when discussing irreducibility, we prefer the term "irreducibility" in this book to emphasize the specific nature of these problems.

The classic reference for hash tables is Knuth (1973), which is still highly recommended. Hashing is also treated in most computer science books on data structures. Early uses of hash tables for the self-avoiding walk problem are Gans (1965) and Jurs and Reissner (1971); a description is also given in Madras and Sokal (1988).

Section 9.2. Two general references on statistics are Silvey (1970) and Cox and Hinkley (1974). References on time series analysis include Priestley (1981) and Brockwell and Davis (1987). Bratley, Fox, and Schrage (1987) discuss time series analysis and other statistical issues in the specific context of simulation. Geyer (1992) and Gelman and Rubin (1992) present two contrasting views on problems of statistical inference for Markov chain simulations.

Proposition 9.2.2, Corollary 9.2.3, and Proposition 9.2.4 are from Appendix A of Caracciolo, Pelissetto, and Sokal (1990). The proof of Equation (9.2.21) in Section 9.7.1 is from Sokal and Thomas (1989), who actually prove a stronger theorem. The proof of Equation (9.2.27) is from Sokal (1989). The exposition of Section 9.2.3 is largely based upon the preceding three papers.

Section 9.3. Strides and biased sampling are reviewed in Hammersley and Handscomb (1964). Kremer and Binder (1988) includes a more recent review of biased sampling, with many references. The dimerization method is due to Suzuki (1968) and Alexandrowicz (1969). The derivation of (9.3.3) is from Madras and Sokal (1988).

Section 9.4. Many local algorithms have appeared in the literature; see Madras and Sokal (1987) and Kremer and Binder (1988) for some references. The failure of irreducibility for local algorithms was noticed early, by Heilmann (1968) (who observed that knots could cause problems) and by Verdier (1969) (who noted the existence of three-dimensional frozen configurations analogous to Figure 9.2). Theorems 9.4.1 and 9.4.2, as well as Proposition 9.4.3, are due to Madras and Sokal (1987); as explained there, the methods also allow one to prove Theorem 9.4.1 in $d = 3$. The proof of (9.4.1) under the stated assumption is due to Caracciolo *et al.* (1990).

Wall and Mandel (1975) commented that the probability of frozen configurations for the slithering snake algorithm did not tend to 0, but expected it to be negligibly small for practical purposes. The rigorous proof of the former assertion [Equation (9.4.2)] is due to Madras (1988).

Reiter (1990) proved irreducibility for a fixed-length algorithm in the spirit of the slithering snake: in this algorithm, a single bond in the walk can be replaced by a 3-bond U while simultaneously removing two bonds from the ends (and of course the reverse of this move can also be done).

The pivot algorithm has been independently rediscovered by many different authors since Lal (1969): Curro (1974), Olaj and Pelinka (1976), and MacDonald *et al.* (1985). Continuum analogues have been used by Stellman and Gans (1972) and Freire and Horta (1976). Except where cited otherwise, the results and discussion of Section 9.4.3 are from Madras and Sokal (1988).

Section 9.5. The rigorous proof of (9.5.3) appeared in Caracciolo *et al.* (1990).

Section 9.6. The BFACF algorithm is due to Berg and Foerster (1981), Aragão de Carvalho, Caracciolo and Fröhlich (1983), and Aragão de Carvalho and Caracciolo (1983); some ambiguities in these papers about the details of the algorithm were clarified in Caracciolo *et al.* (1990), whose presentation we follow here. The irreducibility of the BFACF algorithm in two dimensions (Theorem 9.7.2) is due to Madras (1986, unpublished). The bound (9.6.11) is due to Caracciolo *et al.* (1990).

Janse van Rensburg, Whittington, and Madras (1990) described a nonlocal fixed-length algorithm for polygons on the face-centred cubic lattice, and proved that it is irreducible.

Chapter 10

Related topics

10.1 Weak self-avoidance and the Edwards model

The weakly self-avoiding walk, known also as the self-repellent walk and as the Domb-Joyce model [Domb and Joyce (1972)], is a measure on ordinary random walks in which self-intersections are discouraged but not forbidden. The measure associates to an n-step simple random walk ω the weight

$$Q_n^\lambda(\omega) = \frac{1}{Z_n(\lambda)} \prod_{0 \leq s < t \leq n} [1 - \lambda v_{st}(\omega)], \tag{10.1.1}$$

where $0 < \lambda \leq 1$, $Z_n(\lambda)$ is a normalization constant, the product is over pairs of integers s and t, and $v_{st}(\omega)$ is 1 if $\omega(s) = \omega(t)$ and otherwise is 0. Taking $\lambda = 1$ gives the uniform measure on n-step self-avoiding walks, while $0 < \lambda < 1$ gives a measure in which self-intersections diminish the probability of a walk. Setting $\lambda = 0$ just gives simple random walk. An alternate parametrization of the interaction which appears frequently is to take

$$\lambda = 1 - e^{-\beta}. \tag{10.1.2}$$

Then in terms of β,

$$Q_n^\lambda(\omega) = \frac{1}{\tilde{Z}_n(\beta)} \prod_{0 \leq s < t \leq n} e^{-\beta v_{st}(\omega)}, \tag{10.1.3}$$

where $\tilde{Z}_n(\beta)$ is a normalization constant. Here it is $\beta = \infty$ which corresponds to the strictly self-avoiding walk.

365

It is a (nonrigorous) prediction of the renormalization group method that the weakly self-avoiding walk, for any $\lambda > 0$, is in the same universality class as the strictly self-avoiding walk. This is borne out in the existing rigorous results. For $d = 1$ it was shown in Bolthausen (1990), using large deviation techniques, that there is a $\lambda_0 > 0$ such that for $\lambda \in (0, \lambda_0]$ there is a $c > 0$ such that

$$\lim_{n \to \infty} Q_n^\lambda \{ \omega : c \le n^{-1} |\omega(n)| \le c^{-1} \} = 1. \tag{10.1.4}$$

In fact the same conclusion was obtained in a more general setting than just the nearest-neighbour walk. This shows that if λ is sufficiently small then the one-dimensional weakly self-avoiding walk behaves like the strictly self-avoiding walk, in the sense that $\nu = 1$. For $d > 4$, Brydges and Spencer (1985) used the lace expansion to show that for λ sufficiently close to zero the mean-square displacement of the model defined by (10.1.1) is linear in the number of steps, and the scaling limit of the endpoint of the walk is Gaussian. This could be extended to cover all $\lambda < 1$ for $d \ge 5$, using the methods that handled the $\lambda = 1$ case; see Chapter 6. So also above four dimensions the weakly self-avoiding walk has the same scaling behaviour as simple random walk. There are no rigorous results in two and three dimensions; four dimensions will be discussed below.

The Edwards model is a continuous space and time analogue of the weakly self-avoiding walk, introduced in Edwards (1965). Its relation to the weakly self-avoiding walk is similar to that of Brownian motion to simple random walk. The strength of the self-avoidance interaction for the Edwards model is analogous to the parameter β of (10.1.2). It could be hoped that as this interaction strength goes to infinity the Edwards model would approach a limit corresponding to a continuum limit of the self-avoiding walk; however methods allowing for such a limit to be carried out rigorously remain to be found.

The Edwards model is defined formally as a measure on d-dimensional continuous paths on an interval $[0, T]$, by multiplying the Wiener measure on such paths by a factor suppressing self-intersections. Specifically, if we denote the Wiener measure by dW^T and a typical path by $r(t)$, then the Edwards model is defined by the measure

$$d\mu^T = \frac{1}{Z_T} e^{-gJ} dW^T \tag{10.1.5}$$

where Z_T is a normalization factor, g is a positive parameter measuring the strength of the interaction, and J is a functional on paths defined by

$$J = J(r) = \int_0^T \int_0^T \delta(r(s) - r(t)) ds \, dt. \tag{10.1.6}$$

The quantity $J(r)$ can be interpreted as the amount of time spent by the path r at its double points, and serves in the measure μ^T to suppress self-intersections.

Rigorous sense can be made of the measure μ^T by first replacing the delta function in (10.1.6) by a regularized delta function δ_ϵ, yielding a well-defined interaction J_ϵ and corresponding measure μ_ϵ^T, and then taking the limit removing the regularization ϵ. This procedure can be carried out literally when $d = 1$, but for higher dimensions it is not so simple and a renormalization is required. In fact the situation is quite similar to the construction of the φ^4 quantum field theory [see Glimm and Jaffe (1987)], and methods used to construct these theories in two and three dimensions provide a basis from which to approach the Edwards model. The required renormalization is simplest in two dimensions, and the Edwards model in $d = 2$ was first constructed in Varadhan (1969). In two dimensions the Edwards measure is absolutely continuous with respect to the Wiener measure. For $d = 3$ the construction was carried out in Westwater (1980,1982) for all T and g; here the measure is singular with respect to the Wiener measure, in a dramatic departure from the formal expression (10.1.5). For small g an alternate construction of the three-dimensional measure was given in Bolthausen (1991), which made use of some simplifications in the constructive field theory technology [Brydges, Fröhlich and Spencer (1983), Bovier, Felder and Fröhlich (1984)].

Although for both two and three dimensions the Edwards model has been constructed for all times T and all $g \geq 0$, there is insufficient control to compute the limiting behaviour of the expected value of $r(T)^2$ as $T \to \infty$, and critical exponents such as ν are not currently accessible. The construction of the Edwards model for any finite T can be considered a construction of a subcritical model, and to obtain control of critical exponents a control of the critical $T = \infty$ model is required. However in one dimension a proof has been given that $|r(T)|$ behaves like a multiple of T as $T \to \infty$; see Westwater (1985). It is believed that the Edwards model is in the same universality class as the self-avoiding walk, i.e. that the critical exponents will be the same.

An alternate regularization of the Edwards measure (10.1.5) is to consider a version of the model in discrete time and space. In Bovier, Felder and Fröhlich (1984), such a regularization was given, and the necessary renormalization was performed to construct the continuum limit of Green functions such as the two-point function in two and three dimensions; the continuum measure itself was however not constructed. A natural discretization of the Edwards model is to replace the delta function in the interaction by a discrete version. Specifically, we discretize $r(t)$ to $n^{-1/2}\omega(\lfloor nt \rfloor)$, with ω

a simple random walk, and define

$$\delta_n(x) = \begin{cases} n^{d/2} & \text{if } ||x||_\infty \le \frac{1}{2}n^{-1/2} \\ 0 & \text{otherwise.} \end{cases} \tag{10.1.7}$$

Then we replace J of (10.1.6) by

$$\begin{aligned} J_n &= \frac{1}{n^2}\sum_{s=0}^{n}\sum_{t=0}^{n}\delta_n(n^{-1/2}[\omega(s)-\omega(t)]) \\ &= n^{(d-4)/2}\sum_{s,t=0}^{n} v_{st}(\omega). \end{aligned} \tag{10.1.8}$$

This gives the measure (10.1.3) on simple random walks, with interaction strength $\beta = 2gn^{(d-4)/2}$. From this relation it is clear that in dimensions two and three the discrete Edwards model interaction is weaker than that of the weakly self-avoiding walk. In Stoll (1989) the two dimensional Edwards measure was constructed by taking the continuum limit of this discrete model; this has not yet been carried out in three dimensions.

In four dimensions the discrete Edwards model and the weakly self-avoiding walk are identical, apart from a factor of two in the coupling constants. As this book is being written, rigorous results in four dimensions are beginning to appear. Brydges, Evans and Imbrie (1992) have considered a model of weakly self-avoiding walk on a four dimensional hierarchical lattice, and have proved that a quantity closely related to the critical two-point function decays asymptotically as a multiple of $|x|^{-2}$ if the interaction is sufficiently weak. This work uses an identity to write the two-point function of the model as the two-point function of a quantum field theory, and then performs a renormalization group analysis of the quantum field theory. Arnaudon, Iagolnitzer and Magnen (1991) have announced a proof that the critical two-point function of a continuum four-dimensional Edwards model with fixed ultraviolet cutoff (a regularization analogous to discretization) and sufficiently weak interaction behaves asymptotically like a multiple of $|x|^{-2}$, with $\log|x|$ and $\log\log|x|$ corrections, using constructive field theory methods.

10.2 Loop-erased random walk

During the 1980s considerable progress was made in the study of the loop-erased self-avoiding random walk, which is a model of self-avoiding walk different from the one studied in this book. In this section we give a brief definition of the loop-erased random walk, and state the principal rigorous results which have been obtained for it. Most of the rigorous work is due to Lawler, and is described in his book [Lawler (1991)].

There are two equivalent formulations of the model. The first, from which the name is derived, can be described as follows. Consider the path of an infinite ordinary simple random walk, for the moment in at least three dimensions. We associate to this walk an infinite self-avoiding walk by erasing loops from the path chronologically. In more detail, we begin by looking for the first time that the walk intersects itself and then erase the portion of the walk (the loop) between the (first) two visits to the site where the intersection occurs. Then we erase the first loop from the resulting path, and continue inductively. This leads to an infinite self-avoiding walk. To define a measure on the set of all n-step self-avoiding walks, we assign to each n-step self-avoiding walk ω a weight equal to the probability that the first n steps of the loop-erased walk agree with ω. This family of measures is *consistent*, in the sense of (6.7.2). In particular, a walk which cannot be extended by a single step and remain self-avoiding is assigned weight zero in this measure, and hence the loop erased walk does not define the uniform measure on the set of n-step self-avoiding walks.

The above procedure works in dimensions $d \geq 3$, where simple random walk is transient, but for the recurrent case $d = 2$ more care is needed. In two dimensions it is necessary to use a limiting process to define the model. Roughly speaking a measure is defined on n-step self-avoiding walks by first performing loop erasure as above on simple random walk paths which lie in a finite box of side length $N \geq n$, thereby obtaining an N-dependent measure, and then the limit is taken of this measure as N goes to infinity.

A second (equivalent) formulation of the model, which goes by the name Laplacian self-avoiding walk, provides a description as a "kinetically-growing" walk, i.e. as a stochastic process defined by transition probabilities. To avoid the special difficulties associated with two dimensions, we consider here only $d \geq 3$. The Laplacian walk is defined to be the process whose transition probabilities are as follows. Given an n-step self-avoiding walk ω, the probability that the next step is to a neighbour x of $\omega(n)$ is proportional to the probability that simple random walk starting from x will never intersect ω. To state this more precisely we introduce the following definition. Given a site $x \in \mathbf{Z}^d$ and a set $A \subset \mathbf{Z}^d$, we define $Q_A(x)$ to be the probability that an infinite simple random walk beginning at x never enters the set A. The transition probabilities of the Laplacian walk are then given, for $\omega = \{\omega(0), \ldots \omega(n)\}$ and x a neighbour of $\omega(n)$, by

$$P(\omega(n+1) = x | \omega) = \frac{Q_\omega(x)}{\sum_{y:|y-\omega(n)|=1} Q_\omega(y)}. \qquad (10.2.1)$$

A proof that this is equivalent to the loop-erased walk is given in Lawler

(1991). The name Laplacian self-avoiding walk derives from the fact that $Q_A(x)$ is a harmonic function on the complement of A, with boundary conditions zero on A and one at infinity.

It is now known that the loop-erased self-avoiding walk has upper critical dimension equal to four, and that if (1.1.12) and (1.1.14) accurately represent the behaviour of the mean-square displacement of the self-avoiding walk for dimensions three and four, then the loop-erased self-avoiding walk is in a different universality class (i.e. has different critical exponents) than the self-avoiding walk defined using the uniform measure. We end this section with a statement of the rigorous results, beginning with high dimensions, where the results are strongest.

Theorem 10.2.1 *(a) [Lawler (1980)] Let $d \geq 5$ and let $\hat{S}(n)$ denote the loop-erased walk after n steps. There is a constant b, depending only on the dimension, such that the process $\hat{X}_n(t) = (bn)^{-1/2}\hat{S}(\lfloor nt \rfloor)$ converges in distribution to the Wiener process [normalized as in (6.6.3)]. Moreover, the mean-square displacement of the loop-erased walk is asymptotic to b times the number of steps.*

(b) [Lawler (1986)] Let $d = 4$. There is a sequence b_n such that the process $\hat{X}_n(t) = (b_n n)^{-1/2}\hat{S}(\lfloor nt \rfloor)$ converges in distribution to the Wiener process. The sequence b_n satisfies

$$\frac{1}{3} \leq \liminf_{n \to \infty} \frac{\log b_n}{\log \log n} \leq \limsup_{n \to \infty} \frac{\log b_n}{\log \log n} \leq \frac{1}{2}$$

and the mean-square displacement satisfies

$$\frac{1}{3} \leq \liminf_{n \to \infty} \frac{\log[n^{-1}E(|\hat{S}(n)|^2)]}{\log \log n} \leq \limsup_{n \to \infty} \frac{\log[n^{-1}E(|\hat{S}(n)|^2)]}{\log \log n} \leq \frac{1}{2}.$$

It is conjectured in Lawler (1986) that in (*b*) of the above theorem $\lim_{n \to \infty} \log b_n / \log \log n = 1/3$. This is different behaviour than the correction $(\log n)^{1/4}$ to the mean-square displacement that is predicted by the renormalization group for the self-avoiding walk. In two and three dimensions Monte-Carlo results suggest a more dramatic discrepancy between the loop-erased walk and the self-avoiding walk, namely $\nu = 4/5$ in two dimensions and $\nu \approx 0.616$ in three dimensions [Guttman and Bursill (1990)]. The following theorem proves that in three dimensions the mean-square displacement of the loop-erased walk behaves differently than the $n^{1.18}$ behaviour expected for the self-avoiding walk.

Theorem 10.2.2 *[Lawler (1988)] For every $\epsilon > 0$ there is a positive constant K such that*

$$E(|\hat{S}(n)|^2) \geq Kn^{3/2-\epsilon} \quad for \ d = 2,$$

and

$$E(|\hat{S}(n)|^2) \geq Kn^{6/5-\epsilon} \quad for\ d = 3.$$

These results for the loop-erased self-avoiding random walk are proved using probabilistic methods quite unlike the methods used in this book.

10.3 Intersections of random walks

The critical exponents γ and Δ_4 for the self-avoiding walk are closely related to intersection probabilities for self-avoiding walks. To be specific, assuming that c_n has the asymptotic behaviour specified in (1.1.11) and (1.1.13) [in fact we know (1.1.11) does hold for $d \geq 5$], then the probability that two n-step self-avoiding walks beginning at the origin do not intersect is given by

$$\frac{c_{2n}}{c_n^2} \sim \begin{cases} A^{-1}2^{\gamma-1}n^{1-\gamma} & d \neq 4 \\ A^{-1}(\log n)^{-1/4} & d = 4. \end{cases} \tag{10.3.1}$$

The critical exponent Δ_4 is relevant for intersection probabilities of self-avoiding walks beginning at different sites. A measure of this is the renormalized coupling constant, defined in (1.4.22), which is believed to satisfy

$$g(z) \sim const.(z_c - z)^{d\nu - 2\Delta_4 + \gamma} \quad as\ z \nearrow z_c, \tag{10.3.2}$$

with Δ_4 obeying the hyperscaling relation $d\nu - 2\Delta_4 + \gamma = 0$ in dimensions $2, 3, 4$ (with a logarithmic correction in four dimensions) and $\Delta_4 = 3/2$ for $d \geq 5$ (this is proved for $d \geq 6$; see Theorem 1.5.5 and the Remark below its statement).

While the above conjectures remain unproven for the self-avoiding walk in low dimensions, it is natural to ask if corresponding statements for simple random walk can be proven. In the remainder of this section we give a brief summary of some of the results which have been obtained in this direction.

To discuss the analogue of (10.3.1) for simple random walk, we denote by $f(n)$ the probability that the paths of two n-step simple random walks beginning at the origin do not intersect (apart from the fact that they have a common initial point). For the statement of the next theorem we introduce the notation $f(n) \cong g(n)$ to mean that $\log f(n) \sim \log g(n)$.

Theorem 10.3.1 *For $d > 4$, $f(n) \sim const.$ as $n \to \infty$, for some constant strictly between 0 and 1 which depends only on the dimension. For $d = 4$, $f(n) \cong (\log n)^{-1/2}$. For $d = 2$ or 3 there is an exponent ζ such that $f(n) \cong n^{-\zeta}$, with*

$$\tfrac{1}{2} + \tfrac{1}{8\pi} \leq \zeta < \tfrac{3}{4} \quad d = 2$$

$$\tfrac{1}{4} \leq \zeta < \tfrac{1}{2} \quad d = 3.$$

For $d > 4$ this was proved in Lawler (1980). For $d = 4$ the proof is given in Lawler (1982,1985a,1991), and for $d = 2, 3$ the proof is given in Burdzy and Lawler (1990a,1990b). Nonrigorous conformal field theory arguments predict that in two dimensions $\zeta = 5/8$; see Duplantier and Kwon (1988). Monte-Carlo computations are consistent with this prediction, and also give a value near 0.29 for ζ in three dimensions [Burdzy, Lawler and Polaski (1989), Duplantier and Kwon (1988), Li and Sokal (1990)]. Results for generating functions related to the above theorem are given in Park (1989).

The logarithmic behaviour in four dimensions is the hallmark of the critical nature of four dimensions for random walk intersections. Heuristically this can be seen from the fact that Brownian motion paths have Hausdorff dimension two, and hence four dimensions is marginal for the intersection of two Brownian paths. By the same argument three dimensions is critical for triple points of three paths (two two-dimensional paths in three dimensions will typically intersect in a one dimensional set, and the intersection of this set with a third two-dimensional path will be marginal in three dimensions). Bounds on intersection probabilities of three random walks in three dimensions are obtained in Lawler (1985b,1991), and using rigorous renormalization group methods in Felder and Fröhlich (1985). On a nonrigorous level, results of this type have been considerably generalized using renormalization methods; see Duplantier (1988).

We denote the analogue for simple random walk of the renormalized coupling constant $g(z)$ by $g_0(z)$. For $g_0(z)$ the following theorem gives $\Delta_4 = 3/2$ for $d > 4$, and hyperscaling for $d \leq 4$ (with a logarithmic correction in four dimensions).

Theorem 10.3.2 *Let $t = (2d)^{-1} - z$. Then there are positive constants c_1, c_2, c_3, c_1', c_2', c_3' such that for $z < (2d)^{-1}$*

$$\left.\begin{array}{c} c_1 \\ c_2 |\log t|^{-1} \\ c_3 t^{(4-d)/2} \end{array}\right\} \leq g_0(z) \leq \left\{\begin{array}{ll} c_1' & d < 4 \\ c_2' |\log t|^{-1} & d = 4 \\ c_3' t^{(4-d)/2} & d > 4. \end{array}\right.$$

Results for the probability of intersection of walks of fixed length n, one beginning at the origin and the other at $x \simeq \sqrt{n}$, are given in Lawler (1982,1991). These results effectively yield more detailed information than Theorem 10.3.2. In (and near) four dimensions the above result was proved in Felder and Fröhlich (1985) using a rigorous renormalization group argument; see also Aizenman (1985) for related work on the intersection of Brownian paths. A proof of Theorem 10.3.2 using inclusion-exclusion methods is given in Park (1989).

10.4   The "myopic" or "true" self-avoiding walk

The model of self-avoiding walk discussed in this book is not a random walk in the usual sense, being defined via a measure on paths rather than via transition probabilities as a stochastic process. One model of self-avoiding walk which *is* defined by transition probabilities is the so-called "true" self-avoiding walk; this model is essentially described by the MSAW algorithm[1] of Section 9.1. The epithet "true" is a misnomer, as the paths of this model need not in general be self-avoiding, nor is it the model of self-avoiding walk which is most commonly studied. In Lawler (1991) this model is referred to as the "myopic" self-avoiding walk; this name emphasizes the short-sightedness of the walk in its effort to be self-avoiding. Although the myopic self-avoiding walk has played a relatively minor role in applications [see however Family and Daoud (1984) for an application to polymers under certain conditions], it is interesting to see how it compares to the usual self-avoiding walk.

The transition probabilities for the myopic self-avoiding walk are defined as follows. Consider a walker on the hypercubic lattice $\mathbf{Z}^d$, beginning at the origin and taking nearest-neighbour steps. The first step is to a nearest neighbour of the origin, each neighbour being chosen with equal probability $(2d)^{-1}$. In subsequent steps, if there are neighbours of the current position which have not yet been visited, the next site is chosen uniformly from the neighbours not yet visited. If all neighbours have already been visited (i.e. if the walk is trapped) then the next site is chosen uniformly from among those neighbours which have been visited least often in the past. This leads to paths with self-intersections — looking just one step ahead cannot prevent the walk from becoming trapped, and a step must always be taken to some neighbour. A simple example demonstrates the computation of weights assigned to paths by the myopic self-avoiding walk: the myopic self-avoiding walk assigns to the walk ENWN in two dimensions the weight $\frac{1}{4}\frac{1}{3}\frac{1}{3}\frac{1}{2} = \frac{1}{72}$; for comparison the self-avoiding walk assigns to the same path $c_4^{-1} = \frac{1}{100}$. It is worth noting that the weights associated to the myopic self-avoiding walk are not symmetric with respect to time-reversal: in two dimensions the walk ENWN has weight $\frac{1}{72}$ whereas the time-reversed walk SESW has weight $\frac{1}{108}$.

The above description defines a walk which is prohibited from stepping to neighbours which were visited most often in the past. A less restrictive

---

[1] There is a slight difference between the MSAW algorithm and the model treated in this section, as the former assigns nonzero weight only to self-avoiding walks, unlike the latter.

self-avoidance constraint would merely discourage such steps. This leads us to consider a nearest-neighbour walk, starting at the origin, with transition probabilities

$$P(\omega(n+1) = x + a|\omega(n) = x) = \frac{e^{-\lambda N_{x+a}}}{\sum_{b:|b|=1} e^{-\lambda N_{x+b}}} \qquad (10.4.1)$$

where $|a| = 1$, $\lambda \geq 0$ represents the strength of the repulsion, and N_u denotes the number of visits to the site u up to time n. The case $\lambda = \infty$ then corresponds to the prohibitive model introduced in the previous paragraph.

There are as yet no rigorous results concerning the critical behaviour of the myopic self-avoiding walk. However the nonrigorous results indicate that the myopic self-avoiding walk behaves quite differently from the self-avoiding walk. Both field theoretic methods [Amit, Parisi and Peliti (1983)] and calculations related to those leading to the Flory exponents for the self-avoiding walk [Pietronero (1983)] point to an upper critical dimension of two. The diffusive behaviour $\nu = 1/2$ is expected above two dimensions, logarithmic corrections to diffusive behaviour are obtained in two dimensions, and the exponent $\nu = 2/3$ is found in one dimension. The claim that $\nu = 2/3$ in one dimension clearly does not apply when $\lambda = \infty$, for which the myopic walk behaves ballistically and $\nu = 1$. This indicates that the $\lambda = \infty$ walk belongs to a different universality class than the finite λ version; however the nonrigorous results appear to claim that the upper critical dimension is two for all $\lambda \leq \infty$. A survey, with references to numerical calculations, is given in Peliti and Pietronero (1987).

Appendix A

Random walk

This appendix contains a number of elementary facts about the usual random walk with no self-avoidance constraint, which are used in the book. Further details on random walk can be found in Lawler (1991), Spitzer (1976), and Montroll and West (1979).

We consider a random walk taking steps in a set Ω which does not contain the origin and which is invariant under the symmetries of the lattice $\mathbf{Z}^d$. The two basic examples we have in mind here are Ω equal to the set of nearest neighbours of the origin, and $\Omega = \{x \in \mathbf{Z}^d : 0 < \|x\|_\infty \leq L\}$. As usual we denote the cardinality of Ω also by Ω. The two-point function is defined by

$$C_z(x, y) = \sum_{\omega : x \to y} z^{|\omega|}, \qquad (A.1)$$

where z is a complex parameter and the sum is over random walks of arbitrary length, taking steps in Ω, which begin at x and end at y. Clearly $C_z(x, y) = C_z(0, y - x)$. If $x = y$ then the two-point function includes a unit contribution from the zero-step walk. Since the total number of n-step random walks beginning at x is equal to Ω^n, $\sum_x |C_z(0, x)| \leq \sum_{n=0}^{\infty} (\Omega|z|)^n$ and hence the two-point function and its Fourier transform $\hat{C}_z(k)$ [defined in (1.4.10)] are both finite for $|z| < \Omega^{-1}$. For $z = \Omega^{-1}$, (A.1) is the Green function

$$C_{1/\Omega}(x, y) = \sum_{n=0}^{\infty} p_n(x, y), \qquad (A.2)$$

where $p_n(x, y)$ is the probability that the walk starting at x is at y after n steps. We will show momentarily that (A.2) is finite for $z = \Omega^{-1}$ if $d > 2$.

Fix z with $|z| < \Omega^{-1}$. Extracting from the two-point function the contribution due to the zero-step walk, and taking into account all possible

375

locations of the walk after the first step, we obtain

$$C_z(0, x) = \delta_{0,x} + \sum_{y \in \Omega} z \sum_{\omega: y \to x} z^{|\omega|}. \tag{A.3}$$

This can be rewritten as the convolution equation

$$C_z(0, x) = \delta_{0,x} + z\Omega D * C_z(x), \tag{A.4}$$

where

$$D(x) = \begin{cases} \Omega^{-1}, & x \in \Omega \\ 0, & x \notin \Omega. \end{cases} \tag{A.5}$$

Taking the Fourier transform of (A.4) gives

$$\hat{C}_z(k) = \frac{1}{1 - z\Omega\hat{D}(k)}, \quad k \in [-\pi, \pi]^d \tag{A.6}$$

with

$$\hat{D}(k) = \frac{1}{\Omega} \sum_{y \in \Omega} e^{ik \cdot y} = \frac{1}{\Omega} \sum_{y \in \Omega} \cos k \cdot y. \tag{A.7}$$

Taking the inverse transform then gives

$$C_z(0, x) = \int_{[-\pi, \pi]^d} \frac{e^{-ik \cdot x}}{1 - z\Omega\hat{D}(k)} \frac{d^d k}{(2\pi)^d}. \tag{A.8}$$

Since $1 - \hat{D}(k)$ behaves like k^2 near $k = 0$, it follows from (A.8) and a limiting argument that $C_{1/\Omega}(x, y)$ is finite if $d > 2$.

The critical exponents for random walk are much easier to determine than those for the self-avoiding walk. It follows immediately from the facts that there are Ω^n n-step random walks starting from the origin and that the mean-square displacement after n steps is proportional to n that $\gamma = 1$ and $\nu = 1/2$. Also, $\alpha_{sing} - 2 = -d/2$ follows from the local central limit theorem, which states that the probability of ending at x after n steps (with n and $\|x\|_1$ of the same parity, x fixed, and $n \to \infty$) is asymptotically of order $n^{-d/2}$. The value $\eta = 0$ can be seen in k-space from the fact that $\hat{C}_{1/\Omega}(k)^{-1} = \text{const}.k^2 + O(k^4)$. In x-space the analogous statement follows from the well-known $|x|^{2-d}$ behaviour of the Green function $C_{1/\Omega}(0, x)$. Unfortunately none of these provides a useful means of understanding the corresponding critical exponents for the self-avoiding walk.

However, for the correlation length the proof of mean-field behaviour for the self-avoiding walk in high dimensions given in Section 6.5.1 is modelled on a corresponding argument for random walk (with considerable additional input). Because this argument is not a standard one, we give it below to prove that the nearest-neighbour (or *simple*) random walk correlation length has critical exponent 1/2. This may serve to motivate the approach used in Section 6.5.1. One ingredient is the next lemma, which is of independent interest. The lemma states quite generally that if a two-point function satisfies a certain inequality, known as the *Lieb-Simon inequality* [Simon (1980), Lieb (1980)], then it decays exponentially if the susceptibility is finite.

In the statement of the lemma, B_R is the set of all $y \in \mathbf{Z}^d$ such that $||y||_\infty \leq R$, and ∂B_R is defined to be the set of sites y with $||y||_\infty = R$.

Lemma A.1 *Suppose $K(x,y)$ is nonnegative and translation invariant, and satisfies*

$$K(0,x) \leq \sum_{y \in \partial B_R} K(0,y)K(y,x), \tag{A.9}$$

for all $R \geq 1$ and $x \notin B_R$. If $\sum_x K(0,x) < \infty$, then there are constants $C, m > 0$ such that $K(0,x) \leq Ce^{-m|x|}$ for all sites x.

Proof. Fix any θ with $0 < \theta < 1$. Since $\sum_x K(0,x) < \infty$, there is an R_0 such that for $R \geq R_0$

$$\sum_{y \in \partial B_R} K(0,y) \leq \theta < 1.$$

Fix an x with $||x||_\infty > R_0$, and let $n = \lfloor ||x||_\infty / R_0 \rfloor$. Then $n \geq 1$. It follows from (A.9) that

$$K(0,x) \leq \theta K(0, x - y_1), \tag{A.10}$$

where y_1 is a site in ∂B_R which satisfies $K(y_1, x) = \sup_{y \in \partial B_{R_0}} K(y,x)$. Clearly $||x - y_1||_\infty \geq ||x||_\infty - R_0 \geq (n-1)R_0$. Now we iterate (A.10) $n - 2$ more times to obtain

$$K(0,x) \leq \theta^{n-1} K(0, x - y_{n-1}),$$

for some site y_n. Since $n \geq \frac{||x||_\infty}{R_0} - 1$, this gives

$$K(0,x) \leq \left(\theta^{\frac{1}{R_0}} \right)^{||x||_\infty} \theta^{-2} \sup_y K(0,y).$$

The desired exponential decay now follows from the fact that the supremum on the right side is finite by hypothesis, together with the inequality $||x||_\infty \geq d^{-1/2}|x|$. □

We now apply this lemma to show that the random walk correlation length has critical exponent 1/2. For simplicity we consider only the simple random walk, and write $e_1 = (1, 0, \ldots, 0)$.

Theorem A.2 *Let Ω consist of the 2d nearest neighbours of the origin in $\mathbf{Z}^d$.*

(a) For $0 < z < (2d)^{-1}$ the mass

$$m_0(z) \equiv - \lim_{n \to \infty} n^{-1} \log C_z(0, n e_1)$$

exists, is strictly positive and finite, and for all x

$$C_z(0, x) \leq C_z(0, 0) e^{-m_0(z) \|x\|_\infty}. \tag{A.11}$$

(b) For $0 < z < (2d)^{-1}$, the simple random walk susceptibility $\chi_0(z) = \sum_x C_z(0, x)$ satisfies

$$\chi_0(z)^{-1} = 2z[\cosh m_0(z) - 1], \tag{A.12}$$

and hence $m_0(z)^2 \sim z^{-1}(1 - 2dz)$ as $z \nearrow (2d)^{-1}$, i.e.,

$$\xi_0(z) = m_0(z)^{-1} \sim \left(\frac{1}{2d}\right)^{1/2} \left(\frac{1}{1 - 2dz}\right)^{1/2}.$$

Proof.   (a) Assuming the limit exists, the proof from Section 1.3 that $m(z)$ is finite and nonzero applies also to $m_0(z)$ [see (1.3.14) and (1.3.16)]. We now obtain a subadditivity inequality which will yield existence of the limit. Given three sites w, x, y, any simple random walk ω from w to y which passes through x can be decomposed into two subwalks as follows. Let j be the last time that $\omega(j) = x$, and let $\omega^{(1)}$ be the initial j-step portion of ω and $\omega^{(2)}$ be the remainder. Then $\omega^{(2)}$ does not return to x. The generating function for walks from x to y which do not return to x is equal to $C_z(x, x)^{-1} C_z(x, y)$. Therefore

$$\sum_{\omega : w \to y} z^{|\omega|} I[x \in \omega] = C_z(w, x) C_z(x, x)^{-1} C_z(x, y). \tag{A.13}$$

Using (A.13) and translation invariance,

$$C_z(0, (n + m)e_1) \geq \sum_{\omega : 0 \to (n+m)e_1} z^{|\omega|} I[n e_1 \in \omega]$$
$$= C_z(0, n e_1) C_z(0, 0)^{-1} C_z(0, m e_1). \tag{A.14}$$

This implies that the sequence $- \log[C_z(0, n e_1)/C_z(0, 0)]$ is subadditive, so by Lemma 1.2.2 the limit defining m_0 exists and

$$C_z(0, n e_1) \leq C_z(0, 0) e^{-m_0(z)n}. \tag{A.15}$$

To prove (A.11) we first interchange two components of x if necessary, and then change the sign of x_1 if necessary, to ensure that $x_1 = \|x\|_\infty$. This does not change the value of $C_z(0, x)$. Then symmetry and (A.13) with $w = 0$ and $y = 2\|x\|_\infty e_1$ give

$$C_z(0, 2\|x\|_\infty e_1) \geq \sum_{\omega:0 \to 2\|x\|_\infty e_1} z^{|\omega|} I[x \in \omega] = C_z(0, x)^2 C_z(0, 0)^{-1}.$$

The bound (A.11) then follows from (A.15).

(b) To prove the identity (A.12) we introduce the exponentially weighted susceptibility

$$\chi_t(z) = \sum_x e^{tx_1} C_z(0, x). \qquad (A.16)$$

Inserting (A.3) into (A.16) and using translation invariance gives

$$\begin{aligned}
\chi_t(z) &= 1 + \sum_{y:|y|=1} z e^{ty_1} \sum_x e^{t(x_1-y_1)} C_z(y, x) \\
&= 1 + 2z\,[d - 1 + \cosh t]\, \chi_t(z). \qquad (A.17)
\end{aligned}$$

Therefore

$$\chi_t(z) = \frac{1}{1 - 2z[d - 1 + \cosh t]}. \qquad (A.18)$$

By (A.18), $\chi_t(z)$ is an increasing function of t. By (A.11), $\chi_t(z)$ is finite if $z < 1/2d$ and $t < m_0(z)$. We now argue that for fixed $z < 1/2d$,

$$\lim_{t \nearrow m_0(z)} \chi_t(z) = \infty.$$

To see this, suppose to the contrary that $\chi_t(z)$ is uniformly bounded for $t < m_0(z)$. Then by the monotone convergence theorem $\chi_{m_0(z)}(z) < \infty$. Let $K(0, x) = e^{m_0(z)x_1} C_z(0, x)$. This K satisfies the Lieb-Simon inequality (A.9), as can be seen using the fact that a walk from 0 to $x \notin B_R$ must hit a site in ∂B_R. Hence by Lemma A.1 $e^{m_0(z)x_1} C_z(0, x)$ decays exponentially, which contradicts the definition of $m_0(z)$.

By (A.18),

$$\chi_0(z)^{-1} - \chi_t(z)^{-1} = 2z[\cosh t - 1].$$

Taking the limit $t \nearrow m_0(z)$ gives

$$\chi_0(z)^{-1} = 2z[\cosh m_0(z) - 1].$$

This gives the desired result

$$m_0(z)^2 \sim \frac{1 - 2dz}{z} \quad \text{as } z \nearrow \frac{1}{2d},$$

and completes the proof.   □

We now turn to two lemmas which are needed in Section 6.2 to prove convergence of the lace expansion. These lemmas give estimates for the two-point function for two particular choices of the step set Ω. We begin with the nearest-neighbour walk, for which the set Ω consists of the $2d$ nearest neighbours of the origin. In this case

$$\hat{D}(k) = \hat{D}_0(k) \equiv \frac{1}{d} \sum_{\mu=1}^{d} \cos k_\mu. \tag{A.19}$$

We will denote the two-point function for the nearest-neighbour model by $C_z^{(0)}(0, x)$. The next lemma gives estimates for $\hat{D}_0(k)$, and for $\hat{C}_z^{(0)}(k)$ at the critical point $z = \Omega^{-1}$. Although for our purposes the dimension d is always a positive integer, bounds have been given in terms of $(d-4)^{-1}$ rather than d^{-1} to emphasize critical dimensionality. For a different approach to some bounds of this type, see Hara (1990).

Lemma A.3 *For all $d \geq 1$ and $k \in [-\pi, \pi]^d$,*

$$1 - \hat{D}_0(k) \geq \frac{2}{\pi^2 d} k^2. \tag{A.20}$$

For all nonnegative integers n and all $d \geq 1$,

$$\|\hat{D}_0^n\|_1 \leq \left(\frac{\pi d}{4n}\right)^{d/2}. \tag{A.21}$$

There is a constant K, independent of d, such that for $z = (2d)^{-1}$ and for all $d > 4$

$$0 \leq \|\hat{C}^{(0)}\|_2^2 - 1 \leq K(d-4)^{-1} \tag{A.22}$$

and

$$\left\| \frac{\partial_\mu^2 \hat{D}_0}{[1 - \hat{D}_0]^2} \right\|_1 + 2 \left\| \frac{(\partial_\mu \hat{D}_0)^2}{[1 - \hat{D}_0]^3} \right\|_1 \leq K(d-4)^{-1}. \tag{A.23}$$

The above norms are L^p norms on $[-\pi, \pi]^d$ with measure $(2\pi)^{-d} d^d k$.

Proof. We consider each inequality in turn. The first follows from the fact that $1 - \cos k_\mu \geq 2\pi^{-2} k_\mu^2$, for $|k_\mu| \leq \pi$.

For the second inequality, we proceed as follows. By symmetry,

$$\|\hat{D}_0^n\|_1 = \pi^{-d} \int_{[0,\pi]^d} |\hat{D}_0(k)^n| d^d k \tag{A.24}$$

$$= \pi^{-d} \sum_{m=0}^{d} \binom{d}{m} \int_{[0,\pi/2]^m} dk_1 \ldots dk_m$$

$$\times \int_{[\pi/2,\pi]^{d-m}} dk_{m+1} \ldots dk_d |\hat{D}_0(k)^n|.$$

For k in the domain of integration on the right side,

$$|\hat{D}_0(k)| \leq d^{-1} \sum_{\mu=1}^{d} |\cos k_\mu|$$

$$\leq 1 - \frac{4}{\pi^2 d} \left[\sum_{\mu=1}^{m} k_\mu^2 + \sum_{\nu=m+1}^{d} (\pi - k_\nu)^2 \right]$$

$$\leq \exp \left(-\frac{4}{\pi^2 d} \left[\sum_{\mu=1}^{m} k_\mu^2 + \sum_{\nu=m+1}^{d} (\pi - k_\nu)^2 \right] \right). \quad (A.25)$$

Combining (A.24) and (A.25) and making a change of variables gives

$$\|\hat{D}_0^n\|_1 \leq \left(\frac{2}{\pi} \right)^d \int_{[0,\pi/2]^d} \exp[-4nk^2 \pi^{-2} d^{-1}] d^d k. \quad (A.26)$$

Extending the domain of integration, we obtain

$$\|\hat{D}_0^n\|_1 \leq \pi^{-d} \int_{R^d} \exp[-4nk^2 \pi^{-2} d^{-1}] d^d k = \left(\frac{\pi d}{4n} \right)^{d/2}. \quad (A.27)$$

For the bound on $\|\hat{C}^{(0)}\|_2^2 - 1$, we use a series expansion of $\hat{C}^{(0)}(k)^2 = [1 - \hat{D}_0(k)]^{-2}$ [which follows from (A.6)] to obtain

$$\|\hat{C}^{(0)}\|_2^2 = 1 + \sum_{n=1}^{\infty} \int (n+1) \hat{D}_0^n \frac{d^d k}{(2\pi)^d}. \quad (A.28)$$

The integral here is over $[-\pi, \pi]^d$. It suffices to show that there is a constant K (independent of d) such that for $d > 4$,

$$\sum_{n=1}^{\infty} (n+1) \int \hat{D}_0^n \frac{d^d k}{(2\pi)^d} \leq K(d-4)^{-1}. \quad (A.29)$$

For n odd the integral is zero, while for each even n the integral is positive.

To prove (A.29), we divide the summation into two parts and use (A.27) to obtain

$$\sum_{n=1}^{\infty} (n+1) \int \hat{D}_0^n \frac{d^d k}{(2\pi)^d} \leq \sum_{n=1}^{d-1} (n+1) \int \hat{D}_0^n \frac{d^d k}{(2\pi)^d} + \sum_{n=d}^{\infty} (n+1) \left(\frac{\pi d}{4n} \right)^{d/2}.$$

$$(A.30)$$

Using the fact that for $a > 1$

$$\sum_{n=N}^{\infty} \frac{1}{n^a} \le \int_{N-1}^{\infty} \frac{1}{t^a} dt = \frac{(N-1)^{-a+1}}{a-1},$$

it is not hard to see that the second sum on the right side of (A.30) is bounded by a dimension-independent multiple of $(d-4)^{-1}$. For the first sum on the right side, we use the fact that $|\hat{D}_0(k)| \le 1$ to obtain

$$\sum_{n=1}^{d-1} (n+1) \int \hat{D}_0^n \frac{d^d k}{(2\pi)^d} \le 3 \int \hat{D}_0^2 \frac{d^d k}{(2\pi)^d} + 5 \int \hat{D}_0^4 \frac{d^d k}{(2\pi)^d} + d^2 \int \hat{D}_0^6 \frac{d^d k}{(2\pi)^d}.$$
(A.31)

The right side can be estimated by observing that the integrals on the right side are respectively the probabilities that simple random walk returns to the origin after two, four, or six steps. These probabilities are respectively of the order of d^{-1}, d^{-2} and d^{-3}, which yields (A.29).

For (A.23) we proceed as follows. We first observe that integration by parts gives

$$\left\| \frac{(\partial_\mu \hat{D}_0)^2}{[1-\hat{D}_0]^3} \right\|_1 = -\frac{1}{2} \int \frac{\partial_\mu^2 \hat{D}_0}{[1-\hat{D}_0]^2} \frac{d^d k}{(2\pi)^d}, \qquad (A.32)$$

so it suffices to bound the first term on the left side of (A.23). But for this we simply observe that

$$\partial_\mu^2 \hat{D}_0(k) = -d^{-1} \cos k_\mu, \qquad (A.33)$$

which is bounded in absolute value by d^{-1}, and then use (A.22). $\qquad \square$

The second model of ordinary random walk we consider is the "spread-out" walk in $\mathbf{Z}^d$ with step set $\Omega = \{x \ne 0 : \|x\|_\infty \le L\}$. The cardinality of this set is $\Omega = (2L+1)^d - 1$. In this case we will be interested in $d > 4$ and large L. We write $\hat{D}_L(k)$ in place of $\hat{D}(k)$ of (A.7) for this model. Also, we write $C^{(L)}(0,x)$ for the critical ($z = \Omega^{-1}$) spread-out random walk two-point function. Before proving an analogue of Lemma A.3 for this spread-out random walk, we need the following result.

Lemma A.4 *For any $S \subset \{1, \dots, d\}$, $L \ge 1$ and $k \in [-\pi, \pi]^d$,*

$$|\hat{D}_L(k)| \le 2 \frac{\Omega+1}{\Omega} \left| \prod_{\mu \in S} (2L+1) \sin(k_\mu/2) \right|^{-1} \qquad (A.34)$$

and

$$|\partial_\mu^2 \hat{D}_L(k)| \leq \frac{\Omega+1}{\Omega}(2L+1)^2 \left| \prod_{\nu \in S}(2L+1)\sin(k_\nu/2) \right|^{-1}. \qquad (A.35)$$

Proof. In terms of the Dirichlet kernel

$$M(t) = \sum_{j=-L}^{L} e^{ijt} = \frac{\sin[(2L+1)t/2]}{\sin(t/2)}, \qquad (A.36)$$

we have

$$\hat{D}_L(k) = \frac{1}{\Omega}\left[(2L+1)^d \prod_{\mu=1}^{d} \frac{M(k_\mu)}{2L+1} - 1 \right]. \qquad (A.37)$$

Since $|M(t)|$ is bounded above by both $2L+1$ and $|\sin(t/2)|^{-1}$,

$$\left| \prod_{\mu=1}^{d} \frac{M(k_\mu)}{(2L+1)} \right| \leq \left| \prod_{\mu \in S} \frac{1}{(2L+1)\sin(k_\mu/2)} \right|. \qquad (A.38)$$

This gives (A.34), once we observe that

$$1 \leq \frac{(2L+1)^d}{\prod_{\mu \in S}|(2L+1)\sin(k_\mu/2)|}. \qquad (A.39)$$

For (A.35), we have

$$\partial_\mu^2 \hat{D}_L(k) = \frac{\Omega+1}{\Omega} \frac{\partial_\mu^2 M(k_\mu)}{2L+1} \prod_{\nu \neq \mu} \frac{M(k_\nu)}{2L+1}. \qquad (A.40)$$

In view of the above bounds, it suffices to show that $\partial_\mu^2 M(k_\mu)$ is bounded above by both $(2L+1)^3$ and $(2L+1)^2|\sin(k_\mu/2)|^{-1}$. The first of these bounds follows from

$$\partial_\mu^2 M(k_\mu) = -\sum_{x_\mu=-L}^{L} x_\mu^2 e^{ik_\mu x_\mu} \qquad (A.41)$$

by taking absolute values inside the sum and using $x_\mu^2 \leq L^2$. For the second bound, we use the summation by parts formula

$$\sum_{n=A}^{B}(b_{n+1}-b_n)a_n = -\sum_{n=A}^{B} b_n(a_n-a_{n-1}) + b_{B+1}a_B - b_A a_{A-1}, \qquad (A.42)$$

to obtain

$$
\begin{aligned}
\sum_{x_\mu=-L}^{L} x_\mu^2 e^{ik_\mu x_\mu} &= \frac{e^{-ik_\mu/2}}{2i\sin(k_\mu/2)} \sum_{x_\mu=-L}^{L} [e^{ik_\mu(x_\mu+1)} - e^{ik_\mu x_\mu}]x_\mu^2 \\
&= -\frac{e^{-ik_\mu/2}}{2i\sin(k_\mu/2)} \{ \sum_{x_\mu=-L}^{L} e^{ik_\mu x_\mu}[x_\mu^2 - (x_\mu-1)^2] \\
&\quad +e^{ik_\mu(L+1)}L^2 - e^{-ik_\mu L}(-L-1)^2 \}.
\end{aligned} \tag{A.43}
$$

The desired bound then follows.   □

The next lemma states several properties of the two-point function for the spread-out walk.

Lemma A.5 *For any $d \geq 1$ there is an L_0 depending only on d such that for any $k \in [-\pi, \pi]^d$ and $L \geq L_0$,*

$$
1 - \hat{D}_L(k) \geq \frac{k^2}{2\pi^2 d}. \tag{A.44}
$$

For any $d \geq 1$ there are L_0 and K depending only on d such that for all integers $n \geq 2$ and all $L \geq L_0$,

$$
\|\hat{D}_L^n\|_1 \leq K \frac{(\log L)^{d/2}}{L^d n^{d/2}}. \tag{A.45}
$$

Fix $z = \Omega^{-1}$. For any $d > 4$ and for all L, there is a K depending only on d such that

$$
0 \leq \|\hat{C}^{(L)}\|_2^2 - 1 \leq K \frac{(\log L)^{d/2}}{L^d} \tag{A.46}
$$

and

$$
\left\| \frac{\partial_\mu^2 \hat{D}_L}{[1-\hat{D}_L]^2} \right\|_1 + 2 \left\| \frac{(\partial_\mu \hat{D}_L)^2}{[1-\hat{D}_L]^3} \right\|_1 \leq K \frac{(\log L)^d}{L^{d-2}}. \tag{A.47}
$$

The above norms are L^p norms on $[-\pi, \pi]^d$ with measure $(2\pi)^{-d} d^d k$.

Proof. We shall begin by proving

$$
1 - |\hat{D}(k)| \geq \frac{k^2}{2\pi^2 d}, \tag{A.48}
$$

which in particular implies (A.44). By Lemma A.4 (with $|S| = 1$), there is a universal constant a such that

$$
|\hat{D}_L(k)| \leq \frac{a}{\|k\|_\infty L}. \tag{A.49}
$$

For $||k||_\infty \geq 2a/L$, this together with the fact that $k^2 \leq d\pi^2$ gives

$$1 - |\hat{D}_L(k)| \geq 1 - \frac{1}{2} \geq \frac{k^2}{2\pi^2 d}. \tag{A.50}$$

For $||k||_\infty \leq 2a/L$, we will consider separately the two cases $\hat{D}_L(k) \geq 0$ and $\hat{D}_L(k) < 0$. In either case, it follows from symmetry and the fact that $1 - \cos t \geq 2\pi^{-2}t^2$ for $|t| \leq \pi$ that

$$
\begin{aligned}
1 - \hat{D}_L(k) &\geq \frac{1}{\Omega} \sum_{x \in \Omega : |k \cdot x| \leq \pi} \frac{2}{\pi^2} (k \cdot x)^2 \\
&\geq \frac{2}{\pi^2 \Omega} \sum_{x \in \Omega : ||x||_1 \leq \pi L/2a} (k \cdot x)^2 \\
&= \frac{2}{\pi^2 \Omega} k^2 \sum_{x \in \Omega : ||x||_1 \leq \pi L/2a} x_1^2. \tag{A.51}
\end{aligned}
$$

The sum over x on the right side is bounded below by a (dimension-dependent) multiple of L^{2+d}, for L sufficiently large, and hence there is a positive constant c (depending only on d) such that for $L \geq (2\pi^2 dc)^{-1/2}$ and $||k||_\infty \leq 2a/L$,

$$1 - \hat{D}_L(k) \geq cL^2 k^2 \geq \frac{k^2}{2\pi^2 d}. \tag{A.52}$$

This gives (A.48) for $||k||_\infty \leq 2a/L$ and $\hat{D}_L(k) \geq 0$. If on the other hand $\hat{D}_L(k) < 0$ and $||k||_\infty \leq 2a/L$, then there are positive constants c, c' (depending only on the dimension) such that for large L

$$
\begin{aligned}
1 - |\hat{D}_L(k)| &= \frac{1}{\Omega} \sum_{x \in \Omega} (1 + \cos k \cdot x) \\
&\geq \frac{1}{\Omega} \sum_{x \in \Omega : |k \cdot x| \leq \pi/2} 1 \\
&\geq \frac{1}{\Omega} \sum_{x \in \Omega : ||x||_1 \leq \pi L/4a} 1 \\
&\geq c' \geq cL^2 k^2. \tag{A.53}
\end{aligned}
$$

Taking $L \geq (2\pi^2 dc)^{-1/2}$, it follows from (A.52) and (A.53) that for $||k||_\infty \leq 2a/L$

$$1 - |\hat{D}_L(k)| \geq cL^2 k^2 \geq \frac{k^2}{2\pi^2 d}. \tag{A.54}$$

With (A.50), this gives (A.48) and hence (A.44).

Turning now to (A.45), it follows from (A.50) and (A.54) that if L is sufficiently large then

$$\|\hat{D}_L^n\|_1 \le \int_{\|k\|_\infty \le 2a/L} e^{-cnL^2k^2} \frac{d^d k}{(2\pi)^d} + \frac{1}{2^n} \le \frac{K}{2L^d n^{d/2}} + \frac{1}{2^n}, \quad (A.55)$$

where K depends only on d. Fix $b \in (1, 2)$. Increasing K if necessary, for all $n \ge d \log L / \log b$ we have

$$\frac{1}{2^n} \le \frac{K}{2 b^n n^{d/2}} \le \frac{K}{2 L^d n^{d/2}}. \quad (A.56)$$

Hence for $n \ge d \log L / \log b$,

$$\|\hat{D}_L^n\|_1 \le K L^{-d} n^{-d/2}. \quad (A.57)$$

On the other hand, for any $n \ge 2$, $\|\hat{D}_L^n\|_1 \le \|\hat{D}_L^2\|_1 = \Omega^{-1}$, since $|\hat{D}_L(k)| \le 1$ and $\|\hat{D}_L^2\|_1$ is the probability of return to the origin after two steps. Hence for all $n \le d \log L / \log b$

$$\|\hat{D}_L^n\|_1 \le \frac{1}{\Omega} \le \frac{1}{\Omega} \left(\frac{d \log L}{n \log b} \right)^{d/2} \le K (\log L)^{d/2} L^{-d} n^{-d/2}. \quad (A.58)$$

This completes the proof of (A.45).

The proof of (A.46) now proceeds by writing the left side in terms of a sum of terms of order $(n+1) \int \hat{D}^n$ as in (A.28), and using (A.45).

For (A.47), just as for the nearest-neighbour walk an integration by parts allows the second term on the left to be bounded by the first. The first term is equal to

$$\sum_{n=1}^{\infty} n \int |\partial_\mu^2 \hat{D}_L(k)| \hat{D}_L(k)^{n-1} \frac{d^d k}{(2\pi)^d}. \quad (A.59)$$

By (A.45) and the fact that $|\partial_\mu^2 \hat{D}_L(k)| = O(L^2)$ it suffices to bound the first two terms in the sum, and for these it is sufficient to show that

$$\int |\partial_\mu^2 \hat{D}_L(k)| \frac{d^d k}{(2\pi)^d} \le K (\log L)^d L^{2-d}. \quad (A.60)$$

To prove (A.60), we begin by dividing the domain of integration to obtain

$$\int |\partial_\mu^2 \hat{D}_L(k)| \frac{d^d k}{(2\pi)^d} = \int_{\|k\|_\infty \le 2a/L} |\partial_\mu^2 \hat{D}_L(k)| \frac{d^d k}{(2\pi)^d}$$
$$+ \int_{2a/L \le \|k\|_\infty \le \pi} |\partial_\mu^2 \hat{D}_L(k)| \frac{d^d k}{(2\pi)^d}. \quad (A.61)$$

The first term on the right side is order L^{2-d}, since the integrand is order L^2 and the volume is order L^{-d}. The domain of integration of the second term is the disjoint union over nonempty subsets $S \subset \{1, \ldots, d\}$ of

$$R_S = \{k \in \mathbf{R}^d : 2a/L < k_\mu \leq \pi \text{ for } \mu \in S, |k_\nu| \leq 2a/L \text{ for } \nu \notin S\}. \quad (A.62)$$

By (A.35),

$$\int_{R_S} |\partial_\mu^2 \hat{D}_L(k)| \frac{d^d k}{(2\pi)^d} \leq K L^{2-|S|} \int_{R_S} \prod_{\mu \in S} |k_\mu|^{-1} \frac{d^d k}{(2\pi)^d}$$

$$\leq K L^{2-|S|-(d-|S|)} [\log L]^{|S|}. \quad (A.63)$$

The desired result then follows by summing over S. $\qquad \square$

We conclude this appendix with a theorem about ordinary random walks with general step distribution that is needed in Section 8.1. As explained in that section, this theorem is true under weaker hypotheses than we state here.

Theorem A.6 *Let p be a positive integer. Let $Y_1, Y_2, \ldots$ be independent, identically distributed Z^p-valued random variables with the property that*

$$\beta \equiv \min\{\Pr\{Y_1 - Y_2 = e\} : \|e\|_1 = 1\} > 0.$$

Then there exists a finite constant C (depending on the distribution of Y_1) such that

$$\Pr\{Y_1 + \cdots + Y_m = x\} \leq C m^{-p/2} \quad \text{for every } m \geq 1, x \in Z^p. \quad (A.64)$$

Proof. For $k \in [-\pi, \pi]^p$, let $\phi(k) = E[\exp(ik \cdot Y_1)]$ denote the characteristic function (Fourier transform) of Y_1. Let $e_1, \ldots, e_p$ denote the positive unit vectors of Z^p. Then by independence and symmetry,

$$|\phi(k)|^2 = E[\exp(ik \cdot [Y_1 - Y_2])]$$

$$= E[\cos(k \cdot [Y_1 - Y_2])]$$

$$\leq 2 \sum_{j=1}^{p} \Pr\{Y_1 - Y_2 = e_j\} \cos k_j + \Pr\{|Y_1 - Y_2| \neq 1\}$$

$$= 2 \sum_{j=1}^{p} \Pr\{Y_1 - Y_2 = e_j\}(\cos k_j - 1) + 1$$

$$\leq 2 \sum_{j=1}^{p} \beta(\cos k_j - 1) + 1. \quad (A.65)$$

Since $\cos k_j - 1 \leq -2k_j^2/\pi^2$ for every $k_j \in [-\pi, \pi]$, we have

$$|\phi(k)|^2 \leq 1 - 4\beta k^2/\pi^2 \leq e^{-\delta|k|^2} \qquad (A.66)$$

for every $k \in [-\pi, \pi]^p$, where $\delta = 4\beta/\pi^2 > 0$.

Next let $n = \lfloor m/2 \rfloor$, so that m is either $2n$ or $2n+1$. Let $x \in \mathbf{Z}^p$. Then by Fourier inversion and (A.66) we obtain

$$
\begin{aligned}
\Pr\{Y_1 + \cdots + Y_m = x\} &= \int_{[-\pi,\pi]^p} e^{-ik\cdot x} \phi(k)^m \frac{d^p k}{(2\pi)^p} \\
&\leq \int_{[-\pi,\pi]^p} |\phi(k)|^{2n} \frac{d^p k}{(2\pi)^p} \\
&\leq \int_{R^p} e^{-n\delta k^2} \frac{d^p k}{(2\pi)^p} \\
&= \text{const.} n^{-p/2}.
\end{aligned}
$$

Since $n = \lfloor m/2 \rfloor$, this proves the theorem.   □

Appendix B

Proof of the renewal theorem

In this appendix we prove the version of the renewal theorem that is stated in Theorem 4.2.2, which is all that we need in this book. The probabilistic interpretation of this theorem is discussed briefly just before Theorem 4.2.2. The proof is essentially the standard one [as in for example Feller (1968) or Karlin and Taylor (1975)]. A very different proof using complex analysis appears in Kingman (1972), pp. 10-14.

Theorem B.1 *Assume that $\{f_n : n \geq 1\}$ and $\{g_n : n \geq 0\}$ are nonnegative sequences, and let*

$$f = \sum_{n=1}^{\infty} f_n \quad and \quad g = \sum_{n=0}^{\infty} g_n$$

denote their sums. Assume that $0 < g < +\infty$. Also assume that $f_1 > 0$. Define the new sequence $v_0, v_1, \ldots$ by

$$
\begin{aligned}
v_0 &= g_0 \\
v_n &= g_n + f_1 v_{n-1} + f_2 v_{n-2} + \cdots + f_n v_0, \quad for\ all\ n \geq 1. \quad \text{(B.1)}
\end{aligned}
$$

(a) If $f < 1$, then $\lim_{n\to\infty} v_n = 0$ and $\sum_{n=0}^{\infty} v_n = g/(1-f)$.
(b) If $f = 1$, then

$$\lim_{n\to\infty} v_n = \frac{g}{\sum_{k=1}^{\infty} k f_k} \quad \text{(B.2)}$$

(the limit is 0 if the sum in the denominator diverges). Also, $\sum_n v_n$ diverges.
(c) If $1 < f \leq +\infty$, then $\limsup_{n\to\infty} v_n^{1/n} > 1$.

389

Before proving the theorem, we record an elementary lemma.

Lemma B.2 *Let* $\{\alpha_m : m \geq 0\}$ *and* $\{\beta_{m,k} : m, k \geq 0\}$ *be nonnegative sequences. Assume that* $\sum_{m=0}^{\infty} \alpha_m$ *is finite and that there exists a finite constant* B *such that* $\beta_{m,k} \leq B$ *for every* m *and* k. *Then*

$$\limsup_{k \to \infty} \sum_{m=0}^{\infty} \alpha_m \beta_{m,k} \leq \sum_{m=0}^{\infty} \alpha_m (\limsup_{k \to \infty} \beta_{m,k}).$$

Proof. Since $B - \beta_{m,k} \geq 0$, Fatou's lemma implies that

$$\liminf_{k \to \infty} \sum_{m=0}^{\infty} \alpha_m (B - \beta_{m,k}) \geq \sum_{m=0}^{\infty} \alpha_m (\liminf_{k \to \infty} (B - \beta_{m,k})).$$

The result follows upon subtraction of $B \sum_{m=0}^{\infty} \alpha_m$ from both sides. □

Proof of Theorem B.1. Define the generating functions

$$F(s) = \sum_{n=1}^{\infty} f_n s^n, \quad G(s) = \sum_{n=0}^{\infty} g_n s^n, \quad V(s) = \sum_{n=0}^{\infty} v_n s^n.$$

These are well-defined (possibly infinite) for $s \geq 0$. Observe that $F(1) = f$ and $G(1) = g$. From the defining equation (B.1), we have

$$V(s) = G(s) + F(s)V(s) \tag{B.3}$$

for all $s \geq 0$ (since all terms are nonnegative, the interchange of summation needed to deduce (B.3) is justified).

Consider part (c) first. Assume $1 < f \leq +\infty$ and $\limsup_{n \to \infty} v_n^{1/n} \leq 1$. The latter implies that $V(s)$ is finite whenever $|s| < 1$; but $V(s) \geq F(s)V(s)$ whenever $s \geq 0$ by (B.3), and so $F(s) \leq 1$ whenever $0 \leq s < 1$. Letting s increase to 1 and applying the monotone convergence theorem, we conclude that $f = F(1) \leq 1$, which is a contradiction. This proves (c).

It remains to prove (a) and (b), so we assume now that $f \leq 1$. Observe that $0 \leq v_n \leq \sum_{i=0}^{n} g_i$ for every $n \geq 0$ (by induction), and so $\{v_n\}$ is a bounded nonnegative sequence. From (B.3), we see that $V(s) = G(s)/(1 - F(s))$ whenever $|s| < 1$. Letting s increase to 1, we see that $V(1) < \infty$ if $f < 1$, and part (a) follows. It also follows that $\sum_n v_n$ diverges if $f = 1$.

It remains now to prove (B.2), so we assume $f = 1$. Summation of Equation (B.1) over $n = 0, 1, \ldots, N$ gives

$$\sum_{n=0}^{N} v_n = \sum_{n=0}^{N} g_n + \sum_{m=1}^{N} \sum_{i=1}^{m} f_i v_{N-m}. \tag{B.4}$$

If we now define

$$r_n = 1 - \sum_{i=1}^{n} f_i = \sum_{i=n+1}^{\infty} f_i$$

for $n \geq 0$ (so $r_0 = 1$), then we can rewrite (B.4) as

$$\sum_{n=0}^{N} r_n v_{N-n} = \sum_{n=0}^{N} g_n. \tag{B.5}$$

Define

$$r = \sum_{n=0}^{\infty} r_n = \sum_{k=1}^{\infty} k f_k.$$

Formally, letting $N \to \infty$ in (B.5) gives the desired result $\lim_{N \to \infty} v_N = g/r$. Most of the work in making this rigorous is in proving that the limit exists.

Let $u = \limsup_{n \to \infty} v_n$, which is finite since v_n is bounded, and let $v_{n(j)}$ be any subsequence which converges to u. We claim that

$$\lim_{j \to \infty} v_{n(j)-k} = u \quad \text{for every } k \geq 1. \tag{B.6}$$

To prove the claim, let $u_* = \liminf_j v_{n(j)-1}$. Since v_n is bounded and f is finite, we can apply Lemma B.2 to

$$v_{n(j)} - f_1 v_{n(j)-1} = g_{n(j)} + \sum_{i=2}^{n(j)} f_i v_{n(j)-i},$$

to obtain

$$u - f_1 u_* \leq 0 + \sum_{n=2}^{\infty} f_i u.$$

Since $f_1 > 0$ and $f = 1$, this implies that $u \leq u_*$, and hence $u = u_*$. Thus the claim holds for $k = 1$, and by induction it holds for every positive k.

If we now replace N by $n(j)$ in (B.5) and apply Fatou's lemma, then using (B.6) we obtain

$$\sum_{n=0}^{\infty} r_n u \leq g. \tag{B.7}$$

If $r = +\infty$, then u must be 0 and we are done; so we suppose that r is finite. Let $u' = \liminf_n v_n$, and let $v_{N(j)}$ be a subsequence that converges to u'. Replacing N by $N(j)$ in (B.5) and applying Lemma B.2, and then using the fact that $u \geq \limsup_j v_{N(j)-n}$ for every $n \geq 1$, we obtain

$$r_0 u' + \sum_{n=1}^{\infty} r_n u \geq g.$$

Combining this with (B.7) implies $u' \geq u$; therefore $u' = u$, i.e. v_n converges to u. Now we can apply the dominated convergence theorem to (B.5), which gives $ru = g$ and completes the proof of the theorem. $\square$

Appendix C

Tables of exact enumerations

This appendix contains tables of exact enumerations, for the hypercubic lattices in dimensions $2, 3, 4, 5, 6$, of the number of n-step self-avoiding walks, the sum of squares of endpoints of all n-step self-avoiding walks (this only for $d = 2, 3$), and the number of n-step self-avoiding polygons. We have included all data known to us at the time of publication. The primary use of these tables is to provide partial sequences from which estimates can be made of critical exponents and the connective constant. Methods used are described in the references cited in the tables. An overview of series extrapolation methods is given in the survey Guttmann (1989a).

It is worth noting that there are significant computational problems associated with the generation of these tables. Guttman and Wang (1991) quote a computation time of somewhat less than 700 hours on a Masscomp 5700 computer for the generation of the number of self-avoiding walks and the sum of squares of end-to-end distances for the two values $n = 28, 29$ on the square lattice ($d = 2$). Masand et al (1992) report that the computation of c_n on the square lattice for $30 \leq n \leq 34$ required approximately 100 hours on a Thinking Machine CM-2 massively parallel supercomputer.

393

n	$d = 2$	3
0	1	1
1	4	6
2	12	30
3	36	150
4	100	726
5	284	3 534
6	780	16 926
7	2 172	81 390
8	5 916	387 966
9	16 268	1 853 886
10	44 100	8 809 878
11	120 292	41 934 150
12	324 932	198 842 742
13	881 500	943 974 510
14	2 374 444	4 468 911 678
15	6 416 596	21 175 146 054
16	17 245 332	100 121 875 974
17	46 466 676	473 730 252 102
18	124 658 732	2 237 723 684 094
19	335 116 620	10 576 033 219 614
20	897 697 164	49 917 327 838 734
21	2 408 806 028	235 710 090 502 158
22	6 444 560 484	
23	17 266 613 812	
24	46 146 397 316	
25	123 481 354 908	
26	329 712 786 220	
27	881 317 491 628	
28	2 351 378 582 244	
29	6 279 396 229 332	
30	16 741 957 935 348	
31	44 673 816 630 956	
32	119 034 997 913 020	
33	317 406 598 267 076	
34	845 279 074 648 708	

Table C.1: The number of self-avoiding walks for $d = 2, 3$, from Masand *et al* (1992), Guttmann and Wang (1991), Guttmann (1989b). Values for $d = 2$ are now known at least to $n = 39$ [Conway, Enting and Guttmann (private communication)].

n	$d = 2$	3
1	4	6
2	32	72
3	124	582
4	704	4 032
5	2 716	25 566
6	9 808	153 528
7	33 788	886 926
8	112 480	4 983 456
9	364 588	27 401 502
10	1 157 296	148 157 880
11	3 610 884	790 096 950
12	11 108 448	4 166 321 184
13	33 765 276	21 760 624 254
14	101 594 000	112 743 796 632
15	302 977 204	580 052 260 230
16	896 627 936	2 966 294 589 312
17	2 635 423 124	15 087 996 161 382
18	7 699 729 296	76 384 144 381 272
19	22 374 323 436	385 066 579 325 550
20	64 702 914 336	1 933 885 653 380 544
21	186 289 216 332	9 679 153 967 272 734
22	534 227 118 960	
23	1 526 445 330 900	
24	4 347 038 392 480	
25	12 341 626 847 324	
26	34 940 293 640 400	
27	98 660 244 502 668	
28	277 910 662 983 584	
29	781 060 493 709 204	

Table C.2: The sum of squares of $|\omega(n)|$ over all n-step self-avoiding walks, for $d = 2, 3$, from Guttmann and Wang (1991), Guttmann (1989b) and Guttmann (1987). The mean-square displacement is obtained by dividing by c_n.

n	$d = 2$	3
4	1	3
6	2	22
8	7	207
10	28	2412
12	124	31 754
14	588	452 640
16	2 938	6 840 774
18	15 268	108 088 232
20	81 826	1 768 560 270
22	449 572	
24	2 521 270	
26	14 385 376	
28	83 290 424	
30	488 384 528	
32	2 895 432 660	
34	17 332 874 364	
36	104 653 427 012	
38	636 737 003 384	
40	3 900 770 002 646	
42	24 045 500 114 388	
44	149 059 814 328 236	
46	928 782 423 033 008	
48	5 814 401 613 289 290	
50	36 556 766 640 745 936	
52	230 757 492 737 449 636	
54	1 461 972 662 850 874 916	
56	9 293 993 428 791 901 042	

Table C.3: The number of self-avoiding polygons for $d = 2, 3$. The values for $d = 2$ are taken from Guttmann and Enting (1988). Values for $d = 3$, $n \leq 14$ are from Fisher and Gaunt (1964). Values for $d = 3$, $n \geq 16$ are from Sykes *et al* (1972). Values are now known for $d = 2$ up to at least 70 steps [Enting and Guttmann (private communication)].

n	$d = 4$	5	6
0	1	1	1
1	8	10	12
2	56	90	132
3	392	810	1 452
4	2 696	7 210	15 852
5	18 584	64 250	173 172
6	127 160	570 330	1 887 492
7	871 256	5 065 530	20 578 452
8	5 946 200	44 906 970	224 138 292
9	40 613 816	398 227 610	2 441 606 532
10	276 750 536	3 527 691 690	26 583 605 772
11	1 886 784 200	31 255 491 850	289 455 960 492
12	12 843 449 288		
13	87 456 597 656		

Table C.4: The number of self-avoiding walks for $d = 4, 5, 6$. These values are taken from Fisher and Gaunt (1964), except for $d = 4$, $n = 12, 13$ which are from Guttmann (1978).

n	$d = 4$	5	6
4	6	10	15
6	76	180	350
8	1 434	5 170	13 545
10	32 616	186 856	679 716
12	844 432	4 060 132	17 761 132
14	23 919 864		

Table C.5: The number of self-avoiding polygons for $d = 4, 5, 6$. The values for $n \leq 10$ are from Fisher and Gaunt (1964), while those for $n \geq 12$ are due to Guttmann (private communication).

Bibliography

[1] J. Adler (1984). A second look at a controversial percolation exponent — is η negative in three dimensions? *Z. Phys. B - Cond. Matt.*, **55**:227–229.

[2] R. Ahlberg and S. Janson (1980). Upper bounds for the connective constant. Unpublished manuscript.

[3] M. Aizenman (1982). Geometric analysis of φ^4 fields and Ising models, Parts I and II. *Commun. Math. Phys.*, **86**:1–48.

[4] M. Aizenman (1985). The intersection of Brownian paths as a case study of a renormalization group method for quantum field theory. *Commun. Math. Phys.*, **97**:91–110.

[5] M. Aizenman and R. Fernández (1986). On the critical behavior of the magnetization in high dimensional Ising models. *J. Stat. Phys.*, **44**:393–454.

[6] M. Aizenman and R. Graham (1983). On the renormalized coupling constant and the susceptibility in ϕ_4^4 field theory and the Ising model in four dimensions. *Nucl. Phys.*, **B225** [FS9]:261–288.

[7] M. Aizenman, H. Kesten, and C.M. Newman (1987). Uniqueness of the infinite cluster and continuity of connectivity functions for short and long range percolation. *Commun. Math. Phys.*, **111**:505–531.

[8] M. Aizenman and C.M. Newman (1984). Tree graph inequalities and critical behavior in percolation models. *J. Stat. Phys.*, **36**:107–143.

[9] K.S. Alexander, J.T. Chayes, and L. Chayes (1990). The Wulff construction and asymptotics of the finite cluster distribution for two dimensional Bernoulli percolation. *Commun. Math. Phys.*, **131**:1–50.

399

[10] Z. Alexandrowicz (1969). Monte Carlo of chains with excluded volume: a way to evade sample attrition. *J. Chem. Phys.*, 51:561–565.

[11] S.E. Alm (1992). Upper bounds for the connective constant of self-avoiding walks. Preprint.

[12] S.E. Alm and S. Janson (1990). Random self-avoiding walks on one-dimensional lattices. *Commun. Statist.-Stochastic Models*, 6:169–212.

[13] D.J. Amit (1984). *Field Theory, the Renormalization Group, and Critical Phenomena*. World Scientific, Singapore, 2nd edition.

[14] D.J. Amit, G. Parisi, and L. Peliti (1983). Asymptotic behaviour of the "true" self-avoiding walk. *Phys. Rev. B*, 27:1635–1645.

[15] C. Aragão de Carvalho and S. Caracciolo (1983). A new Monte Carlo approach to the critical properties of self-avoiding random walks. *J. Phys. (Paris)*, 44:323–331.

[16] C. Aragão de Carvalho, S. Caracciolo, and J. Fröhlich (1983). Polymers and $g|\phi|^4$ theory in four dimensions. *Nucl. Phys. B*, 215 [FS7]:209–248.

[17] D. Arnaudon, D. Iagolnitzer, and J. Magnen (1991). Weakly self-avoiding polymers in four dimensions. Rigorous results. *Phys. Lett. B*, 273:268–272.

[18] M.N. Barber and B.W. Ninham (1970). *Random and Restricted Walks: Theory and Applications*. Gordon and Breach, New York.

[19] D.J. Barsky and M. Aizenman (1991). Percolation critical exponents under the triangle condition. *Ann. Probab.*, 19:1520–1536.

[20] J. Batoulis and K. Kremer (1988). Statistical properties of biased sampling methods for long polymer chains. *J. Phys. A: Math. Gen.*, 21:127–146.

[21] B. Berg and D. Foerster (1981). Random paths and random surfaces on a digital computer. *Phys. Lett.*, 106B:323–326.

[22] J. van den Berg and H. Kesten (1985). Inequalities with applications to percolation and reliability. *J. Appl. Prob.*, 22:556–569.

[23] A. Berretti and A.D. Sokal (1985). New Monte Carlo method for the self-avoiding walk. *J. Stat. Phys.*, 40:483–531.

[24] P. Billingsley (1968). *Convergence of Probability Measures.* John Wiley and Sons, New York.

[25] K. Binder and D.W. Heermann (1988). *Monte Carlo Simulation in Statistical Physics: An Introduction.* Springer-Verlag, New York.

[26] E. Bolthausen (1990). On self-repellent one dimensional random walks. *Prob. Th. and Related Fields*, 86:423–441.

[27] E. Bolthausen (1992). On the construction of the three dimensional polymer measure. To appear in *Prob. Th. and Related Fields.*

[28] J.-P. Bouchaud and A. Georges (1989). Flory formula as an extended law of large numbers. *Phys. Rev. B*, 39:2846–2849.

[29] A. Bovier, G. Felder, and J. Fröhlich (1984). On the critical properties of the Edwards and the self-avoiding walk model of polymer chains. *Nucl. Phys. B*, 230 [FS10]:119–147.

[30] A. Bovier, J. Fröhlich, and U. Glaus (1986). Branched polymers and dimensional reduction. In K. Osterwalder and R. Stora, editors, *Critical Phenomena, Random Systems, Gauge Theories.* North-Holland, Amsterdam. (Les Houches 1984.)

[31] R.G. Bowers and A. McKerrell (1973). An exact relation between the classical n-vector model ferromagnet and the self-avoiding walk problem. *J. Phys. C: Solid State Phys.*, 6:2721–2732.

[32] P. Bratley, B.L. Fox, and L.E. Schrage (1987). *A Guide to Simulation.* Springer-Verlag, New York, 2nd edition.

[33] E. Brezin, J.C. Le Guillou, and J. Zinn-Justin (1973). Approach to scaling in renormalized perturbation theory. *Phys. Rev. D*, 8:2418–2430.

[34] P.J. Brockwell and R.A. Davis (1991). *Time Series: Theory and Methods.* Springer-Verlag, New York, 2nd edition.

[35] D. Brydges, S.N. Evans, and J.Z. Imbrie (1992). Self-avoiding walk on a hierarchical lattice in four dimensions. *Ann. Probab.*, 20:82–124.

[36] D.C. Brydges (1986). A short course on cluster expansions. In K. Osterwalder and R. Stora, editors, *Critical Phenomena, Random Systems, Gauge Theories.* North-Holland, Amsterdam. (Les Houches 1984.)

[37] D.C. Brydges, J. Fröhlich, and A.D. Sokal (1983). A new proof of the existence and nontriviality of the continuum φ_2^4 and φ_3^4 quantum field theories. *Commun. Math. Phys.*, 91:141–186.

[38] D.C. Brydges, J. Fröhlich, and T. Spencer (1982). The random walk representation of classical spin systems and correlation inequalities. *Commun. Math. Phys.*, 83:123–150.

[39] D.C. Brydges and T. Spencer (1985). Self-avoiding walk in 5 or more dimensions. *Commun. Math. Phys.*, 97:125–148.

[40] K. Burdzy and G.F. Lawler (1990a). Non-intersection exponents for random walk and Brownian motion. Part I. Existence and an invariance principle. *Prob. Th. and Rel. Fields*, 84:393–410.

[41] K. Burdzy and G.F. Lawler (1990b). Non-intersection exponents for random walk and Brownian motion. II. Estimates and application to a random fractal. *Ann. Probab.*, 18:981–1009.

[42] K. Burdzy, G.F. Lawler, and T. Polaski (1989). On the critical exponent for random walk intersections. *J. Stat. Phys.*, 56:1–12.

[43] M. Campanino, J.T. Chayes, and L. Chayes (1991). Gaussian fluctuations of connectivities in the subcritical regime of percolation. *Probab. Th. Rel. Fields*, 88:269–341.

[44] S. Caracciolo, A. Pelissetto, and A.D. Sokal (1990). Nonlocal Monte Carlo algorithm for self-avoiding walks with fixed endpoints. *J. Stat. Phys.*, 60:1–53.

[45] S. Caracciolo, A. Pelissetto, and A.D. Sokal (1992). Join-and-cut algorithm for self-avoiding walks with variable length and free endpoints. *J. Stat. Phys.*, 67:65–111.

[46] J.T. Chayes and L. Chayes (1986a). Percolation and random media. In K. Osterwalder and R. Stora, editors, *Critical Phenomena, Random Systems, Gauge Theories*. North-Holland, Amsterdam. (Les Houches 1984.)

[47] J.T. Chayes and L. Chayes (1986b). Ornstein-Zernike behavior for self-avoiding walks at all noncritical temperatures. *Commun. Math. Phys.*, 105:221–238.

[48] J.T. Chayes and L. Chayes (1986c). Critical points and intermediate phases on wedges of Z^d. *J. Phys. A: Math. Gen.*, 19:3033–3048.

[49] L. Chayes (1991). On the critical behavior of the first passage time in $d \geq 3$. *Helv. Phys. Acta*, 64:1055–1071.

[50] J. Cheeger (1970). A lower bound for the lowest eigenvalue of the Laplacian. In R.C. Gunning, editor, *Problems in Analysis: A Symposium in Honor of S. Bochner*. Princeton Univ. Press, Princeton.

[51] J. des Cloiseaux and G. Jannink (1990). *Polymers in Solution: Their Modelling and Structure*. Oxford University Press, New York.

[52] D.R. Cox and D.V. Hinkley (1974). *Theoretical Statistics*. Chapman and Hall, London.

[53] J.G. Curro (1974). Computer simulation of multiple chain systems— the effect of density on the average chain dimensions. *J. Chem. Phys.*, 61:1203–1207.

[54] P. Diaconis and D. Stroock (1991). Geometric bounds for eigenvalues of Markov chains. *Ann. Appl. Probab.*, 1:36–61.

[55] M. Doi and S.F. Edwards (1986). *The Theory of Polymer Dynamics*. Clarendon Press, Oxford.

[56] C. Domb (1976). Finite cluster partition functions for the D-vector model. *J. Phys. A: Math. Gen.*, 9:983–998.

[57] C. Domb and M.E. Fisher (1958). On random walks with restricted reversals. *Proc. Camb. Phil. Soc.*, 54:48–59.

[58] C. Domb and G.S. Joyce (1972). Cluster expansion for a polymer chain. *J. Phys. C: Solid State Phys.*, 5:956–976.

[59] L.E. Dubins, A. Orlitsky, J.A. Reeds, and L.A. Shepp (1988). Self-avoiding random loops. *IEEE Trans. Inform. Theory*, 34:1509–1516.

[60] B. Duplantier (1988). Intersections of random walks. A direct renormalization approach. *Commun. Math. Phys.*, 117:279–329.

[61] B. Duplantier (1989). Fractals in two dimensions and conformal invariance. *Physica D*, 38:71–87.

[62] B. Duplantier (1990). Renormalization and conformal invariance for polymers. In H. van Beijeren, editor, *Fundamental Problems in Statistical Mechanics VII*, pages 171–223. Elsevier Science Publishers B.V., Amsterdam.

[63] B. Duplantier and K-H. Kwon (1988). Conformal invariance and intersections of random walks. *Phys. Rev. Lett.*, **61**:2514–2517.

[64] R. Durrett (1988). *Lecture Notes on Particle Systems and Percolation*. Wadsworth & Brooks/Cole, Pacific Grove.

[65] S.F. Edwards (1965). The statistical mechanics of polymers with excluded volume. *Proc. Phys. Soc. London*, **85**:613–624.

[66] R.S. Ellis (1985). *Entropy, Large Deviations, and Statistical Mechanics*. Springer, Berlin.

[67] M.H. Ernst (1988). Random walks with short memory. *J. Stat. Phys.*, **53**:191–201.

[68] F. Family and M. Daoud (1984). Experimental realization of true self-avoiding walks. *Phys. Rev. B*, **29**:1506–1507.

[69] G. Felder and J. Fröhlich (1985). Intersection probabilities of simple random walks: A renormalization group approach. *Commun. Math. Phys.*, **97**:111–124.

[70] W. Feller (1968). *An Introduction to Probability Theory and Its Applications*, Volume I. Wiley, New York, third edition.

[71] R. Fernández, J. Fröhlich, and A.D. Sokal (1992). *Random Walks, Critical Phenomena, and Triviality in Quantum Field Theory*. Springer, Berlin.

[72] M.E. Fisher (1966). Shape of a self-avoiding walk or polymer chain. *J. Chem. Phys.*, **44**:616–622.

[73] M.E. Fisher (1969). Discussion following "Statistics of long chains with repulsive interactions" by J. des Cloiseaux. *J. Phys. Soc. Japan*, **26**, suppl.:42–45.

[74] M.E. Fisher and D.S. Gaunt (1964). Ising model and self-avoiding walks on hypercubical lattices and "high-density" expansions. *Phys. Rev.*, **133**:A224–A239.

[75] M.E. Fisher and M.F. Sykes (1959). Excluded-volume problem and the Ising model of ferromagnetism. *Phys. Rev.*, **114**:45–58.

[76] P.J. Flory (1949). The configuration of a real polymer chain. *J. Chem. Phys.*, **17**:303–310.

[77] P.J. Flory (1971). *Principles of Polymer Chemistry*. Cornell University Press, Ithaca.

[78] P.J. Flory (1976). Spatial configuration of macromolecular chains. In H. Markovitz and E.F. Casassa, editors, *Polymer Science: Achievements and Prospects*. Wiley Interscience, New York.

[79] K.F. Freed (1981). Polymers as self-avoiding walks. *Ann. Probab.*, 9:537–556.

[80] J.J. Freire and A. Horta (1976). Mean reciprocal distances of short polymethylene chains. Calculation of the translational diffusion coefficient of n-alkanes. *J. Chem. Phys.*, 65:4049–4054.

[81] J. Fröhlich (1982). On the triviality of φ_d^4 theories and the approach to the critical point in $d \geq 4$ dimensions. *Nucl. Phys.*, B200 [FS4]:281–296.

[82] J. Fröhlich, B. Simon, and T. Spencer (1976). Infrared bounds, phase transitions, and continuous symmetry breaking. *Commun. Math. Phys.*, 50:79–95.

[83] P.J. Gans (1965). Self-avoiding random walks. I. Simple properties of intermediate-length walks. *J. Chem. Phys.*, 42:4159–4163.

[84] A. Gelman and D.B. Rubin (1992). Inference from iterative simulation using multiple sequences. To appear in *Statist. Sci.*

[85] P.G. de Gennes (1972). Exponents for the excluded volume problem as derived by the Wilson method. *Phys. Lett.*, A38:339–340.

[86] P.G. de Gennes (1979). *Scaling Concepts in Polymer Physics*. Cornell University Press, Ithaca.

[87] C. Geyer (1992). Practical Markov chain Monte Carlo. To appear in *Statist. Sci.*

[88] J. Glimm and A. Jaffe (1987). *Quantum Physics, A Functional Integral Point of View*. Springer, Berlin, 2nd edition.

[89] S. Golowich and J.Z. Imbrie (1992). A new approach to the long-time behavior of self-avoiding random walks. *Ann. Phys.*, 217:142–169.

[90] P. Griffiths and J. Harris (1978). *Principles of Algebraic Geometry*. Wiley, New York.

[91] G. Grimmett (1989). *Percolation*. Springer, Berlin.

[92] R. Grishman (1973). Mean square endpoint separation of off-lattice self-avoiding walks. *J. Chem. Phys.*, **58**:220–225.

[93] A.J. Guttmann (1978). On the zero-field susceptibility in the $d = 4$, $n = 0$ limit: analysing for confluent logarithmic singularities. *J. Phys. A: Math. Gen.*, **11**:L103–L106.

[94] A.J. Guttmann (1981). Correction to scaling exponents and critical properties of the n-vector model with dimensionality > 4. *J. Phys. A: Math. Gen.*, **14**:233–239.

[95] A.J. Guttmann (1983). Bounds on connective constants for self-avoiding walks. *J. Phys. A: Math. Gen.*, **16**:2233–2238.

[96] A.J. Guttmann (1987). On the critical behaviour of self-avoiding walks. *J. Phys. A: Math. Gen.*, **20**:1839–1854.

[97] A.J. Guttmann (1989a). Asymptotic analysis of power-series expansions. In C. Domb and J.L. Lebowitz, editors, *Phase Transitions and Critical Phenomena*, Volume 11. Academic Press, New York.

[98] A.J. Guttmann (1989b). On the critical behaviour of self-avoiding walks: II. *J. Phys. A: Math. Gen.*, **22**:2807–2813.

[99] A.J. Guttmann and R.J. Bursill (1990). Critical exponents for the loop erased self-avoiding walk by Monte-Carlo methods. *J. Stat. Phys.*, **59**:1–9.

[100] A.J. Guttmann and I.G. Enting (1988). The size and number of rings on the square lattice. *J. Phys. A: Math. Gen.*, **21**:L165–L172.

[101] A.J. Guttmann and J. Wang (1991). The extension of self-avoiding random walk series in two dimensions. *J. Phys. A: Math. Gen.*, **24**:3107–3109.

[102] A.J. Guttmann and N.C. Wormald (1984). On the number of spiral self-avoiding walks. *J. Phys. A: Math. Gen.*, **17**:L271–L274.

[103] J.W. Halley and C. Dasgupta (1983). Percolation and related systems in equilibrium statistical mechanics. In B.D. Hughes and B.W. Ninham, editors, *The Mathematics and Physics of Disordered Media*, pages 260–282. Springer, Berlin. Lecture Notes in Mathematics 1035.

[104] J.M. Hammersley (1961a). The number of polygons on a lattice. *Proc. Camb. Phil. Soc.*, **57**:516–523.

[105] J.M. Hammersley (1961b). On the rate of convergence to the connective constant of the hypercubical lattice. *Quart. J. Math. Oxford*, (2), **12**:250–256.

[106] J.M. Hammersley (1991). Self-avoiding walks. *Physica A*, **177**:51–57.

[107] J.M. Hammersley (1992). Corrigendum: Self-avoiding walks. *Physica A*, **183**:574–578.

[108] J.M. Hammersley and D.C. Handscomb (1964). *Monte Carlo Methods*. Chapman and Hall, London.

[109] J.M. Hammersley and K.W. Morton (1954). Poor man's Monte Carlo. *J. Roy. Stat. Soc. B*, **16**:23–38.

[110] J.M. Hammersley and D.J.A. Welsh (1962). Further results on the rate of convergence to the connective constant of the hypercubical lattice. *Quart. J. Math. Oxford*, (2), **13**:108–110.

[111] J.M. Hammersley and S.G. Whittington (1985). Self-avoiding walks in wedges. *J. Phys. A: Math. Gen.*, (2), **18**:101–111.

[112] T. Hara (1990). Mean field critical behaviour for correlation length for percolation in high dimensions. *Prob. Th. and Rel. Fields*, **86**:337–385.

[113] T. Hara and G. Slade (1990a). Mean-field critical behaviour for percolation in high dimensions. *Commun. Math. Phys.*, **128**:333–391.

[114] T. Hara and G. Slade (1990b). On the upper critical dimension of lattice trees and lattice animals. *J. Stat. Phys.*, **59**:1469–1510.

[115] T. Hara and G. Slade (1992a). Self-avoiding walk in five or more dimensions. I. The critical behaviour. *Commun. Math. Phys.*, **147**:101–136.

[116] T. Hara and G. Slade (1992b). The lace expansion for self-avoiding walk in five or more dimensions. *Reviews in Math. Phys.*, **4**:235–327.

[117] T. Hara and G. Slade (1992c). The number and size of branched polymers in high dimensions. *J. Stat. Phys.*, **67**:1009–1038.

[118] G.H. Hardy and S. Ramanujan (1917). Asymptotic formulae for the distribution of integers of various types. *Proc. Lond. Math. Soc.*, (2) **16**:112–132.

[119] K. Hattori (1992). Self-avoiding processes on the Sierpinski gasket. In K. D. Elworthy and N. Ikeda, editors, *Asymptotic Problems in Probability Theory*. Longman, London, to appear.

[120] O.J. Heilmann (1968). Mathematical polymers. I. *Mat. Fys. Medd. Dan. Vid. Selsk.*, **37**:3.

[121] S. Hemmer and P.C. Hemmer (1984). An average self-avoiding random walk on the square lattice lasts 71 steps. *J. Chem. Phys.*, **81**:584–585.

[122] C. Itzykson and J-M. Drouffe (1989). *Statistical Field Theory*, Volumes 1 and 2. Cambridge University Press, Cambridge.

[123] E. J. Janse van Rensburg (1992a). Ergodicity of the BFACF algorithm in three dimensions. *J. Phys. A: Math. Gen.*, **25**:1031–1042.

[124] E. J. Janse van Rensburg (1992b). On the number of trees in Z^d. *J. Phys. A: Math. Gen.*, **25**:3523–3528.

[125] E. J. Janse van Rensburg and S.G. Whittington (1991). The BFACF algorithm and knotted polygons. *J. Phys. A: Math. Gen.*, **24**:5553–5567.

[126] E. J. Janse van Rensburg, S.G. Whittington, and N. Madras (1990). The pivot algorithm and polygons: results on the FCC lattice. *J. Phys. A: Math. Gen.*, **23**:1589–1612.

[127] P.C. Jurs and J.E. Reissner (1971). Efficient Monte Carlo generation of self-avoiding polymers. *J. Chem. Phys.*, **55**:4948–4951.

[128] S. Karlin and H.M. Taylor (1975). *A First Course in Stochastic Processes*. Academic Press, New York, 2nd edition.

[129] H. Kesten (1963). On the number of self-avoiding walks. *J. Math. Phys.*, **4**:960–969.

[130] H. Kesten (1964). On the number of self-avoiding walks. II. *J. Math. Phys.*, **5**:1128–1137.

[131] H. Kesten (1982). *Percolation Theory for Mathematicians*. Birkhäuser, Boston.

[132] H. Kesten (1987). Percolation theory and first passage percolation. *Ann. Probab.*, **15**:1231–1271.

[133] H. Kesten (1990). Asymptotics in high dimensions for percolation. In G.R. Grimmett and D.J.A. Welsh, editors, *Disorder in Physical Systems*. Clarendon Press, Oxford.

[134] J.F.C. Kingman (1972). *Regenerative Phenomena*. Wiley, New York.

[135] C. Kipnis and S.R.S. Varadhan (1986). Central limit theorem for additive functionals of reversible Markov processes and applications to simple exclusions. *Comm. Math. Phys.*, **104**:1–19.

[136] D.A. Klarner (1967). Cell growth problems. *Canad. J. Math.*, **19**:851–863.

[137] D.J. Klein (1980). Asymptotic distributions for self-avoiding walks constrained to strips, cylinders, and tubes. *J. Stat. Phys.*, **23**:561–586.

[138] D.J. Klein (1981). Rigorous results for branched polymer models with excluded volume. *J. Chem. Phys.*, **75**:5186–5189.

[139] D.E. Knuth (1973). *The Art of Computer Programming*, Volume 3. Addison Wesley, Reading.

[140] K. Kremer and K. Binder (1988). Monte Carlo simulation of lattice models for macromolecules. *Comp. Phys. Rep.*, **7**:259–310.

[141] A.K. Kron (1965). Monte-Carlo method in the statistical calculations of macromolecules. *Vysokomolekul. Soedin. [Polymer Sci. USSR]*, **7**:1228–1234.

[142] A.K. Kron, O.B. Ptitsyn, A.M. Skvortsov, and A.K. Federov (1967). A study of statistical globula-coil transition in macromolecules using the Monte-Carlo technique. *Molek. Biol.*, **1**:576–582.

[143] M. Lal (1969). 'Monte Carlo' computer simulations of chain molecules. I. *Molec. Phys.*, **17**:57–64.

[144] A.I. Larkin and D.E. Khmel'Nitskiĭ (1969). Phase transition in uniaxial ferroelectrics. *Soviet Physics JETP*, **29**:1123–1128. English translation of *Zh. Eksp. Teor. Fiz.* 56, 2087–2098, (1969).

[145] G.F. Lawler (1980). A self-avoiding random walk. *Duke Math. J.*, **47**:655–693.

[146] G.F. Lawler (1982). The probability of intersection of independent random walks in four dimensions. *Commun. Math. Phys.*, **86**:539–554.

[147] G.F. Lawler (1985a). Intersections of random walks in four dimensions. II. *Commun. Math. Phys.*, 97:583–594.

[148] G.F. Lawler (1985b). The probability of intersection of three random walks in three dimensions. Unpublished manuscript.

[149] G.F. Lawler (1986). Gaussian behavior of loop-erased self-avoiding random walk in four dimensions. *Duke Math. J.*, 53:249–269.

[150] G.F. Lawler (1988). Loop-erased self-avoiding random walk in two and three dimensions. *J. Stat. Phys.*, 50:91–108.

[151] G.F. Lawler (1989). The infinite self-avoiding walk in high dimensions. *Ann. Probab.*, 17:1367–1376.

[152] G.F. Lawler (1991). *Intersections of Random Walks*. Birkhäuser, Boston.

[153] G.F. Lawler and A.D. Sokal (1988). Bounds on the L^2 spectrum for Markov chains and Markov processes: A generalization of Cheeger's inequality. *Trans. Amer. Math. Soc.*, 309:557–580.

[154] J.C. Le Guillou and J. Zinn-Justin (1989). Accurate critical exponents from field theory. *J. Phys. France*, 50:1365–1370.

[155] B. Li and A.D. Sokal (1990). High-precision Monte-Carlo test of the conformal-invariance predictions for two-dimensional mutually avoiding walks. *J. Stat. Phys.*, 61:723–748.

[156] E.H. Lieb (1980). A refinement of Simon's correlation inequality. *Commun. Math. Phys.*, 77:127–136.

[157] T.C. Lubensky and J. Isaacson (1979). Statistics of lattice animals and dilute branched polymers. *Phys. Rev.*, A20:2130–2146.

[158] B. MacDonald, N. Jan, D.L. Hunter, and M.O. Steinitz (1985). Polymer conformations through 'wiggling'. *J. Phys. A: Math. Gen.*, 18:2627–2631.

[159] N. Madras (1988). End patterns of self-avoiding walks. *J. Stat. Phys.*, 53:689–701.

[160] N. Madras (1991a). On the Ornstein-Zernike decay of self-avoiding walks. In G.J. Morrow and W.-S. Yang, editors, *Proceedings of the Conference on Probability Models in Mathematical Physics*. World Scientific, Singapore.

[161] N. Madras (1991b). Bounds on the critical exponent of self-avoiding polygons. In R. Durrett and H. Kesten, editors, *Random Walks, Brownian Motion and Interacting Particle Systems*. Birkhäuser, Boston.

[162] N. Madras, A. Orlitsky, and L.A. Shepp (1990). Monte Carlo generation of self-avoiding walks with fixed endpoints and fixed length. *J. Stat. Phys.*, **58**:159–183.

[163] N. Madras and A.D. Sokal (1987). Nonergodicity of local, length-conserving Monte Carlo algorithms for the self-avoiding walk. *J. Stat. Phys.*, **47**:573–595.

[164] N. Madras and A.D. Sokal (1988). The pivot algorithm: A highly efficient Monte Carlo method for the self-avoiding walk. *J. Stat. Phys.*, **50**:109–186.

[165] F. Mandel (1979). Macromolecular dimensions obtained by an efficient Monte Carlo method: The mean square end-to-end separation. *J. Chem. Phys.*, **70**:3984–3988.

[166] B. Masand, U. Wilensky, J.P. Massar, and S. Redner (1992). An extension of the two-dimensional self-avoiding walk series on the square lattice. *J. Phys. A: Math. Gen.*, **25**:L365–L369.

[167] F.L. McCrackin (1972). Weighting methods for Monte Carlo calculation of polymer configurations. *J. Res. Nat. Bur. Stand.—B. Math. Sci.*, **76B**:193–200.

[168] E.W. Montroll and B.J. West (1979). On an enriched collection of stochastic processes. In E.W. Montroll and J.L. Lebowitz, editors, *Studies in Statistical Mechanics, Volume VII: Fluctuation Phenomena*. North Holland, Amsterdam.

[169] B.G. Nguyen (1987). Gap exponents for percolation processes with triangle condition. *J. Stat. Phys.*, **49**:235–243.

[170] B.G. Nguyen and W-S. Yang (1991). Triangle condition for oriented percolation in high dimensions. Preprint.

[171] B. Nienhuis (1982). Exact critical exponents of the $O(n)$ models in two dimensions. *Phys. Rev. Lett.*, **49**:1062–1065.

[172] B. Nienhuis (1984). Critical behavior of two-dimensional spin models and charge asymmetry in the Coulomb gas. *J. Stat. Phys.*, **34**:731–761.

[173] B. Nienhuis (1987). Coulomb gas formulation of two-dimensional phase transitions. In C. Domb and J.L. Lebowitz, editors, *Phase Transitions and Critical Phenomena*, Volume 11. Academic Press, New York.

[174] E. Nummelin (1984). *General Irreducible Markov Chains and Nonnegative Operators*. Cambridge University Press, Cambridge.

[175] G.L. O'Brien (1990). Monotonicity of the number of self-avoiding walks. *J. Stat. Phys.*, 59:969–979.

[176] S.P. Obukhov (1980). The problem of directed percolation. *Physica*, 101A:145–155.

[177] O.F. Olaj and K.H. Pelinka (1976). Pair distribution function and pair potential of lattice models under theta condition. I. Numerical evaluation. *Makromol. Chem.*, 177:3413–3425.

[178] S. Orey (1971). *Limit Theorems for Markov Chain Transition Probabilities*. Van Nostrand Reinhold, London.

[179] L.S. Ornstein and F. Zernike (1914). Accidental deviations of density and opalescence at the critical point of a single substance. *K. Akad. Amsterdam*, 17:793–806.

[180] G. Parisi (1988). *Statistical Field Theory*. Addison Wesley, New York.

[181] Y. M. Park (1989). Direct estimates on intersection probabilities of random walks. *J. Stat. Phys.*, 57:319–331.

[182] L. Peliti and L. Pietronero (1987). Random walks with memory. *Rivista del Nuovo Cimento*, 10:1–33.

[183] M.D. Penrose (1992). On the spread-out limit for bond and continuum percolation. To appear in *Ann. Appl. Probab.*.

[184] P.H. Peskun (1973). Optimum Monte-Carlo sampling using Markov chains. *Biometrika*, 60:607–612.

[185] L. Pietronero (1983). Critical dimensionality and exponent of the "true" self-avoiding walk. *Phys. Rev. B*, 27:5887–5889.

[186] N. Pippenger (1989). Knots in random walks. *Discrete Appl. Math.*, 25:273–278.

[187] M.B. Priestley (1981). *Spectral Analysis and Time Series*. Academic Press, London.

[188] D.C. Rapaport (1985). On three-dimensional self-avoiding walks. *J. Phys. A: Math. Gen.*, **18**:113–126.

[189] S. Redner and P.J. Reynolds (1981). Position-space renormalisation group for isolated polymer chains. *J. Phys. A: Math. Gen.*, **14**:2679–2703.

[190] M. Reed and B. Simon (1972). *Methods of Modern Mathematical Physics. I: Functional Analysis.* Academic Press, New York.

[191] J. Reiter (1990). Monte Carlo simulations of linear and cyclic chains on cubic and quadratic lattices. *Macromolecules*, **23**:3811–3816.

[192] M.N. Rosenbluth and A.W. Rosenbluth (1955). Monte Carlo calculation of the average extension of molecular chains. *J. Chem. Phys.*, **23**:356–359.

[193] D. Ruelle (1969). *Statistical Mechanics, Rigorous Results.* W.A. Benjamin, Inc., New York.

[194] S.S. Shapiro and M.B. Wilk (1965). An analysis of variance test for normality (complete samples). *Biometrika*, **52**:591–611.

[195] Z. Šidák(1964). Eigenvalues of operators in l_p-spaces of denumerable Markov chains. *Czech. Math. J.*, **14**:438–443.

[196] S.D. Silvey (1975). *Statistical Inference.* Chapman and Hall, London.

[197] B. Simon (1980). Correlation inequalities and the decay of correlations in ferromagnets. *Commun. Math. Phys.*, **77**:111–126.

[198] G. Slade (1987). The diffusion of self-avoiding random walk in high dimensions. *Commun. Math. Phys.*, **110**:661–683.

[199] G. Slade (1988). Convergence of self-avoiding random walk to Brownian motion in high dimensions. *J. Phys. A: Math. Gen.*, **21**:L417–L420.

[200] G. Slade (1989). The scaling limit of self-avoiding random walk in high dimensions. *Ann. Probab.*, **17**:91–107.

[201] G. Slade (1991). The lace expansion and the upper critical dimension for percolation. *Lectures in Applied Mathematics*, **27**:53–63. (Mathematics of Random Media, eds. W.E. Kohler and B.S. White, A.M.S., Providence).

[202] A.D. Sokal (1979). A rigorous inequality for the specific heat of an Ising or φ^4 ferromagnet. *Phys. Lett.*, 71A:451–453.

[203] A.D. Sokal (1982). An alternate constructive approach to the φ_3^4 quantum field theory, and a possible destructive approach to φ_4^4. *Ann. Inst. Henri Poincaré*, 37:317–398.

[204] A.D. Sokal (1989). Monte Carlo methods in statistical mechanics: Foundations and new algorithms. Lecture notes: Cours de Troisième Cycle de la Physique en Suisse Romande (Lausanne, June 1989).

[205] A.D. Sokal (1991). How to beat critical slowing-down: 1990 update. *Nucl. Phys. B (Proc. Suppl.)*, 20:55–67.

[206] A.D. Sokal and L.E. Thomas (1988). Absence of mass gap for a class of stochastic contour models. *J. Stat. Phys.*, 51:907–947.

[207] A.D. Sokal and L.E. Thomas (1989). Exponential convergence to equilibrium for a class of random walk models. *J. Stat. Phys.*, 54:797–828.

[208] C.E. Soteros, D.W. Sumners, and S.G. Whittington (1992). Entanglement complexity of graphs in Z^3. *Math. Proc. Camb. Phil. Soc.*, 111:75–91.

[209] C.E. Soteros and S.G. Whittington (1988). Polygons and stars in a slit geometry. *J. Phys. A: Math. Gen.*, 21:L857–L861.

[210] C.E. Soteros and S.G. Whittington (1989). Lattice models of branched polymers: effects of geometrical constraints. *J. Phys. A: Math. Gen.*, 22:5259–5270.

[211] F. Spitzer (1976). *Principles of Random Walk*. Springer, New York, 2nd edition.

[212] A.J. Stam (1971). Renewal theory in r dimensions II. *Comp. Math.*, 23:1–13.

[213] D. Stauffer (1985). *Introduction to Percolation Theory*. Taylor and Francis, London.

[214] S.D. Stellman and P.J. Gans (1972). Efficient computer simulation of polymer conformation. I. Geometric properties of the hard-sphere model. *Macromolecules*, 5:516–526.

[215] A. Stoll (1989). Invariance principles for Brownian intersection local time and polymer measures. *Math. Scand.*, 64:133–160.

[216] D.W. Sumners and S.G. Whittington (1988). Knots in self-avoiding walks. *J. Phys. A: Math. Gen.*, **21**:1689–1694.

[217] K. Suzuki (1968). The excluded volume effect of very-long-chain molecules. *Bull. Chem. Soc. Japan*, **41**:538.

[218] M.F. Sykes, D.S. McKenzie, M.G. Watts, and J.L. Martin (1972). The number of self-avoiding rings on a lattice. *J. Phys. A: Gen. Phys.*, **5**:661–666.

[219] K. Symanzik (1969). Euclidean quantum field theory. In R. Jost, editor, *Local Quantum Field Theory.* Academic Press, New York.

[220] H. Tasaki and T. Hara (1987). Critical behaviour in a system of branched polymers. *Prog. Theor. Phys. Suppl.*, **92**:14–25.

[221] C.J. Thompson (1988). *Classical Equilibrium Statistical Mechanics.* Oxford University Press, Oxford.

[222] S.R.S. Varadhan (1969). Appendix to: Euclidean quantum field theory, by K. Symanzik. In R. Jost, editor, *Local Quantum Field Theory.* Academic Press, New York.

[223] P.H. Verdier (1969). A simulation model for the study of the motion of random-coil polymer chains. *J. Comput. Phys.*, **4**:204–210.

[224] P.H. Verdier and W.H. Stockmayer (1962). Monte Carlo calculations on the dynamics of polymers in dilute solution. *J. Chem. Phys.*, **36**:227–235.

[225] F.T. Wall and J.J.. Erpenbeck (1959). New method for the statistical computation of polymer dimensions. *J. Chem. Phys.*, **30**:634–637.

[226] F.T. Wall and F. Mandel (1975). Macromolecular dimensions obtained by an efficient Monte Carlo method without sample attrition. *J. Chem. Phys.*, **63**:4592–4595.

[227] G.S. Watson (1983). *Statistics on Spheres.* John Wiley and Sons, New York.

[228] F.J. Wegner and E.K. Riedel (1973). Logarithmic corrections to the molecular-field behavior of critical and tricritical systems. *Phys. Rev. B*, **7**:248–256.

[229] J. Westwater (1980). On Edwards' model for long polymer chains. *Commun. Math. Phys.*, **72**:131–174.

[230] J. Westwater (1982). On Edwards' model for long polymer chains III. Borel summability. *Commun. Math. Phys.*, **84**:459–470.

[231] J. Westwater (1985). On Edwards' model for long polymer chains. In S. Albevario and P. Blanchard, editors, *Trends and Developments in the Eighties. Bielefeld Encounters in Mathematical Physics IV/V*. World Scientific, Singapore.

[232] S.G. Whittington (1982). Statistical mechanics of polymer solutions and polymer adsorption. *Adv. Chem. Phys.*, **51**:1–48.

[233] S.G. Whittington (1983). Self-avoiding walks with geometrical constraints. *J. Stat. Phys.*, **30**:449–456.

[234] S.G. Whittington and C.E. Soteros (1991). Polymers in slabs, slits, and pores. *Israel J. Chem.*, **31**:127–133.

[235] W.-S. Yang and D. Klein (1988). A note on the critical dimension for weakly self-avoiding walks. *Prob. Th. and Rel. Fields*, **79**:99–114.

[236] W.-S. Yang and Y. Zhang (1992). A note on differentiability of the cluster density for independent percolation in high dimensions. *J. Stat. Phys.*, **66**:1123–1138.

[237] K. Yosida (1980). *Functional Analysis*. Springer-Verlag, Berlin, 6th edition.

Notation

$$\|x\|_p = \left(\sum_{\mu=1}^{d} |x|_\mu^p\right)^{1/p}$$
$$|x| = \|x\|_2$$
$$\|x\|_\infty = \max\{|x_\mu| : \mu = 1, \ldots d\}$$
$$x \cdot y = \sum_{\mu=1}^{d} x_\mu y_\mu$$

$$\partial_z = \frac{\partial}{\partial z}$$
$$\partial_\mu = \frac{\partial}{\partial k_\mu}$$
$$\nabla_k = (\partial_1, \ldots, \partial_d)$$
$$\nabla_k^2 = \sum_{\mu=1}^{d} \partial_\mu^2$$

$$\hat{f}(k) = \sum_{x \in Z^d} f(x) e^{ik \cdot x}$$
$$f(x) = (2\pi)^{-d} \int_{[-\pi,\pi]^d} \hat{f}(k) e^{-ik \cdot x} d^d k$$
$$\|f\|_p = \left[\sum_{x \in Z^d} |f(x)|^p\right]^{1/p}$$
$$\|f\|_\infty = \sup_{x \in Z^d} |f(x)|$$
$$\|\hat{f}\|_p = \left[(2\pi)^{-d} \int_{[-\pi,\pi]^d} |\hat{f}(k)|^p d^d k\right]^{1/p}$$
$$f * g(x) = \sum_{y \in Z^d} f(x - y) g(y)$$

$\bar{z}$	complex conjugate of z
Re	real part of a complex number
$\mathrm{Res}(f(z), z_0)$	residue of $f(z)$ at z_0
$\lfloor x \rfloor$	greatest integer less than or equal to x
$|S|$	cardinality of the set S
$f(x) \sim g(x)$	$\lim f(x)/g(x) = 1$
$f(x) \simeq g(x)$	existence of c_1, c_2 such that $c_1 g(x) \le f(x) \le c_2 g(x)$
$f(x) \approx g(x)$	unproven belief that $f(x)$ and $g(x)$ have similar asymptotic behaviour
$f(x) = O(g(x))$	existence of K such that $|f(x)| \le K|g(x)|$
$f(x) = o(g(x))$	$\lim f(x)/g(x) = 0$

417

In the following list, the second column gives the section number where the notation is defined.

$\circ$	1.2	$\omega^{(1)} \circ \omega^{(2)}$ denotes concatenation of $\omega^{(2)}$ to $\omega^{(1)}$
$\langle \cdot \rangle$	2.3	expectation
A	1.1	amplitude for c_n
α_{sing}	1.3	critical exponent for $c_n(0, x)$
b_n	1.2	number of n-step bridges
$b_{n,L}$	3.1	number of n-step bridges with span L
$b_{n,L}(y)$	4.1	number of n-step bridges ending at (L, y)
$b_n(y)$	8.1	number of n-step bridges ending at (L, y) for some L
$B(z)$	1.5	bubble diagram
B_z	3.1	generating function for bridges
$B_z(L)$	4.1	generating function for bridges of span L
$B_z(L, y)$	4.1	generating function for bridges ending at (L, y)
$\mathcal{B}_\tau[a, b]$	5.2	set of graphs on $[a, b]$ with edges of length $\leq \tau$
c_n	1.1	number of n-step self-avoiding walks
$c_{n,\tau}$	1.2	number of n-step memory-τ walks
$c_n(x, y)$	1.2	number of n-step self-avoiding walks from x to y
$c_{n,\tau}(x, y)$	1.2	number of n-step memory-τ walks from x to y
c_{N_1, N_2}	1.3	number of pairs of intersecting self-avoiding walks
$c_n[k, P]$	7.1	number of n-step walks with at most k occurrences of the pattern P
$c_n[k, (P, Q)]$	7.2	number of n-step walks with at most k occurrences of the pattern-cube pair (P, Q)
$C_z(x, y)$	1.3	two-point function for simple random walk
$C^{(0)}(x, y)$	6.2.1	critical nearest-neighbour simple random walk two-point function
$C^{(L)}(x, y)$	6.2.1	critical spread-out random walk two-point function
$C(x, y)$	6.2.1	either $C^{(0)}(x, y)$ or $C^{(L)}(x, y)$
$C^s(U)$	1.6	set of functions with s continuous derivatives on U
$C_d[0, 1]$	6.1	set of continuous functions from $[0, 1]$ to $\mathbf{R}^d$
$\mathcal{C}_N(x, y)$	5.1	set of N-step self-avoiding walks from x to y
$\mathcal{C}_\tau(L)$	5.2	edges of length $\leq \tau$ compatible with L
$C_X(k)$	9.2.2	covariance of $X^{[t]}$ and $X^{[t+k]}$
$CLEC_X$	9.4.1	cardinality of largest ergodicity class of algorithm X
$\chi(z)$	1.3	susceptibility
$\chi_0(z)$	1.3	susceptibility for simple random walk
D	1.1	amplitude for mean-square displacement
$D_\tau(\epsilon)$	6.8	$\{z : |z| \leq z_\tau[1 + \epsilon\tau^{-1}\log\tau]\}$
$D(x)$	5.1	Ω^{-1} times indicator function of Ω
$\hat{D}(k)$	5.1	Fourier transform of $D(x)$; usually $\hat{D}_0(k)$ or $\hat{D}_L(k)$

$\Lambda_z(L,y)$	4.2	generating function for irreducible bridges ending at (L,y)
$m(z)$	1.3	mass = rate of exponential decay of $G_z(0,x)$
$m_0(z)$	A	simple random walk mass
$M(z)$	4.1	mass for bridges that end on the x_1-axis
$\overline{M}(z)$	4.1	mass for bridges
μ	1.1	connective constant
μ_τ	1.2	connective constant for memory-τ walks
μ_{Bridge}	1.2	connective constant for bridges
$\mu_\Lambda(z)$	4.2	mass for irreducible bridges
ν	1.1	critical exponent for mean-square displacement
$\bar{\nu}$	1.3	critical exponent for $\xi(z)$
ν_p	1.3	critical exponent for $\xi_p(z)$
ν_{Flory}	1.1	Flory values for ν
p_λ	6.4.1	$z_c e^{-\lambda^{1/(1-\epsilon)}}$
$P_{m,n}(\omega)$	6.7	fraction of n-step self-avoiding walks which extend the m-step walk ω
$P(i,j)$	9.1	transition probability from i to j
$P_D(A)$	3.1	number of partitions of A into distinct integers
$\pi(i)$	9.1	equilibrium probability of the state i
Π	9.2.3	projection onto constants in $l^2(\pi)$
$\Pi_z(0,x)$	5.2	irreducible two-point function in x-space
$\Pi_z^{(N)}(0,x)$	5.2	N-loop diagram in x-space
$\Pi_z^{(N)}(0,x;\tau)$	5.2	memory-τ version of $\Pi_z^{(N)}(0,x)$
$\hat{\Pi}_z(k)$	5.2	irreducible two-point function in k-space
$\hat{\Pi}_z(k;\tau)$	5.2	memory-τ version of $\hat{\Pi}_z(k)$
q_n	3.2	number of n-step self-avoiding polygons
s	6.2.2	$= 0$ for nearest-neighbour model; $=$ a small positive constant for spread-out model
st	5.2	edge $\{s,t\}$
S	9.1	state space of Markov chain
$\mathcal{S}$	9.5.1	set of all walks (self-avoiding, starting at the origin)
$\mathcal{S}_N$	9.1	set of all N-step walks
$\mathcal{S}(x)$	9.6	set of all walks that end at x
$\mathcal{S}_N(x)$	9.6	set of all N-step walks that end at x
$\mathcal{S}_N[P,R]$	7.4	set of all N-step walks that begin with the pattern P and end with R
T_X	9.1	average running time of the (static) algorithm X
τ_{exp}	9.2.2	exponential autocorrelation time of Markov chain
$\tau_{exp,g}$	9.2.2	exponential autocorrelation time of observable g
τ'_{exp}	9.2.3	modified autocorrelation time of Markov chain

$\tau_{int,g}$	9.2.2	integrated autocorrelation time of observable g
u	6.2.2	$= 3/2$ for the nearest-neighbour model; $= 5/2 - 5s/2 - 2/d$ for the spread-out model
u_n	5.1	number of n-step self-avoiding loops
$\mathcal{U}_n$	5.1	set of n-step self-avoiding loops
$\mathcal{U}_{st}(\omega)$	5.2	$= -1$ if $\omega(s) = \omega(t)$; $= 0$ otherwise
$|\omega|$	1.1	number of steps in ω
$\langle|\omega(n)|^2\rangle$	1.1	mean-square displacement
Ω	5.1	symmetric set in $\mathbf{Z}^d$, often $\{x : 0 < \|x\|_\infty \leq L\}$; cardinality of same
$\xi(z)$	1.3	correlation length
$\xi_p(z)$	1.3	correlation length of order p
z	1.3	activity
z_c	1.3	self-avoiding walk critical point
$z_c(k;\tau)$	5.2	radius of convergence of Fourier transform of $G_z(0, x; \tau)$
z_τ	6.8	abbreviation for $z_c(0; \tau)$
$z_\tau(k)$	6.8	abbreviation for $z_c(k; \tau)$

Index

Probability and Its Applications

Editors

Professor Thomas M. Liggett
Department of Mathematics
University of California
Los Angeles, CA 90024-1555

Professor Loren Pitt
Department of Mathematics
University of Virginia
Charlottesville, VA 22903-3199

Professor Charles Newman
Courant Institute of
Mathematical Sciences
251 Mercer Street
New York, NY 10012

Progress and Its Applications includes all aspects of probability theory and stochastic processes, as well as their connections with and applications to other areas such as mathematical statistics and statistical physics. The series will publish research-level monographs and advanced graduate textbooks in all of these areas. It acts as a companion series to Progress in Probability, a context for conference proceedings, seminars, and workshops.

We encourage preparation of manuscripts in some form of TeX for delivery in camera-ready copy, which leads to rapid publication, or in electronic form for interfacing with laser printers or typesetters.

Proposals should be sent directly to the editors or to:
Birkhäuser Boston, 675 Massachusetts Avenue, Cambridge, MA 02139, U.S.A.

Series Titles
K. L. CHUNG / R. J. WILLIAMS. *Introduction to Stochastic Integration*, 2nd Edition
R. K. GETOOR. *Excessive Measures*
R. CARMONA / J. LACROIX. *Spectal Theory of Random Schrödinger Operators*
G. F. LAWLER. *Intersections of Random Walks*
H. LINHART / W. ZUCCHINI. *Statistik Eins*
R. M. BLUMENTHAL. *Excursions of Markov Processes*
S. KWAPIEN / W. A. WOYCZYNSKI. *Random Series and Stochastic Integrals*
N. MADRAS / G. SLADE. *The Self-Avoiding Walk*
R. AEBI. *Schrödinger Diffusion Processes*